Issues in Ethics and Animal Rights

Issues in Ethics and Animal Rights

Editor
Manish A. Vyas

2011
REGENCY PUBLICATIONS
Delhi - 110 002

ISBN 9789351240877

Published by : **REGENCY PUBLICATIONS**
(*Division of Daya Publishing House*)
4760-61/23, Ansari Road, Darya Ganj,
New Delhi - 110 002
Phone: 23245578, 23244987
Fax: (011) 23260116
e-mail : info@regency-books.com
website : www.dayabooks.com
website : www.regencybooks.com

Laser Typesetting : **Classic Computer Services**
Delhi - 110 035

Printed at : **Chawla Offset Printers**
Delhi - 110 052

PRINTED IN INDIA

for Pooja

Preface

The issue of ethics and animal rights has always been at the centre-stage in the philosophical, religious and, to some extent, literary discourse. From the Upanishadâs to Mahavira and Buddha, from Pythagoras to Aristotle, from Descartes to Kant and from Albert Schweitzer to Peter Singer, the human-animal relationship has generated a wide range of discussions; some denying the animal world their rights, while some acknowledged their importance in the general schema of the universe and advocated the ban on their exploitation and ruthless killing. Besides, the relation between human beings and animals was believed to belong to the department of philosophy and/or religious studies, rather than to literary discourse. Nevertheless with the introduction of Postcolonial Studies in the English Departments across the globe, students, scholars and critics have begun to locate an alliance between literature, environment and animals. Graham Huggan and Helen Tiffin in their introduction to *Postcolonial Ecocriticism: Literature, Animals and Environment* (2010) argue for the need to club postcolonial and ecological discourses 'as a means of challenging continuing imperialist modes of social and environmental dominance' (2). The hitherto isolated postcolonial literary studies and environmental concerns are now looked at, not as two distinct and unrelated branches of study, but as supplementing each other. Rather, it is through the literary study of nature and animals that social, political and economic discourses emerge in a new light.

This approach, in turn, serves as a catalyst that helps eradicate the traditional belief and practice of human supremacism and anthropocentricism, which are incompatible with justice to other species. Human supremacism in its strongest forms, argues Val Plumwood in *Global Ethics and Environment* (1999), denies ethical recognition to animals and nature marking them as 'Other.' Discarding the Cartesian division of the world, she staunchly advocates a need to break from the anthropocentric traditions in the interest of both justice and prudence. In the present context of ecological destruction, it would be wise for us to adopt philosophical strategies and methodologies that maximize our sensitivity to other members of our ecological communities and openness to them as ethically considerate beings.

This anthology attempts at recording a cogent and researched study of major themes in animal rights. In this collection of thirty-six essays, the writers discuss the foundations of animal ethics in its broadest sense incorporating the Western and Oriental philosophies and religions. Besides philosophy and religion, the anthology further records these diverse voices from the points of view of literary studies and media representations. It, thus, makes essential reading for animal rights advocates and will appeal to all those concerned with the nature, scope and language of contemporary animal ethics. I am hopeful that both students and scholars will find this anthology useful in the pursuit of their study and research.

In outlining and designing this anthology, I am grateful to all the contributors for being a part of this project. It is their cooperation, hard work and trust in me that have resulted into this book. I am also respectfully indebted to Dr. Piyush Joshi and Dr. Rajendrasinhji Jadeja at H M Patel Institute of English Training and Research, Vallabh Vidyanagar, for orienting me towards this challenging field of research. I also thank the Osmania University Centre for International Studies, Hyderabad (formerly American Studies Research Centre) for providing me all the academic inputs during my Post-doctoral Fellowship stay. Many thanks to Professor Marc Bekoff for readily accepting my invitation (his was the first paper to reach me) and Professor Freya Mathews for permitting me to include her short-story (technically it serves as the Introduction to this anthology). It is a privilege to thank Dr. D. Venkataramanan and Dr. Indira Nityanandam for their affection and encouragement. My personal thanks to Pooja and Suhani for being the silent victims of my schedule.

In the end, I thank Daya Publishing House for their meticulous processing of the book and making it reach you in time.

Manish A. Vyas

Contents

Religious Outlooks

Literary and Media Positions

Profile of Contributors

1. **Freya Mathews** is currently Adjunct Professor of Environmental Philosophy at La Trobe University, Australia. She has published widely in the field of environmental philosophy. Her books include *The Ecological Self* (Routledge, 1991), *For Love of Matter: A Contemporary Panpsychism* (SUNY Press, 2003) and *Reinhabiting Reality: Towards A Recovery of Culture* (SUNY Press, 2005).

2. **Marc Bekoff** is Professor Emeritus of Ecology and Evolutionary Biology at the University of Colorado, a Fellow of the Animal Behavior Society, and a former Guggenheim Fellow. In 2009 Marc became a member of The Humane Society University (http://www.humanesocietyuniversity.org/) and a Scholar-In-Residence at the Institute for Human-Animal Connection at the University of Denver (http://www.du.edu/socialwork/faculty/facultypages/bekoff.html) and was also presented with the Saint Francis of Assisi Award by the New Zealand SPCA. In 2000 he was awarded the Exemplar Award from the Animal Behavior Society for major long-term contributions to the field of animal behavior. Marc is also an ambassador for Jane Goodall's Roots & Shoots program, in which he works with students of all ages, senior citizens and prisoners. He has published more than 200 scientific and popular essays and twenty-two books including *Minding Animals, The Ten Trusts* (with Jane Goodall), *The Emotional Lives of Animals, Animals Matter, Animals at Play: Rules of the Game* (an award-winning children's book), *Wild Justice: The Moral Lives of Animals* (with Jessica Pierce), *The Animal Manifesto: Six Reasons for Expanding Our Compassion Footprint,* and the *Encyclopedia of Human-Animal Relationships* , the *Encyclopedia of Animal Behavior,* and two editions of the *Encyclopedia of Animal Rights and Animal Welfare*. In 2005 Marc was presented with The Bank One Faculty Community Service Award for the work he has done with children, senior citizens, and prisoners. In 1986 he became the first American to win his age-

class at the Tour du Var bicycle race (also called the Master's/age-graded Tour de France). His websites are http://literati.net/Bekoff and, with Jane Goodall, www.ethologicalethics.org

2. **Robin Attfield** is Professor of Philosophy at Cardiff University, where he has taught philosophy since 1968. He has also served as Visiting Lecturer in Philosophy, University of Ife, Nigeria (1972-3), and National Research Council (Republic of South Africa) Visiting Research Fellow (July/August 1999). Among his major publications are: *The Ethics of Environmental Concern* (1983 and 1991), *A Theory of Value and Obligation* (1987), *Environmental Philosophy: Principles and Prospects* (1994), *Value, Obligation and Meta-Ethics* (1995), *The Ethics of the Global Environment* (1999), *Environmental Ethics: An Overview for the Twenty-First Century* (2003), and *Creation, Evolution and Meaning* (2006). He is the joint editor of *Values, Conflict and the Environment* (1989 and 1996), of *International Justice and the Third World* (1992), and of *Philosophy and the Natural Environment* (1994), and editor of *The Ethics of the Environment* (2008).

3. **Ralph R. Acampora**, Associate Professor of Philosophy, teaches in the areas of applied ethics and history of (esp. modern) philosophy. He conducts research in the fields of environmental philosophy, bioethics, and animal studies. After earning a B.A./M.A. at CUNY, Acampora gained his doctorate at Emory University (writing a dissertation on inter-species ethics and phenomenology of body). He has authored *Corporal Compassion: Animal Ethics and Philosophy of Body* (University of Pittsburgh Press, 2006), co-edited *A Nietzschean Bestiary* (Rowman & Littlefield, 2003), has published work in a variety of books and journals, referees for *Environmental Ethics*, and is a member of the editorial board for *Anthrozoös*. Recent interests of his include the hermeneutics of spectatorship at zoos, moral issues pertaining to the built, including biotechnical environment, and the ontological status of nature.

4. **Rebekah Humphreys** works as a tutor in Philosophy and has recently been awarded PhD in Philosophy at Cardiff University, South Wales, UK . Her specialised area of research is animal ethics with particular reference to the ethics of factory farming and animal experimentation.

5. **Sandra Swart** is an Associate Professor at the University of Stellenbosch, South Africa. She received her doctorate in History from the University of Oxford and publishes widely in social and environmental history. Her most recent book is *Riding High - Horses, Humans and History in South Africa* (Witwatersrand UNivesity Press, 2010).

6. **Katerina Prudchenko** was born in Kiev, Ukraine and raised in Pittsburgh, Pennsylvania. She holds a B.A. in Mathematics from University of Southern California and an MA from California State University Long Beach. She is currently pursuing PhD in Education at Old Dominion University and lives in Los Angeles. Her short stories are forthcoming in *The Nevada Review* and *Platinum Page*.

7. **Bindi Shah** received her PhD in Sociology from the University of California, Davis, USA and is currently a Lecturer in Sociology in the School of Social Sciences at the University of Southampton, United Kingdom. Her research interests focus on issues of identity, belonging and citizenship among the children of Asian immigrants in the United States and United Kingdom. Her current project is a comparative study of the role of religion and religious institutions in shaping community, belonging and citizenship among young Jains in the two countries. The study also examines how the Jain religious tradition is being translated and/or transformed in new settings, and investigates the kind of faith communities that young Jains are building.

8. **Sebastian, A.J. sdb** is Associate Professor, Dept. of English, Nagaland Central University, Kohima. A Gold Medalist in MA in English Literature from Manipur University (1988), he was awarded PhD in 1992 for his work on the poetry of G.M. Hopkins. The research undertake in collaboration with the Hopkins' Centre at Gonzaga University, Spokane, USA. He also holds a PG Diploma in Psychological Counselling. His major publications include: *Aesthetic and Religious Quest in the Poetry of G.M. Hopkins* (Delhi, Bahri Publications, 1994); *Musings: A First Book of Verse* (Kolkata, Writers' Workshop, 1996); *My Travels My Teacher* (Kolkata, Writers' Workshop, 1998); *Literary Terms in Poetry* (New Delhi, Authorspress, 2001) co-authored with NDR Chandra; *Literary Terms in Drama, Theatre & Cinema* (New Delhi, Authorspress, 2002) co-authored with NDR Chandra; *Literary Terms in Fiction & Prose* (New Delhi, Authorspress, 2004) co-authored with NDR Chandra; *The Poetry of G.M. Hopkins: An Ecological Study* (New Delhi, Adhyayan Publishers 2009). He has written over 80 papers and articles for National and international Journals. He is interested in recent Indian writings in English with special reference to Literature from Northeast India and creative writings.

9. **Steve Russell** is a citizen of the Cherokee Nation born and raised in Oklahoma. He currently serves as a visiting judge for the state of Texas, having had an active career in various appointed and elected courts for seventeen years before entering academia. He is also Associate Professor Emeritus of Criminal Justice, Indiana University, Bloomington. His latest book is *Sequoyah Rising: Problems in Post-Colonial Tribal Governance.*

10. **Sara M. Walsh** is currently Visiting Assistant Professor of Criminology & Criminal Justice in the School of Social Sciences at Indiana University Southeast. Her research interests center on the application of cultural, gender and critical geography studies to public policy and programs.

11. **William Greenway** is Associate Professor of Philosophical Theology at Austin Presbyterian Theological Seminary, where he teaches courses in Christian theology, ethics and spirituality in relationship to hermeneutics, pragmatism, critical theory, Levinas, animals and nature. His essays and review essays have appeared in *Interpretation, Journal of Religion, Scottish Journal of Theology, Religious Studies Review* and *Theology Today.* Most recently he published, "Modern Metaphysics, Dangerous Truth, Post-Moral Ethics: The Revealing Vision of

Bernard Williams" in *Philosophy Today* (2007, 51 [2]: 137-151). In 2000 he published for a general audience, "Animals and the Love and God" in the *The Christian Century* (June 21-28, 2000). In October 2009, he was the Knight Lecturer at Hardin Simmons University and the Logsdon School of Theology, where he lectured on "Modernity and the Betrayal of Creation" and "To Love as God Loves: Animals and the *Imago Dei*," which are central themes in a book-length study now nearing completion.

12. **Pankaj Jain** is an Assistant Professor at the University of North Texas where he is teaching courses related to South Asian Religions and Ecology. Earlier he has also taught Indian Films, Sanskrit, Hindi/Urdu languages and literatures at North Carolina State University, Rutgers, Kean, New Jersey City University, and University of Iowa. He is interested to connect ancient traditions with contemporary issues which he dealt with in his dissertation 'Indic Traditions and Ecology'. He has published portions from his research in journals such as *Journal of Vaishnava Studies, Numen, Religion Compass, Religious Studies Review, Visual Anthropology*, and *Worldviews* and is currently working on his monograph *Dharma and Ecology*. He received his Doctorate from the University of Iowa and Master's degree from Columbia University (both in Religious Studies).

13. **Jeffery D. Long** is Associate Professor and Chair of Religious Studies and Co-Director of Asian Studies at Elizabethtown College in Elizabethtown, Pennsylvania. He is the author of *A Vision for Hinduism: Beyond Hindu Nationalism, Jainism: An Introduction*, and the forthcoming *Historical Dictionary of Hinduism.* Long is currently Co-chair of the North American Hinduism unit of the American Academy of Religion. He previously served as Chair of the Steering Committee of DANAM (the Dharma Academy of North America) and is a regular consultant for the Hindu American Foundation. Long received both his master's and doctoral degrees from the Divinity School at the University of Chicago. He is a member of the Vedanta Society and a regular contributor to *Prabuddha Bharata*, the journal of the Ramakrishna Order.

14. **Lawrence A. Babb** is professor of anthropology and Asian Studies at Amherst College in Massachusetts (U.S.A). He has done field research in central, northern, and western India in both Hindu and Jain communities. His most recent books are *Alchemies of Violence: Myths of Identity and the Life of Trade in Western India* (2004) and *Desert Temples: Sacred Centers of Rajasthan in Historical, Art-Historical, and Social Contexts* (with John Cort and Michael Meister, 2008).

15. **Rod Bennison** is a Conjoint Fellow in Geography and Environmental Studies at the University of Newcastle in Australia. He has a long-standing history within the environmental and animal protection and advocacy movements. He is the Convenor of Minding Animals International and the now highly respected conferences of the same name. Rod lectures at the University of Newcastle on an occasional basis, is an environmental consultant, and also currently holds a position on the Worimi Conservation Lands Board of Management. His main interests in animals lies with the commercial exploitation of wildlife, representations of animals as insignificant others, and the divide between animal protection and environmentalism.

16. **Colette Sciberras** is currently finishing off her Ph.D. thesis in philosophy, entitled *Buddhist Philosophy and the Ideals of Environmentalism,* which she will be submitting at Durham University. She is based in Malta, Europe, where she teaches philosophy at undergraduate level.

17. **Alyce Miller** is an award-winning writer of fiction, poetry, and essays. Her latest book is WATER, winner of the Mary McCarthy Prize for Fiction. She is a professor of creative writing and literature in the English department at Indiana University-Bloomington. She leads a double life as a pro bono attorney specializing in animal rights law and family law.

18. **Ricardo Rozzi** is a philosopher and ecologist, who has contributed to incorporate environmental philosophy into the practices of biocultural conservation and education in Latin America. Currently, he is Associate Professor at the Department of Philosophy, University of North Texas, and associate researcher at the University of Magallanes and Institute of Ecology and Biodiversity in Chile. With these institutions he established the Sub-Antarctic Biocultural Conservation Program (www.chile.unt.edu).

19. **Dhishna. P** is a scholar from Kerala, India. She has recently attained her doctorate in English, with a gold medal for the Best Ph. D scholar in English from Pondicherry Central University, Puducherry, India. Her area of interest is travel literature. Her specialization is on the travel narratives and the encounter of culture with special reference to Western and Indian writers who are all renown travellers. She has done an extensive study on D.H. Lawrence, V.S. Naipaul, Bruce Chatwin and S.K. Pottekkatt. She is working as a lecturer in the Department of English, University of Calicut, Kerala, India.

20. **Brian Deyo** is the Visiting Professor of World Literature in the English Department at Grand Valley State University (Allendale, Michigan). He earned his M.A. in humanities from the University of Chicago and did his doctoral work on J. M. Coetzee in 2008.

21. **Sally Borrell** holds BA (Hons) and MA degrees from the University of Canterbury in New Zealand. She has recently completed a PhD thesis at Middlesex University in London, examining representations of human-animal relations in contemporary postcolonial fiction, with a focus on responses to humanism. She worked as administrator for the former British Animal Studies Network, and is an associate of the New Zealand Centre for Human-Animal Studies.

22. **Indira Nityanandam** has been teaching ELT and English literature for last more than three decades. Being a versatile and prolific scholar, she has published six books and more than twenty-five research papers on ELT, translation studies and women's writing. She is also a recognized M. Phil and Ph. D. guide at the Gujarat University, Ahmedabad. At present, she serves as the Principal, Smt. S. R. Mehta Arts College, Ahmedabad, Gujarat, India.

23. **Anju Bala Agrawal** is Reader and Head in the Deptt. of English, R.C.A.Girls P.G.College, Mathura. She obtained her Ph.D. Degree on "The Critical Study of the plays of Graham Greene". She submitted her thesis of D.Litt on

"Nature and Man in Wordsworth and Seamus Heaney" in 2006 and soon will be awarded the degree. She is guiding several research scholars under her guidance. She has participated and presented papers in various national and international seminars. She has to her credit many published articles in the field of Indian, British, Irish and American literature in different journals and books. She has edited *Post Independence Indian Writing in English* (vol. I and II) and authored *William Wordsworth: A Collection of Critical Essays*. She is an active member in several national, academic and social organizations.

24. **Lisa Heldke** is Professor of Philosophy and Sponberg Chair in Ethics at Gustavus Adolphus College, where she also teaches in the Women's Studies program. She is the author of *Exotic Appetites: Ruminations of a Food Adventurer* (Routledge) and co-editor of two other works on the philosophy of food: *Cooking, Eating, Thinking: Transformative Philosophies of Food* (Indiana) and *The Atkins Diet and Philosophy* (Open Court). A founding member of Convivium: the Philosophy and Food Roundtable, she has devoted much of her philosophical career to advancing the philosophical study of food.

25. **Syam Sudhakar** is a Ph.D Reasearch Scholar at the Department of English, University of Madras. His research interests include Beat Literature, Performance Studies, Gender and Masculinity Studies, Popular Cultural Studies, South Indian Poetics, creative writing, and poetry. He has published two collections of poems, *Irpam* (Dampness) and *Syam Sudhakar Kavithakal* (translated into Tamil by Yumavasuki) in Malayalam. He is a programme director of Cpracsis (Centre for Performance Research and Cultural Studies in South Asia).

26. **Johansen Quijano-Cruz** is a professor of English at the Center of Multidisciplinary Studies in Puerto Rico. He holds graduate degrees in TESOL and English Literature from the University of Puerto Rico. He has published various papers on classroom methodology, digital literacies, and Victorian poetry. He is currently teaching first year English in the University of Texas at Arlington, where he is working on topics ranging from Blakean Ideology the social implications of digitally simulated narratives on society.

27. **S.S.V.N.Sakuntala** is an Associate Professor and Head, Dept of P.G.English, Dr.L.B.P.G.College, Visakhapatnam, Andhra Pradesh. She obtained her Ph.D in Commonwealth Literature in1994 and has 13 years of teaching experience. Her area of research interest is Commonwealth Literature with special focus on African Literature.

28. **Denise Russell** is currently a Research Fellow in the School of English Literatures, Philosophy and Languages at the University of Wollongong in Australia. She had a long association with the University of Sydney where she was Head of the Philosophy Department for some years. She was the founding editor of the Journal *Animal Issues: Philosophical and Ethical Issues Related to Human-Animal Interactions* and the managing editor for five years. She introduced a course on 'Humanity and Animality' at Sydney University and has published widely in philosophy including animal ethics. Her last book *Who Rules the Waves: Piracy, Overfishing and Mining the Oceans*, (Pluto, London, 2010) includes a discussion of the demise

of fish populations and the on-going threats to whales and dolphins. She has a paper on whale songs accepted for a new anthology of work on animals and literature.

29. **Thomas Crowley** is currently working as an editor and writer in Delhi. In 2009, he completed a yearlong Fulbright scholarship at the University of Pune, where he researched intersection of ethics, ecology and culture in the Indian (and more specifically, the Maharashtrian) context. His work has appeared in *Ethics and the Environment*, the *Indian Journal of Ecocriticism*, and the online magazine *Reality Sandwich*.

30. **Wendy Woodword** is a Professor in the English Department at the University of the Western Cape in greater Cape Town. She has published widely in the fields of Southern African literature, and gender and colonialism. In 2002 she co-edited *Deep Histories: Gender and Colonialism in Southern Africa* (Rodopi). Her research on literary representations of nonhuman animals has culminated in *The Animal Gaze: Animal Subjectivities in Southern African Narratives* (Wits University Press 2008)–the first monograph in Human-Animal Studies in Southern Africa. Her next research project will focus on animals in the city–in fiction, the media and photographs. She is also a published poet and her second volume of poetry *Love, Hades and other Animals* (Protea 2008) engages with animals in Greek myth, colonial family histories and daily experience.

31. **Aiswarya S.** is a Ph.D Research Scholar at the Department of English, University of Madras.

32. **D. Venkataramanan** earned his Ph.D. and M.Phil from Madras University, Chennai. He had been an Assistant Editor with Orient Longman, Chennai for four years as well as sub-editor with *The Indian Express* and wrote Book Reviews. In addition to these, he has presented and published papers on Curriculum, Canadian English, Postmodernism and Canadian Literature. At present, he teaches at the department of English, University of Madras. His areas of expertise are Russian Studies, Postmodernism and ELT.

33. **K. Subapriya** is a Ph.D Research Scholar at the Department of English, University of Madras.

34. **Lisa Kemmerer** (B.A. in international studies, Reed College; MTS in comparative religions, Harvard Divinity; Ph.D. in Philosophy, University of Glasgow, Scotland), is the author of *In Search of Consistency: Ethics and Animals* (Brill 2006) and *Religion and Animals: Rightful Relations* (Oxford 2011). She is also the editor of several upcoming books: *Sister Species: Women, Animals, and Social Justice* (U. of Illinois Press, 2011), *Primate People: Personal Stories of Advocacy and Adventure* (U. of Utah Press, 2011), and *Call to Compassion: Reflections on World Religions and Animal Advocacy* (co-edited with Anthony Nocella, Lantern 2011). She has published nearly thirty articles, and a dozen chapters in books and encyclopedias. Kemmerer is a philosopher-activist, artist, and lover of wild places, who has hiked, biked, kayaked, backpacked, and traveled widely. She is currently Associate Professor of Philosophy and Religions at Montana State University Billings.

"Without Animals Life Is Not Worth Living"

☆ *Freya Mathews*

Pookie: A Tale

Let's imagine spending a moment or two in the Middle Ages. What are the images that come to mind? Turreted castles, ladies in rose-coloured gowns and pointy hats, peasants gathering sheaves of wheat, monks in black cassocks.........and pigs! Flick through the pages of any Book of Hours or look at any medieval tapestry and there they are, wandering in alleyways or grubbing for acorns in oak-leafy woods, always in tilted profile, like the animals in bestiaries, and always with that mona lisa smile and those knowing human eyes. The jolly pig, rotund and self-satisfied.

Now set those images of the Middle Ages aside, and take a stroll around your local urban neighbourhood. Any pigs about today? Can you hear the clippety clop of trotters in the shopping mall? Are there snouts in the gutters, muddy wallows where football boots have cut up the soil around the goal posts in the park? Hardly. If you are a city dweller, animals will have pretty much disappeared from your landscape, apart from the odd dog-on-leash or cat-on-porch and a scattering of doughty little street birds. Pigs are still out there somewhere, of course, as witnessed by the ham and bacon in the nation's fridges. But they are out of sight, sequestered in prison-camps that are externally unidentifiable, no longer part of anybody's landscape, anybody's life. Why is this so? Why have pigs, and all the other animals who have figured so essentially in human history, dropped from view? Is this exclusion an integral aspect of modernity? What are the consequences of the rupture of these age-old relations between humanity and its nonhuman confreres? Could these relations be re-established? Could relations of commensality, for instance, replace those of farming or herding in a contemporary urban context? But what would that imply for our eating habits? Could we, with our current levels of reflexivity, continue to eat animals whose conspecifics were cherished by us as domestic companions? And if we found it intolerable to eat such animals, how would we deal with them when they escaped captivity and turned into feral invaders of our indigenous ecosystems? Is there indeed any place for wildlife, feral or indigenous, in the modern imagination? Is making such a place essential to the ecological revisioning of modernity?

I was given a chance to ponder these questions some years ago when my friend, Julia Bell in Western Australia, air mailed me a six week old miniature pig as a birthday present. The arrival of the present was not a complete surprise, though nor was the step to pig ownership a carefully considered one on my part. But after my rash '"yes" to Julia's breathless telephone inquiry as to whether I wanted a piglet for my birthday, I had made the necessary preparations. These had included writing long letters to my local inner-city council requesting a pig permit and justifying that request. It was the first such application this council had ever received, and pig jokes were flying when I went, in trepidation, into the council offices to discuss the matter with the animal by-laws officer. This was, as it turned out, a charming person named Heather. Heather was an animal nut herself: she had only recently moved into town from a country house she had shared with an amiable pack of dogs and goats and a donkey. I assured Heather that my pig would spend most of the day indoors sitting on the sofa and watching television (though I would have to acquire a television for this purpose - and what a worthy purpose for doing so!). This was my honest expectation. I had talked to a local breeder of miniature pigs who had dazzled my imagination with his tales of pigs tripping along obediently, collared and leashed, at the heel of their owner by day and companionably watching television with the rest of the family at night. I didn't doubt that my pig would be just like this. Her breeders had estimated that she would grow to "the size of a kelpie" or about 30kg. Well, our dog, Sashi, was in fact part kelpie, and just about 30 kg. She lived very comfortably in our small innercity backyard, and so would Pookie.

So the day arrived, and I drove out to the airport with my son to fetch the already famous Pookie home. The girls in the air freight depot hummed the tune from the movie Babe – then still current - as they carried the container out to us. I was a bit surprised to find that the piglet was black, not pink, as I had imagined. I had only ever seen black pigs in the bush, where they looked very fierce indeed. But Pookie was quiet and polite all the way home, and when we let her out on the veranda she stood demurely in the corner, unapproachable but unfazed. It was Sashi, our dog, who had hysterics, rolling her eyes and looking aghast at the little black stranger, as if at some unholy apparition. For Pookie, however, Sashi was instantly a mother figure, and though maintaining her distance from us, she eagerly sought to attach herself to Sashi's teats.

It was thus in quite a subdued fashion that our life with Pookie began. I thought to myself, 'what a good little girl. The movie got it right. Pigs really are self-possessed creatures, like Babe, polite and composed though assertive in a wholesome way.' I made up a litter tray and bed with hot water bottles and blankets on the enclosed veranda at the rear of the house. This would be her home for the moment. When she was toilet trained she could graduate indoors, to the sofa and the television.

But Pookie spurned the litter tray. The floor of the veranda, sluiced with hot soapy water several times a day, soon showed signs of woodrot. And if the piglet was initially quiet and self-possessed, this was because she had been taking stock. When she felt sure of her new surroundings, the screaming began. Morning and evening, around feeding time, or as soon as anyone stirred inside the house, Pookie started battering at the kitchen door with her long rubbery snout and front hooves, screaming in a voice that sounded like a massacre in a primary school.

The effect was shattering. The dream of the sofa and the television began to collapse, and after a few weeks of sluicing and being battered, I took a firm decision to relocate the Pookster into the backyard rather than into the front parlour. I set up a small shed under an awning by the back door and into it I moved her bed, still full of blankets and hot water bottles that were recharged several times a day. But now I also added straw, and every day Pookie would artfully make a nest in the straw, lining it with the hot water bottles and somehow managing to spread the blankets over the top. She would pass the time munching the straw down into a chaff that could be conveniently pressed into all the corners of the shed to shut out draughts. She fastidiously remade that bed a couple of times a day.

Pookie was by now very chummy with my son and me. When we sat outside she would spring into our laps and push her head up under our jumpers. This was endearing but hazardous for us: while she would doze for a while in our arms, she would soon startle or rally abruptly and her head would fly up. Accidental contact with that mallet-like skull could dislocate the jaw of anyone making the mistake of cuddling up cheek to cheek with her.

The screaming continued. The breeders had explained that if one wanted a pig to remain "miniature" one would have to restrict its food intake. It seems hard to credit in retrospect that I could have accepted this, but so crucial was scale to the whole venture that I followed the breeders' dietary guidelines to the letter. But how to keep a pig on a diet happy? I added large quantities of bran to her small ration of pig pellets and weetbix; a stack of lucerne hay kept her occupied for at least an hour each morning and night as she licked up every last dried clover leaf with an adroit tongue. But still she protested. I started trawling the neighbourhood for greengrocer waste. What an amazing bounty I found outside nearby supermarkets. Why wasn't somebody collecting all this and converting it into compost? The ecological fallibility of the market, of the law of supply and demand! But such fallibility was in this case, at any rate, working to my advantage, though I had to be persistent. None of the stores could tell me just *when* their skips would be filled. So I had to prowl, maintain a perennial vigilance. Then, there it would be! A truckload of lettuce or cabbage leaves, and on luckier days, rarer treasures: crates of over-ripe tomatoes or mangoes, scatterings of bruised plums or blemished bananas. On occasion, in summer, I would even alight upon an entire watermelon. With what triumph I would bear my bags of booty home and with what satisfaction I would watch my Pookie devour the contents. To witness her demolishing a watermelon would make me laugh out loud. First she would breach the rind with her snout, then she would plant her two front feet inside the shell, sucking the juice noisily before chomping through the whole thing, rind and all, from the inside out. Her enjoyment was unqualified. She would tilt her nose at the sky and close her eyes as she gobbled the food with her mouth open, pink juice and squish running down her cheeks. Watching the Pookster eat was more fun than eating oneself!

I started foraging down at the local creek, joining the Italian ladies with their bundles of wild greens. I gathered dandelion leaves and armfuls of a yellow-flowered weed related to chicory and broccoli, the kind of thing we could eat ourselves if times were lean. I was enjoying my role as a forager and relishing my little forays into feraldom. My activities on Pookie's behalf were placing me in new company, not only that of the Italian ladies by the creek but, at the skips, that of bag ladies, scavenging for their own needs, and all manner of colourful crazies indentured to bantams and rabbits and other, more bizarre pets. Keeping this company didn't do much for my prestige, but the thrill of the treasure hunt outweighed the social costs. Still, pig owning is definitely not for the image-conscious: when shoppers in the carpark started offering to let me wheel their trolleys back to the stores to collect the refund, even I had second thoughts.

Actually, I had never noticed before that bag ladies and people in funny hats with bantams riding on the backs of their bikes frequented the shopping centres. I had assumed that in our commercial age these figures out of Breughel and Bosch had vanished from public view. But foraging for Pookie opened up new windows on reality for me, and there they were, the pre-modern faces. Their teeth uncapped, their warts and moles intact, their heads wrapped in handkerchiefs, they were still laughing to themselves at the margins of the village, just behind the gleaming facades of commerce. I could focus on them and know them now because I was seeing the world through eyes attuned to new registers of salience - to piggy concerns.

Daily life under Pookie's reign was contained within cast-iron routines. As soon as I awoke each morning - and there was no question of sleeping in - I donned my pig gear, which consisted of gum boots and layers of long pink and purple hippie skirts to protect my legs from black hoofprints and mud kisses. In this flowing and voluminous, pre-modern garb, I headed down to the kitchen to prepare the steaming breakfast mash. Hysterical stuck-pig cries accompanied my every move from the other side of the door. But soon the protesting little mouth was stopped, and the round of Pookie chores begun. Manure, which was always deposited in a particular corner of the yard, had to be collected, hay fetched, the water bucket scrubbed and refilled. Pookie would decline the bucket once the water had become sullied, even when it was her own snout that had done the sullying. Both bucket and water had to be pristine before she would deign to drink. After breakfast, before haytime and the daily peanut hunt, it was tummy-time. Just one tickle was enough, and thump, down she would go, ready for her rub. Each fat little leg would be daintily lifted backward or forward to ensure that you could reach the whole expanse of her tummy, and as you rubbed or brushed - sometimes with a scrubber, sometimes with a straw broom - she would stretch out her neck and close her long-lashed eyes with bliss. I would also groom her sides and comb the tassel of her tail, telling her stories and singing nursery rhymes as I brushed. Her favorite was The Owl and the Pussycat. "....and there in a wood a *piggywig* stood, with a ring in the end of his nose, his nose, with a ring in the end of his nose". In the evening, before dark, the whole routine would be repeated. I *had* to be there. There was no question of missing or even of delaying a feed, no matter how pressing my other engagements. Had I done so, the whole of Brunswick, roused by massacre noises, would have been on my doorstep demanding an explanation.

Despite the rigours, life was sweet that spring. To arrive home to a small pig standing on her hind legs at the side gate beside a dog in the same stance invariably lifted my heart. The sound of hoofbeats galloping up the side path lent an edge of brumby wildness to the air of inner-city Brunswick. And Pookie's sauciness, with her black stockings, patent leather high heel hooves and wiggly bottom, could dispel the gloomiest mood. But sweetest of all, and fairest compensation for the trials I endured, were the times when I crawled into her shed at night, and sat in the straw amongst the increasingly flabby hot water bottles, her dozing head in my lap. The city, the pain of the past, all my many disappointments seemed then to fall away……..

Sashi took a less sanguine view of the new arrangements. Pookie adored Sashi, but Sashi continued to be appalled at the intrusion of this black hobgoblin into her domain. At first I was afraid that Sashi might bite Pookles, but these worries soon seemed quaint. As she grew rapidly bigger, comparable in size to Sashi, Pookie's idea of an affectionate tussle was to turn Sashi upsidedown then stomp on her upturned belly with all four hooves. Pookie found this tremendously funny, but Sashi crawled away from these piggy pranks terrorized and bleeding. As Pooks was also digging up the brick paving and ringbarking the trees, having torpedoed everything else in the garden, it seemed that the time had come for an enclosure of some kind. So my son fenced off an elongated area beside and behind the house. The run included her shed, and once she had adjusted to the confinement, she seemed entirely happy again. A large area in which to romp would certainly have been welcome to her, but closer to the top of her hierarchy of needs were food, a cosy bed and a wallow. Companionship did rank somewhere on her list, but she was reticent on this point. She had by now outgrown cuddling, and had never been a sentimentalist. Besides, having been assigned a space of her own, she became territorial about it, and was liable to give me a shove with that lethal head if I lingered too long on her side of the fence. But low-key, pottering-about-in-the-backyard kind of companionship, 'each to their own but doing it together', seemed to make for a contented pig.

Pookie had by this stage become a topic of conversation rather favoured by me. I secretly enjoyed astounding new acquaintances with casual references to my pig. I found I could keep whole tables of people amused for hours with Pookie anecdotes. Pig was, if not chic, at least droll. But I was surprised at how opinionated people were on the subject. I myself had never had anything whatever to do with pigs before Pooks erupted into my life, and I knew that the vast majority of people were in the same position. Yet the moment the subject of pigs was broached, everyone declared, "pigs are so intelligent!". Everyone. This was always delivered as a statement, never a question, and no-one was interested in whether I, as one who consorted with pigs, had a dissenting opinion. I realized I was witnessing an unconscious instance of ideological conviction. Why was everyone so convinced of pigs' superior intelligence when scarcely anyone in the Western world has shared a household with a pig since the Middle Ages? Our language still carries pig lore accrued from the time when folk did cohabit with their porcine charges. "As greedy as a pig", "pig out", "snouts in the trough". All true! In the entire universe no-one could adore food more than Pookie did. "Pig-headed". Yes. Pookie wanted her own way and expected to get it. She butted anyone who obstructed her. "Male chauvinist pig" and "pig cop". Yes. Unflattering as these analogies were to pigs, it was undeniable that Pookie put herself first and showed very little consideration, even by cross-species standards, for the feelings of others. But "smart as a pig"? I hadn't heard that one. "Cunning as a pig"? Not that one either. When Pookie wanted something, she used crash power, rather than a calculus, to get it.

So from whence had this universal attribution of intelligence come? I had found Pookie to be in no way comparable to a dog, despite pig breeders' beguiling assurances. Judging from my own experience, pigs cared nothing for their owner's approval. There was no slinking away, tail between legs, following a scolding. When I scolded Pookie she became infuriated and retaliated with a feisty side-swipe or two. Praise did not even register on her relevance spectrum. There was therefore no way she would be induced to co-operate in an intelligence test - except by bribes. She would, it was true, jump through burning hoops if a jelly bean were in the offing. But it would be difficult to draw inferences from this alacrity. The prospect of a jelly bean would activate every one of her neural pathways right down to the roots of the reptilian brain. For a packet of jelly beans she could probably solve the field equations for the General Theory of Relativity. But was this a case of brain power or motivation, of endowment or of making much of little? How much is "intelligence" a function of appetite in any case? An interesting question.....

I was at this stage contemplating a series of personal growth workshops, with Pookie as workshop leader. This pig was without question a paragon of self esteem. She knew her own mind. Her priorities were in order. It mattered nothing to her what others thought of her. She expected her needs to be met. She was queen of her universe: the sun, moon and stars, together with her loyal human subjects, revolved around her. How she relished life - just the life processes themselves, eating and excreting, sunning and sleeping, snorkelling in mud and sifting through chocolate-crumble soil with that moist and mobile snout. Such unmediated joy of life has long been impossible for our-civilized-selves, and how could we who have lost it disappoint it in one who still possesses it?

I thought that the principal client group for the workshops could be philosophers. Living with Pookie had, in my mind, cast doubt on Socrates' maxim that it is better to be a philosopher dissatisfied than a pig satisfied. Pookie snorted her contempt for all things reflective and speculative. "I want life!" she squealed, "and I want it now! What's more," she added, with a stamp of her hoof, "I *deserve* to get what I want. The world was created specifically for my pleasure. The fish of the sea and the fowl of the air and every creeping thing that creepeth upon the earth was born to wait on me." What purpose did it serve to reflect on things, to spend precious time thinking instead of savouring the inexhaustible delights of the given? What purpose indeed. But of course no misfortune had ever yet befallen Pooks. Nothing had yet transpired to dent her confidence in the established scheme of things. She was unaware, for instance, that three million of her kind existed in steel cages behind closed doors in the countryside. She had no inkling that, far from having the upper hand in her domestic affairs, she could, despite her bossiness, be despatched by me at any moment. Nevertheless, it was salutary to bask in the company of one who was not only pre-copernican but pre-lapsarian; I was sure that my hog therapy for anxious philosophers would be a winner.

Was it coincidental, I wondered, that a creature with such supreme self-esteem should, in recent times, have suffered the greatest abasement at human hands? Is factory farming the revenge of modernity on a species which embodies pre-modern elan and self-infatuation? And mightn't all that political correctness - the pigs-are-so-intelligent mantra - emanate from a still unconscious guilt at the enormity of our human crime against the blithe spirit of the pig?

Not that Pookles was without her achilles hoof. Although fearless in daytime situations - turning head-lowered, glowing-eyed to any perceived threat - her nerves could get the better of her at the fall of night. Burrowed deep into bales of straw inside her shed, with only the flouncy tassle of her tail protruding, she could erupt vertically at a sudden noise: ancestral predators obviously still stalked the pookster's dreams.

Despite all my dietary precautions, this irrepressible pig continued to grow. At six months she was as tall as the projected kelpie, but as long as three kelpies. The narrow run beside the house was now inadequate. Besides, both the house and fence were succumbing to Pookie's alarmingly increasing strength. My son filled the excavations she made around the stumps of the house with bluestone pitchers, but although I could barely lift these, Pookie thought nothing of digging them out with her snout then tossing them contemptuously aside. She was loosening the weatherboards on the walls and working on the gas and water pipes. Gaps were starting to appear between the palings, and one day my startled neighbour, Maria, was joined by Pookie at her washing line. Quick thinking with an apple on my part succeeded in luring her back.

It was becoming obvious that in order to avoid wholesale demolition we would have to construct a proper pig pen. We cleared a corner at the far end of the yard, where a shed on a concrete base already stood, and with bags of cement and lengths of 4x4 red gum, my son built a sturdy post and rail fence. When a shiny new pail was wired to a stake and a couple of bales of straw were strewn in the shed, Pooks was very amenable to her new quarters. The pen was shaded by an apple tree and summer windfalls were already abundant. I was watched closely and entreated with urgent grunting as I gathered the windfalls up for her each morning. There was just room in the yard for a toilet corner, a wallow, a dining area and a sunning spot, in addition to her bed, which it took her hours each day to make and remake, sealing the tiniest cracks with munched straw and contouring the sleeping hollow to her liking.

Our routine was now re-established. I would muck out the pen morning and evening and lay fresh straw. I stored the manure in big compost bins and applied it to my vegetable plot at CERES (Centre for Education and Research in Environmental Strategies), our local environment park. Elixir to vegetables, this pig manure was the secret behind the mysteriously towering broccoli and zucchini in my plot. I brought the leaves and left-overs back home to Pooks. At last I had discovered a place for pigs in sustainable agriculture - not as meat but as compost kits. Feed in at one end all the available green-waste in your neighbourhood and remove, from the other end, pure gold. Nor was this the only role I could envisage for Pookie in the new economy: I schemed endlessly with my inventor friend from CERES, Bill, about setting up a methane plant in the back yard. Bill was the keeper of the methane digesters at CERES. We fantasized about summer nights boiling an outdoor kettle on a cooker rigged up from a 44 gallon drum and run on pookie power.

Although the new pen was a lot more miniature than she was, Pookles now seemed very content. The constellations continued to revolve around her. Visitors to the house were immediately ushered into her presence, where they invariably gasped gratifyingly at her enormous size, her sheer amazingness. She basked in the praise, which was only her due, after all. Ku, our little duck, and Rosie, the galah, kept her company. She would call to Sashi, who sometimes consented to play rough with her, as in the old days, but from the safe side of the fence. Occasionally Sashi would stand with an ambivalent expression on her face and allow Pookie to lick her fur with her greedy little tongue.

Pookie plainly enjoyed my company but because I was her provider she was always impatient when I appeared. The very sight of me seemed to make her angry that the goods and services were not reaching her faster. So I did sometimes feel taken for granted. In the winter of that first year I had to attend a conference in the United States. My son had kindly agreed to stand in for me as Pookie's minder (slave!). While I was away I had a lucid dream of arriving home and going out into the back yard, which was preternaturally bright and flowery in the dream; Pookie ran to greet me with a series of excited little grunts so affectionate they pierced my heart. But no such welcome awaited me on my actual return. If anything, Pookie seemed *dis*gruntled at my absence.

But this surliness soon passed, and with the pig in the sty and the new routines in place, things at last began to settle down. Pookie no longer screamed. The back yard positively purred with her benign and gratified emanations. Had we known that she was going to reach the size of a young hippopotamus rather than a kelpie, we would have penned her from the start and saved ourselves a lot of stress. But now that the difficulties were finally over I began, perversely enough, to have serious doubts. I lived in dread of a break-out. The pookster's size and strength inexorably increased. Our constructions continually turned out to be inadequate, calling for further improvisations. The post and railing fence had to be reinforced with industrial mesh. Gaps started to appear between the palings behind the bamboo hedge. I laced the hedge with barbed wire. But would the old shed stand up to the heavings and hoings now taking place within it? One afternoon I found the yard seemingly vacant - no eager wet snout poking through the rails, no startled snort as a dozing pig awoke to the sound of my approach. When I got closer to the pen there seemed to be half a pig lying in front of the shed – and a very flat half, at that, its tail flicking with excitement. In response to my anguished cry, the whole pig emerged backwards from a gigantic excavation. Large chunks of concrete had been tossed up into the pen. It seemed she had been working on the base of the shed from underneath. I spent a frantic half-hour wrestling bluestone pitchers into the hole before refilling it. One of them jammed my hand, bruising it badly.

I lay awake at night imagining Pookie's snout working its way through the fence palings, her head at last bursting through, unseen, into George and Tessia's yard next door. My mind recoiled at images of Pookie ripping the tops off George's tomato plants, Pookie ploughing the immaculate lawn, tugging at washing, knocking over pot plants. And this was the merest beginning! Now I could see other back yards trashed, Pookie trotting out triumphant onto the street. People were scattering in terror; there were accidents, pile-ups. Pookie was marauding up Lygon Street, derailing trams, upsetting bicycles and prams, storming cafes, ramping up against the cake counter in the pasticceria and gobbling the handmade chocolates from the shelf. Pookie everywhere, amusing herself recklessly, without limit. I saw myself ruined, my house seized, my shoulders stooped before a lifetime of debt. I felt an avalanche of public outrage pouring down on my sleepless head......... For months I wrestled with these secret worries, but eventually I came to the conclusion that, much as we loved Pookalina and enjoyed indulging her, it was necessary to think about alternative arrangements.

Even the prospect of relocating her however posed hair-raising possibilities. How were we to do it? How were we to manoeuvre her through the front gate without her breaking loose? If only she were broken to collar and leash! But our attempts to tie her to the pole of the hills hoist in order to habituate her to the feel of a lead, early in the piece, had led to an ECG for me, such was the volume of her screaming. (Truly. The decibel level of a screaming pig is said to be higher than that of a jet at take-off, and twenty times more stressful in its expressiveness. Pains in the chest had become a regular occurrence for me.) Even if we were somehow to funnel her up the side path to the front gate, how were we to coax her into a waiting vehicle? Wouldn't she just, with a toss or two of that death-dealing skull, breach our feeble defences and launch herself into my nightmare scenario? Wouldn't it be safer and easier simply to keep reinforcing the pen and allow her to live out her fifteen year life span there? After all, there were certain advantages in lodging a pig. Not only was she a compost-making kit; she was also a security device. Any intruder who dropped into our back yard in the dead of night was in for a richly deserved shock! Still, my mind would not rest. I shadow-boxed endlessly with the problem. Meanwhile the pookster basked, demanding more, and yet more, petulantly, never pausing to offer thanks, but giving joy nonetheless.

To find a home for a porcine prima donna whose claim to miniature status could no longer be upheld was no small task. I was willing to finance a pension for the rest of her life if a pig-care establishment could be found. But how to endear her to prospective carers? How to explain to anyone who had *not* changed her hot water bottles at eight weeks and taught her the Owl and the Pussycat that the charm of this piggywig lay precisely in the fact that she had *not* noticed that she was born into a world of human sovereignty.

In any case, wherever a vacancy for a miniature pig might conceivably exist, it was by now filled. I checked out all the children's farms around town. Stories of battered ex-owners abounded. One pig at Bundoora – a fetching little white-haired sow – had been brought in after ripping the door off the fridge. The owner had returned home one day to find the pig standing on its hind legs half inside the smashed fridge, with a yoghurt carton on its head and ice cream and raw eggs dripping from its whiskers.

My secret dream was for Pooks to be adopted by CERES. The CERES site included a children's farm, and the animals were a motley collection of give-aways and foundlings. The current star at the stables was Peggy Sue, a five year old sow who dwarfed the young Pookster. Two rare-breed ginger pigs, Dennis and Rita, only slightly smaller than Peggy Sue, were also in residence, with a brood of piglets. Chris, the animal-farm coordinator, was of the firm view that CERES did not require another pig, but I disagreed. What a great arrangement for a cash-strapped community organization, I argued warmly, to have a pig on a pension! As luck would have it, Dennis and Rita happened to catch the eye of a rare-breed fancier at just that moment and were signed up for a new life, once Rita's piglets were weaned. So the longed-for vacancy had eventuated. I doubled my lobbying efforts with the CERES management committee, and before long Pookie was in! Chris was less than enthused, but I hoped that my piggy's cheeky charm would eventually win him over.

Although relieved and elated at Pookie's acceptance by CERES, I awaited removal day with dread. I phoned Christian, Pookles' personal physician, and asked him to attend. It had taken me a long time to track down an urban vet who was up to pigs, but Christian, who hailed from a clinic on the edge of town, was a young Indiana Jones of a vet. Slight but fearless, with a little black braid at the back of his head and cowboy boots, he was ready to wrestle any pesky beast to the ground. As soon as he arrived he administered a hefty tranquillizer and while we waited for Chris over the next twenty minutes we watched the Pookster gradually capsize. (Chris had been adamant that a tranquillizer was unnecessary, but this was the conviction of a man who had not yet had the pleasure of actually meeting Pooks.) When Chris eventually joined us we erected some makeshift barriers. Paper barriers, to my anxious eyes. My son then opened the gate to the pen. I had a watermelon at the ready, and a back-up supply of mangoes and peanuts. Pookie's out-of-focus eyes widened at the sight of the temptations and she rushed, as fast as her slow-motion state would allow, out of the enclosure. I directed her with the watermelon, and Christian and Chris encouraged her from the rear. She staggered the entire length of the side path, squealing all the while, but when she reached the front of the house she sank, taking my son, who had his arms around her neck, with her. The two of them lay together in a crumpled heap for a while, puffing, until we all managed to revive Pook's interest in the watermelon. Chris had backed a trailer-cage up to the front gate, and now we pushed and petted and begged her to mount the ramp. There was only the smallest space between the gatepost and the trailer, but just as I had feared, Pookie, even in her befuddled state, did her utmost to break through the barrier of arms and legs and make a freedom dash. Somehow, between us, however, we deflected that battering ram of a head and heaved her up the ramp and into the trailer, which was soon dripping with squashed fruit. Along the street people had appeared and gathered in little knots, staring silently.

Oh, the sadness of seeing the unsteady little figure looking back at me as the vehicle, driven by Chris, with my son on board, drew away. The sadness and the relief! For eighteen months that beady-eyed parcel of elemental power had ruled our lives, but now she was launched into the wide world, a world unimaginably unsafe for even the most recalcitrant pig. Her fate was still partly in my hands, but the thread of connection between us would from here on become tenuous. I could soothe her and stroke her ears through the transition period, but her story was no longer tied to mine. Oh Pookie, unmanageable, monstrous but so-vulnerable child.......

Things did not go well for Pook over the first days, weeks and even months of her new life. She was housed in a large but dark stall in the animal stables, and her status sank from queen of the universe, sun at the centre of the firmament, to mere animal amongst animals. No longer did she enjoy fresh straw every day for her bed or muck-outs twice a day. No longer did anyone care that she had sullied her drinking water with her snout and needed a sparkling fresh refill; the same slimy water was topped up day after day. Sackloads of fresh treats from the fruit and vegetable skips no longer dropped at her feet for morning and afternoon tea. She was not even a main attraction. Peggy Sue, being a quiet and well-mannered pig as well as twice Pookie's size, was undisputed box office star, and although Peg died soon after Pookie's arrival, top billing passed to Rita and her red piglets. Both Dennis and Rita were much larger than Pookalina, and the schoolchildren who visited the stables in their scores each day, on educational outings, would hurry past Pookie in their eagerness to see the "really big pig". In fact, a lot of people who passed by did not even notice the little Pookster in her corner – a black pig in a dark poke.

Pookie grieved. Overnight she had undergone her very own copernican revolution. Ejected from the centre of the universe, she was thrust out to ignominy on the edges. Truly she had believed that pigs ruled the world, that humans existed only as devotees of pigs. How rudely that illusion was now shattered. And the world was so much bigger and busier than little Pookles had imagined. There were so many people, so many animals, a cast of thousands, so much going on. No longer did everyone who came into her field of view hurry toward her, exclaiming at her size, her sheer amazingness. Instead, they passed her by, hundreds of them, laughing and joshing amongst themselves, ignoring her, sometimes even making crude sausage jokes at her expense. Oh Pookie! How could they be so heartless! She lay in her stall with her back to visitors. At first she raised her head each time excited voices approached. But eventually she lost any hope that the excitement might be on her account, and she lay unmoving as the voices came and went.

Nor did her new keepers take kindly to her. Her normal bossiness had degenerated into snappiness, and the volunteers were scared of her. A "firm hand" was brought into play. But a firm hand had never made any impression on Pookie. I had tried smacking her in her tantrum days, only to succeed in almost fracturing my hand and enraging her. The way into the Pookster's good graces – the only way – was through shameless bribery. For pig-headed though she was, her free will melted at the prospect of a smartie. The length of black hose that Chris used to guide her around the yard made her sulky and defiant.

I visited her every day, with scaled-down quantities of her former treats, and an array of different brushes and brooms with which to groom and pamper her. Suddenly she was visibly thrilled to see me. The queen of the universe would hurry over to the fence, talking ten to the dozen, when she heard my call. "Oomp, oomp oomp oomp

oomp," she would burble, her face brightening, her wet snout searching my hands for peanuts. If she were out in the paddock she would break into a run at the sight of me and would arrive with a skid, grunting and chattering. As I greeted her, nose to nose through the wire, she would sometimes actually lift her head and open her mouth wide in a kind of excited barking. She was eager for her tidbits, of course, but this was more than tidbit talking; this was serious social excitation. She would continue that strange excited barking even after the tidbit bag was empty. This was exactly the way she had greeted me in the dream in Boston; the dream had been prescient. The heart-melting enthusiasm of that greeting was now my daily reward for eighteen months of indenture. If I were unable to enter the paddock or her enclosure, she would flop down as close as possible to the fence so that I could crouch on hands and knees and reach through the wire to scrub her tummy. She would close her eyes, as of old, and I would recite the Owl and the Pussycat as I stroked, praying that no-one would walk past and surprise me in this eccentric attitude.

But despite my devoted visits, Pookie desponded. The flies worried her. Her less-than-pristine conditions demoralized and depressed her. Her hair fell out. Skin afflictions erupted. She was moved into a more salubrious outdoor pen, but Rita's bullying in the paddock caused her to pull muscles, and soon she was lame as well as bald. In the weekends I sometimes found stones in her pen, thrown by young out-of-hours tormentors.

How the mighty had fallen. How the world had turned, from her oyster to a slough of despond, a vale of disillusionment. Her piggy heart, so ill-adapted to reflection, lay in a ruined state for six long months. Her beady eye grew dull, her condition dirty. I grieved for her, but I knew that so long as she could weather the transition, life at CERES offered the best prospect for her. She could not live in a back yard in Brunswick, and alternative options were scarce. If she could only win people's hearts, attract a new circle of admirers, the moon and the stars could eventually be drawn back into her orbit.

After six months, Demeter, presiding goddess at CERES (Ceres being the Roman version of the Greek deity), took pity on Pookie. Pigs are sacred to Demeter, and there came a point at which the goddess could no longer bear the sight of her little charge, by nature so jovial, reduced to such a state of abjection. A wind of change blew through the stables. Political troubles had long been brewing, and Pook had been caught in the emotional crossfire of undeclared hostilities. Now regime change occurred, amidst political fireworks. Chris and his assistant left. A new animal co-ordinator, part of a larger team with a new and ambitious vision for an organic farm at CERES, took over. Pookie found herself in the care of a beautiful man with silver-fox curls whose background was in dance and biodynamics and Steiner principles rather than the "firm hand" school of farming. Martin believed in animal souls; he understood that Pookie was indeed queen of the universe, that she *needed* to be spoiled and indulged. He pledged himself to her gallantly as loyal subject and squire, while still keeping a handle on the practicalities of her everyday requirements. She was bathed regularly and cleaned, fed an organic diet, and moved to a leafy enclosure with a nice private concrete boudoir to which she could retreat when tired of the attentions and jibes of passers-by. Her skin cleared up and her hair re-grew. A spring returned to her step, and the air of sullen defiance that had settled on her gave way to the old swagger of naughtiness. Pookie was, not in love exactly – an impossibility for a pig – but *gratified*. Even I found myself tangled up in a shadowy oedipal triangle. If Pookie had terrorized her way into my psyche as my child, admittedly monstrous, but all the more consuming for that, then who was this new figure who had entered the unconscious parental contract, if not .....? For months I was in the grip of these strange oedipal undercurrents, and giant pigs and silver goat-men cavorted and shape-shifted in my dreams with an insistency that confounded my reality principle.

However, now that Pookie was back on her feet I could allow her to recede a little from my life. With Martin as her knight and champion, and her own improved morale, she began at last to attract the longed-for circle of admirers. Indeed, she was becoming famous in the neighbourhood. So my visits became less frequent and at the same time more enjoyable. Indeed, there ensued an idyllic phase, during which I dropped into CERES two or three times a week, both to tend my vegetable plot and to visit my beloved child, who was now, if not as puffed up as once she was, at least content and mild. CERES really was mine during that phase. Having Pookie on site, as well as my plot, gave me an inner link to it that made it the mythic centre of my psyche as well as the focus of my bioregional vision. It was the place, it seemed, to which I had been always-coming-home. But such dreams are hard to sustain in the context of modern society, and after this joyous interlude Demeter's gentle gusts of change began to blow again.

Although the new regime at the stables had seen the departure of most of the animals, leaving only Pookie and three of the goats and an assortment of poultry, the stables themselves and one of the two remaining animal paddocks were now ear-marked for "development". An organic produce garden and a Saturday market were proposed. Any animals that remained would have to be integrated into the management-plan – they would have to be organically-certified productive members of the farm. No more taking in of give-aways and foundlings now. A free-range same-breed flock of laying chickens – strict professionals - was envisaged, and a couple of Dexter cows, Dexters being a miniature variety, suitable for urban conditions, with a high milk yield. There was no place in this business plan for deluded pigs whose only claim to permaculture status was the production of little steaming piles of poo. Even the goats and ducks and geese would have to go. Despite my impassioned arguments and interventions, Martin could see the wisdom of this plan, and advertised Pookie in the newspaper as free to a good home.

Martin was confident he could screen out pork-eaters from the prospective carers that came forward in response to the advertisement, but I had a much darker view of Pookie's prospects. Once again I racked my brains for alternatives, and once again fortune, touched by Demeter's hand, smiled. For it just so happened that during the two years Pookie had been at CERES, I had met a man named Peter with a large property up near Hanging Rock. Peter was one of us, an animal nut with a vast menagerie. Roaming the grounds of Bindara Farm were highland cattle, looking like bison from the American wild west, black-pointed suffolk sheep, a pair of ostriches named Tony and Narelle, alpacas, shetland ponies, pheasants and guinea fowl roosting high in gum trees, emus, ferrets, joeys, baby wombats, a pack of dogs.....and pigs. Three pigs, to be precise. Molly in a leafy chook run, and Jean and Wilber in a shady paddock up the hill. I had been staying off and on, writing, in a little flat that Peter had available for rent on the farm, so I was already on fairly familiar terms with Tony and Narelle and the rest of the crew. I didn't like to impose on Peter's extremely good nature, but this was a moral emergency. So after the dreaded advertisement appeared, I popped the question: could Peter possibly find it in his heart to....um..... take in another...um...pig. Peter didn't miss a beat, and quite casually agreed. Oh, the relief! The joy! The eternal gratitude! The horrid ad was abandoned. Pookie was still in the family!

So now another removal was necessary. But this time Martin was in charge. My son and I assisted, but I did not bear the responsibility. Besides, about twenty people were there to help and watch the fun. Martin backed a trailer up to the stable yard gate, made a corridor out of straw bales, opened the stall in which Pook was being held, rodeo-style, then ran ahead of her with pellets and potatoes. Pookie followed helplessly. There was some struggle and stumbling at the straw-bale steps up into the trailer, but then slam, she was in. Martin drove the truck that towed the trailer, and we followed in my car, keeping a close eye on Pookalina all the way. She actually lay down in the copious straw that Martin had provided and seemed quite relaxed for someone enjoying the fresh air of the freeway for the first time.

On arriving at Peter's, we had to entice her through three gates and across an emu enclosure into the pig paddock. The pig paddock stretched up an expansive, granite-studded, wooded slope. At the foot of the hill ran a creek in which Jean and Wilber had made enviable wallows. Little marble angels stood about on tree stumps, relics of goodness knows what funerary adventures on Peter's part. Pookie followed Martin across the emu enclosure, paying no attention whatever to the emus crowding and craning over her. When she reached the pig paddock however, the lure of the wide open spaces proved greater than that of the pellets, and she set out purposefully for the hilltop. A watermelon swiftly re-engaged her attention, and guided her into the little yard where she was to stay until she had successfully completed the protocols with Wilber and Jean. That's where we left her, after a parting tummy rub, lying contently in lush grass in the shade of tall gums. We all went into Peter's kitchen for a cup of tea and a slice of the orange cake that Peter had baked, because it just so happened that it was Martin's birthday.

For a year or so I continued to stay at Peter's now and again, and I witnessed Pookie's transformation into a quieter and happier pig. Peter didn't believe in putting pigs on diets, organic or otherwise, and dumped a trailer load of left-overs from the local bakery in the pig paddock a couple of times a week. Stale cream buns. Wrinkled chocolate cakes. There was a new serene light in Pookie's eyes as her tummy expanded. In time she even stopped ambling up to the fence to greet me. With no more need of me she seemed, to my relief, to have forgotten me, and my visits eventually ceased.

However, early in 2004 I dropped in to visit Peter again. After a cup of tea we strolled up to the pig paddock. Pookie was standing near the fence and shuffled over when we appeared. She was larger than when I had last seen her, larger even than Wilber, her teats sweeping the ground as she walked. Her corpulence was beautiful, not gross, but cuddly and attractively pear-shaped when she sat down. To my astonishment, she seemed extremely pleased to see me, pressing her snout against my hand – the old mud kisses – while oomp oomp oomping to me excitedly the way she used to do at CERES, though peacefully, without agitation. I thought, well, I might as well try the Owl and the Pussycat. So I started reciting, and by the time I reached the bit about the piggywig, clunk, she was down in the dust, her tummy up against the mesh of the fence, ready to be rubbed. Oh Pookie! Who would have thought! Her face wore the old mona lisa smile. Her heart was back in its rightful precopernican slot. We could have been standing in a Book of Hours. All was well with the world. It was a tonic to see.

Living with Animals

'Without animals,' says Peter, a Maasai nomad interviewed once in the *New Internationalist*[1], 'life isn't worth living'.

Even without my love affair with Pookie, I could not be in more heartfelt agreement. I have always lived with animals, and animals have been amongst my closest family members and mentors. I could not imagine my life without them. But how many people today would share this sentiment? For how many would it be football that makes life worth living, or cars, or opera, or ice-skating? Is there anything to ground my conviction that the company of animals is a necessary part of social life, in a way that football, cars, opera and ice-skating manifestly are not, and that we relinquish or forego this aspect of life at our peril?

There are two parts to this question. The first is, is it important for *us*, for our own well-being or the realization of our human potential, that we live in intimate commensal relations with animals? The second is, is it important for the *natural environment* that we live in such relations? Does the *world* need us to continue to live in our ancestral communalism with animals?

Is our present estrangement from both the natural world (as evidenced in the environmental crisis) and from ourselves (as evidenced in the intense neuroticization of life in contemporary 'advanced' societies) due at least in part to the progressive removal of animals from our day-to-day urban reality? Do we then need to find ways of restoring animals to the human household in order to address both the environmental crisis and our own crisis of consciousness?

Our Need for Animal Company

Let's start with the question whether intimate connections with animals are foundational to our human well-being. It is an oft-cited research finding that people who enjoy the day-to-day company of animals are healthier in various respects than people who do not: they tend to visit the doctor less frequently, use less medication, have lower cholesterol and blood pressure levels, recover more quickly from illness and suffer less from feelings of loneliness.[2] Clearly, inexperienced pet pig owners were not included in these surveys, since judging from my case they would have suffered vastly increased rates of stress and heart attack and sudden death, relative to the general populace. But putting the special hazards of consorting with pigs aside, it has been estimated that pet ownership saves the Australian health care system one and a half billion dollars per year[3]. And even pet pig owners, if they survived the tantrums and tyranny, would share in a set of intangible benefits that arise from an obscure, ill-defined and all-but-forgotten form of ....well, love.

What are these benefits, these fillips to our well-being? One of the benefits is that animal company can defuse a lot of the socially generated pressure in our lives. Animals are non-judgmental friends. They do not compete with us. Hence we can relax with them, and enjoy spontaneous affection and cathartic physical closeness: we can 'be ourselves' in the presence of such companions, since they have no socially acquired expectations of us. They offer us emotional and psychological release.

In addition, emotional involvement with creatures who do not share our human goals and aspirations, our systems of values, enables us to gain an external perspective on those values. It enables us to appreciate how odd or arbitrary our human priorities might appear to non-human observers. It is, after all, entirely contingent that there have not evolved on this planet diverse species with mutually communicable forms of intelligence that would enable their members to offer explicit and forceful appraisals of human conduct. As it is, we have to try to *imagine* the cross-species perspectives from which such critiques would emanate. But when we do so, assisted by the intimacy we acquire with animals through commensality, then socially-prescribed imperatives have less hold on us - we can achieve a certain distance from them, a certain detachment. We become less driven, less enslaved to discursive ideals and images, and hence more receptive to our actual bodily and instinctual needs, more self-accepting, with all the implications for health and healing that flow from this.

It does not seem too far-fetched, to me, to speculate that there may even be a direct physiological dependence of humans on animal companionship that would help to explain why people who enjoy that companionship are, by and large (tantrum-induced heart attacks notwithstanding), healthier than others. Some evolutionary theorists argue that our ancestors' early genetic "contract" with certain animals - particularly dogs - enabled us to develop the characteristics that now mark us as human. According to this theory[4], it was our association with dogs - which was initiated at least in part by the dogs themselves, possibly as early as one hundred thousand years ago[5] - which enabled our ancestors to dispense with something that is otherwise mandatory for mammalian predators, namely an acute sense of smell: when dogs agreed to join us in the hunt, they could henceforth do our sniffing for us. The advantage for us of delegating our scenting function in this way was that we could thereby dispense with our *muzzle*. *Sans* muzzle, we could achieve frontal vision, and hence improved hand-eye coordination, where this in turn was a precondition for the development of our tool-making capability. The retraction of the muzzle also entailed the shrinkage and refinement of the tongue, which thereby became capable of the short, highly differentiated sounds required for *speech*. According to this theory then, it was through a functional inter-dependence with dogs that we became human. The deal for dogs, in this scenario, was of course that they received board and lodgings; history has resoundingly vindicated the proto-dogs' evolutionary choice.

Delectable as this theory is, it overlooks the small detail that other primates have also lost their muzzles, without the assistance of dogs. Since it is nevertheless undeniable that nearly all known human communities have included dogs, some co-evolutionary cooperation between the two species does look plausible: human beings may indeed have evolved some kind of physiological need for contact with dogs. If our age-old compact with dogs rests on certain adaptive benefits, then that compact may be wired in by other more direct, physiological forms of inter-dependency. Our bodies may unconsciously respond to subtle canine emanations, just as women's bodies, for instance, unconsciously respond to the subtle menstrual signals emanating from female housemates. If all dogs were banished from our cities - and many indignant citizens are crying out for just such a ban - a massive malaise in the human population might ensue. Such a malaise might take directly physical form, such as immunological decline: the evidence that raising children without exposure to ('dirty') animals tends to weaken their immune systems, where this renders them susceptible to allergies, counts in favour of this kind of interpretation. But the malaise might also take more psychological form - it might be more akin to the depression which is already endemic in our relatively animal-free "advanced" industrial civilizations. It might manifest as a vague sense of incompleteness or meaninglessness, leading to emotional neediness and compensating material acquisitiveness. Or it might be experienced as an existential loneliness which no amount of intra-species socialising can assuage.

Consider the latter possibility for a moment. If we have lived in intimate community with dogs and other animals for anything up to a hundred thousand years, wouldn't it be likely that we would have a distinct psychological need for their company, a need that could not be satisfied by human substitutes? Anyone who habitually walks in open spaces with a close canine friend can testify to the unique appropriateness of dogs as walking companions. Bounding along with infectious interest and joy in their surroundings, they leave us free - free to think our own thoughts and to observe those surroundings keenly ourselves - while nevertheless staying faithfully within our orbit, maintaining an unobtrusive closeness with us. Additionally, anyone who has spent time in Aboriginal settlements can testify to the feeling of comfort that dog packs lend to a community, provided of course that the dogs are not themselves a source of danger. Their constant mingling with the people, their presence at meetings and their forays onto the football field, their barking and carrying on amongst themselves on the margins of human activities, add a safe, convivial and companionable dimension to life. Small pot-bellied pigs (genuine "miniatures", excluded from Australia by quarantine laws) provide this dimension of conviviality in the villages of Papua New Guinea and other parts of South East Asia. Horses do it in Outer Mongolia, camels in northern India, holy cows in India. Even to sit in the garden with an affectionate duck tugging at one's shoelaces can afford a uniquely peaceful interlude in the daily round.

In light of the emotional and psychological satisfactions that we have experienced for thousands of years in the wider social world of the "mixed community"[6] of humans and animals then, isn't it reasonable to assume that, deprived of these satisfactions, we moderns might feel unfulfilled and obscurely lonely, even if we have never experienced these satisfactions at first hand, for ourselves. And mightn't this unfulfilment and loneliness contribute to the social malaise of modern life?

So perhaps it is important for our own well-being to continue the ancient human tradition of living in mixed households or communities. But why might it be important for the natural environment itself that we honour and maintain our ancestral commensal links with animals?

Why Nature Needs Us to Live in Company with Animals

If animal companions do indeed help to make us less driven, competitive and acquisitive, then their presence in our lives works against the world-destroying ethos of contemporary economics, with its competitive individualism and consumerism. If animals help to bring us down to earth, deflating our modern ambitions and pretensions by exposing them to inter-species scrutiny, then we shall be less anxious to remain in the scramble for success, wealth and power, where this scramble, on a mass scale, is driving the engines of the great bulldozer that is our present economy. Indeed, to the extent that we share our lives with animals, we shall not only be less willing but less able to adapt to the regime of efficiency and control which is a prerequisite of capitalist production: animals constantly disrupt our life and work with unpredictable contingencies - escapes, fights, sudden illnesses, injuries, embarrassing lapses. They bring an element of slapstick and anarchy into the cool, smart, self-absorbed world of business and public affairs. They make us miss work; they muss up the perfect clothes, perfect hair, that are needed to assure our "professionalism'", our presentability, in this public world; they strew shit and dirt around the manicured gardens, and leave pawmarks through the tidy houses, that announce our hard-won social status. While this can admittedly be stressful, it does gently lead us back from the obsessive quest which is definitive of the modern ethos and which is at the root of the environmental crisis: the quest to usurp and transcend nature[7], to place ourselves above and beyond its reach, to inhabit a kind of glossy advertiser's version of Plato's heaven, in which moth and rust do not corrupt, because they are kept at bay by chemical warfare, and where thieves do not break in and steal, because the place is patrolled by security guards. In other words, by staying in touch with our animal kin, we stand a greater chance of seeing through the dangerous illusions of a world increasingly dedicated to economic ideals of wealth, power and success that are defined in stark opposition to, or at the expense of, the natural environment.

Another reason why, as environmentalists, we should encourage commensal relations between animals and people, especially children, is that such relations presumably help to engender human empathy towards animals in general, including those in the wild. When people discover the unique personalities and communicative capabilities of their animal friends and familiars, they are logically drawn to credit other animals with such potentialities too, and to extend to them, in principle, a degree of consideration commensurate with that which, they have realised, is due to the animals of their acquaintance. In this way, as is well recognized, animal companions can serve as "ambassadors" for animal life generally, awakening in us new levels of awareness and responsibility vis a vis the natural world.

It must be admitted, however, that this "ambassador" argument is, prima facie, open to objection. In the first place, what of those rural folk, whom we have all encountered, who have been in contact with animals throughout their lives yet who nevertheless treat all animals as totally inconsiderable economic objects? Then there are the people who enjoy family relationships with particular, privileged animals, yet continue to handle the rest with callous indifference. How are we to account for the fact that daily contact with animals has not, in these instances, led to a more considerate attitude towards animals in general?

One way of accounting for this is via the hypothesis that it was the fact of domestication itself, in its more grossly instrumental forms, which led to our cultural objectification of animals. That is, according to some theorists[8], in drawing animals into our domiciliary space, and raising them within the circle of the human clan, then slaughtering them for food or other purposes, we in fact violated the taboo against violence towards kin. The moral gravity of this transgression required that we rationalize our action by denying the moral significance of domestic - and by extension, all other - animals, reducing them to the status of objects that may be produced and consumed without the slightest compunction. In other words, to justify the utilization of animals raised, like kin, within the human circle, we invented an ideology of animals as objects, which effectively closed our eyes to their otherwise manifest subjectivity. Ideology unquestionably *can* blind us to the subjectivity of others, as is plainly attested by the phenomena of slavery, racism and sexism in the human context. So the mere fact that we keep "pets", or come into daily contact with other animals, will not of itself ensure that we develop empathy for them. Communication between self and other can occur only when occlusive ideologies have been exposed and removed. For companion animals to serve as moral "ambassadors" for the animal world at large then, prima facie anthropocentric prejudices have first to be set aside. But when we do set them aside, and begin to discover the rich, subjectivity-charged presence and particularity of familiar animals, then not only will the absurdity of these prejudices be exposed, but the long journey to a new moral universe, a richly variegated universe of moral beings, can begin.

If it is accepted that companion animals do induce in us a new moral seriousness about animals generally, then a question arises concerning the status of domestic animals used for productive purposes. Does our new moral seriousness condemn the utilization of animals for such purposes? If so, is it really in the interests of the species in question, since those species at present owe their very existence to the fact that they are so utilized? How ironical would it be if the dawning of this new moral seriousness led not to an animal renaissance, but to the further retreat of animals both from their present evolutionary strongholds and from our own lives! The question, then, is whether it is possible to reconcile empathy for animals with their domestic utilization.

Such reconciliation of empathy and use appears to be possible to the extent that utilization is of net benefit to the animals concerned. When the animals concerned are considered as species rather than as individuals, it is clear that productive forms of domestication have been of net benefit to them: domestic animals are some of the few animal species still flourishing in a world of declining biodiversity. However, the kind of empathy induced by intimate relationships with animal companions leads us to consider animals as individuals rather than as mere instances of species.

So although reproductive success at the level of species is obviously a necessary condition for an individual's existence, and is in this sense in its interests, it is, equally obviously, not a sufficient condition for the individual's well-being.

To reconcile utilization with empathy, we need to be assured that the existence that our exploitative intentions confer on an individual domestic animal affords both the experiential opportunities and the requisite life span to enable that animal to achieve the form of self-realization appropriate to its particular kind. This implies that the use we may justifiably make of animals will vary according to their species: what may be an acceptable use of one species with a particular set of needs and sensibilities may not be acceptable for a species differently endowed. In particular, while humane killing of animals who lack any consciousness of death may be admissible, the killing of animals who understand and fear death, and who grieve for their own dead (as apparently do elephants and chimps), may be completely inadmissible, involving as it would the systematic infliction of intolerable suffering. Such suffering may very well, from the viewpoint of animals themselves, cancel the benefits of being alive. (This is evidenced by the fact that such animals can pine to death when bereaved[9]).

In short, the fact that domestic utilization affords evolutionary niches for certain species in a world of disappearing niches seems to be a prima facie reason for regarding such utilization as compatible with respect. However, a full-blown attitude of empathy - such as we develop through intimate association with animal companions - requires that the forms of utilization we countenance be compatible with the self-realization of the animals used, where this implies that different forms and degrees of utilization will be appropriate for different species. Moreover, once we acknowledge the moral significance of the animals we use, together with the moral gravity of our practices of utilization, it would seem to become incumbent on us to develop cultural expressions of respect, gratitude and indebtedness for the lives we have thus dedicated to our own ends. In this way, our conduct towards domestic animals could develop more affinity with the familial attitudes of hunter-gatherer peoples towards the wild species that constitute their prey.

However, when domestic utilization of animals is subject to such protocols, it may not only be consistent with empathetic concern for the interests of animals: it might actually be required by such concern. As environmentalists, committed to the maximal preservation of non-human life on earth, yet facing the cold, hard fact that in the 21st century, the processes of urbanization and industrialization that have been synonymous with the disenchantment and tragic devastation of the non-human world are only going to accelerate and intensify, don't we have to admit that one of our best chances for "saving nature" is by bringing nature back into the human domain. For the last few centuries we have witnessed the runaway humanization of nature; now let us inaugurate the wholesale naturalization of human habitat. Our cities are one of the major biological habitats of the future, and our task, as environmentalists, is to ensure that they provide the best opportunities for non-human life that we can devise. We can do this partly by increasing the amount of urban habitat for wildlife. Such habitat can be created by way of indigenous plantings and by permacultural programs of food production in the city. Buildings can also be designed or adapted to create, rather than exclude, habitat opportunities for wild animals (by way roofs that accommodate bats and nesting birds, for instance, as exemplified by the traditional habits of storks and swallows in Europe). However, the urban opportunities for non-human life can also be enhanced by our finding new ways for animals to "earn their living" in the city.

How might some of these new ways be envisaged? The usefulness of sheep as lawn-mowers has been appreciated by a church in my own local neighborhood, and there is no reason why other urban land-holders, including local councils, should not follow suit. Sheep have also been used for traffic calming in the Netherlands, and strategic use of horse-drawn vehicles - for tourist rides or milk deliveries, for instance - could serve a similar purpose. City farms afford educational opportunities for urban schoolchildren increasingly distanced from the realities of food production. The possibilities for reintegrating animals productively into urban life are as limitless as our imaginations. However, the principal way in which animals can "earn a living" in the city is still surely via their companionate role. The exclusive reign of the dog and the cat in this connection needs to be challenged, and the adaptability of other species to the human hearth and home investigated. There is immense scope for the conservation of native species, even endangered ones, in such a program of domestication. Species such as the quoll, or native cat, and the fruit bat, are reputed to make affectionate and contented hearth companions, and the domestic potentialities of many smaller, endangered wallabies, such as quokkas and bettongs, are relatively unexplored. (The quokkas on Rottnest Island, offshore from Perth in Western Australia, have already adapted to the kind of semi-tame, dump-side existence which is, according to certain evolutionary theorists[10], the first step in a species' self-surrender to domestication.) Our reluctance, as "animal lovers", to countenance confinement of wild animals, and the loss of autonomy that domestication entails, may need to be off-set by the recognition that humanity is just another potential niche in the biosphere, one which many species have in the past successfully occupied of their own free will.

The green city of the future, then, would be a mixed community rich in habitat opportunities for a great diversity of animal species. This reintegration of animals into human life would also help to expand human imaginative and empathetic horizons, undermining anthropocentrism and reinforcing commitment to the protection of the non-human world. At the same time, the multiple contacts with animals that it would afford would enhance the health and sanity of the human population.

To envisage the green city of the future as a mixed community in this way would of course involve considerable re-thinking of current urban and environmental planning principles. Restrictions on the ownership of native animals would have to be revised, and new local council regulations allowing for the responsible keeping of a wide range of "pets" would be required. Housing would have to be designed with the needs of both wild and tame non-human occupants in mind. Such demands on design would not in themselves militate against the medium density housing currently favoured by town planners, but they would require that urban consolidation be counter-balanced by large increases in communal green space. Public spaces would also have to be rendered more hospitable to animals, with protection from traffic, and areas designated and set aside for inter-species exercise (dogs would presumably have to be kept apart from quokkas and quolls, for instance!). Urban planners who currently concentrate on high density development for the sake of energy conservation and curtailment of urban sprawl forget that, in excluding non-human beings from the city and creating human ghettos, they are intensifying the anthropocentric mind-set of urban populations. The green city is one which not only conserves energy and utilizes existing infrastructure but also challenges the traditional conceptual division between humankind and nature, making itself a frontier of ecological possibility and opening its people to the degree of contact with non-human life required to awaken their ecological sensibilities.

A Responsive World: Personal Reflections

Living with animals may indeed then be important, for us and for them and for the natural environment. But there is something larger at stake here, something that is difficult to articulate. It has to do with a sense of completeness, a certain texture that somehow holds us and makes our existence cohere with the greater scheme of things. I felt this texture when the back yard hummed with the elemental presence of Pookie during that long summer of our initiation into.....something larger, more demanding, more generous than the scheduled life of our modern civilization. The loss of this texture makes our contemporary animal-exclusive existence curiously "thin", curiously unreal despite our epic impositions on reality. I don't think I can really explain this sense of completeness; it is probably graspable only through experience. So in the remainder of this reflection let me continue the autobiographical thread, and try to communicate, through recounting a little more of my own experience, that feeling of completeness, of barely perceptible cohering.....

I grew up surrounded by loving animals on what today would be described as a hobby farm, situated on the rural outskirts of Melbourne. These animals included dogs and cats, ducks, geese, hens, and, at one stage, a turkey. There were brief episodes with sheep and cows. The main focus of my entire childhood, however, was my ponies. My first pony, and the horses that came after her, were my day-long playmates and confidants. It was to them that I recited my earliest poems, and to them that I ran when I was hurt or excited. They nuzzled me in the same soft, considerate way whatever the occasion. I chose their company not for want of family and friends, but for its own sake. The form of intimacy that grew up between us was qualitatively different from anything that could have developed between myself and human persons. It was a kind of uncluttered closeness, or being-with, which existed despite the fact that our subjectivities were, in terms of content, mutually unknowable. We took it for granted, on either side, that this unknowability did not matter, that our psyches could touch and pervade each other, without need for explanations or self-disclosures, such as those conveyable by language. These animals were, for me, 'primary others', in the psychoanalytic sense; they were not substitutes for, but additional to, significant humans, nor could humans substitute for them. My subjectivity - my sense of self and world - was constituted through my "object relations"[11] with these animals just as fundamentally as it was through my relations with primary human others.

Domestic animals were not the only non-human influences shaping my sense of self and world in those early days. There were also kindly ancient gum trees on our land - we knew they dated from before colonization because they bore canoe scars in their trunks. And there was the creek, steeped in elemental mystery for me, yet at the same time busy and loquacious, swirling with news of other unknown yet connected places. These, together with my animal family, and the wild birds and snakes, all contributed to my sense of a world of communicative presences beyond the circle of human concerns.

Nor was my childhood home the only place which turned my psyche outward in this way. There was also an old sheep station on the vast western plains of New South Wales, which I occasionally visited in school holidays. It was no ordinary sheep station, but, even in those days, a relic of an earlier era. The owner, an oldtimer with eyes as wide as the blue desert sky, had been born in the homestead and raised on the property, and he ran the place in the pre-mechanical style, with the aid of stock ponies, dogs and horse-drawn buggies. We children were out all day in the searing sun on the saltbush plains, lunching out of battered tuckerboxes, racing our ponies, chasing kangaroos, emus and wild pigs with delirious excitement. Back at the homestead, animals filled our every waking moment: there were sheep and lambs, of course, as well as the ponies, most of whom spent the main part of the year in a large herd out on the range, only coming in for a tour of duty now and again, as the need arose. (These tough but happy little horses lived to extraordinary ages. One died many years later at the age of forty-five!) Cattle, pigs, tribes of chooks, ducks and geese, a flock of diminutive long-haired bush goats, an army of dogs, and at different times tame emus and kangaroos all congregated around the homestead. An old white goat named Snowy and a cocoa-coloured hand-reared filly clattered about on the wide back veranda. A sack containing a recently orphaned joey usually hung from the clothes line over the enormous wood-fired stove in the kitchen.

Compassion and fondness for animals jostled, in the daily round, with unabashed slaughter and brutality. From my saddle I witnessed mother kangaroos being torn to shreds by dogs, "for fun"; emus, in flight from our young stockman friends, failing to clear a fence, becoming entangled in the wire instead, and being bludgeoned to death with a fence-post; and back at the homestead, pigs uttering torture-chamber screams as their throats were cut and their still-convulsing bodies dropped into troughs of scalding water. I sat with the other kids in the back of a jeep on a kangaroo-shooting excursion, and as the bodies piled up under our feet, I remember the blood of the kangaroos soaking my green felt boots dark red. The cruelty shocked me to the core - in fact, it was this which first made me aware of my core, a still, silent, inner place of watching, beyond speech. But it did not diminish the overwhelming sense of enchantment that this place awakened in me. (Much, much later, I was to discover that the old station had had a similar effect on many of the people who had been associated with it.) For the enchantment, and the heightened feeling of being alive that accompanied it, arose from the fact that animals - and the uncompromising land which decided their fate - were the almost exclusive focus of everyone's life there, and the carnage, for all its horror, was part of that involvement.

When I was fifteen, my family moved into the inner city, and both my rural life, and my visits to the sheep station, ceased. However our new home overlooked extensive parklands, and I set up house with a dog in an old Victorian loft in our backyard, so the transition was not unduly traumatic. It was not until I was eighteen, when I abandoned my home and my country to live in London, that a keen sense of loss and deprivation at last set in. I moved in with a friend who leased a top-storey studio in the Kings Road in Chelsea, and for various reasons I was soon trapped in the life I had reluctantly chosen to lead there. The apartment was without a garden, without the slightest glimpse of green from its high windows. The grand old building in which it was located was legendary as one of the nerve-centres of the London "underground". Artists, writers and rock musicians congregated there, and every night, till dawn, the entire building was shaken with musical reverberations from the nightclub in the basement. People were embarked on what were for them exciting adventures with sex and drugs. The joint was unquestionably jumping. With comings and goings at all hours, residents and visitors alike were charged to the eyeballs with the fizz of glamour, the intoxication of notoriety and celebrity.

I alone, it seemed, languished. I felt deadened. Without any trees in sight, with all presence and memory of animals expunged from this world, without even a proper sky above me (the London sky appearing more like a low ceiling than the soaring invitation to infinity to which I was accustomed in Australia), I felt truly "underground", buried alive. My spirit, with its lifelong habit of expansiveness, had to submit for the first time to grey urban confinement, to a world built exclusively to human specifications, in which no court of appeal existed beyond socially-prescribed perceptions and perspectives. There was here no turning out to a wider world of subtle voices and signals, a world of myriad, at first indiscernible, but with patient attention increasingly differentiated, responsive presences. Rather, there was a turning in, and a turning up of the volume of human-generated and human-directed self-infatuated cacophony and chatter. This turning-in found its ultimate expression in the essential project of the counter-culture: to transform reality into an inner picture show, a spectacle of hallucinatory images and sexually induced sensations orchestrated for our private entertainment. This project was, in fact, nothing more than a hip rendition of the old transcendental idealism, or solipsistic anthropocentrism, of the Western tradition, which places reality in us rather than us in reality.

I had no words, at the time, to name this human introjection of reality, or to justify my sense of exile from a world that was truly alive, and, unlike the one in which I found myself, a source of true enlivenment. I especially had no words to challenge the high claims of Art on which the counter-culture rested. Instead, I kept some snails and bare twigs in a jar in my room, and gazed at them for months. I retreated into a state of fantasy and intense creativity, writing and drawing obsessively, calling up from my own deep unconscious the images and motifs I needed to survive. I composed song cycles, and stories of origins, before I had heard of Aboriginal dreamings. I hung around old book shops and antique stores, seeking out illustrations and folk tales that could be threaded into my nascent mythologies. I haunted the Natural History Museum in South Kensington, with its layer upon layer, colonnade after colonnade, of magical animal statuary. Whenever I found a numinous image - an old French engraving of a lone seal, for instance, or a Chinese painting of wild geese - I enshrined it, hanging it as a religious icon in the gallery of my mind. Out of such gathered fragments, and out of my own memory, imagination and dreams, I tried to recreate the sense of enchantment that had always been the essence of my experience of the world, and without which I did indeed find life scarcely worth living.

From the viewpoint of Western psychoanalysis, this sense of enchantment is regressive, and signals a failure of individuation in infancy. But to adopt this point of view is, of course, to beg the metaphysical question. Looking back on my early years now, it seems more plausible to me to assume that the ample opportunities for close communion with animals that were available to me throughout my childhood had opened me to a larger world, a world astir with presence or presences that vastly exceeded the human. It was this direct contact with unknowable but pervasive presence which instilled in me a sense of the sacredness or enchantment of the world, and the potentiality for "magic" within it. "Magic" was, in this context, just the possibility of the world's response - the possibility, indeed probability, that the world, when invoked in good faith, *will* respond, though not necessarily in the manner one anticipates or with the results for which one hopes. One should certainly not, in my view, rely on this world to fulfil requests or afford protection, but if one entreats it simply to reveal itself, to engage in an act of communication, then, in my experience, it will generally do so, though in its own ever-unpredictable way. I learned this as a child, through the receptiveness that my animal familiars created in me, and it filled my whole being with a sense of being accompanied, of never being alone, a sense of background love, akin to the background radiation of which physicists speak. This is a "love" which has nothing to do with saving us from death and suffering, or with making us happy. From the viewpoint of the world, death and suffering are just inevitable concomitants of individual life. The point for individuals, from this perspective, is not to seek to evade these inevitabilities, but to reach beyond them - to call into the silence beyond human selfhood in search of a reply. This is the moment for which the world has been waiting, and in which it will rejoice: the moment when we ask it to speak. To receive its reply is to enter a love far greater than the kind of protection and indulgence that our traditional importunate forms of prayer expect, for that reply signifies that we belong to an animate order, a pattern of meaning, from which death cannot separate us, and to which suffering only summons us.

To engage with the unknowable subjectivities of animals and to experience their response to us, particularly in childhood, is perhaps then a principal bridge to communication with the unknowable subjectivity of the wider world. To experience the world thus, as an inspirited thing, will not only direct the course of our own self-realization in a most fundamental way; it will also ensure an attitude of profound mutuality and awed protectiveness towards the world itself.

Pookie Rip

It was only about three weeks after I had finished writing the story of Pookie, reproduced above, that I had a phone call from Peter at Bindara farm. Pookie was dead. She had died in her sleep the previous night, Peter said, apparently from a heart attack. There was no sign of a struggle in the little pig-house she shared with Wilber and she had appeared perfectly normal, and as cheeky as ever, when Peter had last seen her, at bedtime. She was eight years old, which is only middle-aged for a pig, but since there was possibly no tubbier pig on earth, the heart attack hypothesis was extremely plausible.

I was saddened by the news. It was hard to imagine the world without the feisty pookster in it. But I was also disturbed. Why this strange conjunction between the writing of the story and the ending of the life? By finishing her story – which I had begun more than five years earlier – had I, by the mysterious and dangerous magic of writing, finished Pookie herself? I know there are powerful taboos around what can be written, but I hadn't thought they would apply to Pookie - belching, farting, no-nonsense material girl that she was. But of course those taboos have nothing to do with old Christian contrasts between materiality and immateriality. On the other hand however, this was a perfect death, taking Pookie without even waking her, in the midst of her supremely gratified life. Was there really any indication of the untoward in this?

As I pondered these questions, a vanished story from my very early childhood came to me. The story was called The Cat Who Went To Heaven, and it was a favourite of ours, famous for always making all of us, including mummy and daddy, cry. It told of a poor artist living long ago in a Japanese village. (I am telling the story here as I found it, deep on the floor of memory's ocean, worn and whiskered and adulterated by self and time. I prefer to tell it this way rather than correcting it against the original.) With the last of his money the artist sent his servant out to buy food but the servant bought a little kitten instead and brought it home, to his master's despair. However, just when it seemed that starvation was imminent, the young man received an assignment to paint a huge picture for the imperial palace. The painting was to depict the Buddha welcoming all the creatures of earth to Heaven. For months lengthening into years the artist laboured on the canvas, painstakingly detailing every creature in Creation. Every day the kitten, grown into an affectionate and sweet-natured cat, kept her master company, lightening his labour and his loneliness. At each day's end she would come around to the front of the canvas and inspect it carefully. The master knew she was waiting for the expected depiction of a cat, but he was under the strictest imperial orders not to include a cat in the sacred scene, as cats were banned from the Buddhist Heaven on account of their cruelty. Years passed. The Buddha, larger and more beautiful and luminous than any real life figure, stretched out his hand to bless a multitude of sentient beings, all lovingly portrayed down to the finest detail of their whiskers, claws, feathers and scales. Every sentient being was there, except one. The painting was nearly complete. Each day the master's cat still came to inspect the day's work….and each day she went away crestfallen. The master loved his cat more than anything in the world. He feared that her heart would break. But if he included her in the picture, the painting would be seen as blasphemous. It would be destroyed, all his effort would be wasted and ruin would be his lot. Finally the painting was finished. It was one of the most beautiful paintings ever to have been created in the empire. But at the last moment the master, unable to face his small companion's heartbreak, painted a likeness of her inconspicuously in the bottom right-hand corner of the picture. When his cat made her inspection that evening, she was overjoyed. So overjoyed was she, in fact, that she lay down in front of the canvas and died, from pure happiness. The poor master went to bed that night having lost not only his painting and his livelihood, but his beloved companion. When he awoke the next morning however, he found that the figure of the Buddha had turned in the night and was now extending its hand in blessing not to the assembled multitude but directly to the small bowed figure in the right hand corner of the canvas. The miracle became known throughout the empire, and the painting was accepted by the Emperor and universally acclaimed.

I am not suggesting that Pookie died of happiness at being "immortalized" in my little story. She was someone who couldn't have cared a toss about matters immortal. But just as the artist, by including his beautiful cat in the painting, overcame prejudice and opened the gates of Heaven to cats, so, by telling the story of an adorable pig, one might make at least a tiny dent in the monumental prejudice that keeps millions of pookies worldwide in unspeakable subjection and abjection.[12] If telling the tale of Pookie could achieve this, then her death, in the midst of pure happiness, might indeed begin to open a gate for other pigs, if not into Heaven, then at least out of Hell.

May it be so.

Notes

1 Nikkivan der Gaag, 'The Maasai and the Travellers', *New Internationalist*, 266, April, 1995, pp. 24-25.

2 Information supplied by Australian Companion Animal Council.

3 See Jonica Newby, *The Pact for Survival: Humans and their Animal Companions*, Sydney: ABC Books, 2001, Chapter 6.

4 *Ibid.*

5 The theory that many of our present day domestic animals initiated the process of domestication themselves, in pursuit of their own evolutionary advantage, has been explored at length in Stephen Budiansky, *Covenant with the Wild,* William Morrow, New York, 1992.

6 "Mixed community" is Arne Naess' term. See Arne Naess, "Self-realization in Mixed Communities of Humans, Bears, Sheep and Wolves", *Inquiry* 22, 1979, pp. 231-241.

7 Many works could be cited in support of this account of modernity; see, for instance, Carolyn Merchant, *The Death of Nature*, San Francisco: Harper and Row, 1980; Val Plumwood, *Feminism and the Mastery of Nature*, London: Routledge, 1993, Chapters 1,2; Freya Mathews, *The Ecological Self,* London: Routledge, 1991, Chapter 1.

8 James Serpell, *In the Company of Animals*, Cambridge: Cambridge University Press, 1996, chapter 11.

9 See, for instance, the account of elephant's consciousness of death in Joyce Poole, *Coming of Age with Elephants*, London: Hodder and Stoughton, 1996, Chapter 19; for a more ambivalent account of chimpanzee attitudes, see Jane Goodall, *In the Shadow of Man*, William Collins, Glasgow, 1971.

10 See Budiansky, op cit.

11 The term "object relations" is deployed in a branch of psychoanalytic theory, known as object relations theory, to designate the kinds of relations with primary others that an infant internalizes in the process of developing its individual sense of self. It is associated with the work of D.W. Winnicott, and later feminist theorists, such as Nancy Chodorow.

12 Anyone wishing to protest at the plight of pigs in factory farms in Australia can find information at www.savebabe.com.

Ethical and Philosophical Perspectives

Chapter 1

Increasing Our Compassion Footprint: The Animals' Manifesto

☆ *Marc Bekoff*

Our relationships with nonhuman animals (hereafter "animals") are complicated, frustrating, ambiguous, paradoxical, and wide-ranging. When people tell me that they love animals and then harm or kill them I tell them I'm glad they don't love me. We observe animals, gawk at them in wonder, experiment on them, eat them, wear them, write about them, draw and paint them, and represent them in many varied ways. We often dispassionately ignore who they are and what they want and need, moving them from here to there as we "redecorate nature" and making decisions for them without their consent. Far too many of us, including practicing scientists, ignore the fact that we are responsible for our choices about how we interact with animals and that simply doing something "in the name of science" is not an adequate reason for intentionally causing suffering, pain, or death. "Good welfare" usually isn't "good enough" (http://arbs.biblioteca. unesp.br/viewissue.php). Surely we can do better in our relationships with animals and other people.

It is important to give serious attention to the emotional lives of animals (Balcombe 2006; Allen and Bekoff 2007; Bekoff 2007a, b; Bekoff and Pierce in press). Studying animal emotions is exciting, challenging, and frustrating. We already know a lot about animal emotions and sentience, more than we often give ourselves credit for.

Behavioral and neuroscientific research shows that animals' lives are not all that private, hidden, or secret. Numerous lines of what scientists call independent evidence, information from many different disciplines, show that many animals have rich and deep emotional lives. When taken as a whole, the ensemble of data leads to no other reasonable conclusion. Scientific data (science sense), evolutionary biology, and common sense show clearly that animals have a point of view and preferences about what they do and do not like.

Anthropomorphic Doubletalk

Some readers may say, "Oh, you're just being anthropomorphic; we don't really know that other animals experience deep and enduring emotions." I find this to be a poor excuse often used to justify animal abuse and to shirk personal responsibility for one's actions. When someone says he is not sure if dogs, for example, have emotions, if they feel joy or grief, I say I'm glad I'm not his dog.

Many researchers engage in anthropomorphic doubletalk. Over the years I have noticed a curious phenomenon. If a scientist says that an animal is happy, no one questions it, but if he or she says that an animal is unhappy, charges of anthropomorphism are immediately raised. Like the dissonant personal beliefs of scientists, this doubletalk seems mostly aimed at letting humans feel better about themselves and allowing them to continue on their merry way doing whatever they choose to do to animals. Of course, inappropriate anthropomorphism is always a danger, because it is easy to get lazy and presume that the way we see and experience the world must be the only way. It also is easy to become self-serving and hope that because we want or need animals to be happy, they are. In fact, the only guard against the inappropriate use of anthropomorphism is knowledge that results from detailed study of the minds and emotions of animals.

Many researchers now recognize that we must be anthropomorphic when we discuss animal emotions but that, if we do it carefully, what I call biocentrically, we can still give due consideration to the animals' point of view. Being anthropomorphic is doing what comes naturally. No matter what we call it, most agree that animals and humans share many traits, including emotions. This is merely an extension of Charles Darwin's well-accepted ideas about evolutionary continuity, in which it is argued that the differences among species are differences in degree rather than in kind. Thus, we are not inserting something human into animals but are identifying commonalities and then using human language to communicate what we observe.

Anthropomorphism is a much more complex phenomenon than we might expect. It may very well be that the seemingly natural human urge to impart emotions onto animals, far from obscuring the "true" nature of animals, may actually reflect a very accurate way of knowing and the knowledge that is gained, supported by solid scientific research, is essential for making ethical decisions on behalf of animals. I think that some individuals are far too worried about being anthropomorphic without realizing how important it is for gaining access to who animals are and what they feel (Horowitz and Bekoff 2007).

The Importance of Compassion: Minding Animals

Compassion is the key for bettering animal and human lives. Compassion can lead to justice for all. All over the globe people are talking about ways to lighten our carbon footprint and accrue carbon credits. But what about increasing our compassion footprint and accruing compassion credits? A good way to make the world a more compassionate and peaceful place for all animals, to increase our compassion footprint, is to "mind" them. Minding animals means recognizing that they have active minds and feelings. We also mind animals as their caretakers in a human-dominated world in which their interests are continually trumped in deference to ours.

To mind animals it is essential for people with varied expertise and interests to talk to one another, to share what we know about animals and use this knowledge for bettering their and our lives. Figuring out how science and the humanities, including those interested in animal protection, conservation, and environmentalism (with concerns ranging from individuals to populations, species, and ecosystems), can learn from one another is essential.

Think of the way we double-cross animals. Imagine an utterly exhausted polar bear asking "Where's the ice?" as she attempts to swim with her off-spring from one ice floe to another as she had in years past only to discover that the ice is gone due to climate change(http://www.timesonline.co.uk/tol/news/uk/article767459.ece; http://online.wsj.com/public/article_print/SB113452435089621905vnekw47 PQGtDyf3iv5XEN71_o5I_20061214.html). Imagine the anxiety and fear she feels as she, and perhaps her offspring, slowly drown.

We have a long way to go. Existing laws and regulations allow animals living on land, in water, and in air to be treated in regrettable ways that demean us as a species. Indeed, in the eyes of the law animals are mere property and can be treated like backpacks, couches, and bicycles with no legal recourse. The animals' eyes tell us that they do not like this at all. They have a point of view. Viewing animals objectively does not work.

We are only fooling ourselves if we claim that animals are adequately protected from pain and suffering. Over the past five years violations of the federal Animal Welfare Act (AWA) in the United States have increased more than 90 percent (http://www.all-creatures.org/saen/). In 2006 alone there were more than 2,100 violations of the AWA, with the highest level of violations occurring in the areas of Institutional Animal Care and Use Committees (58 percent) and veterinary care (25 percent). It has been estimated that about 75 percent of all laboratories violate the AWA at one time or another.

"In the Name of Science"

Despite global attempts to protect animals from wanton use and abuse, what we have been doing has not been working. As already mentioned, "good welfare" is not good enough. "Good welfare" and allowable research according to existing regulations permit mice to be shocked and otherwise tortured, rats to be starved or force-fed, pigs to be castrated without anesthesia, cats to be blinded, dogs to be shot with bullets, and primates to have their brains invaded with electrodes. Only about 1 percent of animals used in research in the United States are protected by legislation, and the legislation is sometimes amended in nonsensical ways to accommodate the "needs" of researchers. The desperation of science to rob animals of their sentience produces distortions that open the door for egregious and reprehensible abuse. For instance, here is a quote from the federal register:

We are amending the Animal Welfare Act (AWA) regulations to reflect an amendment to the Act's definition of the term animal. The Farm Security and Rural Investment Act of 2002 amended the definition of animal to specifically exclude birds, rats of the genus Rattus, and mice of the genus Mus, bred for use in research. (Vol. 69, no. 108, 4 June 2004)

It may surprise you to hear that birds, rats, and mice are no longer considered animals, but that is the sort of logic that epitomizes federal legislators. Researchers are not allowed to abuse animals, so the definition of animal is simply revised until it refers only to creatures researchers don't need. We know that mice are empathic beings who feel the pain and suffering of other mice, yet this scientific fact has not entered into discussions about the well-being of mice or other animals.

Excuses justifying animal exploitation such as "Well, it's okay, I'm doing this in the name of science" or in the name of this or that usually mean "in the name of humans." We are a very arrogant and self-centered lot. All practicing scientists need to take responsibility for their practices and always use the most humane and noninvasive techniques. This will produce more reliable data and also set an example for future researchers, including young children who might want to pursue a career in science.

Bill and Reno: A Scientist and His Dog

It is all too easy to push aside the animals' point of view and shirk responsibility. Just when I think I've heard it all, something happens that surprises me. One such event took place a few years ago that made me ask myself, "Isn't a dog a dog?" It seems like a no-brainer to me, but a very bright friend and colleague made me keep asking this question and trying to figure out how he and others could argue otherwise.

I was visiting a major American university where I was going to give a talk on the nature of animal emotions. A friend, let's call him Bill, came up before the lecture, and I asked him how his dog Reno was doing. Well, for more than five minutes Bill extolled Reno's deep emotional life: He loved to play with his friends, missed Bill when he was gone, had separation anxiety, and a few days ago was jealous of the attention that Bill was giving his daughter. "Why," Bill exclaimed, "Reno even knows what he can and can't do when he plays, like he has some moral sense!" A bright, emotional, and virtuous dog, Reno was. He was happy, he knew social rules and right from wrong, and he told Bill this without hesitation.

I was thrilled to hear these stories, because Bill always had been one of those researchers who claimed that we really don't know what animals are feeling and in some ways we can't be sure that they are feeling anything. While animals do act "as if" they experience an array of various emotions, it is too early to make any grandiose pronouncements about their feelings.

My excitement was short-lived. During the question-and-answer time following my talk Bill went for my throat, accusing me of being far too anthropomorphic and too sure of myself. He did it in a sort of lighthearted academic way, and when he was done I simply asked him to recall for the audience the conversation that we had had before my lecture, about Reno's emotional life. Bill turned slightly red, then said, "Well, you know what I meant when we were talking before. I was just letting my hair down and telling stories about Reno. I really don't know that he enjoys playing with his friends or that he is depressed when I leave him alone. And I feel rather certain that he wasn't really jealous of the attention I was giving to my daughter; he just behaved as if he were."

Frankly, I do not know what Bill meant. He was quite comfortable telling me about Reno's feelings in one context but not the other–a distancing mechanism that is used by many people who live with dogs, cats, rats, birds, and other animals at home and who also confront them or similar animals in their laboratories. Bill didn't even seem to know that there was a disconnect–dissonance, psychologists call it–between the things he told me in the space of an hour. And he ignored the fact that humans and many other animals share ancestral parts of the brain that are important in positive and negative emotions, so that it is not at all surprising that Reno would experience and express an array of different emotions.

Similar conversations take place at cocktail parties when it is permissible to let one's hair down and not feel a need to sanitize animal feelings with all sorts of "Well, you know what I mean" disclaimers. I feel sorry for dogs who live with humans who wonder if they really have feelings.

There seems to be inconvenient truths that drive people into indefensible situations. Many individuals have vested interests in what they are doing and how they evaluate information. They are resistant to change even in light of solid scientific data and don't seem to realize that they are being inconsistent in their views on what animals know and feel depending on whether they are at work or at home. Conventional wisdom tells them that it's okay to see dogs in different lights as different beings, depending on where they are and to whom they are speaking about their close companions. However, it is conventional wisdom and bad biology that resulted in animals' losing their minds and hearts "in the name of science." Conventional wisdom has been a large part of the problem, so it will not likely be the solution to reconciling how dogs and other animals are portrayed in different venues. The least we can ask for is consistency in different contexts. Isn't a dog a dog wherever we encounter her?

Empathy in Mice

Scientific evidence is mounting that animals, even rodents, have the capacity to feel empathy. In June 2006, researchers reported in the journal Science the first unequivocal evidence for empathy between adult non-primate mammals. Dale Langford and her colleagues demonstrated that mice suffer distress when they watch a cage-mate experience pain (Langford *et al.*, 2006). Langford and her team injected one or both members of a pair of adult mice with acetic acid, which causes a severely painful burning sensation. The researchers discovered that mice who watched their cage-mates in pain were more sensitive to pain themselves. A mouse injected with acid writhed more violently if his or her partner had also been injected and was writhing in pain. Not only did the mice who watched cage-mates in distress become more sensitive to the same painful stimuli, they became generally more sensitive to pain, showing a heightened reaction, for example, to heat under their paws. The researchers speculated that mice probably used visual cues to generate the empathic response–which is interesting because mice normally rely most heavily on olfactory communication.

Other researchers were quick to note the importance of these unanticipated findings. Renowned primatologist Frans de Waal (2006) said of Langford's research, "This is a highly significant finding and should open the eyes of people who think empathy is limited to our species." These data confirm that empathy is an ancient capacity probably present in all mammals. Neuroscientist Jaak Panksepp (2006) similarly remarked, "If it turns out that the 'empathetic' effect in mice is mediated by the same brain mechanisms as human empathy, then the evidence would be truly compelling that Langford's model actually reflects evolutionary continuity in a pro-social mechanism among many different mammalian species."

Sadly, one of the researchers who was part of the team suggested that an opaque barrier be used to separate mice so that they could not know what was happening to another mouse, because mice who observe each other during experiments may be "contaminating" the data. I am sure that if you saw what was going on in the other cage you'd prefer not to see it, either. Here is a good example of a scientist shirking his responsibility to provide the animals he uses with the very best care possible.

The Privilege of Studying Animals

Studying nonhuman animals is a privilege that must be taken seriously and not abused. Although the issues are difficult and challenging, they are not impossible. Certainly we must not let the animals suffer because of our inability to come to terms with difficult issues or to accept responsibility for how we treat them. Questioning the ways in which humans use animals will make for more informed decisions about such use. By making such decisions in a responsible way, we help to ensure that in the future we do not repeat the mistakes of the past and that we will move toward a world in which humans and other animals may share peaceably the re-sources of a finite planet.

There is a continuing need to develop and improve general guidelines for research on free-living and captive animals. These guidelines should be inspirational and aspirational as well as regulatory. In our efforts to learn more about the worlds of other animals, we need to study many different species. We must not be afraid of what learning about other species may bring in terms of knowledge about animal consciousness and intelligence and their ability to feel pain and suffer. We should not continue to view animal suffering from afar, nor should we blind ourselves to the many ways in which we cause harm to the world around us.

What I fear most is that if we stall in our efforts to take animal use and abuse more seriously and fail to adopt extremely restrictive guidelines and laws, still more and even irreversible damage will result. Our collective regrets about what we failed to do to protect animals' rights in the past will be moot. We need to enter into close and reciprocal relationships with all beings in this more-than-human world.

Our starting point should be that we will not intrude on animals' lives unless we can argue that we have a right to override this maxim, that our actions are in the best interests of the animals irrespective of our desires. When unsure about how we influence the lives of other animals, we should err on the side of the animals. Some guiding principles include:

- ☆ Taking seriously the animals' points of view
- ☆ Putting respect, compassion, and admiration for other animals first and foremost
- ☆ Erring on the animals' side when uncertain about their feeling pain or suffering
- ☆ Recognizing that almost all of the methods that are used to study animals, even in the field, are intrusions on their lives–much research is exploitative
- ☆ Recognizing that speciesist assessments based on vague notions of an animal's cognitive or mental "complexity" are misguided
- ☆ Focusing on the importance of individual animals as opposed to species membership
- ☆ Appreciating the variations among individual animals and the diversity of the lives of different individuals in the worlds within which they live
- ☆ Valuing common sense and empathy, even though traditionally these qualities have had no place in science
- ☆ Using broadly based rules of loyalty and nonintervention as guiding principles
- ☆ Informing all potential readers of scientific papers how animals were negatively affected by the research so that others can avoid making the same mistakes
- ☆ Recognizing that humans are necessarily anthropocentric and that the nonconsenting animal's point of view can never be adequately assimilated into the utilitarian calculus regardless of how right-minded individual persons are.

Animals should be used only as a last resort, and the researchers are responsible for the choices that they make as they "do science." Limitations of time, money, and energy are not satisfactory excuses for using animals when alternatives are available or can be developed, which is often the case.

Accruing Compassion Credits

It is time that people began to think about how to accrue compassion credits, as they do carbon credits (see for example http://www.time.com/ time/health/article/ 0,8599,1709186,00.html). Every individual can make positive changes for all living beings by weaving compassion, empathy, respect, dignity, peace, and love into their lives. It is simple to make more compassionate choices about what we eat and wear and how we educate students, conduct research, and entertain ourselves at the expense of animals. Increased compassion for animals can readily lead to less carbon because of the inverse relationship between these markers especially in our consumption of factory-farmed meat from highly abused animals (http:// www.ciwf.org.uk/ globalwarning/index.html).

We also can focus on the value of individual lives when we try to restore animal populations and ecosystems. It is fair to ask whether the life of an individual should be traded off for the good of its species, for example when we try to restore wolves to Yellowstone National Park and individual wolves die so that others might live.

Making every attempt to coexist peacefully and to do so in the most compassionate ways possible is a win-win situation. Compassion for animals will make for more compassion among people, and that is what we need as we journey into the future. Cruelty to animals has serious implications for humans. Studies by Frank Ascione, Phil Arkow, Barbara Boat, and many others show that children who are cruel to animals are significantly more likely to commit violence against humans later in life; the absence of empathy for one indicates lack of empathy for the other (http:// www.psyeta.org/sa/sa10.4/raupp.shtml). Indeed, studies of prison inmates reveal that as many as 75 percent of violent offenders had early records of animal cruelty (http://www.arkonline.com/violence.html). The Humane Society of the United States has a program, called "First Strike," devoted to learning more about the connection between cruelty to animals and to humans (http://www.hsus.org/ hsus_field/first_strike_the_connection_between_animal_cruelty_and_ human_violence/). The Society & Animals Forum and the Human/Animal Violence Education Network have launched similar programs that deserve our support (http:/ /www.psyeta.org/sa/sa10.4/raupp. shtml). Along these lines, Albert Schweitzer (1923) wrote: "Until he extends his circle of compassion for all living things, man will not himself find peace."

It may seem trivial to claim that we are the future, as are our children and theirs, but I think it bears repeating. There is much work to be done, and every human plays a role in fostering more compassion, respect, and love for animals. Each of us is responsible for the choices we make, and each makes a difference. As we learn about animals and come to appreciate them for who they are, this knowledge can produce growth and tolerance. By loving other animals we make the world a better place for all beings, including ourselves. Showing children that animals are emotional and moral beings is critical (Bekoff 2008).

Human relationships with animals are full of contradictions and ambivalence as well as deep connections and love. On the one hand, animals are used and abused in a sickening and morally distasteful array of human-centered activities. On the other hand, animals are revered and worshipped and form an indispensable part of the tapestry of our own well-being. They make us whole and make us feel good. In many situations it's a double-cross, as we welcome animals into our lives and then treat them as if their lives don't count. We slaughter, silence, and squelch sentient beings with little more than a fleeting thought and with reprehensible indignity. The love-hate relationship that humans have with their animal kin makes life difficult for all.

What is thoroughly unacceptable is that animals often are used to define who we humans are in the great chain of being, and the chain is then presented as a hierarchy in which humans place themselves separate from and above other animals, when we are no better than mice or rabbits or dogs or cats. Hierarchical speciesism results in endless harm and is bad biology. Trumping the interests of animals in the name of science really means in the name of humans. We declare that we are special and better and more valuable than our animal kin and go on to close the door on the lives of other animals. We shut down our senses and our hearts to their pleas that we should take them seriously for who they are and not for what we want them to be in our narrow anthropocentric view of the world.

Let us not forget that throughout the world animals have no legal standing. They are merely property, like backpacks or bicycles, and humans are their owners. Animals can be legally abused, dismissed, disenfranchised, moved here and there, bartered, harmed, and killed. Often this happens in the name of education, science, entertainment, clothing, or food–self-serving excuses that boil down to "in the name of humans." Shame on us. Animals are not merely property. Even young children know they're not (http://connecticutforanimals.blogspot.com/2007/11/hartford-courantop-ed-7-yo-ct-ar_12.html).

The Animals' Manifesto

Animals want to be treated better, with more dignity and respect, and we can make their lives easier and more comfortable by paying attention to what they are asking for. Like us, other animals normally seek pleasure and avoid pain. Here are ten overlapping reasons why animals deserve far more than we've been giving them.

- ✰ They exist, and we share Earth with them.
- ✰ This land is their land, too.
- ✰ They are more than we previously thought.
- ✰ We are alienated from them.
- ✰ We need to mind animals and look out for one another.
- ✰ We are powerful and must be responsible for what we do to them.
- ✰ What we are doing now isn't working.
- ✰ "Good welfare" isn't "good enough."
- ✰ We need to increase our compassion footprint.
- ✰ We all can do something to make the world a more compassionate and peaceful place for them and for us.

I offer these ten reasons to stimulate discussion, not because they are the only reasons why we need to unravel the notion of animal welfare and treat animals with more respect and dignity but because reflecting on these (and other) reasons will force us to be more responsible for what we do to other animals and help to expand our compassion footprint. More compassion is always needed in the world. It's easy to make small, simple changes in our interactions with animals. Ultimately, compassion for animals will make for more compassion among people, weaving more empathy, respect, dignity, and love into all our lives.

Animals are asking us to treat them better or leave them alone. So, let us increase compassion. Animals and future generations of humans will thank us for our efforts.

References

Allen, Colin, and Marc Bekoff. 2007. "Animal Minds, Cognitive Ethology, and Ethics." *Journal of Ethics*. 11:299–317.

Balcombe, Jonathan. 2006. *Pleasurable Kingdom: Animals and the Nature of Feeling Good*. London: Macmillan.

Bekoff, Marc. 2007a. *The Emotional Lives of Animals: A Leading Scientist Explores Animal Joy, Sorrow, and Empathy–and Why They Matter*. Novato, Calif.: New World Library.

Bekoff, Marc. 2007b. *Animals Matter: A Biologist Explains Why We Should Treat Animals with Compassion and Respect*. Boston: Shambhala.

Bekoff, Marc. 2008. *Animals at Play: Rules of the Game*. Philadelphia: Temple Univ. Press.

Bekoff, Marc, and Jessica Pierce. In press. *Wild Justice: The Moral Lives of Animals*. Chicago: Univ. of Chicago Press.

de Waal, Frans. 2006. "Message from Mouse to Mouse: I Feel Your Pain." *New York Times*. July 4.

Horowitz, Alexandra C., and Marc Bekoff. 2007. "Naturalizing Anthropomorphism: Behavioral Prompts to Our Humanizing of Animals." *Anthrozoös*. 20:23–36.

Langford, D. J., S. E. Crager, Z. Shehzad, S. B. Smith, S. G. Sotocinal, J. S. Levenstadt, M. L. Chanda, D. J. Levitin, and J. S. Mogil. 2006. "Social Modulation of Pain as Evidence for Empathy in Mice." *Science*. 312:1967.

Panksepp, Jaak. 2006. www.the-scientist.com/news/display/23764/#23829. Schweitzer, Albert. 1923. Kulturphilosophie. http://en.wikiquote.org/wiki/Albert_Schweitzer.

Chapter 2

Intrinsic Value and Transgenic Animals

☆ *Robin Attfield*

Henk Verhoog has argued that one of the key objections to the genetic engineering of animals lies in the intrinsic value of wild species, which such genetic engineering undermines. Respect for the intrinsic value of animal species supposedly constitutes a deontological ground (a non-consequentialist ground, that is) for a qualified "no" to genetic engineering, or rather for a "no, unless". Here "unless" is short for "unless a basic or very serious human or animal interest ... is involved which cannot be met by other means" (1992b:160)[1]. In this paper I set out to explore what might be meant by the suggestion that animal species have intrinsic value, why philosophers seek to introduce here a deontological or non-consequentialist ground of objection to genetic engineering independent of good or bad states of individual living creatures, and whether what needs to be said can be said in consequentialist terms and on a consequentialist basis. If so, I shall suggest, the objections to genetic engineering can be articulated much more cogently, where they stand up at all.

Verhoog's Intrinsic Value Principle

Where the housing systems of food-animals cause them suffering, it is sometimes proposed that, to prevent this suffering, rather than change the housing systems, the animals required to live in those systems be genetically manipulated so as not to suffer. As Verhoog says, this "goes against the moral intuition of many people "(155).

This chapter appeared previously as Robin Attfield, 'Intrinsic value and transgenic animals', in Alan Holland and Andrew Johnson (eds), *Animal Biotechnology and Ethics*, Chapman & Hall: London, 1998, 172-189,

Yet, as he also says, approaches which locate value in the pleasures and satisfactions of individual animals and treat suffering alone as evil cannot give a satisfactory account of this intuition.

Admittedly not everything need rest on this one intuition, however widespread and persistent. Yet there does seem to be something lacking in theories which, as I have argued elsewhere, imply that there is nothing wrong with breeding or generating animals lacking the capacity to suffer pain so that they may live in otherwise painful conditions, and which imply indeed that no harm whatever can be done to creatures lacking sentience[2]. Thus Verhoog's search for a better theory so far deserves applause. And the same applies when the position which he is criticising allows that adjusting animals to their environment might be objectionable, but only on aesthetic grounds; for, while these grounds are often in place, there would also seem, as Verhoog says, to be a moral issue here as well as an aesthetic one. He is criticising Rollin, who holds that being harmed depends on having conscious experiences[3].

Verhoog's search for a better theory takes him to the notion of intrinsic value. Now part of his diagnosis of the defective nature of theories such as that of Rollin consists in his view that their locating of intrinsic value in the experiences and states of individual creatures pays insufficient regard to the creatures themselves. Here he is in agreement with Tom Regan, who holds that this approach makes individual animals mere receptacles of value[4]. Indeed his endorsement of Regan's position extends to his deciding actually to name the view of intrinsic value in which it applies to experiences and states (rather than to individuals) as the "receptacle view" (*EGEA*, 270). Unlike Regan, however, who writes of the "inherent value" of individuals, Verhoog, to avoid confusion, decides to use the term "intrinsic value" for the value borne by individual animals (*Ibid*, 270).

Can individual entities be the bearers of intrinsic value? In the classical sense employed by G.E. Moore and C.I. Lewis, intrinsic value attaches to whatever is valuable or desirable for no reason beyond the intrinsic nature of the thing in question[5]. While the nature of individuals might conceivably underpin their goodness, intrinsic value as intrinsic desirability attaches much more plausibly to experiences or qualities or to the having of such experiences or qualities, rather than to individuals. Individuals as such do not seem to supply reasons for action. Accordingly, the majority of philosophers since Moore have restricted their application of "intrinsic value" to qualities, experiences and states, as opposed to subjects and individuals. It is important, therefore, to investigate Verhoog means by "the intrinsic value of animals."

Verhoog's position is summarised as follows:

> ... when we say that an animal (or any other entity) has intrinsic value, we mean that (a) the animal deserves to be treated as a goal in itself, and not merely as a means towards the well-being of others; and (b) the animal has characteristics which make it justified to consider the animal as morally respectable. These characteristics are constitutive of the nature of the animal (*EGEA*, 207).

(Here "respectable" almost certainly means "worthy of respect".) Shortly, what is meant here by "goal" and "nature" will be reviewed; but it should at once be remarked that some kind of Kantian position is implicit here. Thus it might be suggested that what Verhoog really has in mind, particularly with respect to clause (b), is that animals (or perhaps some animals) have *moral standing, i.e.* have characteristics such that they warrant practical respect, and such that their interests must be taken into consideration by moral agents. The term 'moral standing' would certainly be clearer in contexts concerning the place of individuals within morality (I suggest) than the language of "intrinsic value". But this reading would probably be less than Verhoog means; for it is at least possible that some form of Categorical Imperative is already implicit in his use of "has intrinsic value". Certainly at the end of the same paper he proposes the introduction of a deontological element, with respect to the ethics of the treatment of animals, based on their intrinsic value (*Ibid*, 277). Nor is Verhoog alone in following Kant in such matters; for Alan Holland has written, with at least partial approval, of a Kantian proscription against regarding benefits for one subject as compensation or justification for adverse impacts on another[6].

Now Holland recognises that this would involve an extension of Kant's Categorical Imperative, since most animals lack the kind of rationality which Kant required as the basis of respect. In its extended version, the Kantian principle is grounded in a "common significance ... in every creature's organised form of life", and "the view, which may be traced back to Aristotle, that every creature is possessed of a *telos* (which literally translated in fact means 'end' or 'goal') which constitutes the fulfilment of its particular form of life." Holland expounds the Kantian principle as follows: "According to this view, other things being equal, one ought not willingly and avoidably to subject any creature to conditions which deny it the ability to express its *telos* (1995: 297)." This principle is then ascribed to Michael Fox, who holds that "transgenic manipulation is wrong because it violates the genetic integrity or *telos* of organisms or species" (cited Holland, 1995: 297)[7].

This apparent digression now serves to throw light on what Verhoog probably means by "the animal deserves to be treated as a goal in itself" (1992a: 270); for he wrote in awareness of Rollin's reintroduction of the Aristotelian concept of "*telos*", with a somewhat reduced sense comprising (to borrow Holland's phrase) "the interests and nature of an animal which it possesses as a member of a species" (1995: 298), or, in Rollin's words, "the set of needs and interests, physical and psychological, genetically encoded and environmentally expressed, which make up the animal's nature" (*EGEA*, 272). Though Verhoog quarrels with Rollin's willingness to speak of changing an animal's *telos* by genetic engineering, there is no sign that he dissents from Rollin's sense of the word. Thus in writing of treating an animal as a goal in itself, Verhoog replaces the Kantian phrase "an end in itself" with words ("a goal in itself") which probably convey "the bearer of a *telos*" (1992a: 270). And in adding that the animal is not to be treated as a means only, he may well be voicing the Kantian proscription against treating benefits for human beings as compensation or justification for adverse impacts on another–subject, apparently, to the explicit, rather non-Kantian proviso mentioned above, "unless a basic or very serious human or animal interest ... is involved which cannot be met by other means" (CI, 160).

Verhoog's basis of respect, however, includes "the nature of the animal", and about this he writes that "To have a 'nature of its own' is unthinkable without taking into account the species to which the animal belongs" (*EGEA*, 273). Here he is critical of Rollin for not regarding species as legitimate objects of moral concern. In order to employ "a concept of species which reflects the dynamism and change inherent in the process of evolution" (*EGEA*, 276), Rollin adopts the definition of a species as an "interbreeding population" (cited Verhoog, 1992a: 276), but Verhoog points out that the rival phylogenetic theory, in which species are compared to individuals, is equally eligible as far as these same criteria are concerned. The second definition lends itself much more readily to talk of the integrity of the species being disturbed, and thus to the species itself being an object of moral concern. And if so, an animal's species-specific nature, Verhoog may be taken to imply, strengthens the basis for respecting the animal.

Here it is appropriate to express doubts whether, even conceived as an individual, a species has a good of its own, and thus is of itself morally considerable independently of its members. Species can certainly flourish, but they do not go through inherited patterns of development in anything like the ways in which their members do. Nor can it be assumed that their naturalness, as organised forms of life or as products of evolution, makes it wrong in general to modify them. It is rather their members which have a good of their own (a characteristic which they share with domestic animals and with animals produced by genetic engineering).

Verhoog, however, believes otherwise. He writes of "a degradation of the intrinsic value" of wild species when domestication takes place, at any rate from an ecocentric point of view (CI, 159), and thus presupposes that at least wild species have moral standing. In addition, he holds that species have intrinsic value, and are thus to be respected, even when they have been eroded or nullified. While this may seem unsatisfactory, what he is in any case able to claim with some plausibility is that, however dynamic and changing, species are sufficiently real at any one time as to confer biologically detectable, species-specific natures on their wild (and sometimes on their domesticated) members; and that these natures are a suitable basis for moral respect. And this conclusion is sufficient to raise questions about genetic engineering, independently of his claims about intrinsic value.

Verhoog does not, however, regard all domestication as wrong. "When man is viewed as a part of an ecosystem", he writes, "and a 'partner' of other natural entities, then it is not necessarily a bad thing when animals are domesticated for human survival" (*Ibid*, 159). "In the 'partner-view', the acceptability of human intervention depends on the 'quality' of the new relations arising" (*Ibid*, 159). (As will be seen, this has implications for the acceptability of genetic engineering.)

Meanwhile there are already problems for domestication relating to the *telos* which has been partially lost. For (a) domestic animals are not to be held to have a changed *telos*; at best they have an imposed, new *telos*, and at worst none at all, except in sofar as their wild *telos* remains partially intact. Also, (b) it ceases to be possible to identify psychological suffering, as the nature of this is species-specific and depends on having an intact *telos* (controversial claims these); and (c) the nature of these animals is ceasing to be autonomous and beginning to be artificial. The implications of all this are less than clear; but Holland infers that, for Verhoog, "the process of domestication is open to critical scrutiny" (Holland, 1995: 300). This would mean that existing patterns of domestication cannot be regarded as a precedent liable to justify whatever genetic engineering may be shown to be parallel to them[8]. And here, if this is what he is saying, Verhoog is surely right.

When the focus changes to genetic engineering, the moral issues become yet more acute. For there is now no natural *telos* at all, but only an imposed one. Hence it is no longer possible to discover what the animal's needs and interests are, or when it suffers. (This claim will be contested below.) Again, genetically engineered animals have ceased to be autonomous, or to be involved in relations of partnership with humanity; in these circumstances, it is at least dubious whether they have intrinsic value at all (*EGEA*, 272).

From these implications of his principle of the intrinsic value of species, Verhoog derives his "no, unless" verdict on genetic engineering. He rejects the reflection that, if we cannot identify the interests of genetically engineered animals, then there are no moral objections any longer, however badly they are treated, though he does see this as a problem for Rollin (*EGEA*, 272). Presumably he rejects this by appealing to what is involved in respecting a natural species and its natural *telos*. This is strong enough to prohibit actions which erode species and *telos*, but not strong enough to do so in all circumstances, since "a basic or very serious human or animal interest ... which cannot be met by other means" (Verhoog, 1992b: 160) may be involved.

A Critique of this Principle

One unresolved issue concerns how these precise conclusions can be derived. Thus the recognition of basic or very serious human and animal interests of other creatures suggests that Verhoog does not rule out a consequentialist appeal to interests or to the possibility of comparing them, and this already clashes with the tacit Kantianism of Verhoog's main principle. Relatedly, his main principle seems specifically to prohibit the sacrificing of the interests of *e.g.* a transgenic animal for the sake of others' interests, and yet this is just what his overall conclusion permits.

These problems seem to result from an attempt to combine "the now dominant consequentialist theory of animal ethics" with "a deontological element based on the intrinsic value of animals" (*Ibid*, 272). But, strictly speaking, it is impossible to combine consequentialist and non-consequentialist theories. For the former make obligations and right and wrong actions turn wholly on consequences (whether the consequences of acts, rules or practices), while the latter do not. And even if this is recognised, and it is acknowledged that would-be mixed theories are, strictly, deontological theories, the problem still arises for any particular such theory of supplying satisfactory criteria for the extent and limits of the consequentialist element in the theory; for it is not plausible that there is any basis of comparison between the weight of the consequentialist element and the weight of the deontological element.

To tackle these problems, a fairly precise meaning would be required for the principle of the intrinsic value of species; but the terms chosen make this, too, an intractable problem. For "intrinsic value" has long functioned as a phrase from the consequentialist vocabulary, and applies to states of affairs which supply reasons for action in virtue of their intrinsic desirability. When Verhoog (to avoid the "receptacle view") follows Regan and ascribes value to individuals rather than to their states and experiences, there is already a difficulty, since it is not, by and large, individuals which are desirable. Maybe this problem could be cured if "intrinsic value" were now given a new and unqualifiedly Kantian sense, such as "never to be treated only as a means to the wellbeing of anything else." But Verhoog is too realistic to do this, recognising that such protection, whether for species, wild creatures, domestic animals or genetically engineered animals, is implausibly strong, and perhaps recognising too that it would generate large numbers of contradictions, in cases where the interests of more than one protected creature conflict. Hence his sense of "intrinsic value" turns out to be compatible, as in its consequentialist use, with the subordination of the interests of the intrinsically valuable individual after all.

There is, however, another reason for using "intrinsic value" in a less-than-Kantian sense. For Verhoog wants to ascribe intrinsic value to species with a natural *telos*; he wants to argue from the respect due to a natural species to conclusions such as the wrongness of genetic engineering which adds to the erosion of a *telos* which was no longer fully present beforehand (having been eroded already through prior processes of domestication). Here there is little or no question of using the species as a means only, since it is supposed to be scarcely present any longer at all. Also there is certainly not supposed to be any question of using its members as means only, as by this stage it is supposed to be impossible to identify their interests; and so it has become unclear what it would be like to treat them as means only. The objection lies rather in failing to respect the natural species which there once was, and at the same time undermining the kinds of relations between species which evolve naturally. But, whether or not there is merit in these objections, there is surely some clearer way of conveying them than through a semi-Kantian use of "intrinsic value".

Is Deontology Really Necessary

It is time to explore the possibility that Rollin's position can be supplemented, and grounds found for rejecting the generation of desensitised creatures, without a deontological theory grounded on the supposed intrinsic value of species. Rollin's objection to causing or allowing avoidable suffering should be retained, but should surely be extended to include the causing or allowing of harms whether or not they consist in suffering (physical or psychological) of which the suffering animal is aware. This is because harm in the form of unconscious disease or injury constitutes an evil alongside suffering which is noticed and distresses the sufferer. One further extension is also in place; for, as Derek Parfit (1984: 487-490) has argued, a creature can be harmed by bringing it into being if its life will foreseeably be one of misery, or worth not living[9]. In this extended form, though the wrongness of generating desensitised creatures is yet to be covered, consequentialism begins to look suited to the ethics of animal husbandry and of genetic engineering; for it already prohibits generating chickens destined for the conditions of factory farms, where their lives are lives of misery, without in general prohibiting killing animals for food, and it also prohibits engineering pigs (like the Beltsville pigs) destined to arthritic lives through their genetic endowment, without debarring the generation of *e.g.* Factor IX sheep.

But before consequentialism can be rehabilitated, including the consequentialist claim that it is states and activities of individuals, rather than those individuals themselves, which are of intrinsic value or disvalue, Verhoog's condemnatory classification of this position as "the receptacle view" needs to be reviewed. Does this position really encourage or allow lack of respect to individuals?

The first point to make is that, far from failing to ground or underpin respect for individuals, it presupposes that individuals whose lives are capable of intrinsic value and disvalue have moral standing or are morally considerable - that is that they have properties which qualify them for practical respect and that their interests are, morally speaking, to be taken into consideration. That is not to say that their interests will prevail, but it is (as has been seen) to say that they are not to be treated solely as means. Accordingly nonanthropocentric consequentialism includes provision for respect for individual creatures in its very foundations.

If, further, the objection is that what matters is states of intrinsic value rather than individual creatures, it may be replied that what matters is the having or enjoying of states of intrinsic value by nothing but individual creatures. The relation between pleasure or flourishing and their bearers is not a contingent one, as there can be no pleasure and no flourishing without it being someone's or something's. Free-floating pleasure is a fantasy of consequentialism's deontological opponents.

Sometimes, however, the objection concerns the readiness of consequentialism to aggregate good and bad states of affairs, whoever undergoes them, and to judge by the overall balance of good over evil, rather than safeguarding the interests of each individual concerned. But no one, and certainly not Verhoog, is prepared to reject all comparisons of affected interests; as has been seen, he is prepared for the interests of individual animals to be overridden for the sake of certain interests of human beings. Indeed any theory which made the intactness of every individual's interests an absolute and indefeasible right would rapidly be inundated with contradictory judgements in cases of conflicts of interest, where both action and inaction were forbidden. Since respecting individuals may be supposed not to have these implications, it cannot prohibit comparisons of interests.

But aggregation now becomes the focus of complaint; it might seem that consequentialism is indifferent between diverse distributions of the same good, and disregards *e.g.* greater needs or stronger entitlements. While an adequate consideration of this issue would take us too far afield, two replies may be given. First, if the satisfaction of basic needs is recognised as of much greater value than other goods, then greater needs will not be disregarded; and it is open to consequentialists to recognise this greater value (and indeed it is open to deontologists to ignore it). Second, consequentialists are free to appraise consequences not as the consequences of single acts but as the consequences of rules or of practices; and if practices embodying entitlements turn out to have a better overall outcome than alternative practices, then consequentialists will favour them. Thus accusations of unprincipled aggregation and of distributional indifference are wide of the mark.

What remains of the "receptacle view" objection consists in the impersonal character of consequentialist ethics. It may, by now, be recognised that consequentialism does not authorise the replacement of anyone and everyone by another creature who or which would be happier, but it may still be objected that the willingness of consequentialists to take into account the good of possible creatures which could be brought into being is liable to downgrade the interests of those which are actual. To this, however, there is a forthright reply, stemming, as before, from the work of Derek Parfit. Since most future creatures are at present no more than possible creatures, our responsibilities to future generations would come to nothing if our obligations ran out at obligations with regard to actual rather than possible creatures. Although the numbers and even the identities of future creatures are as yet unknown, we still have responsibilities extending beyond duties to actual, identifiable creatures, to ensure that whichever ones live also lead lives worth living, and not significantly worse than action on our part could have prevented. To put things another way, some spheres of morality are importantly impersonal.

What has been argued is insufficient to develop and defend a nuanced consequentialist theory, but it does suggest that there is nothing amiss with the consequentialist view that reasons for action consist in the good and bad conditions which can befall creatures actual and possible. It is not so much that there is nothing wrong with the "receptacle view" as that this classification is a misnomer. The concern of the consequentialist is for the wellbeing of creatures, and not for the dispassionate aggregation of free-floating value; and the fact that this is a more comprehensive form of concern than any deontological position yet encountered does not make it any the less a form of concern, and any more disposed to treat individuals as mere channels, vehicles or receptacles.

But since the ethics of transgenic engineering is in question, and consequentialism has been found wanting in this very connection by Alan Holland, a brief review of his confessedly brief critique is necessary before further progress can be made. (Strictly Holland was criticising utilitarianism, the version of consequentialism which seeks to optimise the balance of happiness and unhappiness, but parallel points arise when a consequentialism concerned with value and disvalue in general is in question.)

Holland's first criticism is that pleasures cannot be uncritically counted and balanced, since some pleasures may be wrongfully or shamefully gained. As he forcefully puts matters, "the balance of advantages cannot decide acceptability, since it presupposes it" (AI, 296). Here the consequentialist can reply that she is not committed to counting or balancing uncritically. There are easily defensible consequentialist rules against robbery (Holland's example), or against shameful practices such as corruption or nepotism; for such practices are enormously destructive of social cohesion and trust, and of the conditions widely necessary for the flourishing of human life.

The second of Holland's criticisms is that "calculation of cost and benefit cannot ignore the question of the relation between the two. A benefit which is obtained *by* paying a cost cannot be regarded in the same light as one which is not" (*Ibid*, 296). Not just any paying of a cost is intended (or this would be an objection to buying from shops). Holland's example is the moral condition which would characterise motorised transport if its advantages could only "be secured in advance by the sacrifice of a certain number of lives" (*Ibid*:296), like the sacrifice of Iphigenia to gain a fair wind from Greece to Troy. Since Holland acknowledges that currently we appear to think that the advantages of motorised transport outweigh the loss of life involved, a first reply is that this loss of life already includes the loss of life among construction workers before roads and roundabouts have even been built. Not that society's acquiescence in this makes all road-building acceptable. Indeed sometimes the loss of life among construction workers is predictably so great that, at least on a consequentialist basis, construction should not have begun. Further, such are the pollution and accidents resulting from motor vehicles that transport policies in many cities and countries stand in need of drastic change; but this judgement is itself one enjoined by the consequentialist considerations.

Probably, however, Holland envisages cases where death or serious injury is deliberately inflicted on others, as when one country goes to war with another for territorial gain, or, to devise an example closer to motorised transport, where someone is assaulted in order to commandeer their vehicle. These different examples relate to different practices and their infringement, and the consequentialist is as entitled as anyone else to scrutinise distinct practices differently. Thus she may adhere to a consequentialist version of the just war doctrine, which would plausibly outlaw most wars, including, incidentally, the retaliatory war of the Greeks against the Trojans, and thus the sacrifice of Iphigenia; and she may likewise adhere to laws which (on consequentialist grounds) forbid assault, but which allow commandeering and the use of force in exceptional cases of genuine military "necessity". In none of these cases is the consequentialist committed to aggregation unqualified by regard to the distinctive consequences resulting from the practice in question and the diverse circumstances in which it may be invoked; and in all these cases consequentialism plausibly generates defensible principles.

Consequentialism, then, is not a discredited theory, but a defensible one. But it remains to be seen whether it can cope with whatever can be salvaged from Verhoog's talk of the intrinsic value of wild species, and can say what needs to be said about the ethics of generating desensitised creatures. Indeed a consideration of what needs to be said about this is long overdue.

Consequentialism and Transgenic Animals

Verhoog is doubtless correct in holding that domestication and its processes cannot be taken as above criticism. Many animals are prematurely killed, and many more mutilated *e.g.* through sterilisation; while domesticated breeds are often selected for tractability and submissiveness. While, as Holland points out, domestic animals by and large live longer than their wild counterparts, and wild animals lead on average a tougher life (*Ibid*, 296), there is considerable room for more compassionate forms of farming, and a strong case for the termination of certain alienating practices.

But protests and objections targeted at the processes of domestication need not be grounded in lack of respect to the species which preceded domestication or to its *telos*. Most obviously (although not, as will be argued, exclusively) they can be grounded in avoidable suffering and avoidable harm, and thus in states of intrinsic disvalue. Certainly Verhoog is right to hold that to identify some forms of suffering we need to know the needs and interests of the creature concerned, and that these inherited needs and interests determine whether domesticated creatures are deprived of behaviour patterns natural to their kind. But this does not make the grounds of objection to such deprivation respect for their species or its *telos*. Grounds enough are to be found in concern for the wellbeing of the individual animals as they now are, intrinsically valuable as this clearly is, and varying as are the forms which wellbeing may take.

Verhoog's very acceptance of domestication in cases where humanity is viewed as a partner of other natural entities concedes that not all domestication, and thus not all deviations from the *telos* of a wild species, is wrong. As was mentioned above, his criterion of the acceptability of domestication consists in "the 'quality' of the new relations arising"; but this criterion is too vague to serve as such. His further suggestion, that the animals must be autonomous, in that they must not have a humanly imposed nature, seems too strict, as will shortly become apparent; for selective breeding could be advantageous to animals, in virtue, for example, of resistance to disease. One defensible interpretation of the new relations having a suitable quality would consist in the basic needs of the animals (identifiable through the natures of their kind or lineage) being satisfied; certainly nothing less would avert charges of exploitation. This criterion is radical enough to clash with many of the methods of modern farming (as is to be expected of an interpretation of the "partnership"-view). It is also a criterion which can be incorporated within consequentialism, granted a value-theory which attaches a high intrinsic value to the satisfaction of basic needs (or to what they are necessary for), high enough to outweigh the benefits which might accrue to human beings from animals' basic needs being overridden.

Imagine, further, an animal selectively bred to be immune from a painful illness which besets most members of its species, but otherwise leading a flourishing life characteristic of its kind. Since its wellbeing would seem to be greater than that of most of its kind, generating this animal is hardly open to criticism, or so most people, including consequentialists, would say. Yet if showing respect for wild species as such were what matters, we might have to say that generating this animal is wrong. I am assuming here that the corresponding wild species is not in danger of extinction, for if it were in such danger, there would be more factors to take into account, including the value which could have been present in the lives of species-members which will never now be lived; indeed the loss of this value could account for the inclination to hold that the species is to be respected and preserved as such. But where the animal with the immunity is simply brought into existence alongside the population lacking the immunity, there is hardly a ground for complaint, not even if immune animals are bred in large numbers and become more numerous than members of the traditional species. Here respecting the species seems one deontological principle too far, prohibiting what should not be prohibited.

Holland, however, considers that, if the nature of animals is treated as an end in itself, then there is a Kantian objection even to this benign genetic manipulation. His example concerns building into an animal's endowment "a resistance to disease or injury" (*AI*, 303); and I should acknowledge that this more sweeping measure could be open to objection, as it could have side-effects which made the animal, for example, more timid or more placid (how else has it become resistant to injury?), or generally possessed of reduced capacities. And if less fulfilments were open to it, there would plausibly be consequentialist objections to the original manipulation. But as this is not Holland's case, let us imagine the less likely scenario in which there are no such side-effects, and the only difference to the animal's nature consists in its resistance to disease and injury. To convey the Kantian objection, Holland, writing about "changing an animal's nature for the sake of rendering it less susceptible to disease" claims that "Essentially, it puts respect for the states of a subject above respect for the subject" (*Ibid*, 304). The process is, allegedly, disrespectful of the animal's nature, albeit benign to the wellbeing of the individual animal. Yet all the individuals so treated (plus their offspring if this is an inheritable change) will be healthier and probably more vigorous and longer-living, and *ex hypothesi* there are no side-effects. Besides, even if the animal's "whole nature" is "subordinated" to this process (*Ibid*, 304), it must largely remain intact, or the creature's capacities would have been reduced, contrary to our supposition. So, unless changing an animal's nature is wrong in itself, there seems nothing left to object to, even if the process might be held to involve disrespect for the ends implicit in the animal's nature.

Now the ethics of the breeding of the animals with this immunity would be no different whether it resulted from conventional selective breeding or from genetic engineering, if the assumption is retained that the immunity could be genetically engineered without side-effects. (In actual fact there will usually be a risk of such engineering producing unexpected and unintended bad effects, and at least the risk has to enter into appraisals of such processes.) So it turns out that for some kinds of such manipulation, apart from the consideration of risk, there would be no ethical objection at all, despite the fact that all genetic engineering could be represented as showing disrespect for natural species.

Certainly there is a new difficulty which enters with genetic engineering; for it is much less easy to identify the needs and interests of the animals thus produced, and thus to locate cases where basic needs are being infringed. Since genetic engineering is at least typically intended to benefit human beings, then in view of the desirability of seeking ways of avoiding exploitation of the other party affected (the animals), it would be reasonable to require that people with genetically engineered animals in their charge should patiently seek to discover whether their animals carried the behaviour-patterns of their wild or domesticated ancestors, such as rooting in pigs. Where this is the case, the observed behaviour should be treated as an expression of a basic need, at least where the need had been expressed in such behaviour among the ancestors or among other comparable animals. Thus the "partner"-view has important implications for the ethics of genetic engineering, calling, as it does, for special efforts to identify and satisfy the basic needs of transgenic animals. Yet immunity from a painful disease would still count as a benefit, whatever was discovered about behaviour; for, except in very unusual circumstances, diseases are necessarily disadvantageous. All these considerations can clearly be reconciled with consequentialism.

There are, however, other ethical constraints, I suggest, on genetic engineering. Thus avoidable harm is done where animals are engineered whose lives are not worth living; as was argued above, a creature is harmed by being made to live out an existence which it would be better spared. Such animals should not be generated at all.

But what of creatures engineered to lack feelings, and thus not to suffer? Philosophers for whom the only evil is suffering cannot regard generating such creatures as wrong, while philosophers who appeal to *telos* are also hard-pressed, as little of a wild *telos* remains, but only individuals whose quality of life is slightly above zero. Again, if such creatures were to occupy the ecological space previously occupied by their sentient ancestors and thus supplant their natural cousins, then the defensible principle would be infringed, that it is wrong to generate creatures with a lesser range of capacities where creatures with a greater range of capacities would have lived instead; and this will be a consequentialist principle if its ground consists in the value of realising capacities, as plausibly it does. But what is amiss if the alternative would be no animals at all occupying this same space? This is a question put by Holland with regard to animals whose lives are not lives of suffering, but whose capacities are little above the vegetable (*Ibid*).

With regard to animals lacking sentience, it may be replied that the inability to suffer is itself, in animals, a biological disadvantage; thus animals with this dubious endowment would be disadvantaged as compared with both their contemporaries and their ancestors. Also generating such animals would often violate what Holland calls the "reduction of capacity principle" (*Ibid*), as it is often likely enough that creatures with a greater range of capacities would have existed there instead.

But is it wrong to generate creatures which lead lives more truncated than ones which *could* have been brought into existence instead, and can the generation of creatures lacking sentience be condemned on these grounds? Here Verhoog would probably agree that it is wrong to generate them, but advance the implausible ground that this is wrong because it violates the *telos* which their ancestors once had. For his part, Holland suggests that it is wrong because the zygotes used by genetic engineers to produce the transgenic animals would have grown into an individual with an ampler range of capacities if no genetic engineering had occurred, and thus the very same individual would have been better off, and has actually been harmed (*Ibid*). But if this were the only basis of objection, then there would be no objection to the second and subsequent generations of transgenic animals, if we may assume that at least some of the first generation products of genetic engineering are able to reproduce themselves with the genetic changes intact. For individuals of the second and subsequent generations would not have existed with ampler capacities if the original engineering had not occurred; indeed they would not have existed at all.

This returns us to the question concerning the wrongness of generating creatures with reduced capacities (and thus reduced opportunities for fulfilments) where other creatures with traditional capacities could have been generated instead, even if, market economics being what it is, the other creatures just as likely would not have been generated at all. Yet it is quite unacceptable to say that in these circumstances generating creatures of reduced capacities (for example, without sentience) is allowable. For this is like saying the following: it would be wrong to generate creatures of reduced capacities where the alternatives are to generate creatures with traditional capacities and to generate no creatures at all; but it would not be wrong first to decide not to generate creatures with traditional capacities, and then to claim that there is nothing wrong with generating creatures with reduced capacities, since generating no creatures at all is now the only alternative remaining. And what this suggests is that, if action on the part of the agent in question could have generated creatures with traditional capacities, then the likely fact that this alternative would have been rejected out of hand makes no difference to the wrongness of generating creatures with reduced capacities. In other words, the answer to the question "Is it wrong to generate creatures which lead lives more truncated than ones which *could* have been brought into existence instead?" is "Yes".

The fact that the individuals with truncated capacities and the individuals with traditional capacities are not identical thus turns out to make no difference. This finding is in line with the conclusion reached earlier in the context of Parfit's work, to the effect that morality is, in some spheres, impersonal, and does not simply concern whether identifiable individuals are benefited or harmed. Agents are also responsible for the quality of life of whoever or whatever lives, whatever its identity, insofar as this turns on their own action or inaction; and this supplement to the principles of benefiting and of abstention from harm is once again a consequentialist principle. Given also a value-theory according to which an ampler range of realisations of capacities is, other things being equal, of greater value, there is a strong ethical case against bringing into existence creatures with a reduced range of capacities, as creatures lacking sentience would be in comparison with traditional animals.

Holland has, in fact, challenged me to show that reduced capacities are to be regarded as inferior (or to portend an inferior quality of life) (*Ibid*). Yet he himself had earlier written, "I suggest that in general, and within the framework of a biotic community, there should be an objection to any practice which involves taking a form of life with a given level of capacity to exercise options and reducing that capacity significantly."[11] This view is, I suggest, to be supported because an ampler range of capacities facilitates an ampler range of realisations of capacities, and this is of greater value than a more restricted range. Another way of putting this is, to borrow the phrase which Holland employs, that reducing capacities is "life-denying" and letting them be is "life-enhancing" (*AI*, 303).

He also challenges my use of "crippled" to describe a reduced form of life[12]. This too is a little surprising, as I was taking up Holland's own usage: "'Crippling' a virus is, without doubt, harming it, and there is surely something disturbing in the idea of deliberately creating 'crippled' life-forms, that is, creating life-forms which are permanently harmed."[13] Holland's new question concerns "what is undesirable about a case where the genetic engineer has redesigned what counts as a worthwhile life and in which the condition in question is, precisely, no longer 'crippling'[14]. Here, granted the form of the reduced capacity principle defended above ("It is wrong to generate creatures which lead lives more truncated than ones which *could* have been brought into existence instead"), it may be replied that Holland's first thoughts were best, and (whether or not we say that the redesigned species is permanently "harmed") that this principle supplies a sense in which the members of the redesigned species are indeed crippled; for the principle supplies the comparison (with animals of the traditional life-form which could have lived instead) which the analogy demands. This remains true even though the members of the redesigned species have new (reduced) needs and interests, and on this reduced basis may still be capable of leading a (just about) worthwhile life. Imagine a human being redesigned to assemble cars automatically and flawlessly, for just three hours each day. His or her humanity would be reduced, but (s)he might still be capable of some kind of worthwhile life, centring on the rest of the day. If this monstrous transformation were possible, condemnation would be in place because of the reduction of capacities involved, in the form of the loss of choice and control over large segments of an otherwise human life.

Conclusions

My theoretical conclusion is that there is no need to appeal to the supposed intrinsic value of species or of their *telos*, or to their nature being an end in itself, to account for the objectionable nature of much possible genetic engineering, for this is much more clearly objectionable on consequentialist grounds, concerning the wrongness of generating lives which are not worth living, of causing suffering or other harm, or of reducing range of capacities and realisations. Where, however, appeal to the intrinsic value of species or to *telos* is made for cases which are unobjectionable on consequentialist grounds (such as animals genetically manipulated to be immune to disease without the manipulative process having side-effects), the appeal is, I suggest, unpersuasive. Respect for naturally evolved life-forms rather than for their members might even be held to involve a new kind of "receptacle-view" of the latter.

At the same time the practical conclusion is that there are important ethical criteria applicable to genetic engineering and to the husbandry of transgenic animals which extend beyond a concern to prevent avoidable suffering and avoidable harm and to avoid producing them, and which mandate abstention from reducing capacities. Thus it is wrong to generate creatures which lead lives more truncated than ones which the same agents *could* have brought into existence instead. Further, people in charge of genetically engineered animals should take pains to identify and satisfy the basic needs of these animals. When people become responsible both for the quality of life liveable by animals and for the identities and to some degree the natures of the animals who live, this confers strong obligations, and does not give them *carte blanche* to manipulate as they please. Even creation has its ethics.

Notes

1 Henk Verhoog, "The Concept of Intrinsic Value and Transgenic Animals", *Journal of Agricultural and Environmental Ethics*, 2 (1992), Henceforth cited as CI.

2 Robin Attfield, "Genetic Engineering: Can Unnatural Kinds be Harmed?", in Peter Wheale and Ruth McNally (eds.), *Animal Genetics: Of Pigs, Oncomice and Men*, Pluto Press, London, 1994.

3 Verhoog, "Ethics and Genetic Engineering of Animals", in A.W. Musschenga *et al.* (eds.), *Morality, Worldview and Law* (Assen/Maastricht: Van Gorcum, 1992), p.271 Henceforth cited as *EGEA;* Bernard Rollin, "The Frankenstein Thing: the Moral Impact of Genetic Engineering of Agricultural Animals on Society and Future Science", in J.W. Evans and A. Hollaender (eds.), *Genetic engineering of Animals. An Agricultural Perspective* (New York; Plenum Press, 1986), pp.285-297, p.296.

4 Tom Regan, *The Case for Animal Rights* (London: Routledge & Kegan Paul, 1984)

5 G.E. Moore, "The Conception of Intrinsic Value" in *Philosophical Studies* (London: Routledge & Kegan Paul, 1922); and C.I. Lewis, *An Analysis of Knowledge and Valuation* (La Salle, Illinois: Open Court, 1946)

6 Alan Holland, "Artificial Lives: Philosophical Dimensions of Farm Animal Biotechnology," in T.B. Mepham, G.A. Tucker and J. Wiseman (eds.) *Issues in Agricultural Bioethics* (Nottingham: Nottingham University Press, 1995). See sections on "The Kantian principle" and "Kant revisited". Henceforth cited as *AL.*

7 Holland, "Artificial Lives": see section on "The Kantian principle"; Michael Fox, "Transgenic Animals: Ethical and Animal Welfare Concerns", in Peter Wheale and Ruth McNally (eds.),*The Bio-Revolution* (London: Pluto Press, 1990)

8 Holland, "Artificial Lives"; see section on "The natural *telos* principle".

9 Derek Parfit, *Reasons and Persons* (Oxford: Clarendon Press, 1984), Appendix G, pp. 487-490

10 Holland, "Artificial Lives"; see section on "The reduction of capacity principle"; also Holland, "The Biotic Community: A Philosophical Critique of Genetic Engineering", in Peter Wheale and Ruth McNally (eds.), *The Bio-Revolution* (London: Pluto Press, 1990), pp.166-174.

11 Holland, "The Biotic Community", p. 172

12 Attfield, "Genetic Engineering"; Holland, "Artificial Lives", *ibid.*

13 Holland, "The Biotic Community", p. 171

14 Holland, "Artificial Lives"; see section on "The reduction of capacity principle".

References

Attfield, Robin. 1995. "Genetic Engineering: Can Unnatural Kinds be Harmed?" *Animal Genetic Engineering: Of Pigs, Oncomice and Men.* Ed. Peter Wheale and Ruth McNally. London: Pluto Press.

Evans, J.W. and Hollaender, A., eds. 1986. *Genetic engineering of Animals. An Agricultural Perspective.* New York: Plenum Press

Fox, Michael. 1990. "Transgenic Animals: Ethical and Animal Welfare Concerns." *The Bio-Revolution.* ed. Peter Wheale and Ruth McNally. London: Pluto Press. 31-50.

Holland, Alan. 1990. "The Biotic Community: A Philosophical Critique of Genetic Engineering." *The Bio-Revolution*, ed. Peter Wheale and Ruth McNally. 166-174.

Holland, Alan. 1995. "Artificial Lives: Philosophical Dimensions of Farm Animal Biotechnology." *Issues in Agricultural Bioethics.* ed. T.B. Mepham, G.A. Tucker and J. Wiseman. Nottingham: Nottingham University Press. 293-305.

Lewis, C.I. 1946. *An Analysis of Knowledge and Valuation.* La Salle, Illinois: Open Court.

Moore, G.E. 1922a. "The Conception of Intrinsic Value." *Philosophical Studies.* G.E. Moore. London: Routledge & Kegan Paul.

Musschenga, A.W. *et al.* 1992b. *Morality, Worldview and Law.* Assen/Maastricht: Van Gorcum.

Parfit, Derek. 1984. *Reasons and Persons.* Oxford: Clarendon Press.

Regan, Tom. 1984. *The Case for Animal Rights.* London: Routledge & Kegan Paul.

Rollin, Bernard. 1986. "The Frankenstein Thing: the Moral Impact of Genetic Engineering of Agricultural Animals on Society and Future Science." *Genetic Engineering of Animals. An Agricultural Perspective.* ed. J.W. Evans and A. Hollaender. New York: Plenum Press. 285-297.

Rollin, Bernard. 1989. *The Unheeded Cry: Animal Consciousness, Animal Pain and Science.* Oxford and New York: Oxford University Press.

Verhoog, Henk. 1992. "Ethics and Genetic Engineering of Animals." *Morality, Worldview and Law.* ed. A.W. Musschenga *et al.* Assen/Maastricht: Van Gorcum. 267-278.

Verhoog, Henk. 1992. "The Concept of Intrinsic Value and Transgenic Animals." *Journal of Agricultural and Environmental Ethics.* 2 (147-160).

Wheale, Peter and McNally, Ruth eds. 1990. *The Bio-Revolution.* London: Pluto Press.

Wheale, Peter and McNally, Ruth eds. 1994. *Animal Genetics; Of Pigs, Oncomice and Men.* London: Pluto Press.

Chapter 3

Philosophizing as if Animals Themselves Mattered More: A Zoontological Review of *Philosophy and Animal Life*

☆ *Ralph R. Acampora*

Philosophy and Animal Life brings to the table the resources of major figures in the Wittgensteinian and analytic traditions of philosophy and literary theory.[1] From my perspective, however, too much of the primary and secondary texts that comprise the book is devoted to themes other than (at least half) its ostensible subject(s). Time and again several contributors divert or submerge interests in and of animal life toward or below wonderment about affairs other than animality itself–for examples, the self-constitution of philosophy, the status of an ideology (such as humanism or post-humanism), the nature of language (its generic technicity or the inadequacies of certain species of discourse), and so on. As something of a counterweight to this tendency, I should like to make animals themselves the focus of attention for my commentary on the book at hand. In other words, my gambit is that animals are not only good to think with (as Levi-Strauss and others have said and shown for a good while now) but that they are also here to live with and thus, they place a claim upon us to figure out salutary modes of cohabitation, a challenge that entails attending to them for their own sakes as well as ours and the (inter)relationships that evolve between us. With animals in their own right forming a conglomerate sort of touchstone for my remarks, I argue that some forms of philosophy may be better than others to articulate animal experience (against the presumption of multiple contributors to speak of philosophy as a generic or univocal discipline ill-constituted to handle animality), that not all of the commentators on Coetzee's lectures merit the opprobrium that multiple contributors levy upon them (indeed one commentator in particular would seem, ironically, to well pose the very points the critics wish to press), and that the discourse of animal rights (however incomplete on its own) deserves greater credit and usage than multiple contributors appear ready to grant or endorse.

Coetzee/Costello

A, if not the, main pivot around which *Philosophy and Animal Life* turns is interpretation of J.M. Coetzee's Tanner lectures, their protagonist Elizabeth Costello, and their commentators in *The Lives of Animals*. One prominent thread of critique in this interpretive work is the charge that the commentators mistake the ultimate import of Coetzee's narrative because they fail to appreciate the existential crisis of Costello. The latter character is profoundly wounded by the callousness most of her human fellows display as regards the animal holocaust they commit, permit, or subsidize; she is haunted by the sufferings and killings passively ignored or willfully neglected by so many around whom she must live. As initiated by Cora Diamond, the line of criticism built on this observation proceeds to fault the lectures' commentators for turning away from Costello's and the animals' respective plights by misunderstanding Coetzee's discourse as essentially philosophical argumentation about animal rights dressed up in the rhetorical finery of fictional prose (PAL, 46ff.). Stanley Cavell seconds Diamond's charge on this point and adds that the respondents are in deflective accord on treating the question of other animals as an issue of discerning the presence or absence of an interest or capacity that might qualify them for ethical consideration or treatment–an outlook distinctly different from Costello's approach, which treats the matter as an instance of moral terror that dishonors the fellow-feeling we should otherwise register in the company of other animals (PAL, 97 and 103f.). At the tertiary level of commentary, John McDowell concurs that the lecture respondents are fleeing or suppressing the existential gravity of Costello's crisis by translating it into a propositional game of what could be called concept-chess on the academic board of animal ethics (PAL, 135).

I confess having been taken aback by these impressions of the responses in *The Lives of Animals* – I felt as if I had not read quite the same book as the co-authors of *Philosophy and Animal Life*. The commentaries in the former volume appear to me a good deal more sensitive and penetrating than is granted by the latter writers.[2] About the only response that perhaps deserves their criticisms would be that of Peter Singer, but even he is subtler and more playful than they would lead one to believe. Another respondent, Wendy Doniger, may also be fair game – yet she does admit at the outset that her approach is not entirely faithful to the original: "It seems somehow reductionistic to respond to these deeply moving readings as if they had been dry academic arguments" (LA 93). For her part, Marjorie Garber cannot justly be accused of superficiality when she at least sets her agenda partially to excavate the subtextual significance(s) of Coetzee's discourse by asking, "What does the form of these lectures displace, repress, or disavow? What is striking in its absence here?" (LA 74) – precisely not the sort of query that disposes one to being duped by deflection. And the last respondent, Barbara Smuts, is the least likely target of the critique I have surveyed above (for reasons that require greater and separate attention, as given below). In criticizing these commentators, Cavell mentions that there are five of them; since there are only four actual respondents, I take him to be including the introductory voice of editor Amy Gutmann. Nor is she, however, insensitive to the emotional undertow of the lectures at stake: witness her taking note of Costello's point that (as I would add, a scholastic kind of) philosophy is "relatively powerless ... because it lags our sympathies" (LA 4).

I want now to take up and spotlight the case of Smuts, since doing so will bring into greater relief a point or two I wish to stress. Recall that she is, according to Diamond, "a primatologist [who] describes the Coetzee lectures as a text containing a 'discourse on animal rights'" and thereby misses the deep crisis suffered by the protagonist, Elizabeth Costello (PAL 48f.). So far as I can see, however, that is not Smuts' sense of *The Lives of Animals* as a whole, nor is she insensitive to the main character's affective/relational predicament: she says that Costello "debates the issue of animal rights" with another character (a bioethical exchange that provides evidence the lectures do contain, even if they are not entirely or even primarily about, forays into the philosophy of inter-species morality); yet, in a telling footnote Smuts adds that "when Coetzee writes about animal rights, I take him to be writing, in the first place, about animals and our relations with them" (LA 107, n.1). This last observation raises an interesting possibility–namely, that Smuts is onto something significant deflected by Diamond's own discourse: it could be that *The Lives of Animals* is centrally concerned neither with Costello's angst as such nor with its repercussions for a critique of philosophy, but rather that these elements subserve (even if often or only implicitly) an overarching concern with the promise, power, and pitfalls of cross-species conviviality (what Merleau-Ponty once called "interanimality").

A skeptical audience might at this juncture wonder why, on such a reading, would there be so little about actual animal (inter)relationships in either Costello's or indeed Coetzee's lectures? Smuts herself flags this curious absence when she notes that we learn only indirectly (in the allusive context of another character's self-addressed musing) that Costello cohabitates with cats (LA 107). My hypothesis is that the very condition Diamond fixes upon, viz. Elizabeth's ethico-existential wounding, is precisely what keeps Costello from dwelling upon those with whom she literally dwells. The magnitude and intensity of animal exploitation on the macro-level is so overwhelming that mention of micro-level cohabitation seems either beside the point or apt to be dismissed by company who perpetrate, allow, or ignore the atrocities at stake. Or it could be that the protagonist wants to cherish unto herself the company of feline fellows as an emotional buffer against the maelstrom of immorality beyond her house-hold. Whatever the reason may be, if the general thrust of this interpretation makes sense, then it is not an exaggeration for Smuts to float the possibility that we are here "entering territory where, perhaps, Costello (and maybe even Coetzee) feared to tread" – since, "for the [sympathetic] heart to truly share another's being, it must be an embodied heart, prepared to encounter directly the embodied heart of another" (LA 108) and that is to venture into emotional or existential exposure (a la Wolfe) at a very risky depth of vulnerability indeed.

And what is it that Smuts relates in narrating her own encounters of this sort? We learn that in various contexts–wild, domestic, and in-between–she has spent much of her life in the company of animal others (chiefly primates, cetaceans, and canids). These experiences, largely through imaginative and physical trial and error, teach her the art of inter-species relationship building. One setting in particular seems to have been profoundly revelatory: with a certain Kenyan troupe, Smuts says, "I found myself sharing the being of a baboon because other baboons were treating me like one"; this conviviality was mediated by both patient observation of individuals and somatic habituation to the others' etiquette for routine matters of social life, which enabled Smuts to note and feel their "appreciation for being a baboon body in baboon-land" (LA 110). Not an isolated or ephemeral epiphany, the lesson at stake is carried onward into other scenarios that rekindle in her bodily consciousness "a joyful intersubjectivity that transcends species boundaries" (LA 114). With an adopted dog-wolf hybrid named Safi, for instance, Smuts lives out a companionship that contributes to a psychosomatic insight remarkably relevant for our own discussion: "while we normally think of 'personhood' as an essential quality that we can 'discover' or 'fail to find' in another, in [my] view espoused here person-hood connotes a way of being in relation with others" (LA 118, italics in original).

Ironically, this last point of Smuts' approximates fairly closely what some contributors to *Philosophy and Animal Life* believe is missing from the commentary on *The Lives of Animals*: Cavell says of Diamond that "in contrast to the idea shared by the commentators accompanying Coetzee's stories ... her different notion [of zoontic existence] is one that is not the concept of an animal possessing this or that interest or capacity in common or at variance with our human interests or capacities, but one that 'means a being ... which may be sought as company [Diamond's emphasis]'" (PAL 103f, first brackets mine, second Cavell's). The parallel with Smuts is not exact, because she was speaking of personality while Cavell cites Diamond on animality; nonetheless, they are essentially talking about the same phenomenon of being somebody (or someone else) in and through the experience of what we might call "fellow creatureliness". Nor is this overlap limited to some marginal aspect of the discourses at hand–for Diamond appears deeply impressed by the idea that "Coetzee's lectures ask us to inhabit a body" and, related to that, she seems profoundly concerned with the challenge of how "to express the extraordinary felt character of animal life in relation to our own" (PAL 59,61). Inhabiting the bodily horizon of animal others and expressing the felt character thereof are the very substance of Smuts' research and writing, and in my estimation she does it better than the contributors to *Philosophy and Animal Life*, better than the rest of the commentators on *The Lives of Animals*, or indeed than Coetzee himself (at least in his Tanner lectures).

So far, then, I may have succeeded only in showing that there is more to the commentary on Coetzee's lectures than met the eyes of some critics. A rejoinder could still be lodged to the effect that the example of Smuts does little to counter a central contention of those critics, namely that philosophy is ill-equipped to capture or convey an authentic sense of animality or somatic otherness. Smuts is, after all, a scientist and as such does not represent the philosophic discipline or its characteristic discourse(s). One might yet maintain, as Diamond seems to in echoing Costello, that philosophy's standard mode of argumentation or debate "may make unavailable to ourselves our own sense of what it is to be a living animal" and "may have built into it a distancing of ourselves from ... our capacity to respond to and to imagine the bodily life of others" (PAL 53). Again, if we follow Wolfe in this vein of skepsis, it would be a deflective delusion to think that the lifeworld's resistance to assertoric articulation "could be dissolved or overcome by ever-more ingenious or accomplished propositional arguments, ever-more refined philosophical concepts" (PAL 5). The position at stake reminds me of the poet's protest in the finale of the early 1990s film MindWalk–where, after listening to another character's attempt at a precise yet comprehensive ontology of organic and even cosmic vitality, he contends that life is just not condensable that way (if at all).[3]

It seems to me that this point can be overstated or even misplaced. Biting the bullet, Ian Hacking elsewhere in *Philosophy and Animal Life* admits that what Diamond and Cavell call deflection is tantamount to emotional escapism, but counters that sometimes it constitutes "an escape to better health" (PAL 162). Dodging the bullet, a recent reviewer of the book asks rhetorically and pithily, "is all argumentation really a sign of deflection? ... Can it not be a form of perception, which seeks to reveal prejudices that have gone unnoticed?"[4] To be fair, Diamond herself hedges on this matter, in an endnote: that "there can be a non-deflecting practice of philosophy ..., and that argument may have an essential role in it, is not something I would wish to deny" (PAL 86, n.22). This qualification comes as a rather roundabout way of acknowledging greater power(s) for philosophy than would appear to be recognized in the main body (shall we say argument?) of Diamond's essay. In a more straightforward and sustained way, I would like now to provide evidence not merely that there could be a mode of philosophizing appropriate to the articulation of animality and its discontents but that such actually exists already in the works of several thinkers including myself.

Somatic Sympathy

Since I am obviously most familiar with my own project, I'll refrain from importing too much of anyone else's writing and instead simply furnish an overview of the former–to wit.[5]

Over the past few decades a number of trans-human moral theories have arisen in opposition to the dominant ideology of anthropocentrism. Yet the more generally popular of these ethical systems tend to "elevate" nonhuman beings into our still all-too-humanist sphere of moral concern, often by drawing attention to the "higher" (*i.e.* humanoid) mental capacities of other organisms.[6] Whether contingently or essentially, consequentialist appeals to sentience and deontological definitions of subjectivity function effectively as standards of moral significance (perhaps only) by utilizing positive comparisons to human exemplars of consciousness.[7]

Hence, accidentally or necessarily, they attract the charge of personification from reactionary anthropocentrists. To avoid the real or rhetorical label of anthropomorphism, I have proposed that animal ethicists shed anthropocentric hierarchy altogether–even (the risk of) its appearance–and place their moral thought and political activity behind the truly post-humanist task of reappreciating bodily animacy as such. Thus we may extend the range of caring regard in the very gesture of recognizing our own vital status as animate zoomorphs.

I have undertaken this task by engaging a bio-existential hermeneutic of body in view of Edith Wyschogrod's observation that "classical phenomenology's account of the body subject [can be] recontextualized so as to highlight the body's receptive capacities, its vulnerability, its patience; it is thus replete with ethical significations", and with respect to Carol Adams' conclusion that "A relationship exists between reclaiming the body and its full range of feeling, and reclaiming animals' bodies, including women's. A feminist care ethic for the treatment of [other] animals offers the possibility of such reclamation."[8] Interpreting phenomenal bodiment along these lines, we enter into a mode of philosophizing that is fruitful for inter-species ethics. For the live body is the primary locus of existential commonality between human animals and other organisms, and the appreciation of commonality undergirding differentiation enables the growth of moral relationships.

In order to illustrate the concrete import and philosophic plausibility of these claims, I have sought to appropriate and look to de-anthropocentrize Werner Marx's neo-humanist meditations on phenomenologizing moral reflection.[9] Marx attempts to delineate a path for developing non-metaphysical morality by demonstrating that it is possible–without resorting to transcendent theology or rationalistic systems–to find "a measure on earth" for ethical conduct. He believes that this earthly standard is discoverable in the capacity for compassion, which manifests itself preeminently through one's encounter with mortality. Basically, the pattern suggested is that confrontation with death and dying produces a sense of life's transience, contingency, dependence; this feeling, in turn, breeds an affection of solidarity with those beings who share the condition of mortality. As against our present concerns, however, Marx believes that the concomitant transformation of attunement from solitary horror under the shadow of fatality to altruistic joy in the company of others–is "an unfolding of what characterizes man as a human being above all."[10]

This anthropocentrist interpretation rests on two premises: (1) the conviction that the catalyst of compassion, genuine mortality, stems from self- reflexive awareness and elicits rational insight; (2) the assumption that both these forms of consciousness (*i.e.* self-reflection and intellectual intuition) are exclusively human phenomena. As Marx's mentor, Martin Heidegger, would have it, only humans truly die–other animals, by counter-example, merely perish. One way of challenging such an interpretation would be to marshal ethological evidence for nonhuman modes of "being-unto-death."[11] Another tack, one I have followed instead, is to shift the platform of comparison by suggesting that we may ground moral compassion for other animals in the sensation of sharing carnal vulnerability (rather than mortality under the aspect of mental phenomena).[12]

If what is meant by "sharing" such a sensibility is explicated, the moral connections with compassion are made manifest. On a somatic level, then, it can be held that we are aware of our own physical vulnerability–susceptibility to injury and illness–just in virtue of being sentient entities of animate flesh. We might share this sort of somatic sensitivity with another (kind of) organism in the minimal sense of becoming aware that our susceptibility to suffer harm is like that of the other organism. My claim on this construal has been that such minimal mutuality of common carnal nature suffices phenomenologically to establish compassionate concern for the other–in the mode of its being the proper object or "patient" of ethical consideration. In another, stronger sense of sharing, the second party might also become aware of our vulnerability being similar to its own; this richer form of reciprocity is requisite, it appears to me, for interspecific compassion to take on the aspect of respect – whereby both parties appropriately regard each other as moral subjects, agents or actors.[13] Some of our relationships with other primates (particularly apes), with cetaceans (*e.g.* dolphins), and most especially with domesticated companion and work animals (*e.g.* dogs and horses), feature reciprocally cognizant compassion grown into moral respect.

Yet, I have concentrated not so much on these examples as on the former (and broader) sort of morally compassionate concern, that variety sufficiently based merely on one party's sensing of bodily vulnerability felt (immanently) to be similar to another's. Here we move beyond, or rather behind, Marx–to the founding generation of phenomenologists. For it is Edmund Husserl himself and his student, Edith Stein, who provide more illuminating accounts of relevance. Husserl spoke of what he called the science of somatology, whose "foundation is finally the direct somatic perception that every empirical investigator can effect only on his own body and then the somatic[clarification] that he performs in the interpretive apprehension of perceived alien animate organisms as such."[14] With respect to such somatology's concernful application, Stein's extension of her mentor's thought inclusively incorporates transpecific cases: "The type 'human physical body' does not define the limits of therange of my empathic objects," she says, "For example, I may sense-in pain when [a nonhuman] animal is injured."[15] Although Stein remains skeptical about the possibility of understanding certain gestures of another (kind of) animal, latter-day phenomenologists such as Kenneth Shapiro have pointed to the promise of "kinesthetic empathy" in becoming familiar with (the significance of) the positions and movements of other animate organisms.[16] From the perspective of another tradition, Hacking has made a similar point in a different context (separate from *Philosophy and Animal Life*): "We sympathetically experience an emotion akin to another by picking up cues from the body. ... We perceive the emotions of animals when we attend to their bodies. ... We do not infer [as by analogy] that animals have emotions from the[ir] movements and dispositions".[17]

What is the ethical import of cross-species somatology? I have held that it provides the experientially primordial basis of what we sometimes refer to as the "moral sense"–this being a point that comes into sharper relief through interspecific illustration. Historically, of course, moral sensibility has been the province of a largely Scottish tradition of moral sentiment. Thinkers such as Shaftesbury, Hutcheson, Hume, and Smith conceived of it as an innately altruistic ethical faculty.

Reinterpreted, "moral sense" signifies a more densely physical pattern than does the relatively rarified notion of psychic sympathy to which traditional moral sense theorists appealed.[18] Unhappily, since the eighteenth-century theories leave phenomena of moral sentiment up in the air, more recent models of empathetic identification can be used to (try to) ground those phenomena in a "psycho-ethics" of projective imagination (*i.e.*, stirring moral sentimentality by taking an other's role in the "mind's eye").

However, to the contrary of these psycho-ethical views, it has been my contention that (especially the transpecific) moral life is primarily rooted–as a matter of phenomenal fact–in corporal symphysis rather than in mental maneuvers in the direction of typical sympathy.

As I have used it, "symphysis" is meant to designate the felt sense of sharing with somebody else a live nexus as experienced in a somatic setting of direct or systemic (inter)relationship.[19]

My position has been that speaking of symphysis is the best way to describe the (proto-ethical) feeling that prepares us to take other animal beings into moral account.[20] Inferential reasoning by analogy may rationally justify that assurance, and psychology of imagination may scientifically explain it via empathic projection, but only somatology of genus-being and alien specificity can properly (phenomenologically) articulate its actual experience of conviviality.[21]

From this last perspective, then, an ethos sculpted somatically by symphysical encounters would constitute a character or culture morally sensitive to the existential element or being-in-the-world of flesh, including that camal vulnerability shared with any live body as such.[22]

Rights and Liberation

So far, then, I hope to have shown that there is more in *The Lives of Animals'* commentaries than is explored by contributors to *Philosophy and Animal Life*, and that a certain mode of philosophizing close to somatic phenomena can and does treat animality helpfully whilst avoiding the deflective pitfalls into which analytic philosophy has appeared to land–and thus philosophy is not necessarily to be abandoned or sidelined when it comes to the moral ontology of inter-species relations, because "philosophy" is not methodologically univocal (like "the animal", as per Derrida, it is not really singular but multiform). Indeed the existential/hermeneutic phenomenology that I and others have mobilized gets right into what Diamond (like Coetzee) takes to be the heart of the matter, namely that "the awareness we each have of being a living body, being 'alive to the world', carries with it exposure to the bodily sense of vulnerability to death, sheer animal vulnerability, the vulnerability we share with them" (P&L, 74). Now, following on this, the last point I would like to make is that even the more established ethical discourses of animal rights and animal liberation–the very ones that seem to take so much heat in the book under discussion–can still be useful for understanding and promoting the interests of our fellow creatures.

Not too long ago, *The New York Times* published an essay adapted from a new book entitled *Eating Animals*. In it, Jonathan Safran Foer reminisces about the dietary and psychological effects of a certain sort of collegiate experience:

> ... when, at the end of my sophomore year, I became a philosophy

> major and started doing my first seriously pretentious thinking, I became a vegetarian again. The kind of active forgetting that I was sure meat eating required felt too paradoxical to the intellectual life I was trying to shape. I didn't know the details of factory farming, but like most everyone, I knew the gist: it is miserable for animals, the environment, farmers, public health, biodiversity, rural communities, global poverty and so on. I thought life could, should and must conform to the mold of reason, period. You can imagine how annoying this made me.[23]

Remember how aggravating Coetzee's character, Elizabeth Costello, could be to her family and associates? That was because she did not conceal the wound of her sympathy with those whom she took to be the victims of an ongoing animal holocaust. Foer, by contrast, became a nuisance by strict allegiance to a blood-lessly philosophic ethos of transparency and consistency. (Singer and Regan, it can be noted in passing, irritate many people in a similar fashion.) Yet listen to his next paragraph: "When I graduated, I ate meat–lots of every kind of meat–for about two years. Why? Because it tasted good. And because more important than reason in shaping habits are the stories we tell ourselves and one another. And I told a forgiving story about myself to myself: I was only human."

Ring any bells? Recall Cavell's confessional remarks on his own meat-eating: "I have sometimes felt vegetarianism to be a way of declaring a questionable distance from the human animal", he muses–and then offers his central apology for omnivory, "What I would like to say is simply, 'I am human'" (P&L 122f.). Fittingly, Cavell admits on the same page that he is "embarrassed by, and not yet ready to repair, my ignorance of the general state of philosophical argument concerning vegetarianism" (italics added). One can be forgiven, at this juncture, for wondering which is the more deflective discourse–propositional ethics of deontology and utilitarianism on the one hand, or on the other the linguistic meditations and literary reflections in which Wittgensteinians like Cavell revel?[24]

Contemplating that query inclines this reader at least to agree with Elisa Aaltola (in her review of the book before us and related works) that "animal ethics in its standard form does not, by a long shot, deserve all the criticism that the 'new wave' is at times eager to level at it".[25]

More affirmatively, I would like to say that the rights view and liberation theory –certainly as practiced by their leading proponents–have had the merit of forcing certain matters to the fore that were receiving scant notice previously. They made other animals visible as members of the moral community, bringing nonhuman interests into the arena of ethical consideration. At a time when animal advocacy needed rhetorically to overcome denigration as sentimentalist, theorists such as Singer and Regan gave rational respectability to regard for the sufferings of and injustices against (members of) other species. Their kind of approach, its shortcomings on other counts notwithstanding, could and did travel discursively into the fields of science and law–where it is changing practices and regulations for the better. One reason for these accomplishments is that argumentation requires the marshaling of evidence, and the first generation of animal ethicists did not shy away from revealing and spreading unadorned and detailed descriptions of institutions and procedures of animal exploitation. Here I side with McDowell (PAL 128f.) in discounting a suggestion raised by Cavell, namely "that the extreme variation in human responses to this fact of civilized existence [sic–the mass production of animals for food] is not a function of any difference in our access to information; no one knows, or can literally see, essentially anything here that the others fail to know or can see" (PAL 93). On the contrary: within a decade or two after *The Jungle* was published, Upton Sinclair's muckraking masterpiece about stockyards and slaughterhouses was all but forgotten in mainstream culture–and so, for about half a century until Singer and Regan wrote their seminal works in animal ethics, few knew and even less witnessed the industrialized process by which living individuals were and are transmogrified into supermarket commodities for mass consumption. The growth and acceptance of vegetarianism over the past generation have been phenomenal; hence, echoing Hacking's remark about deflection, it would not be untoward to insist that we knock neither information nor argument overmuch. In other words, for the sake of most everybody associated with factory farming, we can afford to let the rights-bashing subside.

* * *

I began this commentary by promising to stay close to the concerns of actual animals; looking back over it, I am less than satisfied that my promise has been fulfilled. So, in moving toward conclusion, let me try to indicate the ways in which I think my critical comments can be tied together by attention to animality in its various forms. First, I asserted that some of the contributors to *Philosophy and Animal Life* would have found fodder for portions of their own agenda had they scrutinized more carefully the "Reflections" on Coetzee's *The Lives of Animals*. In particular, I flagged the work of Barbara Smuts–and I think it is obvious, even from my own all-too-brief survey, that her comparative psychology is rich in zoographic detail and an especially acute sensitivity to her subjects and/as fellow creatures. Second, I claimed there are modes of philosophy outside the Anglophonic mainstream of analytic method that can and do articulate and appreciate structures and interests of animal experience–and I hope the foray I provided into the moral ontology of inter-species somatology (aka phenomenology of body) demonstrated at least one promising example of such. Third, I defended animal rights and liberation theory against misologistic critique by pointing out their efficacy for consciousness-raising and behavioral change in respect of zoological others. Finally, by way of implication rather than recapitulation, I want to state that though I do believe there is something of a retro-humanistic narcissism running through large stretches of the book here considered, and though I have sought to serve up some counterweights against that tendency, ultimately the best animal philosophy will transcend us/them dualism in order to forge more inclusive senses of cross-species conviviality such that–as anthropocentrism recedes and species apartheid melts away–the experience of we eventually shall feel very multifarious and come to stand for something much more capacious than is currently standard: as a complement (and compliment) to deep ecology, call it maybe "broad biosophy", a wisdom and conscience fit for ethical earthlings.

Notes

1 Which table is being set–and who is invited or unwelcome–is ambiguous: some of the authors, notably Cora Diamond (in her rejoinder to my initial presentation of the present paper at a meeting of the American Philosophical Association in New York City, December 2009), insist that the book is primarily or exclusively a set of meta-philosophic meditations on human life as such and its uneasy relation to reality. I maintain that, authorial intentions notwithstanding, the book's title, editing, promotion, and reception all bespeak an interpretation on which animality generically and human relations to other animals are central objects of concern.

2 To be fair, some of the critics are more cautious than others: Cavell, following Diamond, finds "each and all" of the respondents to be essentially obtuse (PAL 97); McDowell is more nuanced in laying down the charge of deflection, as he allows that it might be accurately descriptive of the commentators only "to varying extents" (PAL 135) – what follows in my main text can be taken as an indication of just how much is hidden by that hedge.

3 Mind Walk, dir. Bernt A. Capra (Atlas, 1990).

4 Elisa Aaltola, "Philosophy and Animal Studies: Calarco, Castricano, and Diamond", *Society and Animals* 17.3 (2009): 285.

5 The next few pages of text derive largely from my essay "Toward a Properly Post-Humanist Ethos of Somatic Sympathy", presented at the Minding Animals conference (Newcastle, Australia – July 2009), translated into Italian (in Nell'albergo di Adamo, ed. M. Filippi, in press), and English revision forthcoming (ed. Gregory Zucker, Lexington Books).

6 Probably the most trenchant critique on this point has come from feminism's contribution to inter species moral philosophy–see, *e.g.*, *The Feminist Care Tradition in Animals Ethics*, J. Donovan and C. J. Adams eds. (New York: Columbia University Press, 2007). Cf. Neil Evemden's suspicion that "the very idea of animal 'rights'– an idea that seems to presuppose a humanistic interpretation of justice and individuality– amounts to a kind of subterfuge to draw certain nonhumans into the protected circle of human self-definition." Quoted from his The Social Creation of Nature (Baltimore: Johns Hopkins University Press, 1992), p. 141 n.33.

7 Attend to the transmission of philosophic theories into political practices. Cf. Peter Singer's *Animal Liberation* (New York: Avon, 1975) and Tom Regan's *The Case for Animal Rights* (Berkeley:U. of California Press, 1983) with socio-historical accounts of movement-building in Susan Sperling's *Animal Liberators* (Berkeley: U. of California Press, 1988) and James Jasper and Dorothy Nelkin's *The Animal Rights Crusade* (New York: Free Press, 1992).

8 "Does Continental Ethics Have a Future?," in *Ethics and Danger*, A. B. Dallery et al. eds. (Albany: S.U.N.Y. Press, 1992), p. 236 ; "Caring about Suffering: A Feminist Exploration", in Donovan and Adams (2007), ch. 8, p. 221.

9 See especially *Towards a Phenomenological Ethics: Ethos and the Life-World*, S. Heyvaert trans.(Albany: S.U.N.Y. Press, 1992).

10 Phenomenological Ethics, p. 56 (italics added).

11 Cf. David Farrell Krell's phenomenological contestation of Heidegger's dying/ perishing distinction, in *Daimon Life: Heidegger and Life-Philosophy* (Bloomington: Indiana University Press, 1992), esp. pp. 317ff.

12 These two approaches are not entirely exclusive of each other. Diamond herself intimates a provocative amalgam of both: "The awareness we each have of being a living body, 'being alive to the world', carries with it exposure to the bodily sense of vulnerability to death, sheer animal vulnerability, the vulnerability we [humans] share with them [other animals]." (From PAL 74.)

13 Mental factors (*e.g.*, second-order consciousness of each other's mutual recognition) must also come into play for respect in this sense to emerge fully as a moral phenomenon.

14 Ideas III, sec. 2 b. *Collected Works*, vol. 1, T. Klein and W. Pohl trans. (The Hague: Martinus Nijhoff, 1980), p. 7.

15 *On the Problem of Empathy*, W. Stein trans., 3rd ed. (Washington, D.C.: I.C.S., 1989), p. 59. The term "sense-in" (as in "Einempfindung") suggests what I would like to stress, viz, that the operation at stake is primarily somatic rather than purely psychic (a point confused by continued use of the words "empathy" and "sympathy," which are encumbered by mentalistic connotations.)

16 "Understanding Dogs through Kinesthetic Empathy, Social Construction, and History," *Anthrozoos* v.3, n.3, 184-95.

17 "On Sympathy: With Other Creatures", Tijdschrift voor Filosofie 63 (2001): 710. For Hacking, this point is under-girded by evolutionary biology: "As Darwin insisted, the [emotion-expressive] musculature, like that of the rest of our [human] bodies, has much in common with that of [esp. mammalian] animals" (711).

18 Compare Max Scheler's *The Nature of Sympathy*, P. Heathtrans. (London: Routledge, 1954).

19 David Seamon uses the word in his "Different Worlds Coming Together: A Phenomenology of Relationship"(*Dwelling, Seeing, and Designing: Toward a Phenomenological Ecology* [Albany: S.U.N.Y. Press, 1993]); as he notes, it originated in medical usage and means in ancient Greek, "the state of growing together" (p. 23). A Latinate cognate would be "concrescence", but that term –at least in philosophic discourse–has acquired Whiteheadian overtones that are not necessarily implied by my usage. It is also important to note that symphysis is not a state of total fusion: as per Kuang-Ming Wu, " 'the twain' osmotically interpenetrate, mutually infuse, while keeping intact their respective integrities" (*On the Logic of Togetherness: A Cultural Hermeneutic* [Leiden: Brill, 1998], p. 12 [italics added]).

20 Hacking seems to be indicating symphysical awareness when he profiles the moral experience "that is a relation of felt identity of body. It is a relation of

sympathy between [hu]man and beast [sic], which works through the living bodies of the two" ("On Sympathy", 691, italics added).

21 Greco-Germanic philosophy from Aristotle to Hegel and Marx has evolved a tradition of philosophical anthropology on the premise of studying the human animal's species-being strictly; the trans-humanizing task before us now is to go beyond that tradition's homo-exclusive bounds into the ontology both of generic animality (if there be any) and of other-than-human speciations.

22 Compare Sue Cataldi's *Emotion, Depth, and Flesh: A Study of Sensitive Space: Reflections on Merleau-Ponty's Philosophy of Embodiment* (Albany: S.U.N.Y. Press, 1993); Shannon Sullivan's *Living Across and Through Skins: Transactional Bodies, Pragmatism, and Feminism* (Bloomington: Indiana University Press, 2001); and Richard Shusterman's *Body Consciousness: A Philosophy of Mindfulness and Somaesthetics* (Cambridge University Press, 2008).

23 "Against Meat", *The New York Times Magazine*, 11 October 2009: 72.

24 Curiously, besides Cavell and Diamond, animal trainer and zoographer Vicki Hearne also displayed a certain antipathy or aversion to animal rights discourse (in her case it was rather pronounced). It might repay further investigation to figure out whether and how the influence of Wittgenstein may have set up their allergic reactions; it plausibly has something to do with the Austrian's later eschewal of atomistic (and aggregative) approaches, or at least so I believe after hearing Nicolas Delon's "Comments on Jeff McMahan's 'Our Fellow Creatures': A Wittgensteinian Critique of Moral Individualism", paper delivered at the New School for Social Research, 3 April 2010). We do not, however, have the luxury in the present context to pursue this line of inquiry and speculation.

25 "Philosophy and Animal Studies", 286; Aaltola offers another essay on this topic in "Coetzee and Alternative Animal Ethics", in A. Leist & P. Singer eds., *J. M. Coetzee and Ethics: Philosophical Perspectives on Literature* (New York: Columbia U. P., 2010), chp. 6 – because this work is still forthcoming, I have not yet been able to peruse it for comparison.

Chapter 4

The Moral Status of Sentient and Nonsentient Creatures

☆ *Rebekah Humphreys*

In recent years, there has been a growing concern for the well-being of animals reared in factory farms and used in scientific experiments. Most people have learned to accept that they have a moral responsibility to their pets, and there appears to be some recognition and reflection of the fact that farm and experimental animals are animals that may be much like our pets.

However, legislation and reform to improve the welfare of animals used in experiments have been virtually non-existent. This is not to say that efforts to improve animal welfare have not been made in the case of animal experimentation. Campaign groups fight tirelessly to improve the lives of animals used in experiments. Rather, the reason why the practice of animal experimentation does not welcome legislation and reform seems, in part, connected to an idea of science as a special sort of field, with a different status to other fields of work. Seen as such, it may be seen as separated from certain moral values. The laboratory is often seen as a place where great knowledge is attained at the expense of animals, and therefore may be somehow seen as a place 'insulated from certain kinds of moral criticism' (Diamond, 1995, p.361), particularly in respect of animals. Indeed, there appears to be a widespread conviction that animals used for scientific purposes should be perceived or at least treated differently from animals in other contexts. Indeed, this conviction is reflected in law, where there exists separate legislation for the treatment of pet animals and experimental animals.[1]

Legislation and reform for factory farm animals have been more successfully introduced, largely through the efforts of non-governmental organisations. However, in spite of increasing uneasiness about the suffering of animals in factory farms, the majority of people in the West eat factory-farmed meat. Consistently standing in opposition to factory farming would mean having to restrain one's eating habits and change one's lifestyle, something that many people are just not willing to do. In the West there seems to be a food culture, or way of life, which involves being able to eat what one wants, whenever one wants, even to the detriment of one's health. Whereas meat was once seen as the luxury part of the meal, meat is now so commonplace in Western diet that it could be seen as the staple food for many people.

In spite of existing new technologies, animal experiments have actually increased and, with the Western food culture as it is, the practice of factory farming looks to be a permanent fixture and the main system of farming in the future. In the light of this and considering the immense suffering animals endure in the practices of factory farming and animal experimentation the future looks bleak for animals destined for the factory farm and the laboratory, and it is arguably our urgent duty to interrogate, 'What is the moral status of animals reared in factory farms and used in experiments?'

In order to answer this question one firstly needs to ask 'What are the necessary and sufficient conditions for X to have moral standing?' (see Goodpaster's discussion of this question in 'On Being Morally Considerable', 1978, p.309). Reflections concerning anthropocentric attitudes towards animals necessitate concerns for environmental ethics, as animals, the environment and humans are interrelated. Indeed, answers to questions concerning the extent of our moral responsibilities towards animals may also provide answers to questions concerning the extent of our moral obligations towards the environment, and vice versa, since both involve ethical reflections that attempt to throw light on the extent of moral standing. In this respect, at a more fundamental level, environmental ethics and animal ethics merge, and, rather than being viewed as two distinct fields of ethics, should form one coherent ethic. Thus, they are both based on issues arising from answers to the question as to what kinds of entities can be said to have moral standing.

Sentient Creatures, Nonsentient Creatures and the Life Criterion

So, let us return to the question, 'What then are the necessary and sufficient conditions for X to have moral standing?' Indeed, for Peter Singer (Singer, 1995b, ch.3) only those beings that can suffer can have interests and only those beings with interests have moral standing. Since nonsentient beings cannot suffer they are deemed not to have any interests and are, therefore, excluded from moral consideration (Singer, 1995b, pp.57-8). For Singer then sentience is a necessary and sufficient condition for X to have moral standing.

Singer's sentientism calls for the principle of equality to be extended to sentient animals. When considering human interests only, it is thought that factual differences in human beings do not justify any difference in the consideration we give to their interests. *'The principle of the equality of human beings is not a description of an alleged actual equality among humans: it is a prescription of how we should treat human beings'* (Singer, 1995a, p.5). Similarly the fact that animals are of a different species, who possess different capacities and intelligence which are lower than or, at least, different from those of humans, does not justify any difference in the consideration we give to their interests. The principle of equal consideration of interests requires us to take the interests of nonhuman beings seriously:

The argument for extending the principle of equality beyond our own species is simple, so simple that it amounts to no more than a clear understanding of the nature of the principle of equal consideration of interests (Singer, 1995b, p.56).

Singer argues that we should extend the principle of equality of consideration to all sentient animals and that their interests be given equal consideration as the like interests of human beings.

In respect of human beings, like or similar interests should be given equal consideration. Similarly, in cases where the like interests of humans and animals are at stake the interests of humans and animals should be given equal consideration. It should be said that the principle of equal consideration of interests, properly understood and applied, does not require equal treatment. Rather, it requires that *like* interests are *considered equally*. All sentient beings have interests in not suffering comparable with human interests in not suffering, and it is reasonable to suppose that where those interests are similar they should be given similar or equal consideration irrespective of species.

In the words of Jeremy Bentham, 'the question is not, Can they *reason*? nor, Can they *talk*? but, Can they *suffer*?' (1982, p.283). Bentham argued that the sufferings of animals should be given equal moral weight to the sufferings of humans. Both Singer and Bentham claim that it is the suffering of a being that is to be taken into account in our moral considerations. That is, sentience, or the capacity to suffer, is the characteristic deemed necessary for inclusion in the moral sphere.

However, it is not just the avoidance of pain and suffering that beings have an interest in. All sentient animals have an interest in life or living, even if they are not aware of such an interest. But it does not follow that those lives have equal value. A human being may stand to lose more from death than, say, a cat, and a dog may lose more from death than, say, an ant. And what is of varying value may not only be life:

> If the goods that my freedom permits are, on the whole, more prudentially valuable than what a bird's freedom permits, then equal consideration does not confer equal moral weight on my freedom and the bird's (DeGrazia, 1996, p.234).

As David DeGrazia says, 'For A's interest X and B's interest Y to be relevantly similar, A and B must have, in some sense, the same thing at stake with X and Y' (*Ibid.*, p.233). Comparative claims can be made whilst taking into account the principle of equal consideration of interests. Those comparative claims or judgments can be not just about life, but about other interests, such as an interest in freedom, and an interest in being free from pain.

Applying sentientism to the ethics of factory farming and experimentation would have massive implications for the future of such practices, and giving equal consideration (or even just serious consideration) to the interests of factory-farm and laboratory animals would mean that these practices would have to be, at the very least, partially abolished. However, although sentience is a sufficient condition for moral standing, sentientism is not sufficient as an ethical theory, as it does have a narrow understanding of what is good and bad, and should, at best, be incorporated into ethical theory, or, more specifically, into an animal ethics theory.

Indeed, on David DeGrazia's sentientism, it follows that we can treat nonsentient animals as we please, as the lack of their being sentient qualifies them to be sans interests. For DeGrazia, while we may be able to speak of what is harmful to nonsentient things, we cannot say that such things have interests that should be taken into consideration:

> It sounds fairly natural to speak of harm to a plant, but not to speak of its welfare. If the sentience requirement is right, while it might not be incorrect to speak of what is good or bad, beneficial or harmful, to plants and nonsentient animals, these terms are not correctly applied to these beings *in any sense relevant to morality* (DeGrazia, 1996, p.227).

However, we do feel that we can wrong nonsentient things in a moral sense. Consider humans who are in a vegetative state or comatose. Many people in these states lack the capacity to feel pain or pleasure, and are also not conscious or concerned about their life. Whether they are sentient or not they do seem to have interests. We can affect them for better or for worse by our actions. And what about entities such as trees and plants? They also have some interests that we can further or damage by our actions. From a moral point of view the good of entities must be taken into account whenever those entities can be affected for better or worse by the actions of an agent.

Indeed, Joel Feinberg suggests that to have interests a being must be capable of being harmed or benefited. A being that has no interests does not possess this capability. As he says, 'a being without interests is a being that is incapable of being harmed or benefited, having no good or "sake" of its own' (Feinberg, 1974, p.51). Feinberg also suggests that only those things that have interests and a good of their own have moral standing (*Ibid.*, p.51). Kenneth Goodpaster also points out that there is an essential link between beneficence and morality. Those things that can be harmed or benefited will naturally deserve moral consideration. Living things are capable of being beneficiaries. Inanimate objects though cannot be benefited or harmed, and so have no place in morality. It can, thus, be argued that all that has interests deserve moral consideration, and that all and only living things (present and prospective) have interests.

Goodpaster suggests why there is a reluctance to see nonsentient creatures as candidates for moral standing. If one's conception of morality or the good is essentially hedonistic, that is one views only pleasure as of intrinsic positive value and only suffering as of intrinsic negative value, one will recognise only sentient beings as capable of being beneficiaries and will not see nonsentient things as having any interests, and therefore as deserving of moral consideration (Goodpaster, 1978, p.321).

Nonsentient things may have less moral significance than sentient beings, but this does not bar them from having moral standing. Moral significance 'aims at governing comparative judgments of moral "weight" in cases of conflict' (ibid., p.311).

There will obviously be cases where the interests of sentient beings will override the interests of nonsentient ones. However, the opposite may also be the case. That is, the interests of nonsentient beings may override the interests of sentient ones. Comparative judgments of moral significance can also be made when giving moral consideration to the conflicting interests of nonhuman beings and human beings.

The interests of animals and of humans will obviously be different so they will require different moral treatment. However, the basic interests of humans are thought to have greater moral significance than trivial or less basic interests of humans, and there is little reason to suppose that this should not apply when weighing animal and human interests. Indeed, it is reasonable to suppose that a being's basic interest in not being subjected to suffering should always override, or be given more moral significance than another's less basic interest.

Sentience is a characteristic that confers strong positive obligations. But sentience on its own is not morally significant. Sentience is not a free-floating faculty. It has to belong to something that is living, and it is that living entity that has moral standing. Being sentient just makes that life more morally significant.[2] So, we may have stronger or greater moral obligations to some things than others in virtue of the kind of things they are, or in virtue of the kind of life they lead as living things and in virtue of our own moral relations to those things.

Biocentrism: The Good of Beings and Independent Value

In extending moral consideration from sentient creatures to nonsentient creatures, and, further, to all organic life we have thus embraced biocentrism. Those who object to this biocentrism ask the question of why the boundary for moral considerability is drawn at life. Why not include rocks, rivers, mountains, and other nonliving entities? Why not also include other inanimate objects, such as cars and machines? If we accept the life criterion how far do our commitments extend?

The reply to these questions is that inanimate objects and nonliving objects do not have a good of their own. They have no interests of their own. To say that living things have a good of their own is to say that they have 'a nonderivative independent good, typically involving their healthy functioning as living systems' (Attfield, 1995, p.20). The good of nonhuman living things will be 'distinctive and peculiar to the capacities with which their own kind is endowed' (ibid., p.20). The good of machines is not an independent good but the good of their users or manufacturers. Rocks and mountains also do not have a good of their own. Trees by contrast do have a good of their own independent of the value ascribed by humans. An analogy can be made between non-conscious living things, like trees and plants, and those conscious things that are widely held to have moral standing, like the so-called 'higher' mammals. Trees have interests and capacities similar to those beings that we consider to have moral standing. They have the capacities for reproduction, growth, self-repair, and self-preservation just as sentient things do. There is then an analogical argument for holding that trees do have moral standing. The analogical argument only applies to those things that have interests.

There are however disanalogies that can be made between conscious things and non-conscious things. Trees have no desires or feelings. They are not interested in what we do to them in the sense that they do not have thoughts. However, the disanalogies do not hinder the fact that they do have interests and a good of their own. As Paul Taylor says,

> The idea of a being having a good of its own, as I understand it, does not entail that the being must... take an interest in what affects its life for better or for worse... It may, indeed, be wholly unaware that favourable and unfavourable events are taking place in its life. I take it that trees, for example, have no knowledge or desires or feelings. Yet it is undoubtedly the case that trees can be harmed or benefited by our actions (Taylor, 1998, pp.72-3).

Rather than the disanalogies excluding such nonsentient things from having standing in the moral sphere, they point to the different moral significance of such things, not moral standing.

We can then talk of what is in a being's interests or good, irrespective of whether that being can reflect upon its interests or good. As said above, the good of living things lies in their flourishing in a way that is natural or distinctive to their species (Attfield, 1995, p.20). That a living thing can thrive or flourish does not depend on that living thing being able to take an interest in its own life.

While the good of living entities in general can be seen in terms of their flourishing, the good of sentient creatures may best be seen in terms of their well-being. Indeed, Attfield argues that well-being and flourishing are 'states of affairs' that are independently or intrinsically valuable (Attfield, 1995, p.36-39). Such states are good-in-themselves, rather than good-in-relation to other things (Vilkka, 1997, p.21). That is, the good, well-being or flourishing of living things is good or of value in itself (Attfield, 1995, pp.29-39).

Richard Routley's Last Man thought experiment allows us to reach conclusions about intrinsic value by referring to common judgments which most people would form in the face of such examples. Imagine a case where there is one surviving man left on the earth. (One could imagine that all other sentient beings have died out due to a nuclear bomb or their death is certain due to nuclear fallout.) Does the survivor do wrong if he needlessly destroys every living thing left on the planet? (Routley, 1973, p.207). Further imagine that there is one tree left. Does the survivor do wrong if he needlessly chops down the last tree? The last tree could multiply if left alone (see Attfield, 1994, p.168).[3] Most people would suggest that he does do wrong. It must be then that there is something about the tree that counts. There must be something good that can happen to the tree that counts as a reason against chopping it down. The judgment that it is wrong to destroy the tree unnecessarily implies that there is something about the tree that is of nonderivative value:

> [O]ur judgments about the Last Man example suggest that the good or the flourishing of trees... is of intrinsic value, and embodies... one of the reasons why they should not be needlessly destroyed, and why such destruction would be wrong (Attfield, 1995, p.33).

It appears then that trees count because they have a good of their own and interests of their own. Their good lies in their thriving in a way that is natural to their own species. Such thriving is of independent value.[4]

Both non-conscious and conscious living things then have a good of their own. The good of nonsentient or non-conscious things, likes trees, lies in the flourishing of their capacities. The good of sentient or conscious things lies not just in their flourishing, but also in their well-being. Flourishing and well-being can be seen to be states that are intrinsically valuable (Attfield, 1995, pp.36-9); that is, they are valuable for their own sake.

Anthropocentrism: Against the Value of Nonhuman Beings

John Benson discusses anthropocentric responses to Routley's Last Man example (Benson, 2000, pp.21-3). There are some anthropocentrists who claim that the last man does nothing wrong and destroys nothing of value (Benson, 2000, p.23). However, such claims contradict our judgments about the last man. They seem to be implausible and until supported they should ideally be rejected. Indeed, such claims would permit us to destroy all life on earth if such life was of no benefit to humans, something that most of us would regard as appalling.

Benson's own response to the last man is that in destroying all nonhuman life he does harm human interests, for he harms himself. In this respect he undermines his own interests (ibid., p.23) and this is why he does wrong. Benson then denies that the last man's action harms no human, and denies that the last man destroys nothing of use or value to humans. Benson appeals to well-being as a reason why the last man undermines his own interests. The appreciation of the natural world, Benson argues, is essential for his well-being (ibid., pp.33-35). Bryan Norton makes a similar claim. He argues that humans have an ecological need to interact with the natural world, and that nature is important to us as humans as it increases the quality of our lives (Norton, 2000, 143-51). To destroy the natural environment and its inhabitants is wrong because of the 'loss of richness from the human experience' (Norton, 2000, p.149). For Norton then bio-diversity enriches the quality of our lives.

However, Benson's appeal to well-being is very dubious. It is not certain that we all have a need to appreciate the natural world. Likewise, the claim that experience of the natural world is an evolutionary need, passed on through generations, is unconvincing. Even if some humans do have such a need it is not clear that all humans do, especially when we consider city dwellers who hardly ever see any forests or other uncultivated natural environments and do not feel the need to do so either (Attfield, 2003, pp.69-71). As Routley says, 'ethical principles if correct are universal' (Routley, 1973, p.207), and it is unlikely that Norton's evolutionary need is a universal need. It is even more improbable that the last man would have such a need. For Norton and Benson to plausibly suggest that the last man does wrong because he undermines his evolutionary need, this specific need has to be watertight as a universal need. However, can we be confident that the last man does undermine his interests by destroying nature and thus destroying the potential for fulfilling his evolutionary need? I do not think we can. It seems then that it is not the value of human interests that make the last man's act wrong, but rather the independent value of the good of nonhuman entities themselves.

Benson does discuss extending his anthropocentric ethic to include all sentient beings, not just humans. On this extension the last man does wrong as he harms (human and nonhuman) sentient beings, and destroys what is of use and value to those beings (Benson, 2000, p.24). This is an improvement to Benson's objectionably human-centred ethic. However, it is still too narrow as an adequate environmental ethic as it excludes all nonsentient life.

Actually, Benson himself recognises that some environmentalists would object to this extended ethic on the ground that it is indeed too narrow. He notices that Routley could provide a counter-argument (as an objection) to this extension by arguing that the last man is the last sentient being and he goes about destroying all nonsentient life on earth. In the light of this counter-argument, it could now be objected that on the extended anthropocentric ethic what the last man does is permissible,[5] whereas on an environmental ethic what the last man does is still wrong (ibid., p.24). The extension of the human-centred ethic could also be objected to because its extension to include all and only sentient life excludes nonsentient human life, such as human beings who are comatose, and in this way it undermines itself if it supposed to be a variant on a human-centred ethic.

It should be said that in discussing extending the anthropocentric ethic to include all sentient life, Benson does not actually wish to extend *his* ethic in such a way that it would recognise the moral standing of all sentient beings. In his further discussions it becomes apparent that, in discussing the extension, he seeks to show that sentience-based arguments can be founded on anthropocentric values. As he says, supporters of sentientism 'argue that the reasons we have for granting independent moral status to (most) human beings oblige us, in consistency, to grant it to (some) nonhuman animals'. Benson further claims that sentience-based arguments 'rely upon drawing out unexpected consequences from moral ideals that we already have' (Benson, 2000, p.24), moral ideals which relate to why we value human beings. However, contrary to Benson, sentientism can recognise the value of all sentient beings independent of considerations of what we value about humans. That is, supporters of sentience-based arguments need not derive their conclusions from premises based on the value of *human* beings.

Benson believes that anthropocentric values alone can provide adequate protections for the natural world and its inhabitants, and we do not need to embrace an extended or a new ethic. For Benson, adequate environmental principles can be founded on anthropocentric ones. However, even if we go some way with Benson and suppose that the nonhuman world can be protected by an anthropocentric ethic (though not sufficiently so), the practical consequences of adopting an ethic which is more inclusive and not derived solely from human value would be abundantly different from the consequences of adopting an anthropocentric ethic such as Benson's (that is, one which attempts to provide protections for the nonhuman world by deriving its value from the value of humans only). An ethic which recognises the value of, say, all nonhuman animals, independent of what we value about humans, would have massive implications for animals. For example, adopting such an ethic (in practice) may mean that sometimes animals' interests would take moral precedence, even when there are human interests at stake, or that species that are unknown to us (and have little or no value for humans) would be given due consideration.

So, while Benson may seek to show (through his discussion of the extension of anthropocentrism to include all sentient life) that sentience-based arguments for protecting the environment are founded on anthropocentric principles, it is just not true that these arguments are always founded on such principles. Further, the consequences of adopting an ethic which is more inclusive would be vastly different from adopting an ethic which is derived solely from what we value about human beings and what is of value to human beings.

Perhaps in discussing the extension and particularly the anticipated counter-argument from Routley, Benson also means to reveal the inadequacies of sentientism (which supposedly fails to take account of nonsentient life) as opposed to anthropocentrism (which he believes can indirectly take account of nonsentient and nonhuman life). However, sentientism could take account of nonsentient life by arguing that its survival and protection is of paramount importance for the flourishing of sentient beings. Besides, the extended ethic that Benson discusses is certainly a more inclusive ethic than anthropocentrism, since its adoption would afford genuine protections for all sentient beings, whether or not there are human interests at stake. (But it is not without its problems. For example, we could imagine that there are no sentient beings left on earth, but there exists much nonsentient life. On sentientism, directing a nuclear bomb at the earth that was powerful enough to kill all life on the planet would be permissible, whereas on an environmental ethic this would be wrong. Sentientism then still poses problems in being too narrow as an environmental ethic, in that it only considers sentient beings and their experiences as having value.[6])

Further Objections

Returning to biocentrism then, one objection to this view is that it is not even possible to live if you accept such an ethic. One would have to morally consider all life, even the life of, for example, mosquitoes and bacteria. One would have to live by Albert Schweitzer's ethic of reverence for life (Schweitzer, 1994, p.65). According to Schweitzer all life is sacred and should be treated as such.

The above objection can be met by reference to Goodpaster's distinction between moral considerability and moral significance. As said earlier, moral significance 'aims at governing comparative judgments of moral "weight" in cases of conflict' (Goodpaster, 1978, p.311). Where interests conflict the interests of nonsentient beings may have less moral significance than the interests of sentient beings (although both these kinds of being have moral standing). We can pay respect to the interests and well-being of animals, or the interests of, say, trees (that is, we can give these things moral considerability) even if we do, for example, eat animals, or cut down trees for our own use. The objection to biocentrism then seems to mistake moral significance with moral standing. Plants could have moral standing and at the same time have very little moral significance in comparison to, say, sentient creatures.

There is also the distinction to be made between regulative and operative moral consideration (a distinction made by Goodpaster, 1978). We all have our own moral thresholds and limitations:

> An agent may, for example, have an obligation to grant regulative considerability to all living things, but be able psychologically and in terms of his own nutrition to grant operative considerability to a much smaller class of things (Goodpaster, 1978, p.313).

Moral considerability of an entity is operative when this moral considerability is psychologically possible for an individual. Moral considerability of an entity is regulative when this moral considerability is defensible on all grounds independent of operativity (or independent of whether it is psychologically possible for an individual) (Goodpaster, 1978, p.313). (We could also say, in addition to Goodpaster, that (in respect of some entities) *practically* operative considerability and nothing more may be possible.) Plants and other non-conscious living things do have moral standing and this should be taken into account as a regulative consideration:

> The regulative character of the moral consideration due to all living things asks, as far as I can see, for sensitivity and awareness, not for suicide (psychic or otherwise) (Goodpaster, 1978, p.324).

Although we have an obligation to take into account the moral standing of all living things as a regulative consideration, in reality, it may be no more than psychologically possible for us to grant operative considerability to non-conscious living things, like trees, because otherwise the life criterion would be impossible to live by.

We have reached a conclusion then that moral considerability belongs to whatever has interests. Interests belong to those things that have a good of their own or are capable of being harmed or benefited, and only living things (including future actual and possible ones) have interests.[7] Further, although some living things may have more or less moral significance than others, it appears that all have moral standing; all living things are capable of being harmed or benefited and, as such, have interests. Life[8] then (as a criterion that includes future actual and possible life) is a sufficient and necessary condition for something to have moral standing (see Goodpaster, 1978, pp.310, 313 and 316).[9]

Conclusions

Through an assessment of the question 'What are the necessary and sufficient conditions for X to have moral standing?' we discover that having interests is why animals used in current practices have moral standing – and not just sentient animals. Indeed, all animals (and all living things) have moral standing in virtue of the fact that they have interests. Therefore, a coherent ethical theory about our moral obligations to animals (or towards the environment) will have to take into account and include all living creatures, but animals used in farms and experiments are strong candidates for moral standing in virtue of their sentience.

Considering that all living things have moral standing and that animals with greater capacities arguably carry greater significance just makes our obligations to animals used in current practices stronger than our obligations to many other living things. Animals used in factory farming and in experiments are sentient, conscious, and aware beings. They not only have interests, and can be affected for better or for worse, but are also beings that have a life, rather than just being alive. So, considering that all living things deserve moral consideration, these animals may well have greater moral significance than many other animals. Being sentient just gives them stronger claims on our moral obligations.

That all living things have interests and, therefore, moral standing is what accounts for many of our intuitions and reflections that we can harm nonsentient beings, whether they are human beings, such as comatose humans, or nonhuman things, such as trees. Indeed, looking at the Last Man example it seems that trees do deserve moral consideration. How much moral significance we give to different living things is a different matter. But it is hard to justify giving more moral significance to a human embryo than to a sentient animal used in a factory farm or an experiment. Here is where we can make comparative judgments of interests.[10]

We have found then, through a critical examination of ethical theories, that suffering is not the only thing that informs our moral relations with animals. Animals have other interests, one obvious one being an interest in life or living. Undoubtedly, it is not just sentient animals that have interests. Nonsentient and non-conscious creatures have interests too. Indeed, it has been found that interests belong to all living things. Having interests is a sufficient and necessary condition for moral standing, and since all and only living things (present and prospective) have interests it follows that being a living thing (whether present or prospective) is a necessary and sufficient condition for having moral standing (see Goodpaster, 1978, pp.310, 313, and 316).

Not only do factory farm and experimentation animals have interests that are morally important, but our use of these animals and the considerable suffering we inflict upon them generate strong responsibilities to take their interests seriously and give them due moral consideration. Recognising the moral standing of animals involves seriously considering the interests of the animals used in the practices of factory farming and experimentation, and this in turn would involve taking seriously the fact that they are made to endure immense suffering for human benefits, and recognising that those benefits are not always weighty enough to justify inflicting extensive suffering on *millions* (if not *billions*) of animals *every day*. Indeed, if the interests of these animals were properly considered and weighted against the interests of humans then massive changes would need to be made to the way animals are treated in current practices and certain aspects of these practices (or the practices themselves) would be seen to be unjustifiable.

Notes

1 Under the *Protection of Animals Act 1911*, it is an offence to cause unnecessary suffering to any animal, ill-treat, torture, or frighten an animal, but animals used in experiments are excluded from the act (*Protection of Animals Act*, 1911, cited by

British Union for the Abolition of Vivisection (BUAV), 'Campaign to End Animal Experiments: Legal Cruelty' [article by BUAV accessed at www.buav/law/cruelty, on 13/07/06]. More recently, the *Animal Welfare Act 2006* lists welfare needs that pet owners have a legal responsibility to meet, but states that 'Nothing in this Act applies to anything lawfully done under the Animal (Scientific Procedures) Act 1986' (*Animal Welfare Act*, 2006).

2 The life criterion does not commit one to the view that mere existence has value.

3 This further imaginary scenario is one similar to that presented by Robin Attfield in his 'The Good of Trees', in *Environmental Philosophy: Principles and Prospects* (Aldershot: Avebury, 1994), p.168 (though Attfield's scenario is derived from Routley's Last Man example) and for present purposes it will be used to examine some judgments about the value of trees. Routley's original example will be returned to shortly.

4 One of the best ways for addressing intrinsic value is by thought experiments. See Robert Nozick's Experience Machine (Nozick, *Anarchy, State, and Utopia* (Oxford: Basil Blackwell, Basic Books, 1974), pp.42-45). See also Donald Scherer's Planet Lifeless and Planet Flora (Scherer, 'Anthropocentrism, Atomism and Environmental Ethics', in *Ethics and the Environment*, eds. Donald Scherer and Thomas Attig (Englewood Cliffs, New Jersey: Prentice-Hall, 1983), pp.73-81).

5 In discussing this anticipated counter-argument of Routley's, I do not think that Benson is thereby conceding that the last man's act here is permissible (though his reasons for not conceding this would have nothing to do with the wrong caused to nonsentient beings themselves). On Benson's anthropocentric ethic, as said above, in killing all life on earth (both sentient and nonsentient) the last man does indeed do wrong, for he harms what is of value to human beings or, at least, to the last man himself.

6 For another thought experiment that discovers value in the flourishing of nonsentient things see Donald Scherer's Planet Lifeless and Planet Flora (Scherer, 'Anthropocentrism, Atomism and Environmental Ethics', in *Ethics and the Environment*, eds. Donald Scherer and Thomas Attig (Englewood Cliffs, New Jersey: Prentice-Hall, 1983), pp.73-81).

7 It has been said that companies or corporations can have interests (see Robin Attfield, *Value, Obligation and Meta-Ethics* (Amsterdam: Rodopi, 1995), p.24). However, I am inclined to agree with Joel Feinberg that the interests of companies should be seen in terms of the interests of the people who work for those companies. As Feinberg says, 'corporate interests... are... analyzable into the interests of its numerous flesh and blood members' ('The Rights of Animals and Unborn Generations', in *Philosophy and Environmental Crisis*, ed. William T. Blackstone (Athens: University of Georgia, 1974), p. 57).

8 Here, 'life' should be interpreted to mean being a living thing, rather than having a life. I make this distinction since the latter interpretation could suggest having consciousness or non-vegetative life. (It should be noted here that there is a difference between being alive and having a life.) If this latter interpretation were

given then the life criterion would exclude much nonsentient life, like trees. Kenneth Goodpaster says that 'being alive', 'life' or 'being a living thing' is a necessary and sufficient condition for moral considerability (Goodpaster, 'On Being Morally Considerable', in *Journal of Philosophy*, 75, 1978, pp.309-25). It seems to matter though whether one uses the words 'being alive', 'life' or 'being a living thing' to state the criterion for moral standing. The use of these words is relevant here. If one uses the words 'living thing' to state the necessary and sufficient condition for moral considerability then (following Goodpaster's view) this would mean that all and only living things are capable of being beneficiaries and, as such, of having moral standing. This criterion could include future possible beings, by stipulating that they are included (as I have done). Future possible beings will or may be, at some point, living things. If, however, one talks in terms of 'being alive' or 'life' as a necessary and sufficient condition for moral considerability then this could suggest that all and only beings that *are alive* or *have a life* are capable of being beneficiaries and, as such, of having moral standing. The words 'are alive' could imply that future beings are excluded from moral considerability, while the words 'have a life' could imply that vegetable life is excluded, although I do not think it was Goodpaster's intention to do so, and he certainly wanted to include all living things (including vegetative ones) in the moral sphere. However, I have stipulated above that by 'life' I mean being a living thing (whether present or prospective).

9 Goodpaster argues that 'being alive', 'life', or 'being a living thing' is both a necessary and a sufficient condition for moral standing. Only those beings that satisfy this criterion are capable of being beneficiaries and of being harmed.

10 That all living things have moral standing is why we say that abortion or using human donor eggs or cloning are moral issues. We are dealing with things that are living, and, as such, they have interests (although they may not have a life as such). However, it does seem premature to talk of the sanctity of human life, whatever its kind or quality, in a world where the practices of factory farming and animal experimentation, that cause immeasurable and widespread suffering on a global scale, are accepted as the norm. What about the lives of these animals? Their lives not only matter in respect of having moral standing, but matter to the animals living them.

References

Attfield, Robin. 2003. *Environmental Ethics*. Cambridge and Malden, MA: Polity.

_____. 1995. *Value, Obligation and Meta-Ethics*. Amsterdam: Rodopi.

_____. 1994. 'The Good of Trees', in *Environmental Philosophy: Principles and Prospects*. Aldershot: Avebury. 53-71.

Benson, John. 2000. *Environmental Ethics: An Introduction with Readings*. London: Routledge.

Bentham, Jeremy. 1982. *Introduction to the Principles of Morals and Legislation*, eds. J. H. Burns and H. L. A. Hart. London and New York: Methuen.

DeGrazia, David. 1996. *Taking Animals Seriously: Mental Life and Moral Status* Cambridge: Cambridge University Press.

Diamond, Cora. 1995. *The Realistic Spirit.* Cambridge, Massachusetts and London: MIT Press.

Feinberg, Joel. 1974. "The Rights of Animals and Unborn Generations." *Philosophy and Environmental Crisis.* ed. William T. Blackstone. Athens: University of Georgia Press. 43-68.

Goodpaster, Kenneth. 1978. "On Being Morally Considerable." *Journal of Philosophy.* 75 (308-25).

Norton, Bryan. 2000. "The Cultural Approach to Conservation Biology." John Benson. *Environmental Ethics: An Introduction with Readings.* London: Routledge. 143-51.

Routley, Richard. 1973. "Is There a Need for a New, an Environmental, Ethic?" *Proceedings XV, World Congress of Philosophy.* Bulgaria: Varna. 205-10.

Schweitzer, Albert. 1994. "Reverence for Life." *Environmental Ethics: Readings in Theory and Practice.* ed. Louis P. Pojman. Boston and London: Jones and Bartlett. 65-71.

Singer, Peter. 1995a. *Animal Liberation.* 2nd edition with a new preface by the author London: Pimlico Press.

_____. 1995b. *Practical Ethics.* 2nd edition. Cambridge: Cambridge University Press.

Taylor, Paul. 1998. "The Ethics of Respect for Nature." *Environmental Philosophy: From Animal Rights to Radical Ecology.* eds. Michael E. Zimmerman *et al.* Englewood Cliffs, New Jersey: Prentice Hall. 71-86.

Vilkka, Leena. 1997. *The Intrinsic Value of Nature.* London: Rodopi.

Chapter 5

Animal Rites, Rights and Whites: Traditional African Beliefs and Animal Slaughter in South Africa[1]

Sandra Swart

There are many ways to kill a cow. In modern post-Apartheid South Africa the slaughter of animals is key not only in the physical consumption patterns of a famously carnivorous society, but also in its spiritual life. In religious terms, it plays a role in Jewish and Muslim faiths, and it is central in the ancestor belief systems of many southern African black communities.[2] Of course, the ritual slaughter of animal is entrenched in many of the world's cultures and religions. Indeed, there is a theory that such rituals had their origins in the pre-neolithic hunt and such ritualised chases and killings have been depicted in paleolithic rock art.[3] In some modern Zulu belief systems, *amadlozi* (the ancestors) have direct influence on the living and can intercede with God, much like household *lares et penates* or some Christian saints.[4] The living and the dead are thus connected in an enduring relationship. Some of the spirits, however, wield a singular influence, continuing to commune frequently with the living. Communication demands the action of certain rituals, characterised as orthopraxis rather than orthodoxy.[5] Ancestors are reached through animal sacrifice at the homestead at times of both crisis and celebration. Ancestors remain part of syncretic belief systems – Christianity and traditional beliefs are not exclusionary, but rather complementary.[6] Nearly all South Africans hold religious beliefs, but by far the largest majority now identify with Christianity: 86% of the population (38 million people) belong to a Christian denomination.[7] Only a miniscule 0.3% of the population identify exclusively with African traditional religion. However, Africans self-identifying as Christians often still retain communication with ancestors.[8] Thus the ceremonial killing of cattle, goats and sheep takes place in the rural homesteads, in urban townships and increasingly in the suburbs to mark life transitions: births, coming of age, marriages, death or for specific individual needs, like healing or cleansing.[9] The sacrifice of animals has occupied a great deal of media attention, legal intervention, lobby-group involvement and public interest in the community arenas of the young South African democracy.

A recent controversy cries out for an historicized understanding of the idea of both cruelty and ethical treatment of animals in African belief systems in South Africa. In the annual bull killing ritual, young Zulu men beat a bull to death, stabbing its eyes, and twisting its genitals. This offers us a useful lens into longer societal processes and the traditional notions of slaughter of animals, which is integral to the spiritual life of many people in South Africa. Post-Apartheid, with the end of racialised residential restrictions, the white middle class has been confronted with the visceral evidence of African beliefs on their doorstep. The focus of this research is on how the discourse of animal rights and animal welfare over the issue of cruelty has been pitted against cultural values, particularly African traditional beliefs. This chapter is careful to differentiate between different ceremonies which involve slaughter within traditional African belief systems, through a case-study of Zulu ritual. Since colonial times, African religions have been elided, over-generalised and oversimplified by outsiders, and reduced to animism and ancestor worship. Any sweeping statement about a subject as multifaceted as religion and particularly in a continent as diverse as Africa, is clearly an audacious – or foolhardy – enterprise. So the chapter focuses on one aspect of the religious beliefs of one ethnic group, not as representative of all those in South Africa, but rather as a way in to start asking questions about animal-human relationships in African religions.

In such a diverse society as South Africa notions surrounding animals historically have been fissured with tensions generated by class, race, gender and generation. At the same time there is constitutionally-entrenched right to practice one's religious beliefs freely. Thus some political and religious leaders defend their practices as both traditional and authentic and resist the imposition of (they argue) white western notions of humane treatment. Indeed, Zulu leaders defended the bull-killing ceremony in court by contending that it demonstrates a different *kind* of respect for animals inherent in Zulu religion and culture. This chapter will show, however, that there is nothing straightforward about the adjectives "traditional" or "authentic". The discussion shows that – far from the static diachronic representation of African belief systems presented by experts on tradition – African beliefs are synchronic and always adapting to socio-historical conditions. Concomitantly, African animal rites and rights have changing trajectories of their own and are evolving in contemporary South Africa.

Ways of Dying

The method of slaughter varies. Nineteenth century travellers and later ethnographers described the ceremony in which the men of a particular regiment fought a bull before a "special Doctor" despatched it by severing its vital cord in the neck.[10] Yet, already by the 1960s, there were suggestions that the slaughter method could vary and that the manner had changed in some instances from stabbing in the side to in the neck, leading to a painless and silent death.[11] Yet, for some the bellowing of the beast shows that the ancestors are satisfied and will help the supplicant. At least a moment of real pain must therefore be experienced by the animal "for this bellowing is the method employed by the spirits of expressing their acceptance of the offering".[12] As a cow "needs" to cry – or the ancestors will not be satisfied – the stun gun approved by the Society for the Prevention of Cruelty to Animals (SPCA) is eschewed.[13]

Currently, ritual sacrifices occur daily. Many receive no comment, other than approbation and acceptance, from participants and neighbours. In other cases, white neighbours objected to ritual slaughter of a cow but linked the complaint – either openly or more implicitly – to the fact that their black neighbours were "Other" to their religion and lifestyle. For example, one coupled objections to religious animal sacrifice to the fact that the black neighbours' pool was filthy, they had not mowed the lawn, they partied late into the night and that taxis parked on their verges. Others asked only that animal killing should be discreet and not in front of children.[14] Animal sacrifice was even blamed for attracting Zambezi sharks to the beach, resulting in the death of lifeguard and a surfer.[15] Others staunchly ignored religious relativism, maintaining simply that "animal cruelty is animal cruelty."[16] Ballard, who conducted an ethnographic survey into white attitudes to the newly multiracial suburbs, captures it neatly with "Cattle slaughter... is an example of precisely this kind of intrusion of 'otherness' for some established white residents."[17]

Although now increasingly routine occurrences, some cases of sacrifice catch the public imagination, eliciting both outrage and defensiveness from the two extremes, and a range of responses along the gamut from various individuals and groups. For example, there was public outrage after the (early) release of African National Congress[18] (ANC) former chief whip and disgraced convicted fraudster, Tony Yengeni, who slaughtered a bull to celebrate his release from prison to act purportedly as a cleansing ritual.[19] The SPCA laid a charge of animal cruelty under the Animal Protection Act of 1962. Ironically, it is a charge which carries the possibility of a prison sentence.[20] In another highly publicised incident, during the recent 2010 Soccer World Cup, traditional African leaders and the Traditional Affairs Minister urged animal slaughter prior to the global event, although it was refused by organisers after international condemnation. The idea was that an *inkomo* (a cow) would be slaughtered at each stadium "so we would be inviting our African ancestors to be with us throughout the soccer event."[21] Adding, 'It shows "we are Africans"'.[22]

Taking the Bull by the Horns

Another, more protracted controversy erupted over the *Umkhosi Wokweshwama* ceremony, or Festival of the First Fruits[23], which incorporates *Umkhosi Wezinsizwa* (the ceremony in which a regiment is appointed, is said by its promoters to date back three millennia.[24] It revolves around the killing of a bull at the Zulu king's palace at KwaNongoma. Cultural experts explain that the when the bull is killed, its vigour and potency are transferred to the king and to his army. "Young warriors" – pre-pubescent boys – who are perceived to possess both the "innocence and maturity", members of the *amabutho,* slaughter the bull which is "especially chosen for its strength" (some suggest the earliest ceremonies involved the sacrifice of a buffalo).[25] This provided an arena in which the young warriors could prove their bravery. The aim was to overpower the animal, close its airways and then break its neck. There are mixed reports: some maintain the bull is killed as "humanely as possible", given herbs to "desensitise it" and, should it be seen to suffer, an elder supposed to "put a knife to the spinal column." An ex-participant contended that the bull only suffered for ten minutes and another said it is explicitly stated the bull must be killed *ngokuphazama kwehlo* [in the blink of an eye].[26] Others report that the bull is killed cruelly: its eyes are gauged out, it is choked with sand in its nose and mouth, its testicles are pulled and his tongue pulled out.

The young warriors are then said to become heir to the bull's power, and, then, through their deference, they transfer it to their king, who channels it into defending the realm. No blood-letting or dismemberment of the bull is allowed as "the symbolic representation of the king [is] in the form of the bull". And, as a warrior who once participated in such a killing, bare-handed killing is mandatory as a "sign that no violence should engulf the nation."[27] A culture-broker contended: "From ancient times it was believed the power of the king wanes. While in olden times the king was killed and a new king was installed, the modern practice developed of killing the king symbolically so that his power may be regained and he would be revitalised."[28] The royal house argues that the killing and, significantly, the *manner* of the killing are both unalterable as "the Zulu people can't separate the killing of the bull from the ceremony because they need to have a healthy king, as well as a powerful king." It has also been described as a celebration of a "co-operative society. The youth's work as a team to ensure the bull is brought down is a continuation of the process of co-operation."[29] A slightly different interpretation of the ritual is that, by killing the beast, the young men develop a bond of brotherly comradeship, which they then transmit to others in their age cohort, which strengthens "manhood" in their kingdom and dissuades them from fighting each other.[30]

After the bull is killed, a traditional healer removes its gall and mixes it with secret herbs. This is given to people to drink and its flesh is given to the young warriors; the rest is burnt. Purportedly, citizens were not allowed to consume their first corn or fruit harvest before they had offered them to their king and at the same time the festival was said to serve as a thanksgiving to God (*uMvelinqangi*) and to the *amadlozi*.[31] The king was supposed to accept the first fruits from God on behalf of all his subjects. They say this ceremony, like many others, was "suppressed during colonial times, first under government by the British and then later under apartheid." The ceremony incidentally featured in the Apartheid-era, state-funded television series *Shaka Zulu* and in the film *Zulu Dawn*. The current king, Goodwill Zwelithini, apparently revived the old ceremony, in the last decade which had lapsed for a generation before that, to jump-start an "agrarian revolution" among his subjects who face "poverty, starvation and joblessness".[32] (The two, initially distinct, festivals were combined by Zwelithini when he revived the traditions.) One learns this from the King's website.[33]

A Bull Fight?

The event evoked international outrage.[34] A Facebook campaign and online petition were initiated.[35] A lobby-group Animal Rights Africa (ARA) brought a court action to the Pietermaritzburg High Court Judge to halt the manner of the killing. They argued it was a contravention of the Animal Protection Act and South Africa's international obligation in terms of the Terrestrial Animal Health Code of the World Organisation for Animal Health, to which South Africa is a signatory.[36] According to the Act, any person who "ill-treats, infuriates, tortures or maims or cruelly beats, kicks or goads or terrifies any animal" is guilty of an offence.[37] Other protestors included PETA and representatives in various African countries, including Somalia, Uganda, Egypt, DRC, Nigeria, Ethiopia, Tanzania, and Zimbabwe, calling on parliament to denounce it as "unbecoming to the modern face of Africa".[38] They even petitioned the ultimate moral authority in the modern globalised world – arguably the Buddha and Jesus of twenty-first century consumer society – Oprah Winfrey.[39]

International outcry remained in the virtual cyber-realm, however, but locally the ARA contended that their action was taken "on behalf of the victim, in this case the innocent bull which has moral agency outside of human institutional or cultural arrangements". They lamented that their legal action had been "subverted" and that that their approach was "not anti-culture, not anti-freedom, not anti-human."[40] They made the point that the King's own website acknowledged that the ritual was resuscitated and thus it could not be claimed that the killing was a "necessary part of Zulu culture that must be observed at all costs."[41] They maintained that it recognised the need for respect for other cultures, but wanted to "move the discussion of an inclusive justice and the recognition of animal rights."[42] ARA asked the court to grant an interim interdict to stop this year's ritual on grounds of cruelty. Alternatively, they asked that the bull not be "tortured or maimed, beaten, kicked or terrified" or have "its tongue pulled out of its mouth, its eyes gouged or its reproductive organs mutilated, twisted or severed, or have any substance forced down its throat."[43] The organisers refused to allow the event to be filmed and the ARA appealed to the court to allow representatives to attend and record the ceremony, disputing Zwelithini's contention that the bull dies in a quick and painless fashion by "breaking its neck".[44]

Interestingly, counsel representing the king, KZN Premier Zweli Mkhize and Traditional Affairs MEC, Nomusa Dube, appealed for protection of ancient monarchical beliefs by invoking the new democratic constitution. It is certainly true that the 1996 constitution provides for religious and cultural rights, reinforced by the 2002 Commission for the Promotion and Protection of the Rights of Cultural, Religious and Linguistic Communities Act.[45] In terms of the Act, one group of society cannot assert its beliefs over another. Significantly, that is "particularly" true if those beliefs involve "religious custom and ritual".[46] In fact, the monarch's supporters even appealed to the South African Human Rights Commission.[47]

Bulldozing Bullies?

South African President Zuma's spokesperson in the ANC, Zizi Kodwa, noted that those that opposed the ceremony were attempting to force their own racist agenda: the very existence of debate on this issue carried "strong neo-colonial and racial undertones"[48]; "[i]t is a neo-colonial mentality that suggests the African must be civilised, because it is a dark heathen continent [sic]; "the criticism... is nothing but sheer, unbridled arrogance and the imposition of the mores of western civilisation." Indeed, as Ballard has shown, there is some truth to the assertion that colonists treated animal slaughter as "paradigmatic of [African] cruelty and racial inferiority".[49] Perceptions of the ritual were historically filtered through chauvinistic binaries of "humane-cruel" and "religion-superstition. So Europeans could imagine their own civilization and modernity in contrast to the barbaric and superstitious African.[50]

But Kodwa went further, conflating religion, culture, and hereditary absolute rule, asserting "the king is the custodian of the cultures of his subjects, the people. He is the ultimate voice of reason" on "their behalf."[51] Letter writers to the press urged those who questioned the custom leave:"If Animal Rights Africa feels we are barbaric and cruel to animals, the exits are all open, from airports to harbours, Cape Town to Joburg and Durban."[52] Another asked: Must we forfeit our rights so animals can live? "How [else] are young Zulu mean supposed to come of age? By being thrown a 21st birthday party?."[53] "Where do *izifikanamthwalo* (outsiders) get off telling us what to do in our own back yard?" [54]

A Lot of Bull?

There were hypocrisies and lapses in logic on both sides. Public opinion became polarized.[55] Public discourse ran a continuum between openly racist critique about barbaric cruelty and tolerant acceptance of ritual slaughter. Many secularists who dismissed animal sacrifice as an outdated superstition nevertheless conceded it was not more cruel than state- and public-sanctioned industrial meat production. They noted that the uninitiated were on some level simply expressing visceral horror at confronting reality of animal-human relations that are normally hidden from view, behind the veil of the abattoir. [56]

Others contended that the ritual of slaughtering can be traced to Abraham, and Muslim and Jewish community practice their own forms of slaughter so the interest in this case must have been generated by selective racism. In fact, the possibility of remaining ignorant of other religions and cultures is tied to Apartheid, it was a luxury uniquely available to many white South Africans. Whereas blacks had to learn how to be bicultural, "able to play by the rules of their own cultural community and able to play the game according to the rules established by the former masters – the whites."[57] As one commentator observed: "the noise that persisted was about black versus white, European versus African, colonist versus "freedom fighter".[58] An Inkatha Freedom Party (IFP) member, Blessed Gwala, noted ARA had not objected to Spanish bull-fighting, which he dubbed "a game of teasing a bull until it dies."[59]

Others argued that the festival was crucial, as cultural festivals contribute to self-esteem, especially among the youth.[60] Rituals maintain and create social structure and a world view. "Rituals support change and transition. Because rituals themselves have a beginning and an end, they give people tools to make changes or life transitions." Others conceded openly that "the ritual is cruel in nature is not in doubt, "but one must see "the bigger picture"; it is not intended as sadistic and not be "narrow-minded"."Just "like Jesus died cruelly on the cross as the sacrifice for

A newspaper cartoon illustrating similarities between (Spanish) white, cultural practices and African religious ritual.[61]

mankind in order for us to be cleansed of our sins and be able to return to live with God, the same principle (here in the name of culture, not Christianity) ... applies to the bare-handed killing of the bull. Call it barbaric or whatever, *Ukweshwama* does enjoy constitutional protection. In this case, rights of humans to practise their culture trumps the rights of animals."[62]

Attempts to discuss the matter were boycotted by the king.[63] The ANC leadership in the province issued a thinly-veiled threat that court intervention could cause race tension to flare up.[64] They warned of outpouring of anger and widespread unrest, "vehemently opposing this court order so as to protect the rights of our people to exercise their cultures... as protected [by the] constitution."[65] The King's spokesperson: "We view this court challenge as nothing but a religious persecution of the Zulu people...".[66]

The (white) judge appealed to parties to reach a compromise, then, when that failed, dismissed the action. He stated that it was simply unrealistic for "one judge to put a stop overnight" to an ancient ritual. The judge said if the killing were interdicted the symbolic transfer of power to the king would be prevented and the "Zulu nation left with a powerless king."[67] Civil disobedience, uprising, unrest would result. Disarmingly, he asked: "What if the king then lost his powers or was struck by lightning." What would happen then? "I am sitting on a powder keg here...". He even compared the issue to the Soweto Uprising.[68] He reiterated that he was himself an animal lover but he had to consider the importance of this tradition to the Zulu nation. He maintained he was damned either way.[69]

When the Cows came home

After the judgement, the Zulu king himself allegedly likened the court action to "the destruction of the Zulu nation characterised by the burning of Ulundi and the banishing of Zulu kings by colonialists after 1879."[70] In his speech to thousands at the ceremony he said: "But we won." He called for cultural unity among Zulus, castigating *izimpimpi* (traitors, quislings) who photographed the slaughter and handed pictures to "enemies".[71] Journalists were instructed to leave just before the slaughter.[72] Even cell phone cameras were banned. He warned that the traitors would suffer the wrath of the ancestors. The king warned his warriors that it should only take two minutes to kill the bull and promised a prize to the most skilful slaughterer. Once the bull had been separated from the herd, members of the *Izinyosi* regiment gave chase – but the bull eluded them, escaping the kraal. This enraged the king, who chastised them as *iminqolo* (mama's boys). The bull was caught and brought back in, he was grabbed by the horns and brought down after which the forty youths encircled him, chanting, as the closest tried to break his neck. This continued for about ten minutes.[73]

The king denied that the tongue was pulled out, maintaining it was impossible as "the animal has teeth and can easily bite you." And he added: "I do not know what these people who want to say that we are cruel to animals are talking about. We respect animals to the extent that we even give them names. How can we then be cruel to them?"[74] Indeed, it is true that owners know the genealogy of each animal and care about its family-tree. "Reverence for cattle is symptomatic of the deep ecological consciousness that pervades African culture. Consider the totems... clans adopt animals [which]... makes it taboo for a person to harm an animal that represents his or her totem....it arouses the wrath of the ancestors – *ingqumbo yeminyanya*." He added: "Surely, this cannot be the culture of a people predisposed to animal cruelty?"[75]

A group of Zulu warriors killing a bull, December 9, 1995, at the Royal Palace near Nongoma. [76]

There was only one attempt to see it from the animal's perspective. A commentator, Justice Malala, remarked: "The bull is now dead. In the aftermath of the noise and the anger in the week prior to his slaughter, I wish to speak for the dead bull.[77] He pointed out that it was risible to say that the bull must suffer because his ancestors had made animals suffer: "ancestors had not read the work of J.M. Coetzee and were not on facebook." He strongly criticised the ANC for cruelty to animals in the name of culture. Asking simply, why not kill the bull swiftly?[78] He contended that in silencing debate over religion, Zuma's ANC was going back to old-style cultural chauvinism, taking the once progressive party backwards, "I mourn for an ANC that has abandoned science for superstition." He added with hauntingly poignant simplicity: "I have spoken for the bull."[79]

Where's the Beef?

An academic can ask several questions about this story. One can probe the crude, deterministic language used to describe "culture", "tradition" and "religion" as though these human social constructions were homogenous, synchronic and static. One can ask who makes someone a "cultural expert"? Why elide Zulu language and "ethnic identity"? One could ask whether a festival that had to be "revived" (and in the process altered) is necessary for cultural identity and religion? One could ask why "culture" and "religion" appear to be used interchangeably? (Culture and religion are hard to de-couple but it is clearly necessary for human rights litigation, as the Constitution specifies two separate rights and also modern human rights discourse appears to take religion more seriously than culture.) Instead of being treated equally, as dictated by the Constitution, traditional religions are relegated to manifestations of "culture", with the sub-text that they are not 'true' religions.[80] If it is true, as is increasingly recognised, that violence against animals is a forerunner to violence against vulnerable humans why would this revived tradition be tolerated?[81] Does inhumane killing of animals desensitise one to violence against humans?[82] One could ask if it were really sacred to "all" Zulus if so many appeared to dissent: some asking vehemently to the "barbarity of the ritual and the assertion that it encompasses Nguni culture."[83] There are still hypocrisies, slippages and contradictions within the modern democratic state. African traditional practices and human rights are enshrined in the new constitution, arguably the most progressive in the world but so are polygamy and patriarchy, for example, in "tribal law" and chiefly authority. One could question how such so-called religious festivals reinforce centralised authoritarian leadership.[84] One could expose weakness of the cultural argument, can be used to excuse "sacrificing a human" or genital mutilation – it is not justified simply because it is a cultural practice. As philosopher David Benatar points out "there is nothing inconsistent in opposing greater cruelty while not opposing lesser cruelty". Once could condemn the smug, barely concealed bigotry or secular dismissal of religious beliefs manifested by some of those who opposed the ceremony.[85] Should relationship with animals be private, not open to the scrutiny of other people and other 'cultures' or religious faiths?

Human Rites and Animal Rights – Conclusions

Some key points have emerged from these narratives of slaughter. Some might argue that secular slaughter for mass meat consumption is arguably just as cruel as *Umkhosi Wokweshwama*. Here it is not only the death but the life imposed on the animal that is cruel. A bloodless packet of chicken in white South African supermarket appears innocuous but is the end-product of ruthless factory farming. Urbanites assume they will have sympathetic, non-utilitarian and non-violent animal relationships. Yet factory farming is arguably more brutal. There is just a different expedient: the ritual bull killing appears to be going out of its way to be cruel, whereas factory farming is merely incidentally cruel as a cost-saving device. In the public imagination there is a great distinction as to whether the animal is slaughtered on the altar of the ancestors or on the altar of capitalism.

Animal-human relations serve as proxies for relations between different human groups – castes, classes and ethnicities. Beliefs about animals' rights and treatment of animals are presented as seamless in homogenised cultures. Unitary reading of religious belief: traditions are cultural constructions. In a culture of dehumanising and of violence towards the other, violence towards animals requires our urgent attention. Animals are still fundamental in the human search for meaning and purpose in understanding our relationship with gods, ancestors, and other people. Reactive deference for culture and religion closes down critical debate. Equally, multiculturalism, asserting the moral principle of respect for the principle of the rights of others to believe and practice diverse things, is not the same as relativism.[86] Respect for their right to exist neither translates neither into automatic respect for the beliefs themselves nor into wholeheartedly support for all of the beliefs right to exist. Another key point is the relationship between cultural constructions of masculinity and animals: what is the effect on a boy who tortures a beast so as to become a man?

The final point to take from these narratives is the most significant for our purpose: the purportedly static nature of human-animal relations in religious context fundamentally untrue, as this final vignette reflects. There is a possibly apocryphal story about "traditional" killing that is nevertheless widely told and believed. One of the so-called young "Black Diamonds", the new generation of upwardly mobile African elite, was about to get married recently.[87] This son of famous, exiled ANC activists – scion of a notable lineage – desired a glittering society wedding. He wanted to combine a white wedding with traditional African beliefs and the high point of the ceremony to be his ritual killing of a bull. At the last minute, however, his nerve failed him. Instead, he turned to one of his security guards, who mowed the animal down with a sub-machine gun. There are many ways to kill a cow.

Notes

1 *This chapter was made possible by a Writing Fellowship from the American Humane Society. The opinions expressed here are my own, however.

2 There a numerous African groups with different languages and cultural histories in South Africa. The biggest group is Zulus at about 21% of the population. The Apartheid regime tended to over-emphasise cultural differences between the various ethnic groups, for example between Xhosas and Zulus, in a divide-and-rule ploy to turn them against each other rather than against the white government.

3 W. Burkert quoted in Michael Lambert, "Ancient Greek and Zulu Sacrificial Ritual: A Comparative Analysis", Numen, 40, 3, 1993, p.312; Harald Pager,

"The Ritual Hunt: Parallels between Ethnological and Archaeological Data", *The South African Archaeological Bulletin*, 38, 138, 1983, pp.80-87.

4 In an interesting analysis, Zulu and ancient Greek sacrifice have been explicitly compared by classicist. Lambert, "Ancient Greek and Zulu Sacrificial Ritual: A Comparative Analysis", pp.293-318.

5 In other words, "right action" rather than "right belief".

6 BA Pauw 'Ancestor beliefs and rituals among urban Africans' (1974) 33 *African Studies* 99 103.

7 Statistics South Africa *Census 2001: Primary tables* http://www.statssa.gov.za: 8,2% of the population is Pentecostal/Charismatic, 24,4% belong to established churches and 36% to 'other Christian' denominations; 1,5% of the population identify with Islam and minorities profess such religions as Hinduism and Judaism.

8 Zulu faith includes belief in a creator God (Unkulunkulu), although it is possible that this conviction originated from efforts by nineteenth century Christian missionaries to frame the idea of the Christian God in vernacular idiom.

9 Richard Ballard, "Slaughter in the suburbs": livestock slaughter and race in post-apartheid cities", *Ethnic and Racial Studies*, 33, 6, 2010, pp.1069-1087, p.1072; Bergland, *Zulu Thought-Patterns and Symbolism* p.220; A.T. Bryant, *The Zulu People: as they were before the white man came* (Pietermaritzburg: Shuter and Shooter, 1949, 1967), pp.265-6; Krige, *The Belief Systems of the Zulus*, p.188; *see also Luc de Heusch* (trans by L. O'Brien and A. Morton, *Sacrifice in Africa: A Structuralist Approach* (Bloomington: Indiana University Press, 1985).

10 R.C.A. Samuelson, *Long, long Ago* (1929, Durban: 1974) p.382.

11 Axel-Ivar Bergland, *Zulu Thought-Patterns and Symbolism* (London: Hurst and Company, 1976) p.228.

12 Eileen Krige, *The Belief Systems of the Zulus* (Pietermaritzburg: Shuter and Shooter, 1936, 1957) p.291.

13 Mvuyo Mati, "The day the bellow visited a culture", *Sowetan*, 25 January 2007; Khanyi Ndabeni, "African traditions have tough time in suburbs", *The Herald*, 9 January 2009.

14 Doreen Premdev, "Neighbours upset at ritual animal slaughters", *Sunday Tribune*, 6 April 2008.

15 Guy Rogers, "Beach slaughter feeding shark frenzy", *The Herald*, 20 May 2009.

16 Letter, Ross McClelland, *The Herald*, 14 January 2009.

17 Ballard, "Slaughter in the suburbs", p.1073-1074.

18 The African National Congress has been South Africa's governing party since the establishment of majority rule in 1994 with the democratic transition.

19 Biénne Huisman, Nashira Davids, Ndivhuho Mafela and Buyekezwa Makwabe, "Yengeni hosts cleansing party", *Sunday Times*, 21 January 2007.

20 A person convicted faces a maximum prison sentence of 12 months or a R20 000 fine. Russel Molefe, "To slaughter or not to slaughter", *City Press*, 20 July 2008.

21 Puleng Mashabane, "Slaughter at 2010 soccer showpiece?", *Citizen*, 23 October 2009.

22 Tshifhiwa Shonisani, "Minister wants 2010 slaughter", *Citizen*, 6 November 2009.

23 Of course it also echoes The Feast of First Fruits from Leviticus 23:10-11.

24 Sandile Waka-Zamisa, "King defends bull ritual, urges circumcision", *Witness*, 7 December 2009; Bheki Mbanjwa, "Is ritual bull killing cruel?", *Witness*, 22 December 2007.

25 Sherlissa Peters, "Bull-killing judgement reserved", *Daily News*, 2 December 2009.

26 Professor Pitika Ntuli, quoted in "Bull: 'Minimal suffering", *Citizen*, 7 October 2009; Bheki Mbanjwa, "Is ritual bull killing cruel?", *Witness*, 22 December 2007..

27 Thami Magubane, "'Bull dies quickly, painlessly", *Witness*, 28 November 2009.

28 Professor Simon Maphalala quoted in "What the Ritual Stands For: The Dying Bull's Power Passes to the Nation's King", *Witness*, 2 December 2009.

29 Professor Pitika Ntuli, quoted in "Bull: 'Minimal suffering", *Citizen*, 7 October 2009.

30 Fred Khumalo, "African culture? Not in my name", *Sunday Times*, 11 October 2009; Vusi Shongwe, "Taking the bull by the horns", *Sowetan*, 2 December 2009.

31 Ndela Ntshangase, "Ceremony revives pride and self-confidence", *Sunday Tribune*, 29 November 2009.

32 Bull Killing, Witness, 26 November 2009; Stephen Coan, "The bull ritual in history", *Witness*, 3 December 2009.

33 www.zuluroyals.com/zulu.htm

34 Aleisha Tissen , "Minister Gandhi wants ritual to end", *Citizen*, 6 November 2009; Aleisha Tissen, "Aussies call on Zuma to denounce bull ritual", *Citizen*, 23 October 2009.

35 Aleisha Tissen, "Bull slaughter outcry grows", *Citizen*, 8 October 2009.

36 Ingrid Oellermann, "Bull ritual 'falls foul of code'", *Witness*, 25 November 2009.

37 Sherlissa Peters, "King blasts culture critics", *Daily News*, 7 December 2009.

38 Aleisha Tissen and Katlego Kalamane, "Outrage over bull-killing ritual", *Citizen*, 6 October 2009; Bongani Mthethwa, "Animal activists, Zulu chiefs lock horns over bull", *Sunday Times*, 1 November 2009; Elise Tempelhoff, "Dierewelsynorganisasie begin petisie om 'doodmarteling van bul' te keer", *Beeld*, 3 November 2009; André Le Roux, "Afrika veroordeel Zoeloe-ritueel", *Burger*, 30 September 2009.

39 Bongani Mthethwa, "Oprah asked to save bull from ritual", *Sunday Times*, 15 November 2009.

40 Aleisha Tissen, "Bull-killing decision lauded", *Citizen*, 3 December 2009.

41 Ingrid Oellermann, "Bull ritual 'falls foul of code'", *Witness*, 25 November 2009.

42 "'ARA action isn't to disrespect Zulus'", *Witness*, 3 December 2009.

43 Ingrid Oellermann, "Deadlock. Parties unable to compromise.", *Witness*, 2 December 2009.

44 Ingrid Oellermann, "Deadlock. Parties unable to compromise.", *Witness*, 2 December 2009.

45 Sandile Memela, "Why undermine our culture?", *Sowetan*, 24 January 2007.

46 Ingrid Oellermann, "Deadlock. Parties unable to compromise.", *Witness*, 2 December 2009.

47 Aleisha Tissen, "'Kee hands off our culture", *Citizen*, 27 November 2009.

48 Zizi Kodwa, "Constitution protects our practices", *Star*, 3 December 2009.

49 Ballard, "Slaughter in the suburbs", p.1072.

50 Ballard, "Slaughter in the suburbs", p.1073; M. Mutua 'Limitations on religious rights: Problematising religious freedom in the African context', *Buffalo Human Rights Law Review*, 1999, 75, pp96-97.

51 Letter, Cuan Mills, "'Ukweshama' debate fast becoming irresponsible", *The Times*, 4 December 2009.

52 Letter, T. Mkhize and M. Moholoa, "Respect our culture or leave country", *The Times*, 4 December 2009.

53 Duma Pewa, "Must we forfeit our rights so animals can live?", *Witness*, 21 November 2009.

54 Duma Pewa, "Must we forfeit our rights so animals can live?", *Witness*, 21 November 2009.

55 Martin Prozesky, "Culture, conscience and a call for mutual understanding", *Sunday Tribune*, 29 November 2009.

56 Ballard, "Slaughter in the suburbs".

57 Rich Mkhondo, "We must respect others' rituals", *Daily News*, 21 February 2007.

58 Justice Malala, "Culture no excuse for cruelty", *The Times*, 7 December 2009.

59 The Inkatha Freedom Party is a South African Zulu political organization founded during the Apartheid period in 1975 to counter the Xhosa-dominated ANC. Thanda Mgaga, "ARA will continue 'bull' fight", *Witness*, 26 November 2009; Bheki Mbanjwa, "IFP enters ritual bull killing fray", *Daily News*, 26 November 2009.

60 Zapiro, *The Times*, 1 December 2009.

61 A. Adams Kendall, "The role of Izangoma in Bringing the Zulu Goddess Back to her People", *TDR*, 43, 2, 1999, PP.94-117.

62 Cedric Mboyisa, "All tied up in knots", *Citizen*, 4 December 2009.

63 "King boycotts talks on bull", *Witness*, 25 November 2009,

64 Bheki Mbanjwa and Sherlissa Peters, "Bull-killing: be cautious, warns ANC", *Daily News*, 1 December 2009.

65 Bheki Mbanjwa and Sherlissa Peters, "Bull-killing: be cautious, warns ANC", *Daily News*, 1 December 2009.

66 "King's Office: Court will rule in favour of ritual", *Witness*, 1 December 2009.

67 Ingrid Oellermann, "Bull killing ritual goes ahead", *Witness*, 5 December 2009.

68 African students protested against the *Afrikaans Medium Decree* of 1974 which forced all black schools to use *Afrikaans* as a language of instruction. When high-school students in Soweto began demonstrating for better education on 16 June 1976, police responded with both teargas and live bullets.

69 The judge then offered his hope that parliament would intervene. Ingrid Oellermann, "'Ek is gedoem, 'sê regter in saak oor doodmaak van bul", *Beeld*, 2 December 2009.

70 "King slams bull-killing ritual furore", *Star*, 7 December 2009.

71 "King slams bull-killing ritual furore", *Star*, 7 December 2009.

72 Sherlissa Peters, "King blasts culture critics", *Daily News*, 7 December 2009.

73 Bheki Mbanjwa, "Controversial ritual defended", Daily News, 7 December 2009.

74 Bheki Mbanjwa, "Controversial ritual defended", *Daily News*, 7 December 2009.

75 Sonwabile Mancotywa, "Outcry over slaughter is prejudice masked as piety", *Sunday Times*, 28 January 2007.

76 *Photograph by: JOAO SILVA, Credit: ASSOCIATED PRESS* http://www.timeslive.co.za/news/article218055.ece/ANC-joins-row-over-bull-killing-ritual

77 Justice Malala, "Culture no excuse for cruelty", *The Times*, 7 December 2009.

78 He spoke wistfully of a conference that occurred on the cusp of the democratic transition two decades ago in 1990 of the SA Youth Congress in KaNyamazane. Clause 10 of the draft constitution, which proclaimed that one of its goals was to fight all forms of superstition". The (far North) Polokwane delegates resisted as they contended witches existed and they had the right to burn them. This was rejected by other delegates argued that scientifically it could not exist.

79 Justice Malala, "Culture no excuse for cruelty", *The Times*, 7 December 2009.

80 Jewel Amoah and Tom Bennett, "The freedoms of religion and culture under the South African Constitution: Do traditional African religions enjoy equal treatment?", *African Human Rights Law Journal.*

81 "'ARA action isn't to disrespect Zulus'", *Witness*, 3 December 2009.

82 Aubrey Matshiqi, "Solidarity with nature more than just bull", *Business Day*, 27 November 2009.

83 Fred Khumalo, "African culture? Not in my name", *Sunday Times*, 11 October 2009.

84 Mary de Haas, "Lessons emerging from the first fruits ceremony debate", *Daily News*, 3 December 2009.

85 David Benatar, "Culture does not justify cruelty", *Star*, 18 November 2009.

86 Eusebius McKaiser, "Even a multiculturalist must sometimes say no", *Business Day*, 27 November 2009.

87 South Africa's "Black Diamonds" are the upwardly mobile Africans who are join the ranks of the country's wealthy elite.

Chapter 6

Animals and Ethical Responsibilities: A Utilitarian Perspective

☆ *Katerina Prudchenko*

Introduction

The status of animals has been an important topic in the history of philosophical thought. The Greek philosopher Porphyry (232-305) was one of the first to dismiss Aristotle's argument that humans are the only beings with language and reason and therefore deserve a different moral standard (Legge 8). He stated that the animal's ability to reason is a matter of degree similar to the degree in which men vary in their reasoning abilities. According to Porphyry,

> No one, however, of sound mind, will say that brutes are unreceptive of the reasoning power, on account of the difference between their habit of body and ours, when he sees that there is a great variety of habit in men, according to their race, and the nations to which they belong and yet, at the same time, it is granted that all of them are rational (101).

His position however was not accepted and many philosophers continued to frame their arguments within Aristotle's basic premise. In the 17th century, Rene Descartes (1596-1650) made a famous rationalist argument that led to widespread animal abuse for a substantial period of time (Legge 8). Contending that animals are unable to feel hot, cold, or pain, Descartes said "...there are no men so dull and stupid, not even idiots, as to be incapable of joining together different words, and thereby constructing a declaration by which to make their thoughts understood" (45).

His position led him to the conclusion that animals are like machines and therefore cruelty to animals, just like to machines, is not possible.

Instead of addressing Descartes' rationality argument, British philosophers John Locke (1632-1704) and David Hume (1711-1776) constructed contrary positions which focused on the similarities between humans and animals (Legge 11). Even though they did not go so far as to give animals moral consideration, they did argue against cruelty out of concern for its effects on humans. In *Some Thoughts Concerning Education,* Locke wrote "the custom of tormenting and killing beasts will, by degree, harden their [people's] minds even towards men; and they who delight in suffering and destruction of inferior creatures, will not be apt to be very compassionate or benign to those of their own kind" (171).

German philosopher Immanuel Kant (1724-1804) made a similar argument against cruelty to animals. He maintained that tormenting animals is a violation of the morality of the individual since "our duties toward animals are indirect duties toward mankind" (213). An exemplar of Continental philosophy, Kant's arguments maintained that rationality is the basis of moral worth and therefore animals are not rational beings (Stratton-Lake 2-10). Furthermore, this position led him to the conclusion that animals, children and the mentally retarded, lack moral status and rights and actions toward non-rational beings are not morally wrong.

Another German philosopher, Arthur Schopenhauer (1788-1860) famously critiqued Kantian ethics and his notions of morality. He argued against Kant's position that "cruelty to animals is contrary to man's duty to himself" (96) and instead stated that "boundless compassion for all living beings is the firmest and surest guarantee of pure moral conduct, and needs no casuistry" (172). Schopenhauer maintained that animals are capable of feeling pain and suffering no less than humans. He was extremely critical of the treatment of animals throughout Europe and became a hero of the animal-welfare lobby in Germany (Legge 14). In *The Basis of Morality,* Schopenhauer wrote, "In the European systems of morality they [animals] are badly provided for, which is most inexcusable. They are said to have no rights...[and be] without moral significance. All of this is revoltingly crude, a barbarism of the West" (111).

Utilitarianism

Similarly to Schopenhauer's position, the British utilitarian argument is also constructed in opposition to Kant's stand on morality. Utilitarianism as an ethical doctrine was formed in the eighteenth and nineteenth centuries by philosophers Jeremy Bentham (1748-1832) and John Stuart Mill (1806-1873). It designates an action as being right if that action produces more pleasure or happiness than its alternative. In particular, according to Mill's Greatest Happiness Principle, the rightness of an action is proportional to the degree that the action promotes happiness and the wrongness of an action is proportional to the degree that the action promotes unhappiness (Chopra 333). As one of the first to critique Kant's concept of morality, Bentham redirected the argument on the status of animals from their ability to reason to their ability to suffer.

Utilitarian philosophers consider that moral status is not dependent upon the rationality of the being in question. Bentham was one of the first philosophers to reject the idea that animals should not be given moral consideration because they may lack the ability to reason. Along with Mill, Bentham argued that the purpose of morality is to minimize pain and suffering and maximize happiness and pleasure (Pascalev 210-211). Since all humans and animals share the desire to avoid pain or suffering it is therefore necessary to expand the definition of morality beyond rationality. In particular, moral consideration does come from a being's ability to reason but rather from his ability to feel pain. In one of the first philosophical arguments in defense of animal rights, Bentham wrote:

> The day may come, when the rest of the animal creation may acquire those rights which never could have been with-holden from them but by the hand of tyranny. The French have already discovered that the blackness of skin is no reason why a human being should be abandoned without redress to the caprice of a tormentor. It may come one day to be recognized, that the number of legs, the villosity of the skin, or the termination of the *os sacrum*, are reasons equally insufficient for abandoning a sensitive being to the same fate. What else is it that should trace the insuperable line? Is it the faculty of reason, or perhaps, the faculty for discourse? But a full-grown horse or dog is beyond comparison a more rational, as well as more conversable animal, than an infant of a day, or a week, or even a month, old. But suppose the case were otherwise, what would it avail? the question is not, Can they *reason*? nor, Can they *talk*? but, Can they *suffer*? (143)

Unlike Kant's morality, utilitarianism gives moral consideration to all sentient beings and obligates people to diminish suffering and increase happiness of all those who suffer. This was a revolutionary philosophical position because it placed animals alongside people in the moral universe. In 1871 Charles Darwin, took another revolutionary position when he argued that man and the "lower" animals are both capable of "feeling pleasure and pain, happiness and misery" (104). In *Descent of Man vol. 1*, Darwin concluded that since man "possesses the same senses as the lower animals, his fundamental intuitions must be the same" (96). These developments in thought removed rationality as a requirement for moral consideration and opened the discussion on the moral status of animals.

When evaluating the moral permissibility of an action, utilitarian thinkers proposed that all beings are taken into account, whenever an action affects them directly or indirectly (Pascalev 211). Actions that result in more pain than pleasure are not morally permissible while actions that result in more happiness than suffering are morally permissible. For example, it is morally permissible to kill an animal in order to feed a starving family because the benefit to the family is greater than the pain and death of one animal. Likewise, the deaths of many animals are therefore not morally permissible if all the pain results in the happiness of one person. It is vital to point out that this moral theory does not automatically condemn the use of animals for research, food and other human purposes. Instead, utilitarianism considers and weighs the consequences of pain and pleasure in a way that would result in greater overall happiness in the world (Pascalev 211). In particular, utilitarianism requires a cost-benefit analysis in which the cost is the pain or negative effects associated with all involved beings while the benefit is the pleasure or the positive effects associated with all involved beings.

Since utilitarianism is a philosophy which is concerned with the consequences of actions, the state of a being's awareness is of particular concern. Utilitarianism presupposes that pain is a basic biological function for all organisms. Therefore to deny the most basic pain consciousness is to deny the entire mental state of suffering (Rollin 269-270). Furthermore, if one denies an animal's ability to feel pain on any conscious level then one is logically bound to deny more complex mental states of sophistication as well.

An animal's ability to feel pain is of particular concern to the moral implications of scientific study. The scientific community spent much of the twentieth century denying animal pain consciousness, both in objective documentation and in personal experience. The hypocrisy is expressly clear in the action of the scientists who use animals as "pain models" in their research while simultaneously denying that animal's ability to feel pain (Rollin 270). Until recently, the International Association for the Study of Pain indicated that "language is a necessary precondition for the ability to feel pain," and this definition justified the performance of open heart surgery, without anesthesia, on neonatal humans until the 1990's (Rollin 269-271). Some laws to control pain of research animals did pass in 1985.

Utilitarian philosophy is not limited to the works of Bentham and Mill. In fact, it is one of the guiding principles of animal welfare in the late twentieth century. Peter Singer (1946-present), an Australian utilitarian philosopher, argues that since sentient beings have an interest in not-suffering, they have the right to have that interest be considered (Francione 254). Singer also argues that people should take into account animal interests on the same level as human interests when performing the cost-benefit or utilitarian analysis of a specific action. He notes that not taking the animal's interest into account is a form of discrimination, called speciesism, similar to sexism and racism.

Singer does not go so far as to say that the animal's moral consideration is the same thing as a moral right. In fact, he rejects the entire notion of rights for both humans and animals (Francione 254). Philosophers like Tom Regan (1938-present) who consider an animal's moral claim to be a moral right are not utilitarian. Instead his position states that an action that does not treat an animal like a being of inherent worth violates the animal's non-human rights and is not morally permissive. According to Regan, "Fundamentally, what we want to know is whether the harm we visit upon them [animals], and the freedom and life we take from them, violates their rights" (25). This position rejects any treatment of animals for purely human purposes such as eating and experimenting. It considers all sentient beings with inherent worth and inherent rights that should be protected.

While the animal rights position is absolutist, the utilitarian position is more flexible. Singer argues that the animal's moral significance competes with other morally significances and should be evaluated accordingly. In particular, the focus of the utilitarian perspective is the strength and the nature of the competing interests and not necessarily the particulars of the interests. As a result, the saving of one morally considerable being's life while causing harm but not death to another is morally permissible. Likewise, less suffering, a shorter life and a painless death is morally preferable to a lot of suffering, a limited extension of life, and ultimate death.

Within Bentham's tradition of utilitarianism, moral decision making is based on two criteria. The first criterion indicates that every being's interests carry the same significance (Legge 91). The second criterion indicates all competing interests and the best possible equilibrium between satisfaction and frustration of interests are seriously considered prior to action. In particular, utilitarianism perpetuates a morality in which "society feels able to say that the benefits we gain from using animals for food, experimentation or whatever outweigh the disadvantages to the animals themselves" (Legge 91). As a result, utilitarianism does not purely advocate for animal rights but rather for a balance between competing interests.

Factory Farming

One important focus of the ethical debate is factory farming, a practice that involves large scale production of livestock for human consumption. Factory farming started in the 1920, with the discovery of vitamins A and D, which allowed farmers to raise animals indoors without access to sunlight (Trapp 94). As a result, since the 1950's, meat production around the world increased more than fivefold and has more than doubled since the 1970's. Today factory farms or confined animal feeding operations account for more than 40 percent of the world meat production, up 30 percent since 1990 (Nierenberg 5). They were once limited to North America and Europe but are now found in Africa, Latin America and Asia.

While factory farms bring animals to the market in a relatively quick and inexpensive way, they also pose many problems for animals, people, and the environment. According to the US Department of Agriculture, approximately 900 million or 10 percent of animals raised for food die from injury and disease, prior to reaching the slaughterhouse (HFA, pars 9). The animals are kept in crowded and unhygienic conditions which often result in the outbreak and spread of diseases such as spongiform encephalopathy (BSE), avian flu, and foot-and-mouth disease (Nierenberg 5). These illnesses are not limited to animals and often spread to humans through the contamination of food, milk or water with animal fecal material. In particular, nearly 80 percent of randomly sampled ground beef and 80 percent of poultry contain microbes from animal feces (HFA, pars 20). According to the General Accounting Office, food-borne illnesses cost the nation more than $22 billion per year in medical expenses as a result of 81 million Americans getting sick and 9,000 dying (HFA, pars 21).

Meat and fish from factory farms contain a great amount of unnatural ingredients such as arsenic, hormones, persistent organic pollutants (POPs), polychlorinated biphenyls (PCBs), and other chemicals. Animals are also consistently injected with an abundance of antibiotics and other antimicrobials which eventually undermine their effectiveness in human use. For example, in 1954, US farmers injected animals with about half a million pounds of antibiotics per year. Today, farmers inject animals with 25 million pounds of antibiotics, 80 percent go directly into the feed to increase growth (HFA, pars 17). This widespread overuse is resulting in new strains of antibiotic resistant bacteria, posing a great threat to both humans and animals.

The transformation of farming from small family businesses to large corporate factory farms has industrialized meat production and taken away the livelihoods of small family operated farms. Today, much of the meat industry is consolidated and owned by only a handful of multinational corporations (Nierenberg 6). Over the past 15 years, the number of hog farms in the US decreased from 600,000 to 157,000 while the number of hogs sold increased (HFA, pars 6). As a result of corporate consolidation, just 3 percent of hog farms now produce more than 50 percent of hogs and only 2 percent of cattle feed operations now produce more than 40 percent of all US cattle (HFA, pars 6). Smaller family farms are unable to compete with corporate factory farms and an increase in factory farms has forced thousands of family farms out of business. For example, between 1979 and 1998, the number of American farmers dropped by 300,000 (HFA, pars 5). Smaller farms with better practices, such as grass fed livestock, cannot compete and as a result suffer along with the animals.

When considering the moral permissibility of factory farming, these are the factors that have to be accounted for in the utilitarian cost-benefit analysis. The primary benefit of factory farms is that they allow many people to consume and enjoy large quantities of inexpensive meat. Another benefit is the financial reward for the individuals who are employed by these multinational corporations. Nevertheless, the above mentioned costs of factory farming far outweigh the benefits. The utilitarian perspective seeks balance and is therefore more concerned with the type of interest rather than whose interest it is. Because moral consideration is given to all sentient beings, it is therefore important to weigh their interests comparatively. In particular, crucial interests are given more weight than important interests and important interests are given more weight than replaceable interests (Gruen). Similarly, crucial, important, and replaceable interests are all weighted heavier than trivial interests. For example, an animal's interest in not suffering is a crucial interest which carries appropriate weight. A person's interest in eating the animal for a particularly cheap price, such as a 99 cent hamburger when other sources of food are available, carries far less weight. As a result, the animal has a considerably stronger interest and it is therefore not morally permissible to consume the hamburger originating from a factory farm.

Animal in Research

Besides factory farming, another important area of consideration is animal research. In the 18th century, England was one of the first places to form an animal welfare lobby in order to exert pressure on scientists and improve practices of research on animals. These attitudes were largely influenced by Jeremy Bentham. The battle continued throughout the 19th century and eventually led to the passing of the Cruelty to Animals Act of 1876. Continental Europe was less responsive and some European nations only introduced legislation in the 1980's, in order to gain membership in the European Union (Legge 112).

Experiments on animals have led to the development of variety medical treatments, including heart-lung surgery, kidney transplants, polio vaccine, insulin, whooping cough vaccines, asthma drugs, and anesthetics for surgery. As a result, many scientists argue that research on animals may contribute to development of treatment for cancer, AIDS, as well as other psychiatric, genetic, and neurological conditions (Legge 114). Nevertheless, other scientists like British medical doctor Robert Sharpe take an opposing view. He indicates that since 1876, scientists used 170 million animals in experiments in the UK and of these 85% were experimented on since 1950. According to Sharpe,

> The rapid fall in deaths that commenced more than a hundred years earlier, had largely leveled out. So the vast majority of experiments can have had little impact and today we know that society's control of the infectious epidemics rests primarily in efficient public health service and a good standard of living...Even now infectious diseases like TB and whopping cough are far more common among working class people than in the professional workers, despite the availability of modern drugs and vaccines. And despite the huge increase in animal experiments since 1950, our overall health could now be deteriorating (238-9).

These are serious considerations when evaluating the utility of animals in research. The utilitarian perspective grants moral permissibility of animals in research under specific conditions. Animal research is generally morally permissible in experiments that address crucial and important interests of those who suffer from debilitating or fatal diseases. However, it is important to point out that the probability of satisfying many crucial interests should be high and the number of animals whose crucial interests are violated should be low (Gruen). Conditions of confinement as well as the particulars of the methods of experimentation are also important factors to consider.

On the other hand, animal research is not morally permissible when the experiments are conducted for the purposes of pursuing replicable or trivial interests of humans or other animals. Examples of trivial interests include testing for cosmetics and household products because non-animal tested alternatives and other options are already available for human use. While the use of animals in cosmetic research is still legal in the US, the practice was already banned in the UK in 1998 and made illegal throughout the European Union in 2009 (Jha and Lewis). The future of animal experimentation does appear brighter with the advent of new technologies such as computer simulation and stem cells. This research is promising in not only developing treatment in a variety of debilitating diseases but also as an alternative to animal testing in drug development.

Voices of Opposition

In 1899, Mark Twain wrote a letter to the London Anti-Vivisection Society, in which he said,

> I believe I am not interested to know whether Vivisection produces results that are profitable to the human race or doesn't. To know that the results are profitable to the race would not remove my hostility to it. The pains which it inflicts upon unconsenting animals is the basis of my enmity towards it, and it is to me sufficient justification of the enmity without looking further. It is so distinctly a matter of feeling with me, and is so strong and so deeply-rooted in my make and constitution, that I am sure I could not even see a vivisector vivisected with anything more than a sort of qualified satisfaction. I do not say I should not go and look on; I only mean that I should almost surely fail to get out of it the degree of contentment which it ought, of course, to be expected to furnish (26).

His position is shared by those who tend to scrutinize the utilitarian perspective for its moral flexibility, including Franco-German philosopher and medical doctor Albert Schweitzer (1875-1965). Schweitzer does not aim to strike a balance between competing interests but rather maintain that ethics are a responsibility without limits toward all that live (213-215). Schweitzer received the Noble Peace Price for his most famous work, *The Reverence for Life,* in which he stated that the ethics of reverence for life are absolute. Unlike utilitarianism which opposed Kant's absolute ethics, Schweitzer's position is similar in that he makes no claims in the potential achievement of his ethics (Cicovacki 20). He instead argues that the morality that all men should attain is to be concerned for all creation. As a result, destroying a life in order to survive does not "absolve us of moral responsibility toward the life destroyed. A lesser of two evils is still an evil" (20).

Conclusion

Jeremy Bentham's position redefined the criteria for moral consideration in Western philosophy and the ability to feel pain replaced reason and language as prerequisites for moral status. While animals are granted moral consideration, reverence for life is not absolute. Instead, utilitarianism accepts relativity in ethical values. It proposes a balance in weighing competing interests and consequences of an action with the goal of bringing about greatest happiness to all involved.

References

Bentham, Jeremy. 1859. *The Works of Jeremy Bentham.* Vol 1. Edinburgh: William Tait. Web. 19 Aug. 2010.

Chopra, Ramesh, ed. 2005. *Academic History of Philosophy.* India: Isha Books.

Cicovacki, Predrag, ed. 2009. *Albert Schweitzer's Ethical Values: A Source Book.* New York: Oxford.

Craig, Edward & Floridi, Luciano, ed. 1998. *Routledge Encyclopedia of Philosophy.* Vol 3. London: Routledge.

Darwin, Charles. 1901. *The Descent of Man and Selection in Relation to Sex.* N.p.

_____. 1902. *The Descent of Man. Vol 1.* New York: American Dome Library.

Descartes, Rene. 2008. *Discourse on the Method and Meditations*. Trans. John Veitch. New York: Casimo.

Francione, Gary Lawrence. 1983. *Animals, Property, and the Law*. Philadelphia: Temple.

Gruen, Lori. "Moral status of animals." *The Stanford encyclopedia of philosophy*. 2003. Web. 16 Aug. 2010.

<http://plato.stanford.edu/archives/spr2009/entries/moral-animal/>.

"HFA [Factory Farming]." Web. 20 Aug. 2010. <http://www.hfa.org/factory/index.html>.

Jha, Alok & Lewis, Paul. *Scientist backs animal testing for cosmetics*. 2006. Web. 17 Aug. 2010.

http://www.guardian.co.uk/science/2006/mar/04/businessofresearch.research

Kant, Immanuel. 1997. *Lectures on ethics*. Trans. Peter Lauchlan Heath & Jerome B. Schneedwind. United Kingdom: Cambridge.

Legge, Debbie. 1997. *Law Relating to Animals*. London: Cavedish.

Locke, John. 1693. *Some Thoughts Concerning Education*. London: Strand.

Nierenberg, Danielle. 2005. *Happier Meals: Rethinking the Global Meat Industry*. Massachusetts: Worldwatch Institute.

Pascalev, A. K. "We and they: animal welfare in the era of advanced agricultural biotechnology." *Livestock Science* 103 (2006): 208-220.

Porphyry. *Select works of Porphyry*. Trans. Thomas Taylor. London. 1823. Web. 18 Aug. 2010.

<http://books.google.com/books?id=tj0jAAAAMAAJ&pg=PA101&dq=#v=onepage&q&f=false>.

Regan, Tom. 2003. *Animal Rights, Human Wrongs: An Introduction to Moral Philosophy*. Maryland: Rowman & Littlefield.

Rollin, Bernard E. "Animal mind: science, philosophy, and ethics." *The Journal of Ethics* 11 (2007): 253-274.

Schopenhauer, Arthur. 1998. *On the Basis of Morality*. Trans. E. F. J. Payne. Providence: Berghahn Books.

_____. 2007. *The Basis of Morality*. New York: Casimo.

Schweitzer, Albert. 1961. *Civilization and Ethics*. London: Unwin Books.

Sharpe, Robert. 1988. *The Cruel Deception: The Use of Animals in Medical Research*. Wellingborough: Thorsons.

Stratton-Lake, Philip. 2000. *Kant, Duty, and Moral worth*. London: Routledge.

Trapp, Robert. 2009. *The Debateabase Book: A Must Have Guide for Successful Debate*. New York: Idebate.

Twain, Mark. 2010. *Mark Twain's book of animals*. Shelley Fisher Fishkin & Barry Moser, ed. Los Angeles: University of California.

Chapter 7

Vegetarianism and Veganism: Vehicles to Maintain *Ahimsā* and Reconstruct Jain Identity among Young Jains in the UK and USA

☆ *Bindi Shah*

Almost 600 young Jains, aged between fourteen and twenty-nine, are sitting under a marquee at the Lincolnshire Marriot Resort and Convention Centre in the suburbs of Chicago, listening intently to Will Tuttle giving his keynote speech at the 8th Biennial Young Jains of America Convention in July 2008. Tuttle, the author of 'The World Peace Diet' and a long time practicing vegan and animal rights activist, believes that most people in the West are born into a culture that has a mentality of commodification of animals, of predatoriness and a mentality of exclusion, to exclude other creatures from the sphere of compassion. With the aid of visual images of 'factory farms' he tells the captivated crowd that this mentality was captured in the words of one dairy farmer who argued that animals were put on earth for the four 'Ms' – meat, milk, manure and money. Tuttle suggests that this mentality is obsolete and has to be changed if humans are going to survive on earth. He advocates a culture that is based on the idea that "compassion is the heart and soul of not only spiritual health, but also social harmony, and freedom and justice and sustainability." He reminds the young Jains in attendance, who have travelled from all corners of North America and a handful from the United Kingdom, that they have had good *karma* to be born into Jain families and to understand the principle of *ahimsā*, and urges them to become "Jain millionaires" by each persuading one million other people to become vegans ((Field notes on Will Tuttle's Keynote Speech, Young Jains of America Convention July 4, 2008).

Introduction

In this speech Tuttle, who does not identify as Jain, conjoined two Jain beliefs, that all life is interdependent -*parasparopagraho jivanan* - and that non-violence is the supreme path - *ahimsā paramo dharma* – to highlight its expression in a vegan lifestyle and dietary practice to achieve compassion and ecological sustainability *in* the social world. Most young Jains are unlikely to become "Jain millionaires" in Tuttle's sense, but such an infusion of conversations, ideas and images that link vegetarianism and veganism to the ethic of *ahimsa* provide the inspiration for young Jains in the West to affirm their commitment to express the ethic of non-violence through their dietary practices. Following a brief discussion of the concept of non-violence from the Jain perspective, particularly its emphasis on the interdependence of all living beings and the theory of karma, in this essay I argue that for lay young Jains, growing up and living in the social and cultural milieu of the West, observing a vegetarian diet has become a significant defining feature of their religious practice and identity. Additionally, I suggest that for a small but committed number of young Jains leading a vegan lifestyle and advocating for the adoption of veganism within the Jain community represents a reconstruction of Jain religious practice in the West. I argue that in this way they are both negotiating what counts as the practice of Jainism within the Jain communities in the West and contributing to wider conversations about justice and sustainability.

This paper is based on research conducted with the sons and daughters of Indian immigrants from Jain communities in the UK and USA. I interviewed twenty-five young Jains between the ages of 17 and 30 in each country during 2008-09. My respondents are all middle-class or upper middle-class, have attended or plan to attend university or are working in professional occupations or are self-employed. I also observed Jain conventions in the USA organised by the Federation of Jain Associations in North America (JAINA) and its sub-committee, the Young Jains of America (YJA), and in the UK I observed a Jain convention and smaller social and religious events organised by Young Jains UK. I have also analysed pamphlets, newsletters, internet sites and conference/workshop materials produced and/or consumed by young Jains.

Migration histories of Jains differ in the two countries; in the USA the majority of my respondents were the children of Jain immigrants who had travelled to study in the 1970s. In the UK the majority of my respondents were the children of Jain immigrants who had arrived from East Africa during the 1970s and early 1980s though there were a small number of families who had migrated directly from India during the same time period. As in India, Jains are a minority ethno-religious community in both countries, with estimates of over 30,000 in the UK (Sanghrajka 2007) and over 75,000 in the USA (Pluralism Project 2007), and includes members from both the Shvetambar and Digambar sects and a range of castes.

Principle of *Ahimsā* and Theory of Karma in the Jain Tradition

In examining the role of the Jain tradition in the lives of young Jains in the West I draw on Orsi's (1997) and Joshi's (2006) idea of "lived religion" to examine what counts as religion and how it is constructed, negotiated and contested. The lived-religion approach enables us to go beyond dominant understandings of religion as a stable and unchanging ideology or rule-system and acknowledges not only ritual practices but also the principles and tenets that are internalized and applied in daily interactions, behaviours and practices. The lived religion approach enables us to ask not "How Jain is she? But rather "How is she Jain?" (Joshi 2006:20).

One of the fundamental principles of Jainism is the concept of *ahimsā* or non-violence. All Jain beliefs, rituals, and traditions emanate from this central theme. In Jain ontology a distinction is made between sentient beings (*jiva*) and non-sentient beings (*ajiva*).[1] Sentient beings have a soul and there are two classes of souls, the liberated souls that are pure souls, and the worldly souls that are impure and therefore caught in the continuous cycle of transmigration. Worldly souls are further divided into four main birth categories, often represented in the form of a svastika, those of gods, humans, hell beings, and animals and plants. While even the most lowly forms of worldly existence have a soul and the potential to break free of the endless cycle of birth, death and re-birth, it is only in the human state that an embodied soul can achieve purity and liberation from the cycle of transmigration[2] because humans have a developed mind and can make choices in their conduct.

Almost all religions have faith in the basic doctrine "what you sow, so you reap". The doctrine implies that a person bears the consequences of his/her deeds. In the Jain tradition, this doctrine, specifically the theory of *karma*, explains why the soul becomes embodied and remains caught up in a cycle of transmigration and a worldly existence. All activities, whether desirable or undesirable, produce vibrations in the soul, which in turn activate the dormant *karmas* and induce an influx of *karma* particles made up of non-living matter[3]. If the activity is conducted with equanimity, achieved when the worldly soul has awareness that every circumstance or condition is momentary, then the following result emerges: *karmic* particles will either not bind to the soul, are partly shed and partly calmed down, or calmed for the time being but not shed. However, if the thoughts, speech and actions of worldly souls are accompanied with emotions resulting from attachment and aversion, which can give rise to irrationalism, non-restraint, carelessness, and passions such as anger, greed, ego and deceit, then bondage of *karma* particles to the soul occurs. While all kinds of *karmas* are undesirable positive speech, thought and action such as compassion, friendship and goodwill, generate meritorious (*punya*) *karma*, while negative thought, speech and action, such as anger, pride, ego desire for status, greed and so on, cause demeritorious (*paap*) *karma*[4].

Since *karmic* particles associated with the worldly soul carry the imprints of the consequences of past thought, speech and actions, and travel with the soul from one life to the next, a worldly soul must endeavour to shed bonded *karma* particles and stop further binding to achieve liberation from the cycle of transmigration or *moksha*. Stoppage of *karmas* binding to the soul can be achieved through rational belief, life led with awareness, discipline and guided by vows, absence of passions, and ideally, absence of all activities. *Karmic* particles can be shed without effort when their duration has ended, though this could take many lifetimes, or through austerities, meditation and penance before their expiry date. However, since it is impossible for worldly souls to avoid all activities, a central aspect of a lay Jain's religious practice is to carry out activities with awareness and without desire, and so avoiding the accumulation of *paap karma* and allowing *punya karma* to be shed without effort (Jaini 1979:167).

The Jain theory of *karma* elucidates why the killing of animals is reprehensible not only for the suffering produced in the victims, but even more so because it involves intense passions on the part of the killer, passions which will bind him or her more firmly in the grip of a worldly existence (Jaini 1979: 167). In this sense, the principle of *ahimsā* is directly linked to the theory of *karma*. In everyday life lay Jains make a distinction between necessary and unnecessary violence and accept that a certain amount of violence is unavoidable, especially in the preparation and consumption of food. Jains classify living beings according to the level of development of its sensory capabilities and ability to experience the world. At one end of the scale are one sense microscopic beings which inhabit every corner of the universe. Similarly, while plants have a longer life span and a more complex physical structure, are also classified as one sense beings as they possess the sense of touch. Animal species possess from two to five senses (Jaini 1979:109-110). Since human beings cannot live without eating, the consumption of plants and vegetables causes the least violence. However, there are further restrictions on the consumption of plants and Jains are famously associated with avoiding root vegetables.[5]

For Jains, the practice of vegetarianism enables them to maintain *ahimsā* and bind the least *karma* while living in the social world. Jaini (1979: 169) notes that Jains "became the primary exponents of vegetarianism in India, and their strong stance contributed much to the eventual triumph of vegetarianism throughout the subcontinent."

Being Vegetarian, Being Jain

If vegetarianism triumphed in the subcontinent and became widespread so that Jains were one of many religious communities that practiced vegetarian diets, in the social and cultural milieu of the West vegetarianism is the exception rather than the norm. Thus in India, vegetarianism is only one of many practices that define a Jain.[6] In the West, where few young Jains participate in regular ritual practice austerities, meditation or self-modification, observing a vegetarian diet takes on much greater significance as an expression of what makes a person Jain. This is especially so in the United States where there is a "predominantly meat-eating and meat-advocating culture" (Vallely 2003:13).

All the young Jains in this study emphasised the incorporation of the principle of *ahimsā* in their physical actions as well as in their speech and thoughts, and living by Jainism broadly encompassing moral code as key aspects of their religious practice. Among those living in the UK only those who followed a vegan lifestyle specifically mentioned their dietary practices. For the rest, while they were all vegetarian, their dietary practices were an everyday part of their lives and therefore not a conspicuous aspect of their practice of Jainism. During interviews diet was rarely discussed; when it was mentioned it was often seen as only an "outwardly demonstration of ahimsa" as 23 year-old Jaimini noted, and not one that she particularly emphasised as part of her Jain identity. For others it was a vehicle to cultivate self-control. Twenty-five year-old Mihir believed that "the diet thing is a discipline rather than...it's teaching you a discipline rather than telling you what to do, so I wouldn't say it is an essential part [of Jainism]."

Young Jains living in the United States also understood *ahimsā* as a core aspect of Jain philosophy but many tended to highlight dietary rules and their practice of vegetarianism as an important avenue through which they practiced *ahimsā* and maintained compassion towards other living beings. When asked 'what are the key elements of Jainism for you?' twenty-four year-old Vaishali mentioned vegetarianism and went on to castigate her Jain friends who eat meat for identifying as Jains: "I make fun of them 'You're not Jain; you're really not Jain, why do you call yourself Jain. Call yourself anything, but don't call yourself Jain." For thirty year -old Parita being vegetarian was also "one of the biggest elements of being Jain... because I just think that that's like the core of our belief system is having compassionate... being compassionate towards all living beings and you can't, I don't think one can be compassionate ... if they are not vegetarian."

Despite an extensive focus on dietary practices, actual habits varied greatly among young Jains in both countries. Some young Jains included eggs in their diet while others would eat cakes and biscuits that contained eggs but not in any other form. Others had omitted eggs of any kind in their diet, while others still did not eat root vegetables (potatoes, garlic, onions and carrots) in the home to comply with parents' wishes but did so outside the home. Some young Jains in the study had decided themselves to not eat root vegetables in the home but were flexible when eating at restaurants or when visiting friends as they did not want to impose their dietary preferences on others. Almost all the participants observed *paryushan parva* or *das lakshan*, one the key religious festivals in Jain tradition that focuses attention on cleansing the mind and soul through penance and through giving and seeking forgiveness, by refraining from eating green vegetables for the duration of the festival.

This difference in the place that diet holds in the construction of a Jain identity and practice among Jain youth in the diaspora stems from the differing socio-cultural contexts in which they live. According to the Vegetarian Society of UK in 2000 almost seven percent or four million people in the UK were vegetarian and this figure rose to twelve percent among the young.[7] More recent figures from the Food Standard Agency's Public Attitudes to Food Issues Survey conducted in 2009 suggest a smaller proportion of the UK population, at three percent, is strictly vegetarian, while five percent were partly vegetarian. Nevertheless, as the Vegetarian Society notes, vegetarianism had been growing throughout the 1990s and vegetarian dishes are the norm on menus at any restaurant, fast food outlet, cafe and supermarkets in large and small towns in the UK. More importantly for young people who are more easily influenced by peer pressure, at least one, and generally two, vegetarian options are available for school lunches. Young Jain students in the UK find that they are not the only ones standing in the vegetarian queue for lunch in the school cafeteria; they are joined by Hindus and white British peers, as well as Jewish and Muslim peers wanting to refrain from eating non-kosher and non-halal food. In other words being a vegetarian is no longer considered a 'fad' in the UK but a legitimate dietary choice.

In the United States the Vegetarian Times reports that approximately 3.2 percent of the population or 7.3 million adults were vegetarian in 2008.[8] A poll conducted by market research company Hart Interactive in 2005 found that three percent of young Americans, aged 8 to 18, are vegetarian.[9] In the last decade vegetarianism has become popular in the United States and this is especially evident in large metropolitan areas such as New York, Chicago, San Francisco and Los Angeles. Most restaurants, cafes and specialty supermarkets in these cities offer a range of vegetarian choices. But this is not the case at schools across the United States. During the 1990s, when many of the participants in this study were attending primary or secondary school, they brought lunch from home while friends and peers ate hot lunches offered at school. There were also likely to be very few other Jains or Hindus attending the same school, thus making these young Jains even more conspicuous when consuming vegetarian food. These dietary habits set them apart from their friends and required them to address reactions that ranged from healthy curiosity to outright hostility. Vaishali who went to school in a wealthy suburb of New Jersey, described having multiple conversations with her friends about "why I'm vegetarian, why I believe in it," and answering questions such as "Oh what is it like being vegetarian? "How do you live without eating red meat?" Parita, who grew up in Los Angeles observed that her friends were cognisant of her dietary restrictions and never offered her food that was not vegetarian. However, twenty-five year old Nikhil, who was raised in a small town outside of Chicago, found that in his early teens a lot of his school mates would question the logic of his beliefs:

> "[they] actually compare consuming meat to consuming plants and kind of create some equality there, that you're not eating meat, but then you're eating plants, and plants are also living beings, so um, aren't you going against your belief?

Nikhil's explanation for being vegetarian in terms of the Jain principle of *ahimsā* and not causing harm to other four and five sense beings was perhaps not seen as legitimate because it did not have parallels with Christian concepts. For others such as twenty-four year-old Meghna, who went to middle-school in a working class urban neighbourhood of greater New York City and who often took Indian/Gujarati vegetarian food to school as part of her lunch, encountered negative experiences:

> I can remember very distinctly, um, I had a situation when I was a kid where this one kid, he called me curry [laughter] and he called me curry in front of all of my friends.

Meghna recalled that Gujarati/Indian food "was always alien to other people." For these young Jains growing up in the United States being vegetarian set them apart from their friends and peers. However, the experience of having to negotiate this aspect of their difference at an impressionable age did not lead them to abandon a vegetarian diet; rather these conversations with and/or challenging remarks from peers served to highlight their dietary practices as a key aspect of what made them Jain in the USA.

Veganism as a Greater Ahimsa Dietary Practice

Drawing on the writings of Saman Srutaprajna, who argues that *himsā* is not only caused through physical violence but also through anger, force, harsh words, deceit, accumulation, negligence and so on, Long (2009:103) highlights the notion in Jainism that "knowledge of the true nature of existence brings a burden of responsibility." Throughout the 1990s Jains in the United States have become increasingly aware of the impact of the modern commercial dairy industry on animals. For a growing number of Jains knowledge of commercial mild production techniques means that simply being vegetarian is no longer adequate to minimise violence when consuming food; this would constitute negligence, a form of violence, and therefore accumulate negative karma. Such awareness brings responsibility to act ethically and maintain compassion for other living beings. Saurabh Dalal, Deputy Chair of the International Vegetarian Union and an active member of JAINA for the last twenty years, estimates that between 500 to a thousand Jains in the USA have become vegan, with the vast majority being youth (Interview 3rd July 2009). In addition to meat and eggs, vegans do not consume dairy products such as milk, ghee, and ice-cream and avoid the use of silk, pearls, fur and leather.

Twenty-three old Bhavini described herself as a "fully-fledged vegan." At the time of the interview it had been more than a year since she had omitted all dairy products from her diet and was avoiding wearing silk or wool clothes or using leather purses in an attempt to find "as many non-violent ways" to live her life. She linked her lifestyle to her concern for the treatment of animals. Twenty-five year old Aakash, who had been a vegan for several years at the time of the interview, believed that veganism was a key element of Jainism, "not just vegetarianism but also understanding where milk comes from, what it is..." For Aakash a "general awareness of animal rights issues as well as environmental issues because that plays very much into human rights issues" was also important and understood their interconnections through the Jain maxim of preventing or minimising suffering for all five to one sense beings.

Both Bhavini and Aakash had been inspired to become vegans after participating in one of the biennial YJA conventions that are organised under the auspices of JAINA. As Dalal noted, vegetarianism and veganism have been a particular focus of all YJA convention, beginning with the first one that took place in 1994. For example, Ingrid Newkirk, the founder of People for the Ethical Treatment of Animals (PETA) was invited as a keynote speaker at the 2000 YJA Convention in Los Angeles, California. As noted in the opening paragraph, Will Tuttle, a long-time vegan and animal rights activist delivered the keynote address at the 2008 YJA Convention, and at the 2010 YJA Convention artist and ecologist Michael Tobias' keynote speech was also concentrated on the importance of compassion for animals.

Promotion and education of vegan lifestyles is further reinforced though vegan meal options alongside vegetarian meals at these conventions, and through the JAINA publications included in the souvenir bag given to every young attendee. For example, included in the 2008 YJA Convention pack were several pamphlets that emphasised diet and lifestyle as a way to practice the Jain principle of *ahimsā*. One colourful pamphlet, from the Vegetarian Resource Group based in Baltimore, titled 'Vegan Diets in a Nutshell' identifies dairy and protein substitutes as well as lists food items that could provide a nutritionally sound vegan diet. A vivid and dramatic booklet from PETA2, PETA's young adult division, presents fact and images of the meat and dairy industries. But more importantly for young people eager to emulate their heroes, it lists actors and musicians who have adopted a vegetarian/vegan diet. Also included in the pack are several books published by JAINA that address animal rights. A book compiled by Pramoda Chitrabhanu, director of PETA-India, and Pravin K. Shah of JAINA, 'the Book of compassion: Reverence of All Life', begins with a 'Universal Declaration of the Rights of Animals' and includes numerous articles on the dairy industry in the United States and in India, as well as on pearls and silk production and urges all "concerned and compassionate people" to reduce the use of these products (Chitrabhanu and Shah 2006: 7). A second book, titled 'English Pratikraman: Observance of Self-Reflection', is especially dedicated to people practicing a vegan and alcohol/drug free lifestyle. Throughout the convention workshop sessions also directly address diet and animal rights. At the 2008 YJA convention there were sessions with titles such as 'The power of Vegetarianism', 'Applied Jainism: The Vegan Lifestyle as Greater Ahinsa for Today's World', 'Veganism in a Nutshell', 'Getting Active for Animal Rights'.

In recent years a small number of British young Jains who have attended Jain conventions in the USA have also been inspired to become vegan. One of them is twenty-two year old Amal living in London. He converted to a vegan diet after participating in a JAINA convention in 2007 in New Jersey, USA. There he had heard talks presented by a young organiser from PETA2, and had numerous conversations with other Jains who had been vegan for many years. Through these conversations Amal not only realised the ease with which it is now possible to access vegan food, especially in large cosmopolitan cities such as London, New York, and Chicago, but he gained a philosophical understanding of the perceived barriers to being vegan:

> I just realised, it's not that is difficult, it's me being fussy. About expectations about how my food should be...and about this thing with taste: that taste is indulgence...

For Amal becoming a vegan has allowed him to both cultivate greater self-control and non-attachment to his culinary preferences, as well as passionately educate others about the greater compassion and non-violence embedded in a vegan diet. During the most recent *Paryushan* festival in September 2010 he posted a message on the Jain Vegan Yahoo Group, established by a small but committed group of Jain vegans in the UK a few years ago, urging Jains who traditionally refrain from eating green vegetables during this festival to also consider removing dairy and other animal products from their diet "to reduce their *himsā* footprint".[10] On the other hand, Jaimini was unusual in that she was introduced to the animal rights movement at the tender age of 13 by her father, a founder of the Indian Vegetarian Society in London. She recounted that at home "it was kind of a house obsessed by vegetarianism" and she had participated in numerous demonstrations for the protection of animal rights.

Jaimini is vegan but regarded her dietary preferences as only the external expression of her commitment to *ahimsā*. Her spiritual practice also involved an internal focus on controlling her passions and living life with awareness.

Such a focus on compassion for animal and animal rights more generally suggests that many Jains in the West believe that a vegan lifestyle is more in alignment with Jain ethical and moral values in the contemporary era. In fact Jaimini and Amal both observed that it was easier to follow a vegan diet in the West rather than in India. When visiting India, both relatives and a Jain nun criticised Jaimini for her dietary choice:

> The sadhiviji [Jain nun] said that it wouldn't make any difference. You are one person being a vegan, will that make any difference to cows...I said, but there isn't just me. For a start there are thousands in the UK, and second of all, you do what you have to do that's right and everyone else will follow...my relatives [in India] they said Mahavir Bhagwan never said anything about milk, and if you care so much about that why do you eat potatoes...

Amal perceived Jains in the UK "to be quite open-minded, and they appreciate different ways about thinking about things and won't reject them for not being Jain." Thus both point to the prescriptive nature of the Jain religion in India where omitting root vegetables is equated with being a good Jain, and the flexibility in creating a Jain practice in the West. The shifting notion of what makes a person a good Jain in the West is given further weight through a recent initiative launched by JAINA to promote veganism throughout the Jain community. This project is guided by Gurudev Chitrabhanu, Jain spiritual leader in North America and who became vegan in the mid-1990s. He clearly believes that in the West the best way to be a good Jain is to follow a vegan lifestyle and urges lay Jains to "refrain from food that comes from suffering and violence, and then that suffering will not come back to us like a boomerang – the universal law of cause and effect from which there is no escape. In this way, Jains can continue to be at the forefront of Ahinsa, each of us showing great compassion and unconditional love."[11]

Reconstructing Jain Identity through Dietary Practice

Young Jains growing up in the UK and USA use a language that conveys 'spirituality' and 'ethics' rather than 'rituals' and 'worship' to reflect their religiosity. They place emphasis on Jain philosophy and on understanding *karma* theory, and endeavour to internalise these tenets and apply them in their everyday lives. To maximise their chances of a meritorious birth in the next life young Jains work towards preventing intentional violence and minimising unintentional violence. Among those growing up in the USA, maintaining a vegetarian diet as an aspect of their everyday religious practice has become highly significant. Despite a pluralistic religious environment, this is no small matter in a society where meat-eating is the dominant cultural norm. An emphasis on vegetarianism and its explicit link to the ethic of ahimsa becomes a means through which to define what makes a person Jain in the USA. For young Jains living in the UK, where vegetarianism has become more acceptable and widespread in the last twenty years, their dietary choices constitute simply an external expression of their desire to minimise the binding of negative *karma* and as an aid in their effort to cultivate self-control, discipline and living life with awareness.

A small, committed, and slowly growing group of young Jains in the two countries believe that there is greater *ahimsā* gained from following a vegan diet and lifestyle than simply being vegetarian. For these young people not acting on knowledge of the violence and suffering enacted by the modern dairy industry is itself a form of violence and results in the accumulation of negative karma. Vallely (2003) interprets this as "socially-motivated *ahimsā*", aimed at social change rather than furthering one's spiritual progress and eventual liberation of the soul from the cycle of transmigration. We can discern the themes of justice and compassion in the reasons why Bhavini and Aakash became vegans, themes which resonate with social currents in the wider society. The language of compassion, harmony and respect for all life including that of animals is also the language used by organisations such PETA and long-time animal rights activists such as Tuttle, Tobias and Newkirk. At the same time this commonality with similar minded non-Jain identified individuals and groups can have the effect of maintaining, developing and deepening Jain spiritual practices among young Jains because such common threads provide recognition and legitimacy to Jain beliefs. Dundas (2002: 273)) has identified the rise of a neo-orthodox approach to Jainism amongst Jain in the West. Such an approach promotes Jainism as a "modern and progressive" religion, with "an emphasis on those aspects of Jainism which can be interpreted as scientific and rational and therefore be accommodated to and encompass 'western' modes of thought" (single quotation marks added). Amal and Jaimini, both also committed to promoting animal rights, demonstrate that a vegan diet and lifestyle not only generates greater compassion and love for all living beings, which is the essence of Jainism, but also enables them to lead life with awareness, discipline and self-control, and therefore further their own spiritual progress. In these ways, young Jains are negotiating and reconstructing what it means to be a Jain in the diaspora.

Notes

1 Non-sentient substances (*ajiva*) are divided into five elements –the medium of motion, the medium of rest, space, physical matter, and time (see Jaini 1979: 98-102) for a detailed description of these five elements.

2 Jaini, 1979:111

3 Jain at http://www.jainstudy.org/jscl.03How KarmaThReltoMS.htm.

4 Jain, 1997:122

5 Root vegetables, such as potatoes, carrots, onions, garlic and so on, are believed to have many souls within one body and therefore consumption of these would cause greater violence than those possessing one soul. Additionally, fruits with multiple seeds such as figs serve as hosts for microscopic beings and are forbidden, as is the consumption of honey, the collection of which results in the death of millions of bees.

6 Cort (2001:128) notes that it is the second level of dietary restrictions – avoiding

root vegetables, honey and those fruits or vegetables that are thought to have multiple souls such as figs and pomegranates – are perceived as specifically Jain practices in India.

7 http://vegsoc.org/news/2000/21cv/introduction.html, downloaded on 30th September 2010.

8 http://www.vegetariantimes.com, downloaded on 30th September 2010.

9 http://www.usatoday.com/news/health/2007-10-14-veggie-kids_N.htm, downloaded on 30th September 2010.

10 Downloaded from http://groups.yahoo.com/group/Jain Vegans/messages on 14th September 2010.

11 Posted by Saurabh Dalal on the Jain Vegan Yahoo Group on 10th September 2010, downloaded from http://groups.yahoo.com/group/Jain Vegans/messages on 14th September 2010.

References

Chitrabhanu, Pramoda and Shah, Pravin K. (eds) 2006. *The Book of Compassion: Reverence for All Life*. Raleigh, NC: Federation of Jain Associations in North America (JAINA).

Cort, J. E. 2001. *Jains in the world: religious values and ideology in India*. Oxford: Oxford University Press.

Dundas, P. 2002. *The Jains*. 2nd edition. London: Routledge.

Jaini, Padmanabh S. 1979. *The Jaina Path of Purification*. Berkeley, CA: University of California Press.

Joshi, Khyati. 2006. *New Roots in America's Sacred Ground*. New Brunswick and New Jersey: Rutgers University Press.

Long, Jefferey D. 2009. *Jainism: An Introduction*. New York, NY: I.B. Tauris.

Orsi, Robert. 1997. "Everyday Miracles: The Study of Lived Religion", in David D. Hall (ed), *Lived Religion in America: Toward a History of Practice*, pp. 3-21. Princeton, NJ: Princeton University Press.

Pluralism Project. 2007. Statistics by Religious Tradition: Jain statistics. Retrieved 30th January 2007 from www.pluralism.org/resources/statistics/tradition.php#Jainism

Sanghrajka, H. 2007. Institute of Jainology, London UK. Personal communication.

Vallely, Anne. 2003. 'The Jain Plate: The Semiotics of the Diaspora Diet", in Knut A. Jacobsen and P. Pratap Kumar (eds.), *South Asians in the Diaspora: Histories and Religious Traditions*, pp. 3-22. Leiden, NLD: Brill

Chapter 8

Eco-philosophy of St. Francis of Assisi

☆ *A.J. Sebastian sdb*

The 193 nation Copenhagen summit in 2009, considered the most important meeting in the history of the world to combat global warming, nearly came to a total collapse on 19 December 2009, as their talks merely took note of an accord, which was a non-binding deal for combating global warming. The plan does not specify greenhouse gas cuts required to achieve 2 Celsius goal, to ward off more floods, droughts, mudslides, sandstorms and rising seas. UN sealed the climate deal by agreeing to limit global warming to two degrees till 2050 without having set any target for carbon cuts. The global effort to save Planet earth ended in a whimper with a face saver proposal by US led group of five countries including China, India, Brazil and South Africa.

Ecologists are alarmed by the "awareness that we have reached the age of environmental limits, a time when consequences of human actions are damaging the planet's basic life support systems...Either we change our ways or we face global catastrophe, destroying much of beauty and exterminating countless fellow species in our headlong race to apocalypse" (Glotfelty xx). We are continually challenged by various green movements with their plea to save planet earth.

Considering "The World as Sanctuary," Eco-philosophy sees humanity as one with nature, carrying the universe onward from inanimate matter of life, to consciousness, and ultimately to the Divine. This new worldview emphasizes the unique precious and sacred nature of our planet. The five key tenets of eco-philosophy are: 1) The world is a Sanctuary. 2) Reverence for life in our guiding value. 3) Frugality is a precondition for inner happiness. 4) Spirituality and rationality do not exclude each other, but complement each other. 5) In order to heal the planet, we must heal ourselves (Skolimowski, home.cogeco.ca). It is Arne Naess who defined eco-philosophy as "a philosophy of ecological harmony or equilibrium. A philosophy as a kind of sofia (or) wisdom, is openly normative, it contains both norms, rules, postulates, value priority announcements and hypotheses concerning the state of affairs in our universe" (qtd. Drengson & Y. Inoue 8).

Hence, in the realm of ideology we can speak of the need for Ecological Humanism which points towards social relationships based on the idea of sharing, and stewardship. It sees world as a Sanctuary in which we temporarily dwell, and of which we must take the utmost care. It speaks of human life having a transcendent dimension, with its eschatology, concerned with the ultimate end and meaning of life. Ecological humanism calls for ecological spirituality that takes the Cosmos to its creator (Skolimowski, home.cogeco.ca).

Recently I happened to read "Circle of Life," an eco-poem, which draws our attention to the fact that we are part the cycle of life being one with the cosmos and all the elements of nature.

> For all that can be, really is round.
> Sun, Earth, & Moon rotate around
> with Water & Air cylinders spouting down
> true too of the rhythmic beat of sound
> and the distant lights of a city or town
> where animals graze from the ground
> above where you too shall be found
> fuel for the soil upon now drowned
> as the Circle of Life keeps going round
> (DarnRick.http://www.healthandfitness.com).

We form part of today's ecological crisis and are conscious of our environmental responsibility to protect the earth and its resources. Various ecological positions keep surfacing giving vent to man's response to the environment. Various eco-philosophies confront us with their diversified and some times radical approaches. Those who subscribe to the different approaches seek to solve the environmental crisis in their own ways, some being very subversive and revolutionary. These include Deep Ecology, Social Ecology, Earth First!, Greenpeace, Ecocriticism, Anthropocentrism, Ecofeminism, Ecosophy, Friends of the Earth, etc. As William Rueckert opines:

> The problem…is to find ways of keeping the human community from destroying the natural community, and with it the human community. This is what ecologists like to call the self-destructive or suicidal motive that is inherent in our prevailing and paradoxical attitude toward nature. The conceptual and practical problem is to find the grounds upon which the two communities – the human, the natural – can coexist, cooperate, and flourish in the biosphere (Rueckert 107).

In the context of global warming and the subsequent environmental crisis facing mankind, a reading of the eco-philosophical vision of St. Francis of Assisi based on his love of all creatures being part of God's sacred creation, make us visualize the universe in a different light.

Ecocritic like Lynn White points out that St. Francis believed in the "virtue of humility – not merely for the individual but for man as a species. Francis tried to depose man from his monarchy over creation and set up a democracy of all God's creatures... His view of nature and of man rested on a unique sort of pan-psychism of all things animate and inanimate, designed for the glorification of their transcendent Creator..." (White 13).

Everything around us forms our environment and our lives depend on keeping its vital systems intact. We depend on nature and it is imperative for us to protect the earth's environmental resources for our survival. We need to understand that the environmental problems are human or social problems such as pollution, conservation and sustainable use resources, preservation of endangered species etc. All these call for the protection of biodiversity and ecosystems on which all life on earth depends.

Francis' love for his brothers and sisters as other Christs is a known fact. It is recounted that once a poor man annoyingly kept begging from him. When a friar treated the beggar roughly, Francis ordered him to lay aside his habit and fall at the feet of the beggar, to seek his forgiveness: "My brother, when thou seest a poor man, behold in him a mirror of the Lord and His poor Mother. In the sick, in like manner, consider that He bore our sicknesses" (Bonaventure 66).

Francis manifested great tenderness for all things, believing in the common origin of all creatures. It is said that he often saved lambs that were being led to the slaughter house, being reminded of Jesus the Lamb of God. While in Rome, he used to keep a lamb with him in reverence for Christ the Lamb of God. Later he kept the lamb in the care of a noble lady whom the lamb accompanied to the church as though it had been trained in spirituality by the saint (68). At Grecio, once, a hare was brought to the man of God. He placed it on the ground for it to fly to safety, however, the bird leaped into his bosom. He held it with tender affection and let it fly off into the sky (69). While passing beside the Lake of Rieti close to the hermitage of Grecio, a fisherman brought him a water-fowl, which he accepted and allowed it to fly away. Finding the bird refusing to fly, the saint raised his eyes to heaven in prayer and commanded the fowl to fly away. It did obey instantly and flew to freedom. (Ibidem, p.69). From the same lake someone brought a fish for Francis who addressing it as his brother, set it free into the lake waters. But the fish kept leaping around his boat in great affection until he blessed it to depart (69-70).

While walking near the Lagunes of Venice, he noticed a multitude of chirping birds on a tree. Taking the cue from the birds, he told his companion, "Our sisters, the birds, praise their Creator; let us therefore go into the midst of them, and sing the Canonical Hours to the Lord" (70). At St. Mary of the Angels, Francis was constantly reminded of prayer by a chirping grasshopper. Being inspired by such an insignificant creature about God, he asked the grasshopper, "Sing , my sister grasshopper; rejoice and praise the Lord thy Creator" (70). When Francis was sick at Siena, a nobleman sent him a pheasant. The bird always remained with him and refused to be separated from his presence. Whenever the friars took it out to the vineyard, it flew back to the saint (71). All kinds of birds came chirping melodiously around his cell, when he was at the hermitage of Alvernia. He declared, "I perceive, brother, that it is the Will of God that we should abide here awhile, seeing that our sisters, the birds, thus rejoice at our presence" (71). A falcon is said to have built its nest where Francis lived. It became so friendly, it began to cry every night to wake him up for the Divine Office (71).

When the holy man lived at the hermitage of Grecio, the inhabitants were threatened by a pack of ravenous wolves that devoured people and animals, besides destroying their corn and vines. He asked the people to make amends for their evil lives and repent. When they did penance seeking the mercy of God, Francis asked the wolves to be off from the village (72-3).

Once when he was on his way to Bevagna, he found a multitude of birds of all kinds assembled together. As they flocked to welcome the saint, he admonished them: "Oh, my brother birds, you are bound greatly to praise your creator, Who has clothed you with feathers, and given you wings wherewith to fly; Who has given you the pure air for your dwelling-place, and governs and cares for you without any care of your own" (102). Hearing his exhortation, the birds spread their wings and expressed their great joy, swelling their throats and opening their beaks. The saint covered them with his tunic and blessed them. On another occasion when he came to preach to people at Alviano, he found some swallows building their nests, making a great deal of noise. When he couldn't be audible, Francis bid them, "My sisters, the swallows, it is now time that I should also speak, for you have spoken more than enough. Listen to the word of God, and keep silence until the preaching is ended" (103). The message was taken with great reverence by the birds who remained still in obedience.

Francis believed in the universal brotherhood of all creation. He found a unique relationship between men, animals, birds, plants and the universe. He took great joy in beholding the sun, the moon and the stars. He was delighted to contemplate the beauty of flowers and was mesmerized by their fragrance (Englebert 135). When he walked over stones, he did it with reverence, remembering Jesus the rock. He even avoided trampling over water out of sheer respect. He never let smoking firebrands be tossed aside as he respected it as 'brother fire.' He forbade his friars from chopping down trees exhorting that every thing should be allowed to grow (135). He never trampled upon worms, instead would pick them up to prevent them from being crushed underfoot. During winter he used to give warm wine and honey to the bees. He built nests for doves to lay eggs and multiply (135-6). The legend of the wolf of Gubbio vouches for his love for a ferocious creature. That particular ferocious wolf attacked and devoured men and animals. Francis went out to meet the demonic creature. With the sign of the cross he ordered the beast, "Come here, brother wolf!... In Christ's name, I forbid you to be wicked" (137). The wolf obeyed and surrendered at his feet. He exhorted the beast, "Brother wolf...I am very sorry to hear of the dreadful crimes you have committed in these parts, going even so far as to kill creatures created in God's image... But I want you to reconcile you with them...If you agree to make peace, brother wolf, I will tell the people to feed you as long as you live" (137). The wolf bowed in agreement and sealed the pact by placing its paw in the saint's hand and wagging its tail.

It was in the midst of ill health, blindness and the stigmata he suffered, that the saint sang his *Canticle of Brother Sun.*

> Most High Almighty Good Lord,
>
> Yours are praise, glory, honour and all blessing.
>
> To you alone, Most High, do they belong,
>
> And no man is worthy to mention You.
>
> Be praised, my Lord, with all Your creatures,
>
> Especially Sir Brother Sun,
>
> Who is daylight, and by him You shed light on us.
>
> And he is beautiful and radiant with great splendor.
>
> Of You, Most High, he is a symbol.
>
> Be praised, my Lord for Sister Moon and the Stars.
>
> In heaven You have formed them clear and bright and fair.
>
> Be praised, my Lord, for Brother Wind
>
> And for air and cloud and clear and all weather,
>
> By which You give Your creatures nourishment.
>
> Be praised, my Lord, for Sister Water,
>
> For she is very useful, humble, precious and pure.
>
> Be praised, my Lord, for Brother Fire,
>
> By whom You light up the night,
>
> For he is fair and merry and mighty and strong.
>
> Be praised, my Lord, for our Sister Mother Earth,
>
> Who sustains and rules us
>
> And produces varied fruits with many-colored flowers and plants.
>
> Praise and bless my Lord
>
> And give Him thanks and serve Him with great humility (251-2).

Francis speaks of a vibrant relationship of interdependence in the universe by which the various elements contribute to sustain the cycle of life. He addressed brother Sun as the symbol of a life-giver. As Ian Bradley has very aptly said:

> The sacramental approach to nature of Teilhard de Chardin... new insights gained from quantum physics and process of philosophy; the increasing sense of awe and wonder with which

> scientists gaze on the universe; recovering traditional Christian themes like the great chain of being, the dance of creation and the music of the spheres: all of these may help us to see… that God is engaged in a continuous and reciprocal relationship with all his creation (Bradley 50).

Man cannot stay apart from creation as an onlooker, exploiting its resources since he came from its dust and shall return to that very dust of the earth (Genesis 3:19).

I find an excellent parallel to the *Canticle of Brother Sun* in Gerard Manley Hopkins' "God's Grandeur" where the poet projects his sacramental vision of nature. It gives further insights into the chain of being and the music of creation which St. Francis portrayed. The natural world glorifies God constantly. But it is only man, the apex of God's creation, that can render Him glory consciously. He can render God glory seeing nature in a sacramental way. He can attain union with God through creation, which is the sacrament of God's presence. Hopkins is concerned about the two modes of the divine impact on mankind, beginning with the grandeur of storm and ending with the reassuring beauties of sunrise.

> The world is charged with the grandeur of God.
> It will flame out, like shining from shook foil;
> It gathers to a greatness, like the ooze of oil
> Crushed. Why do men then now not wreck his rod?
> Generations have trod, have trod, have trod;
> And all is seared with trade; bleared, smeared, with toil;
> And wears man's smudge and shares man's smell: the soil
> Is bare now, nor can foot feel, being shod ("God's Grandeur").

The grandeur of God is manifested in the universe, and the energy and beauty in it reveals the face of God. We come across the poet's reference to all things charged with God's grandeur: " All things therefore are charged with love, are charged with God and if we know how to touch them give off sparks and take fire, yield drops and flow, ring and tell of him." (Devlin 195). The presence of God in the world is like the electric power. Sometimes it flashes out brilliantly' like shining from shook foil'. The image of the foil brings out the message powerfully.

Again, the grandeur of God, though great and impressive, begins to show itself in little ways and 'it gathers to a greatness, like the ooze of oil / crushed'. The world is full of Divine power, love and beauty, but the industrial man has utterly lost touch with nature. The poet is puzzled when man does not become subservient to God's supreme authority. He is convinced of man's ultimate destiny in God. His life is to give glory to God. But sinful man does not acknowledge the Maker of the universe. Hence, the poet laments about the evils of Industrialism as 'all is seared with trade; bleared, smeared with toil'. Nature has been disfigured and 'the soil / Is bare now'. But man cannot destroy the essence of God, which is expressed by the inscape of each thing in nature.

Man withers and burns everything in his greed; unlike nature, which 'flames out' God's glory. For the 'dearest' gift by which nature itself, as well as man, is kept from becoming bankrupt is to be found in Christ's Redemption, literally, his buying-back-again of man and nature with his life (Mariani 96-7).

> And for all this, nature is never spent;
>
> There lives the dearest freshness deep down things;
>
> And though the last lights off the black West went
>
> Oh, morning, at the brown brink eastward, springs-
>
> Because the Holy Ghost over the bent
>
> World broods with warm breast and with ah! Bright wings ("God's Grandeur").

Despite the insensitive destruction of the beauty of nature, 'nature is never spent', as there is constant renewal and growth in it. The poet believes in the renewal of life, just like the way sunset and darkness of the night brings in the daybreak every morning. The final couplet strengthens the poet's hope for the world. In the Holy Spirit, there is regeneration and life for the warped world in sin. The Spirit of God hovers over the world hence ' broods with warm breast and with ah ! bright wings'. The lines evoke reference to Genesis 1:2: '...and the Spirit of God was moving over the face of the waters' and Luke 13:34: 'How often would I have gathered your children together as a hen gathers her brood.' Increasingly Hopkins was drawn to acknowledge shapes of natural force as vessels of God's finger - the Holy Ghost sustaining the universe...(Heuser 36). Hopkins celebrates the world of nature with its varieties manifested in manifold ways, like St. Francis.

Francis' eco-philosophy is based on Biblical principles. Bible being basically an ecological book, several parallels between ecology and Bible can be drawn: a) Both view the world in a long-range time frame. b) Both see the natural world as one interconnected whole. c) Both focus on the significance of land. d) Both present us with an awareness of limits. e) Both see the natural order as subject to decay. Both show that all behaviour has consequences (Snyder 45-51). Bible refers to all of creation – the heaven and earth, the constellations, and nature beaming with life in all its splendor.

In the book of genesis we read: "'In the beginning God created the heavens and the earth" (Gen 1:1). "And God said 'Let the earth put forth vegetation...And it was so" (Gen 1:11). "The earth brought forth vegetation...And God saw that it was good" (Gen 1:12). "And God said, 'Let the waters bring forth swarms of living creatures, and let birds fly above the earth across the firmament of the heavens' "(Gen 1:20). "And God said, 'Let the earth bring forth living creatures according to their kind: cattle and creeping things..." (Gen 1:24). "Then God said, 'Let us make man in our image, after our likeness; and let them have dominion over the fish of the sea, and over the birds of the air, and over the cattle, and over all the earth. So God created man in his own image, in the image of God he created him; male and female he created them. And God blessed them, and God said to them, 'Be fruitful and multiply, and fill the earth and subdue it; and have dominion over the fish of the sea and over the birds of the air and over every living thing that moves upon the earth' " (Gen 126-28). Man is thus made the apex of God's creation and given authority and dominion over the earth and God found everything he had made "very good" (Gen 1:31). "The Lord God took the man and put him in the Garden of Eden to till it and keep it" (Gen 1:15). He was thus given power to sustain the earth. But man broke the harmony of nature by his disobedience to God.

The kingdom of God is the central theme running through the New Testament which has significant ecological implications. The kingdom entails the renewal of all creation, human and natural. This expectation is holistic as it affirms the spiritual-physical unity of the person; it relates personal and social renewal; it links human and cosmic aspects of redemption; it affirms the interconnectedness of the spiritual and material dimensions of life; and it means the ultimate unity of all things, including heaven and earth, so that God is all in all. The kingdom unites creation and redemption–redemption as recreation focuses back on the original creation. Both are expressions of God's lordship leading to a redeemed earth. The kingdom is a new order of salvation and relationships. Hence Christians are led to care for creation and stop all its degradation as faithful stewards (Zerbe, Gordon. http://www.directionjournal.org). As Wendell Berry writes, "The first principle of the Kingdom of God is that it includes everything… everything in the Kingdom of God is joined both to it and everything else that is in it" (Berry 44).) St. Paul speaks of the importance of the 'the new creation' in Christ (2 Cor. 5:17), redeeming both the human and the non-human. 'Nature is God's tool of reward and punishment, and its beneficence depends on human morality' (Kay 214). In an article " A Biblical Perspective on Environmental Stewardship" the editors focus attention to 1) Theological and ethical foundations of stewardship, 2) The marvels of human achievement, 3) How economic and environmental trends relate, 4) Some human and environmental concerns for present and future, and 5) Environmental market virtues. Summing up they state, 'On the basis of a biblical worldview and ethics, as well as of sound science, economics, and public policy principles, we believe sound environmental stewardship celebrates and promotes human life, freedom, and economic development as compatible, even essential for, the good of the whole environment.

Biblically it is man's responsibility to protect the earth and sustain it as master of the earth. Several green groups have been advocating for the protection of the environment and to 'think globally and act locally.'(Bate 260). In 1979, proclaiming Francis of Assisi as patron saint of ecology, John Paul II declared:

> Today world peace is threatened not only by the arms race, regional conflicts and continued injustices among peoples and nations, but also by a lack of DUE RESPECT FOR NATURE due to widespread destruction of the environment as he offers a genuine and deep respect for the integrity of creation. He invited all of creation - animals, plants, natural forces, even Brother Sun and Sister Moon, to give honor and praise to the Lord… The poor man of Assisi gives us striking witness that when we are at peace with God we are better able to devote ourselves to building up

> that peace with all creation which is inseparable from peace among all peoples... Saint Francis will help us to keep ever alive a sense of "fraternity" with all those good and beautiful things which Almighty God has created. And my he remind us of our serious obligation to respect and watch over them with care, in light of that greater and higher fraternity that exists within the human family.(John Paul II. http://www.ncrlc.com).

In the backdrop of the growing global ecological concerns, an examination of St. Francis of Assisi makes great sense, having been declared patron of ecology. A re-reading of his life and works will inspire us to respect and love nature in all its bounty. As it has been observed by Jonathan Bate that Nature is a term that needs to be contested, not rejected. It is profoundly unhelpful to say 'there is no nature' at a time when our most urgent need is to address and redress the consequences of human civilization's insatiable desire to consume the products of the earth. We are confronted for the first time in history with the possibility of there being no part of the earth left untouched by man.

> Human civilization has always been in the business of altering the land, whether through deforestation or urbanization or mining or enclosure or even the artificial reimposition of 'nature' through landscaping... When there have been a few more accidents at nuclear power stations, when there are no more rainforests, and when every wilderness has been ravaged for its mineral resources, then let us say '*There is no nature*' (Bate 171).

How are we to save our fragile world from ecological disaster through global warming, pollution and deforestation? World over environmental awareness is created to make people live more eco-friendly and to bring about environmental conservation through protection of flora and fauna and by providing clean energy and sustainable development. Today world peace is threatened not only by warfare and conflicts, but also through lack of due respect for nature by its irresponsible and reckless exploitation and destruction.

St. Francis draws attention to Ecological Ethics with its various approaches examined from anthropocentric, ecocentric and theocentric approaches. The anthropocentric approach places humans at the centre of concern. Conservation of nature is primarily for human benefit and that all species and natural resources should be utilized for human progress. The ecocentric approach claims that humans are of equal value to all other life forms. Rejecting the Christian worldview, the ecocentrics are usually pantheistic, ignoring man's unique dignity within creation. The Theocentric approach claims God at the centre of value, continuously and dynamically involved in creation (Stassen & Gushee 435).

It is imperative to make an assessment of the major ecological crises that we face today such as 1). Massive reduction in biodiversity caused by deforestation, poisons and poor farming methods; 2). Climatic changes caused by pollutants, including carbon monoxide and methane; 3) Pollution by industrial, chemical and post-consumer wastes; 4)and Soil erosion and desertification, caused by deforestation and poor farming methods (Northcott 209). The crux of the problem lies in the deterioration of human values and the loss of the sense of the sacred in man's reckless pursuit of wealth in a consumer society, spiraling unethical exploitation of nature. In the midst of present environmental crisis, the eco-philosophy propounded by St. Francis is a clarion call to respect all things animate and inanimate. His teachings will certainly help us to introspect on our bounden duty to safeguard creation as responsible stewards.

References

Bate, Jonathan. "From 'Red' to 'Green'". *The Green Studies Reader*. Laurence Coupe. Ed. London and New York: Routledge, 2000. rpt. 2004.

Beers, Michael. et al. Eds. "The Catholic Church and Stewardship of Creation." *Environmental Stewardship in the Judeo- Christian Tradition*. Ed. Action Institute. Michigan: Action Institute. rpt. 2008. 33-62.

Berry, Wendell. *Home Economics*. San Francisco: North Point Press, 1987.

Bible, The Holy. Revised Standard Version. London: Catholic Trust Society, 1966. All Biblical references are taken from this edition.

Bonaventure, St. *The Life of St. Francis of Assisi*. Ed. Cardinal Manning. Illinois: Tan Books and Publishers, 1988.

Bradley, Ian. *God is Green: Ecology for Christians*. New York: Doubbleday, 1990.

"Catholicism."http://www.guardian.co.uk/world/2007/apr/27/ catholicism.religion.Accessed on 21 January 2010.

"Conservation."http://conservation.catholic.org/ecologicalcrisis.htm. Accessed on 15 January 2010.

DarnRick.http://www.healthandfitness.com/darren/poetry.html. Accessed on 10 January 2010.

Devlin, Christopher, S.J. Ed. *The Sermons and Devotional Writings of Gerard Manley Hopkins*. London: O.U.P. 1959.

Drengson, Alan & d Yuichi Inoue, Editors. *The Deep Ecology Movement: An Introductory Anthology*. Berkeley, North Atlantic Publishers, 1995.

Englebert, Omer. *St. Francis of Assisi: A Biography*. Trans. Eve Marie Cooper. Michigan: Servant Books, 1979.

Glotfelty, Cheryll. "Introduction." *The Ecocritism Reader: Landmarks in Literary Ecology*. Glotfelty, Cheryll and Harold Fromm (Eds). Athens and London: University of Georgia Press, 1996, xv-xxxvii.

Heuser, Allen. *The Shaping Vision of Gerard Manley Hopkins*. London: O.U.P, 1958.

John Paul II. http://www.ncrlc.com/ecological_crisis.html. *The Ecological Crisis: A Common Responsibility*. 1990. Accessed on 22 January 2010.

Kay, Jeanne. "Concepts of Nature in the Hebrew Bible." *Environmental Ethics: Divergence and Convergence*. Eds. R.G. Botzler and S.J. Armstrong. Boston, MA: MacGraw-Hill,1998.

Mariani, Paul L. *A Commentary on the Complete Poems of Gerard Manley Hopkins*. Ithaca: Cornell University Press, 1970.

Northcott, M.S. *Ecology and Christian Ethics*. (In Gill, R. *ed*. *The Cambridge Companion to Christian Ethics*). Cambridge: Cambridge University Press, 2001.

Rueckert, William. repr. "Literature and Ecology: An Experiment in Ecocriticism." *The Ecocritism Reader: Landmarks in Literary Ecology*. Glotfelty, Cheryll and Harold Fromm (Eds). Athens and London: University of Georgia Press, 1996, 105-23.

Skolimowski, Henryk. home.cogeco.ca/~drheault/ee_readings/Environmental_Ethics_Society/Skolimowski"What is ecophilosophy?"Some Founding Principles. Accessed on 18 January 2010.

Snyder, Howard. *Liberating the Church: The Ecology of Church and Kingdom*. Downer's Grove, IL: Inter-Varsity Press, 1983.

Stassen, G.H. & Gushee, D.P. *Kingdom Ethics*. Downers Grove, IL: IVP, 2003.

White, Lynn. repr. "Historical Roots of Our Ecological Crisis." *The Ecocritism Reader: Landmarks in Literary Ecology*. Glotfelty, Cheryll and Harold Fromm (Eds). Athens and London: University of Georgia Press, 1996, 3-14.

Zerbe, Gordon. http://www.directionjournal.org/article/?7632008. "Ecology According to the New Testament". Accessed on 26 January 2010.

Chapter 9

Animal Rights, Interests and Moral Standing: A Critical Examination of Dialogue Between Regan and Frey

☆ *Rebekah Humphrey*

Introduction

The aim of this paper is to assess R.G. Frey's analysis of Leonard Nelson's argument (that links interests to rights). Frey uses Nelson's argument in an attempt to refute the claim that animals have interests or rights. His discussions are directed against the claim that animals have moral standing in virtue of them having rights or interests.

Frey criticises claims that animals have rights or interests, arguing that such claims have not been established, and that, therefore, a case for the moral standing of animals based on such claims has not been established either. I will discuss Frey's claims that animals have not been shown to have rights nor interests in turn, but the main focus will be on Frey's claim that animals have not been shown to have interests.

One way Frey analyses the claim that animals have interests is by considering H. J. McCloskey's denial of this claim and Tom Regan's criticism of this denial. While Frey's position on animal interests does not depend on McCloskey's views,[1] he believes that a close consideration of McCloskey's views will show that it has not been established that animals have interests and thus also show that Nelson's argument (linking interests to rights) has not been established as sound. My discussion (of Frey's scrutiny of Nelson's argument) will centre only on the dialogue between Regan and Frey in respect of McCloskey's argument. I will endeavor to update the dialogue by providing a re-interpretation of 'rights' in Nelson's argument.

Rights

In his book *Interests and Rights* Frey begins his attack on the belief that animals have rights by discussing Leonard Nelson's argument for animal rights (Frey, 1980, p.5). Nelson's argument links interests to rights:[2]

> All and only beings which (can) have interests (can) have moral rights;
>
> Animals as well as humans (can) have interests;
>
> Therefore, animals (can) have moral rights (Nelson, 1956, pp.136-44).

While this argument is undeniably valid, for the conclusion to be true the premisses must also be true, and Frey disputes that they are true. Therefore, it is the soundness of the argument that Frey is disputing. He objects to both the major and minor premise of Nelson's argument.

Before we continue with his objections we should distinguish here between a weak sense of rights and a strong sense of rights. If a being has a right in the weak sense this can be said to refer to nothing more than a moral reason to take its interests into account in the appropriate way and give that being due moral consideration. Talking of a being's moral rights in this sense is talking of a being's moral standing or talking of an obligation to take that being into moral consideration. (Perhaps Nelson would use the term 'moral consideration' or 'moral standing' if he were writing his argument today. 'Rights' was the only term that was available to him at the time he was writing.[3]) By contrast, if one says that a being has a right in the strong sense one can be said to be referring to an *overriding* reason to take that being's interests into account in the same situation (Attfield, 1995, pp.142-44).[4] Talking of moral rights in this sense is talking of interests that should always take moral precedence; rights, in this sense, are inviolable. While there are rights in the weak sense, I am sceptical about rights in the strong sense (see below).[5] But it is rights in the weak sense that I will interpret Nelson as referring to, and it will be this sense of rights that I will refer to when assessing Frey's claims against Nelson.

Let us now turn to Frey's attack on Nelson's argument. Firstly, Frey objects to the major premise (that all and only beings which (can) have interests (can) have rights), as he believes that it is not true. Frey believes that Nelson merely assumes it to be true. However, disputing this is highly implausible. As Regan says, 'Most of us do not think that *things* can have rights, and this seems to be because we do not think that they can have any interest in what happens to them' (Regan, 1976, p.253). It does seem that one can only have rights if one has (or can have) interests. Joel Feinberg cites two reasons for this. Firstly, 'a right holder must be capable of being represented and it is impossible to represent a being that has no interests.' Secondly, 'a right holder must be capable of being a beneficiary in his own person, and a being without interests is a being that is incapable of being harmed or benefited' (Feinberg, 1974, p.51).

Indeed, it seems plausible that all 'beings capable of being beneficiaries can deserve moral consideration', or can have moral standing, and that all 'beings who have (or can have) interests are capable of being beneficiaries', and that, therefore, all 'beings who have (or can have) have interests can deserve moral consideration', or can have moral standing.[6] It does seem then that all and only beings with interests have rights,[7] and, taking 'rights' to mean rights in the weak sense, Frey appears to be wrong in his denial of the major premise.

Secondly, Frey argues that Nelson's major premise assumes that there actually are rights. Frey doubts whether there are rights, whether these be human or animal rights. However, doubting whether there are rights does not make the major premise false. Even if one doubts that there are rights it can still be true that only beings with interests can have rights (if there are rights), and this appears to be a conceptual truth. (Frey's objection to the Nelson's minor premise will be discussed under the title 'Interests' below.)

Besides, Frey cannot disallow our use of rights in the weak sense, and he provides no argument as to why rights should not be interpreted in this way or an argument against the view that there *are* beings with rights in this sense (where 'rights' should be taken to mean 'moral standing').[8] Construed in this sense, and contrary to Frey's doubts about rights, it does seem that there are beings that have rights, and that all beings with interests have rights.

Indeed, I am inclined to share Frey's doubts about rights (interpreted in the strong sense), though not his doubts about the truth of the major premise (since this premise should be construed to refer to rights in the weak sense.) Saying that a being has a right does not mean that there are overriding reasons to take that being's interests as having moral precedence over another being's interests.

It should perhaps be said here that entities can have moral standing without anything about their having rights being held to follow, where (as Kenneth Goodpaster argues) moral standing should be interpreted 'broadly to include the most basic forms of practical respect (and so is not restricted to "possession of rights"...)' (Goodpaster, 1978, p.309). When we speak of beings having moral standing we are speaking of beings to which 'principles of morality [should] apply' (Warnock, 1971, p.148, quoted by Goodpaster, 1978, p.308), irrespective of particular rights that they may or may not have. The case for extending moral consideration to animals is independent of rights. In particular the case against factory farming and animal experimentation argues for the moral standing of animals, and no reference to rights is required for this case to be made. Factory farming and animal experimentation involve the massive suffering of animals, and this, in itself, requires that we take animals' interests seriously and give them due consideration. Such moral consideration may require that factory farming and animal experimentation be abolished or phased out. Even if animals do not have rights it would not mean that current practices that involve the immense suffering of animals are justified.

I tend to agree with Raimond Gaita that the claim that something is unjust because it violates rights simply tells us that an action is unjust:

> To say that an action is unjust because it violates someone's right adds nothing, I believe, to saying that it is unjust, neither by way of explaining why it is unjust, nor by making it more objectively or more stringently binding on the will... Elaboration which reveals what it means to be cruel to a human being or to an animal, and whose appreciation makes some actions impossible and others necessary, need not, I believe, mention rights (Gaita, 2002, p.200).

Gaita points out that rights talk has the function of making our relations with humans and animals more understandable (Gaita, 2002, p.200). Consideration of a being's interests, of what it means to cause a human being or a non-human being to suffer, or to degrade a human being, is independent of rights talk. Reflection on interests and obligations can tell us why actions are wrong, or what makes our actions unjust. Rights talk just tends to tell us that they are unjust.

An appeal to rights in the face of an unjust action need not necessarily prevent that action, or stop it occurring or continuing. The success of an appeal always depends upon the state of mind, or tolerance, of the person to whom the appeal is made. The appeal need not provoke feelings of duty or obligation.

Indeed, an action may violate a being's rights without actually being wrong. In some instances it may be necessary to violate someone's rights, if they exist that is. For example, we would think it right to violate a rapist's right to freedom. Sometimes it may be more of an injustice not to violate a being's rights. In some cases, then, rights talk cannot tell us whether an action is just or unjust, and it cannot always tell us if an action is right or wrong. Whether an action is just or unjust, then, is to some degree independent of rights.

'Rights' may be enshrined in law in the United Kingdom, but they are readily overridden depending on the context or situation, and this could be held to indicate that they cannot tell us much about whether an action is justified or not. Rights tend to be either revered or discounted depending on what will best promote the interests of the individuals involved. Frey, then, shares this scepticism about rights talk. (In the UK there is increasing scepticism about 'rights talk'. Although appeals to human rights are, *legally*, invaluable in the context of genuine human abuses, many people believe that rights talk is increasingly used as a means to the unfair claiming of huge compensation rewards.)

In spite of this scepticism surrounding rights talk and, in particular, scepticism about rights in the strong sense, Nelson's major premise is still true, since 'rights' in Nelson's argument should be meant to refer to rights in the weak sense. Such scepticism does not make Frey's objection any less implausible. The proposition 'All and only beings which (can) have interests (can) have moral rights' should be upheld as true.

Interests

However, it is not so much the major premise of Nelson's argument that concerns Frey, but rather the minor one that states that animals have interests. Frey turns is argument to a discussion of H. J. McCloskey's criticism of this premise and Regan's response to this criticism (Frey, 1980, p.18).

For McCloskey, to say that something is in a being's interests has both evaluative and prescriptive connotations:[9]

> [I]nterests are distinct from welfare, and are more inclusive in certain respects–usually what is dictated by concern for a man's welfare is in his interests. However, interests suggest much more than that which is indicated by the person's welfare. They suggest that which is or ought to be or which would be of concern to the person/being. It is partly for this reason–because the concept of interests has this evaluative-prescriptive overtone–that we decline to speak of the interests of animals, and speak rather of their welfare (McCloskey, 1965, p.126).

I agree with Regan that what McCloskey means by 'evaluative-prescriptive overtone' is far from clear. Frey says that the evaluative meaning of 'X will be in A's interests' can be conveyed as 'X will contribute (or is thought to contribute) to A's welfare, or well-being', whereas the prescriptive meaning of 'X is in A's interests' can be conveyed as 'A ought to care, or be concerned about X' (Frey, 1980, p.19). Regan interprets McCloskey similarly to Frey. For Regan, the evaluative meaning of 'X will be in A's interests' can be characterised as 'X will (or we think X will) benefit A' and the prescriptive meaning can be characterised as 'X ought to be of concern to A', or 'A ought to care about X' (Regan, 1976, p.255).

It must be said here, before we continue, that 'X will contribute to A's welfare, or well-being' is not a judgment of value or an evaluation. But it is not clear what judgments of value are for McCloskey and he does not define what the evaluative element of interests is. That 'X will contribute to A's welfare' is, in fact, Regan's and Frey's interpretation of what McCloskey means when he says that talk of what is in a being's interests has evaluative connotations.

Indeed, McCloskey himself does not say what the evaluative connotations are and finding this out turns into an interpretive exercise. But it does appear, from McCloskey's quote above, that the evaluative meaning of what is in a being's interests is (somehow) related to talking of 'that which is indicated by the person's welfare' and that the prescriptive meaning of interests is related to 'that which... ought to be or which would be of concern to the person/being'. So, while Regan and Frey seem to be right in thinking that, for McCloskey, the evaluative meaning of 'X will be in A's interests' can be characterised as 'X will contribute to A's well-being' it should be pointed out that, although McCloskey may consider this an evaluation, it is not what is usually called an evaluation or value judgment.

Let us return to Regan's attack on McCloskey's view. For McCloskey we cannot speak of what is in an animal's interests because talk of what is an animal's interests has a prescriptive and evaluative meaning. Regan firstly points out that talk of what is an animal's interests need not have a prescriptive meaning. That is, we can ask 'what is in A's interests?' without issuing a prescription. Regan cites John Searle (Searle, 1962, pp.423-32, cited by Regan, 1976, p.255), who argues that we can say 'X is good' (or we can make an evaluation about X) without thereby saying 'I commend X' (or issuing a prescription related to X). At first glance it seems that this reference to Searle may give an explanation as to why 'X will contribute to A's welfare' is supposedly evaluative. Saying that we do not have to issue a prescription when we say 'X is good', or that the evaluative statement 'X is good' does not necessarily have a prescriptive meaning, is a valid point in response to McCloskey and Regan is correct in saying that a prescriptive meaning need not be given to the statement 'X is in A's interests'.

At first glance it seems that this reference to Searle may give an explanation as to why 'X will contribute to A's welfare' is supposedly evaluative. However, on closer inspection, it is less than clear what relevance Regan's reference to Searle has to McCloskey's view and, in particular, to his belief that saying 'X will be in A's interests' has an evaluative meaning, characterised as 'X will contribute to A's welfare, and a prescriptive meaning, characterised as 'A ought to be concerned about X', because, unlike the statement 'X is good' (the evaluative statement that Searle uses in his argument), the statement 'X will contribute to A's welfare' is not an evaluative statement in the ordinary sense of the phrase. While 'X is good' is obviously evaluative, 'X will promote A's welfare' is not; so alluding to Searle's argument is fruitless until it is known why McCloskey believes that the latter statement is evaluative (since Searle argues that we need not attach a prescription to an *evaluative* judgment).[10]

However, the main points that Regan is trying to make, by referencing Searle, is that a prescription need not follow an evaluation, and interests need not have a prescriptive element (even if we go along with McCloskey and assume that they have an evaluative element). So, if we apply this to the context of animal interests, animals can have interests even though they (or anyone else) do not have to do anything about those interests; that is, animals can have interests without there necessarily being a prescription relating to those interests.

Despite Regan's fair point that talk of what is in a being's interests need not have a prescriptive element, for the sake of argument Regan goes on to assume that such talk does indeed have this element. Regan interprets McCloskey as believing that animals do have a well-being, and so it must be the prescriptive meaning of 'X is in A's interests' which casts doubt on the claim that we can talk of what is in an animal's interests (Regan, 1976, pp.251-57 [p.254], cited by Frey, 1980, p.19). Since animals are not the sort of beings which ought to do something about X, and they are not the sort of beings which ought to be concerned about anything, it follows that we cannot speak of what is in an animal's interests (Regan, 1977, p.335 (interpreting McCloskey, 1965)). However, why do we have to say, 'A (the being with the interests) ought to be concerned about X?' Instead, why not say, 'B (another being) ought to be concerned about X?'

If X is conducive A's welfare or A's interests and A is not the sort of being that ought to do anything about X (say, for example, a baby, an animal or a severely mentally disabled person), then we should say that B (another being, say, a moral agent or a relevant someone) ought to do something about X, or 'B ought to be concerned about X', not 'A ought to do something about X', or 'A ought to be concerned about X'.

So, if, say, eating food is in a baby's interests, then, we should say that someone ought to do something about that food. That is, if X (food) is in A's (the baby's) interests, then, B (the guardian or some other relevant person) ought to do something about X (food). It is not the case that if we judge that X is in A's interests, then the judgments for action concerning those interests should always be prescribed to A.

Prescriptions for action are prescribed to moral agents. They are not always prescribed to the being whose welfare is being evaluated. That being might be incapable of taking any action to safeguard its interests. The being might be a baby, or a senile person. It might be an injured animal, or an animal trapped in a cage. So, instead of saying 'A ought to do something about X', we should say 'B (another being) ought to do something about X'. Indeed, if one were very sick in hospital, say, with an amputated leg, and pain relief was in one's interests, no one would expect one to administer the pain relief oneself, but another person would be expected to do it.

Regan also rejects McCloskey's argument. He too points to the fact that there are some humans who are not the sort of beings that ought to do something about X; they are not the sort of beings which ought to be concerned about anything. Examples include marginal cases, such as babies and the severely mentally disabled. Thus, on the prescriptive meaning of 'interests', as interpreted by McCloskey, we cannot speak of what is in these humans' interests. Both these humans and animals in general will be ruled out as beings that can have interests (Regan, 1977, p.335). However, most would reject the idea that we cannot speak of what is in a baby's interests, or in the interests of a severely mentally disabled human being. Regan argues that although animals and these humans are not in a position to do anything about securing their interests we can nevertheless speak about what is in their interests. For instance, we can say that it is in the interests of a baby that he or she sees a doctor when showing signs of illness. We can speak of something being in a being's interests whether or not that being is capable of doing anything about their interests. When we say this 'the prescription *can* only be directed to some *other* competent person' (Regan, 1977, p.336).

Frey believes Regan's case to be implausible, and argues that for animals and such humans to have interests ultimately depends on the existence of other humans, that is, other competent people to see to it that their interests are provided for. For Regan, when we say that it is in the interests of a baby that he or she sees a doctor the prescription (if any) applies to other people. But Frey believes that it follows from this that if no other people exist then the baby has no interests, and, for Frey, this conclusion is highly implausible (Frey, 1980, p.23). For animals and these humans to have interests then depends upon the existence of other people implausible as this may be.

Frey interprets Regan's position as implying that if others do not exist then there can be no interests at stake:

> Regan's argument to show that animals satisfy McCloskey's requirement either yields an unwelcome result or... requires support of a kind quite beyond anything he provides... animals have not been shown to have interests, with the result that the minor premise in the Nelsonian argument for animal rights is not established (Frey, 1980, p.27).

However, Frey's interpretation of Regan's argument is not correct. That other people need to exist for animals to have interests is not what Regan implies. If no humans exist then we can still talk of what is in an animal's interest; it is just that we can no longer give this talk a prescriptive meaning. Since animals are not the sort of beings which ought morally to do anything about their interests any prescription would need to be directed at someone else and, in Regan's own words, 'there is no "someone" who ought to do anything' (Regan, 1977, p.336); but, it does not follow that, in the absence of someone else, the interests of animals cease to exist. It is just that if no one else exists there will be no one else to prescribe or direct the obligation or action to. It follows that speaking of what is in an animal's or a baby's interest still makes sense. Therefore, on the criterion that interests are necessary for the possession of rights 'both will qualify as possible possessors of rights' (Regan, 1977, p.337).

McCloskey's claim then that we cannot speak of animals' interests should be rejected. Regan has shown that even if we assume that talk of what is in a being's interests has a prescriptive meaning, we can still speak of the interests of animals. Frey has attempted to uphold McCloskey's claim by providing a counter-argument to Regan, but Frey's argument should be rejected as he completely misinterprets Regan with the result that his (rather than Regan's) counter-argument reaches some unwelcome conclusions. Frey's discussion of McCloskey's argument has failed to show that Nelson's minor premise (that animals as well as humans (can) have interests) is not true. Frey's denial that animals have interests is implausible, as is McCloskey's. It seems then (on the basis of the above discussions only[11]) that Frey is wrong in his denial of both Nelson's major premise (all and only beings which (can) have interests (can) have rights), and his minor one (animals as well as humans (can) have interests). Contrary to Frey, Nelson's argument is sound and Frey's doubts about the truth of the premisses, and, therefore, his doubts about the truth of the conclusion are unfounded.

Conclusions

For Frey then the argument that animals have moral standing or rights fails because the claim that animals have interests has not been established (Frey, 1980, p.27). However, Frey's attempt to show that this claim has not been established, through reference to McCloskey's argument, failed. Also, Frey's denial of the truth of the claim that all and only beings which (can) have interests (can) have rights is highly implausible. Nelson's argument should be construed to refer to having rights in the weak sense or something like warranting moral consideration, and construed in this way Nelson's proposition, that all and only beings which (can) have interests (can) have rights, is true.[12] Contrary to Frey's belief then, both Nelson's minor premise (animals as well as humans (can) have interests), and his major premise (all and only beings which (can) have interests (can) have moral rights) are true, and, therefore, Nelson's conclusion (animals (can) have moral rights) is also true.[13]

Notes

1 There are other reasons why Frey does not think it has been shown that animals have interests; reasons that have nothing to do with McCloskey's argument. (See his Interests and Rights (Oxford: Clarendon Press, 1980), cps.3-11.). But, since

this paper aims to focus on, and give an analysis of, the dialogue between Regan and Frey, in respect of McCloskey's argument, these reasons are beyond the scope of this paper. This paper looks at just one reason why Frey thinks it has not been shown that animals have interests.

2 While the argument focuses on moral rights, one could replace the word 'rights', in Nelson's argument, with the word 'standing', and the argument would then stand as one that would probably be endorsed by those people who link animal interests to the moral standing of animals. It may be appropriate then if we consider that the words 'moral rights' could be replaced by 'moral standing'. In refuting the claim that animals can have interests, Frey's discussion of Nelson's argument is directed at not only those philosophers who link interests to rights, but those who link interests to moral standing, or moral consideration, too.

3 Robin Attfield usefully made this relevant comment about Nelson after a paper I presented at Cardiff University of Wales on 26th November 2008.

4 Attfield implicitly employs the distinction between a weak and strong sense of rights. See Attfield, *Value, Obligation and Meta-Ethics*, Value Inquiry Book Series, Vol.30 (Amsterdam: Rodopi, 1995), pp.142-44, and Attfield, 'The Place of Rights in Normative Ethics', lecture handout for Cardiff University, 2002. The two senses of rights I have referred to are based on Attfield's distinction.

5 For discussion about moral and legal rights see R. M. Hare, *Moral Thinking: Its Levels, Method, and Point* (Oxford: Clarendon Press, 1981), pp.150-57.

6 This argument is adapted from Kenneth Goodpaster's interpretation of Joel Feinberg's argument (linking interests to rights). I have replaced the word 'only' (in Goodpaster's interpretation of Feinberg's argument) with the word 'all', though I am not thereby claiming to be interpreting Feinberg, but merely using the argument differently (Goodpaster, 'On Being Morally Considerable', in *Journal of Philosophy*, 75, 1978, p.318; Feinberg, 'The Rights of Animals and Unborn Generations', in Philosophy and Environmental Crisis, ed. Blackstone (Athens: University of Georgia, 1974), p.51.).

7 See James Rachels, 'Do Animals Have a Right to Liberty?', Donald VanDeVeer, 'Defending Animals by Appeal to Rights', and James Rachels, 'A Reply to VanDeVeer', in *Animal Rights and Human Obligations*, eds. Tom Regan and Peter Singer (New Jersey: Prentice Hall, 1976). See also H. J. McCloskey, 'Rights', *The Philosophical Quarterly*, Vol.15, No.59, Moral Philosophy Number, (Apr., 1965), p.115-27.

8 Indeed, Frey suggests that there *are* beings to whom we have obligations to take their interests into moral consideration. He suggests that we have moral obligations to some beings, specifically human beings. See Frey, *Interests and Rights* (Oxford: Clarendon Press, 1980), pp. 10-13 and 145-50.

9 Since McCloskey does not give a definition of what he means by 'evaluative' and 'prescriptive', I will assume that, for McCloskey, 'evaluative' concerns judgments of value and 'prescriptive' concerns by contrast (a) judgments about what ought to be done and (b) directives for action or judgments of obligation.

10 For a fuller discussion of Searle's argument see Searle, 'Meaning and Speech Acts', *The Philosophical Review*, Vol.71, No.4, (Oct., 1962), pp.423-32. Also see R. M. Hare's attack on Searle's argument in 'Meaning and Speech Acts', *The Philosophical Review*, Vol.79, No.1, (Jan., 1970), pp.3-24. Both these articles are cited by Frey in *Interests and Rights* (Oxford: Clarendon press, 1980), p.19. The former article is cited by Regan 'McCloskey on Why Animals Cannot Have Rights', in *The Philosophical Quarterly*, Vol.26, No.104, (Jul.1976), p.255.

11 As said earlier, Frey gives other reasons for his denial of Nelson's premisses. (See his *Interests and Rights* (Oxford: Clarendon Press, 1980), cps.3-11.). This paper has focused only on the dialogue between Regan and Frey in respect of McCloskey's argument.

12 Even if one doubts there are rights, as Frey does, or even if there were no beings having rights, or no beings having interests, it can still be true that only beings that (can) have interests (can) have rights. Of course, this is not enough to uphold the 'all' variant of the major premise, which says that *all* beings which have interests have rights. However, this variant was supported and upheld as true earlier in the paper (see the section on 'Rights').

13 As said at the beginning of this paper, we should be aware here that we could replace the word 'rights' in Nelson's argument with 'standing' and the argument would go through as before and the same conclusions about Frey's claims would apply.

References

Attfield, Robin. 1995. *Value, Obligation and Meta-Ethics*. Value Inquiry Book Series, Vol.30. Amsterdam: Rodopi. 142-44.

Feinberg, Joel. 1974. "The Rights of Animals and Unborn Generations." *Philosophy and Environmental Crisis*. ed. Blackstone. Athens: University of Georgia. 43-68.

Frey, R. G.. 1980. *Interests and Rights*. Oxford: Clarendon Press.

Goodpaster, Kenneth. 1978. "On Being Morally Considerable." *Journal of Philosophy*. 75 (308-25).

McCloskey, H. J. "Rights." *The Philosophical Quarterly*. Moral Philosophy Number. Apr., 1965. Vol.15.59 (115-27).

Nelson, Leonard. 1956. *A System of Ethics*. trans. Norbet Guterman. New Haven: Yale University Press. Part I, Section 2, Ch.7. 136-44.

Raimond, Gaita. 2002. *The Philosopher's Dog*. London: Routledge.

Regan, Tom. 1977. "Frey on Interests and Rights." *Philosophical Quarterly*. 27 (335-37).

Regan, Tom. 1977. "McCloskey on Why Animals Cannot Have Rights." *The Philosophical Quarterly*. Vol.26.104. Jul., 1976. 251-57.

Searle, John R. 1962. "Meaning and Speech Acts." *The Philosophical Review*. Vol. 71.4. 423-32.

Warnock, G. J. 1971. *The Object of Morality*. New York: Methuen.

Chapter 10

Reverence for Life: An Ethical Evaluation of Human-Animal Relationship

☆ *Manish A. Vyas*

In most of the Western religious and philosophical tradition, the non-human world is thought to exist for the sake of human beings. The assumed superiority of human beings over the rest of the world has religious expression in the first account of creation in the Genesis. There is a hierarchy based on the theological concept of "dominion," where Adam is given "dominion over the fish of the sea, and over the fowl of the air, and over the cattle, and over all the earth, and over every creeping thing that creepeth upon the earth (1: 20-28)." As a result, most of the Western cultural, religious and philosophical traditions are committed to a value hierarchy in which human beings stand at a pinnacle of creation or evolution. This gave birth to a widespread belief that human existence is valuable in itself, while animal life is there for human use and, therefore, has only instrumental value. In this regard, many prominent Christian theologians have offered an instrumentalist understanding of life. Deeply entrenched in a theological tradition of the orders of Creation, St. Augustine (1877) states in relation to the Old Testament prohibition against killing:

> When we say, "Thou shalt not kill," we do not understand this of the plants, since they have no sensation, nor of the irrational animals that fly, swim, or creep, since they are dissociated from us by their want of reason, and are therefore by the just

> appointment of the Creator subjected to us to kill or keep alive for our own uses; if so, than it remains that we understand the commandment simply of man" (32).

Augustine, in his rejection of fellowship with non-human life, is not alone in the Christian tradition. St. Thomas Aquinas also feels that "dumb animals and plants are devoid of the life of reason," which is a sign that they are naturally enslaved and accommodated to the uses of others (1918, 64: I)." Influenced by the Aristotelian notion of ordering, Aquinas endorses the notion that lower creation exists to serve the higher. Addressing the question whether it is lawful for humans to kill living beings, he concludes:

> There is no sin in using a thing for the purpose for which it is. Now the order of things is such that the imperfect are for the perfect...it is not unlawful if man uses plants for the good of animals, and animals for the good of man...(*Ibid*).

In *Summa Contra Gentiles*, Aquinas expounds an instrumentalist and hierarchical understanding of Creation:

> Hereby is refuted the error of those who said that it is sinful for a man to kill dumb animals: for divine providence they are intended for man's use in the natural order. Hence it is not wrong for man to make use of them, either by killing or in any way whatever (1950, Bk 3. 12).

Aquinas understands the role of nonhuman creation almost strictly in terms of its capacity to serve human wants. Creation, he insists, was created solely for "intellectual creatures": "The Divine Ordinance of animals and plants is preserved not for themselves but for man," and hence, "as Augustine says...both their life and their death are subject to our use" (*Ibid*). In his theology, the value of non-human life is measured by its utility to human interests. Since it lacks *rationality*, nonhuman creation can be considered as an instrument for humans who alone possess such a capacity.

A similar instrumentalist understanding of nonhuman life is found in the Reformers, notably John Calvin and Martin Luther. Addressing the issue of the subjection of animals to human dominion in Genesis I, Calvin remarks: "Hence we infer what was the end for which all things were crated; namely, that none of the conveniences and necessities of life might be wanting to men" (1847: 96). Calvin believes that "men may render animals subservient to their own convenience, and may apply them to various uses, according to their wishes and necessities" (*Ibid* 9:2: 291). Martin Luther follows suit. After the fall and flood, "the animals are subjected to man as to a tyrant who has absolute power over life and death" (1958: 132). Such understanding of nonrationality and nonfellowship has largely dominated Christian discussions of animal and plant life and have served as the justification for excluding other life forms from moral consideration.

This anthropocentric behavior found its chief proponent in the French philosopher Rene Descartes, who made the distinction between man and animals a philosophical and ethical problem. For Descartes, animals are essentially machines, completely lacking in reason and in fact in any kind of experience. As such they are due no moral obligations at all; we may experiment on them and we may kill and eat them without moral scruple. He used the faculty of reason / rationality as the dividing line between human beings and animals. In *Discourse on Method,* Descartes maintains that non-human animals lack reason and language, and hence are ultimately undistinguishable from machines:

> ....although many animals show more skill in some of their actions than we correspondingly do, it is nonetheless clear than the same animals show none at all in many others, so that what they can do better than us does not prove that they have any mental powers, for it would follow from this that they would have more intelligence than any of us, and would surpass us in everything. Rather, it shows that they have no mental powers whatsoever, and that it is nature which acts in them, according to the disposition of their organs; just as we see that a clock consisting only of ropes and springs can count the hours and measure time more accurately than we can in spite of all our *wisdom* (2006: 124).

Along with Aquinas, Descartes identifies the rational soul as an immortal soul, and he makes a sharp distinction between the souls of rational beings and those of dogs and trees.

Following Descartes, Aquinas and other philosophers, German philosopher Immanuel Kant maintained that animals are there "merely as a means to an end. That end is man" (1963: 239). For Kant, cruelty towards animals was wrong not because it was improper. It was wrong solely on the grounds of that it was bad for mankind. He argued in 1785 that humans have duties only toward other humans, and that "cruelty to animals is contrary to man's duty to *himself,* because it deadens in him the feeling of sympathy for their sufferings, and thus, a natural tendency that is very useful to morality in relation to other humans is weakened" (1785: 17). Four years later, the English philosopher Jeremy Bentham, one of the founders of modern utilitarianism, argued that it was the ability to suffer, not the ability to reason, that should be the benchmark of how we treat other beings. If rationality were the criterion, many humans, including the babies and disabled people, would also have to be treated as though they were things. He argued:

> The day *may* come when the rest of the animal creation may acquire those rights which never could have been witholden from them but by the hand of tyranny. The French have already discovered that the blackness of the skin is no reason a human being should be abandoned without redress to the caprice of a tormentor. It may one day come to be recognized that the number of legs, the villosity of the skin, or the termination of the *os sacrum* are reasons equally insufficient for abandoning a sensitive being to the same fate. What else is it that should trace the insuperable

> line? Is it the faculty of reason or perhaps the faculty of discourse? But a full-grown horse or dog, is beyond comparison a more rational, as well as a more conversable animal, than an infant of a day or a week or even a month, old. But suppose the case were otherwise, what would it avail? The question is not, Can they *reason*?, nor can they *talk?* But, Can they *suffer*?" (1996: 283)

Bentham was, thus, the first philosopher who was outspoken in advocating a need to reframe our relationship to the natural world completely and to cease to see the natural world merely as raw material for the fulfillment of human needs. The emphasis on rationality and the human supremacism have been responsible for the large scale devastation of the ecological resources world-wide; so much so that the very existence of the Earth and humans seems to be in peril. Talking about human-made destruction of ecology, Richard Ryder notes:

> Man is just an animal, one species among many species. He is certainly a very clever animal, but he is also an exceptionally destructive, aggressive and polluting animal. During the twentieth century the species of man has grown to overpopulate many areas of the planet and has begun to expand outwards into space, taking with it a potential for creativity as well as the capacity to spoil and to destroy…This clear and proud animal has already destroyed many other species; he now also has the power to destroy his own…and so the probability of total doom for his species grows (1991: 35).

Bruce Johansen in his book *Global Warming in the 21st Century* echoes similar concerns for the unprecedented mass extinctions of flora and fauna. In contrast with the mass extinctions on the Earth in the past which were driven by natural catastrophes, the present-day human driven devastation of native habitats of animals and plants is a product of "human kind's increasing dominance of the Earth" (2007: 533).

This anthropocentric attitude has been one of the prime reasons for the cruelty towards animals world-wide. In the twentieth and twenty-first centuries, humans indifference towards animals based on the former's claim on rationality has further resulted into what is popularly called factory-farming. It is premised on the view that animals are machines incapable of experiencing pain or pleasure. The business of factory-farming is estimated to be more than $ 100 billion world-wide. Animals in these factories are over-crowded in "barren environments, restricted, stressed and maintained on drug-laced, unnatural diets (Ryder 2006: 107). Virtually all poultry products and meat in most of the countries come from animals produced on a massive scale in these factories. Gaverick Matheny in her article "Utilitarianism and Animals" reveals that around 17 billion land animals were raised and killed during 2001 for food; somewhere between 50 and 100 million other animals were killed in laboratories, while another 30 million were killed in fur farms (2006: 22). On the other hand, Miyun Park, Director of the Farm Animals and Sustainable Agriculture Program in Washington, describes the cruel and unhygienic treatment of hens in one such factory:

> ...life for egg-laying hens in battery cage facilities is harrowing. Hens stand on wire-mesh flooring so unlike the earth that their nails, which would normally wear down while scratching the ground, curl around the bars. Feather loss is common as hens rub against cages until many appear to have been plucked, their bodies raw with sores. They cannot roost at night, dust-bathe to clean themselves, feel sunlight, breathe fresh air, build a nest, raise their young, or even freely stretch their wings, let alone exercise or roam (2006: 176).

The vast majority of these animals are forced to live and die in condition most of us would find morally repugnant. Yet their use has been justified by the belief that animals do not deserve significant moral consideration.

This, in turn, calls for a need to re-evaluate the basis of human-animal relationship from the point of view of ethics. Mark Twain in his essay "The Lowest Animal" (first published 1962) goes to the extent of denying the Darwinian theory of Ascent of Man from the Lower Animals. He argues:

> I have been studying the traits and dispositions of the "lower animals" (so-called), and contrasting them with the traits and dispositions of man. I find the result humiliating to me. For it obliges me to renounce my allegiance to the Darwinian theory of the Ascent of Man from the Lower Animals; since it now seems plain to me that that theory ought to be vacated in favor of a new and truer one, this new and truer one to be named the *Descent* of Man from the Higher animals (1972: 153).

One of the greatest faults of all ethics has been that they were centered only on the relations of man to man. Having taken the ecological systems which support us entirely for granted, we have lost a touch with ourselves as beings who are not only cultural but also natural, embedded in the earth and just as dependent on a healthy biosphere as other forms of life. There is, therefore, a need to break from the anthropocentric traditions in the interest of both justice and prudence. In the present context of ecological destruction, it is necessary to adopt philosophical strategies that maximize our sensitivity to other members of our ecological communities as ethically considerate beings. In this regard, German philosopher and Nobel laureate Albert Schweitzer argues that the life of human beings cannot be viewed in isolation for the simple reason that: "we are part of life–we are born of other lives; we possess the capacities to bring still other lives into existence...So nature compels us to recognize the fact of mutual dependence, each life necessarily helping the other lives linked to it (1965: 47)." In short, our relationship with animals is one of those nodal points at which our relationship with Other is formed. Val Plumwood in her article "Ecological Ethics from Rights to Recognition" extends the reach of anthropocentricism to include conceptual strategies of injustice within the human sphere to support supremacism of nation, gender and race (1999: 189). She identifies Otherization as chief among those strategies of injustice which constructs a major gulf between the elite and non-elite, between the black and white, between man and woman.

Among the Western philosophers it was Albert Schweitzer who first noticed the inadequacy of humanocentric thought as a rational critique of the world and sought to widen the scope of ethics to incorporate concern for all manifestations of life. In contrast to the Western philosophers, Schweitzer locates the value of beings not in any specific faculty or capacity limited to a certain species, but rather in the "will-to-live" which is common to all life. Schweitzer's ethic commands responsibility for all the various life forms with which one comes into contact. His ethic of Reverence for Life is universal in that it does not limit itself strictly to human relations. According to him, a man is truly ethical only when he "obeys the compulsion to help all life which he is able to assist, and shrinks from injuring anything that lives (1987: 310)." Schweitzer's call for reverence for life comes quite closer to the Buddhist and Jain concept of *ahimsà* (non-violence), which is not simply the absence of something (violence). On the contrary, it is practiced on the basis of a deeply positive feeling of respect for living beings. It places great emphasis on concern (dayà) and sympathy (anukampà) for living creatures, and an increasing empathy with them based on the awareness that others dislike pain and death just as much as oneself (Keown 2005: 14).

Rejecting the idea of a hierarchy of beings in nature with humans at the top, Schweitzer tells us that our relation to the non-human world should not be one of moral hierarchies or instrumentality. For him, what is deeply objectionable in Christian theology is the exclusive preoccupation with the human species. He offers his thought: "True philosophy must start from the most immediate and comprehensive fact of consciousness, which says: 'I am life which wills to live in the midst of life which wills to live" (1987: 309). Schweitzer's view that all living things are of the same will serves as the basis of his ethics. Thus he reminds us that:

> The poor fly which we would kill with our hands has come into existence like ourselves. It knows anxiety, it knows hope for happiness, it knows fear of not existing anymore...The beetle that lies dead in your path–it was something that lived, that struggled for its existence like you, that rejoiced in the Sun like you, that knew anxiety and pain like you (1988: 10).

Schweitzer's ethics of reverence for life, thus, makes no distinction between higher and lower, more precious and less precious lives for the simple reason that human beings are part of life; they are born of other lives, as Mark Twain observes:

> Providence leaves nothing to go by chance. All things have their uses and their part and proper place in nature's economy: the ducks eat the flies–the flies eat the worms–the Indians eat all these–the wildcats eat the Indians–the white folks eat the wildcats–and thus all things are lovely (1983: 177).

In this way, true ethics encompass all life and treat it as sacred. Human beings come again and again into the position of being able to preserve their own life only at the cost of other life. If he has been touched by the ethic of Reverence for Life, he injures and destroys life only under a necessity which he cannot avoid. Hitherto ethics have faced the problem of man and animals uncomprehendingly. Even when sympathy with the animal creation was felt to be right, it could not be brought within the scope of ethics because ethics were really focused only on the behavior of man to man. While the truth is that ethical concern should encompass all life. We should respect "the earth, the plants and animals for themselves, before weighing up their utility for human beings" (Kung & Moltman 1990: 132). Speciecism is as great a prejudice as racism. It is the idea that it is justifiable to give preference to beings simply on the grounds that they are members of the species *Homo sapiens*. Those of us who count ourselves as Humanitarians, Muslims, Hindus or Christians should extend our ideologies to include other species. The welfare of animal-citizens should be as much our concern as is that of other humans. If we accept that it is wrong to cause suffering to innocent humans, then, logically, it is also wrong to do this to non-humans. There is no sound evidence for believing that other animals cannot suffer like us, and indeed the evidence points the other way. For these reasons I consider that animal welfare is a serious subject worthy of further research and deserving the support and respect of the world community in an age when animals are increasingly being exploited.

References

Aquinas, St. Thomas. 1918. "Summa Thedogica." *The Summa Theologica of St. Thomas Aquinas*. Trans. Father of the English Dominican Providence. New York: Benzinger Brothers. Pt. I, Question 64:I.

_____. 1950. *Summa Contra Gentiles*. Trans. Joseph Rickaby. London: Burns & Oates. Bk. 3,12.

Augustine, St. 1877. *The City of God*. I.20. Trans. Marcus Dods. Edinburgh: T & T Cark.

Bentham, Jeremy. 1996. Introduction to the Principles of Morals and Legislation. Chapter 17. J. H. Burns and H. L. A. Hart (eds). *The Collected Works of Jeremy Bentham*. New York: Oxford University Press.

Calvin, John. 1847. *Commentaries on the First Book of Moses Called Genesis*. Vol. I (I:26). Trans. John King. Edinburgh: Edinburgh Printing Co.

Descartes, Rene. 2006. *Rene Descartes: A Discourse on the Method*. Trans. Ian Maclean. New York: Oxford University Press.

Johansen, Bruce. 2007. *Global Warming in the 21st Century: Plants and Animals in Peril*. Vol. 3. New Delhi: Atlantic Publications.

Kant, Immanuel. 1785. *Groundwork of the Metaphysic of Morals*. Part II (The Metaphysical Principle of the Doctrine of Virtue), paras 16 and 17.

_____. 1963. *Lecture on Ethics*. Trans. L. Infield. New York: Harper Torchbooks.

Keown, Damien. 2005. *Buddhist Ethics: A Very Short Introduction*. New York: Oxford University Press.

Kung, Hans and Moltman, Jurgen (eds.) 1990. *The Ethics of World Religions and Human Rights*. London: SCM Press.

Luther, Martin. 1958. *Luther's Works*. Vol. 2. Jaroslav Pelikan (ed.). St. Louis, Mo: Concordia Publishing House.

Matheny, Gaverick. "Utilitarianism and Animals." *In Defense of Animals: The Second Wave.* Peter Singer (ed.). MA: Blackwell. 13-25.

Park, Miyun. "Opening Cages, Opening Eyes: An Investigation and Open Rescue at an Egg Factory Farm." *In Defense of Animals.* 174-180.

Plumwood, Val. 1999. "Ecological Ethics from Rights to Recognition Multiple Spheres of Justice for Humans, Animals and Nature." *Global Ethics and Environment.* Nicholas Low (ed.) London: Rutledge. 188-212.

Ryder, Richard. 1991. "Speciecism." *Animal Experimentation: The Moral Issues.* Robert M. Baird and Stuart E. Rosenbaum. New York: Prometheus Books. 35-42.

_____. 2006. "Speciesism in Laboratory." *In Defense of Animals.* 87-103.

Schweitzer, Albert. 1965. *Reverence for Life by Albert Schweitzer: An Anthology of Selected Writings.* Thomas Kiernan (ed.). New York: Philosophical Library.

_____. 1987. *The Philosophy of Civilization.* Trans. C. T. Campion. New York: Macmillan, 1950, reprint New York: Prometheus Books.

_____. 1988. *A Place for Revelation: Sermons on Reverence for Life.* Martin Strege and Lothar Stiehm (eds.) trans. David Larrimore. New York: Macmillan.

Twain, Mark. 1972. "The Lowest Animal." *Mark Twain on Man and Beast.* Janet Smith (ed.) Westport: Lawrence Hill. 152-161.

_____. 1983. "Lowbrow Animals: Mark Twain." *Animals in American Literature.* Mary Allen (ed.) Urbana: University of Illinois Press.

Religious Outlooks

Chapter 11

Chains and Circles: Hierarchy and the Status of Animals

☆ *Steve Russell and Sara M. Walsh*

The debate on the legal protection of animals has, before the modern upsurge in rights discourse, centered on interpreting the Genesis paradigm, answering whether human beings have a right of dominance or a duty of stewardship. Either way, human beings are awarded one of the higher links in Genesis on what came to be known as the "Great Chain of Being," a metaphor that could as easily justify cruelty to lesser humans as to lesser animals. Lord Devlin (1965), in his famous debate with H. L. A. Hart about criminal law and morality, proposed that animal abuse cannot be a crime without reference to morality. Hart (1963) replied that in accord with the harm principle, it should be illegal to inflict pain on a sentient being.

While analogies between the oppression of animals and slavery may be offensive, the arguments for both human and animal slavery are premised on the same assertion of rightful dominion. Without a hierarchy based on the human/animal dichotomy a fundamental basis of 'othering' and of both human and animal mistreatment does not exist. Eliminating the animal/human hierarchy as a gauge for rights is particularly pertinent in criminal justice today since real or imagined animal-like qualities and behaviors (often defined by 'wildness') are frequently used to justify general mistreatment[1] and the reduction/infringement of individual rights. An examination of the hierarchy as it informs both human and animal slavery can be useful to lay bare the justification mechanism that underlies human and animal oppression without offensively likening oppressed persons to non-human animals.

We contrast the Judeo Christian metaphor with the circle of life or sacred hoop recognized by most American Indian cultures and suggest that dominance and stewardship both beg the question of the right relationship of human beings with the rest of the world. Modern insights from feminist theory and the young science of ecology support the Indian view that Hart's reply to Devlin might better have been grounded in rational self-interest.

Fois gras, a delicacy in France and therefore much of the world, is produced by force-feeding ducks or geese until their livers swell to ten times the normal size. Like some methods of producing veal, this force-feeding, "galvage," often accompanied by immobilization, had been targeted for abolition by people who are opposed to meat-eating generally and because many meat-eaters do not support animal torture even when it allegedly improves the meat. In 2006, the city of Chicago joined the state of California and several European nations (other than France) in banning the sale of foie gras. The ban resulted in public outrage, litigation, and civil disobedience (Associated Press). While some opposition to the ban was couched in terms of showing that galvage is not really so uncomfortable for birds, the nub of the controversy was over limiting human exploitation of animals at all. If this ban stands, where will the debate focus next? The ban did not stand; it was repealed in 2008.

Resolved, the criminal law should proscribe torture of non-human animals. This statement is not greatly controversial, the contretemps over foie gras in Chicago notwithstanding. All states criminalize animal abuse (Parmenter 818) while excepting blood sports, just as all states criminalize child abuse while excepting corporal punishment (Pollard 2003). Given that U.S. animal protection laws are weak in the international context (Engelsman 2005) and even what laws exist are, like laws against wife-beating were for so many years (Russell 1992), honored in the breach, the subject seems to us still very much open for discussion. To draw lines between behaviors that should be the concern of the criminal law and behaviors that should not, it is useful to ask why animal abuse (defined in some manner[2]) is commonly the subject of enough social opprobrium to invoke criminal sanctions. Is animal abuse a moral issue or, more precisely, *only* a moral issue?

Lord Patrick Devlin, in his debate with H. L. A. Hart on the relationship between morality and law, took the position that "(r)eal crimes are sins with legal definitions"(27). Devlin was comfortable with crimes defined without regard to John Stuart Mill's (1869) "harm principle." Social opprobrium was for Devlin self-justifying. Animal abuse in public would answer to the harm principle by breaching the public peace, but Devlin would show that we criminalize animal abuse in private as well because our basis for doing so is not harm but morality:

> I suppose that there is hardly anyone nowadays who would not be disgusted by the thought of deliberate cruelty to animals. No one proposes to relegate that or any other form of sadism to the realm of private morality or to allow it to be practiced in public or in private. It would be possible no doubt to point out that until a comparatively short while ago nobody thought very much of cruelty to animals and also that pity and kindliness and the unwillingness to inflict pain are virtues more generally esteemed now than they have ever been in the past. But matters of this sort are not determined by rational argument. Every moral judgement,

> unless it claims a divine source, is simply a feeling that no right-minded man could behave in any other way without admitting he was doing wrong (Devlin 17).

Hart responded that the harm addressed by animal abuse law is not the discomfort of people who witness the abuse but rather the suffering of a sentient being (34). Of course, Hart could have argued that Devlin missed the point. There is a big difference between not being able to harm animals without "admitting…wrong" and justifying application of state power against somebody else who has harmed animals in private. There is at least as much difference as between butchering a cow and ordering a hamburger.

Neither is Hart's chosen argument self-evident. How do we know that non-human animals are "sentient beings" that "suffer?" Generally, we know that other beings are sentient and capable of suffering if they communicate that to us. That is true of humans as well as non-human animals. People who live on farms or keep companion animals probably find the communicative abilities of non-human animals to be self-evident.

If, as Hart said, the harm of animal abuse is in the suffering of a sentient being, and it is true that fewer people in modern times live close to animals, then can Devlin be correct that the law is becoming more protective of non-human animals? We concede that this counterintuitive proposition appears to be true. Our early treatment of animals was probably based less on experience than on archetypal stories about experience. In our time, archetypal stories travel electronically but are more ephemeral. Our early understandings were anything but ephemeral and magnitudes more powerful than they are today. They came from oral tradition, from myths that shaped first religion and then philosophy and even shaped the limits of scientific inquiry.

Myths are, said Joseph Campbell, "...the world's dreams. They are archetypal dreams and deal with human problems. I know when I come to one of these thresholds now. The myth tells me about it, how to respond to certain crises of disappointment or delight or failure or success. The myths tell me where I am." (Campbell 1988: 15). For much of the world, the relationship between human beings and non-human animals has been framed by the Genesis story, common to all three Abrahamic faiths: Judaism, Christianity, and Islam.

God, we are told in the Genesis, made fruits and herbs on the third day of creation.[3] Fish and fowl came on the fifth day.[4] All the other animals, including man, were created on the sixth day.[5] Man was created, said God, to "have dominion…over all the earth, and over every creeping thing that creepeth on the earth."[6] In case we didn't hear the first time or "creepeth" was too restrictive, God quickly repeats that man is to "have dominion…over every living thing that moveth upon the earth"[7].

Apparently, God creates man again on the seventh day and puts him "into the garden of Eden to dress it and to keep it."[8] Then God creates the beasts and birds again and has Adam name them.[9] Adam gives "names to all cattle,"[10] presumably not individually. After the woman commits her *faux pas* with the forbidden fruit, God charges Adam to rule over her, establishing patriarchy.[11]

We then learn that God intends man, as well as non-human animals, to wear the animals' skins.[12] The story that follows is an arguable endorsement of meat as superior to vegetables, because Cain makes a vegan offering[13] while Abel brings fat lambs.[14] God prefers Abel's offering.[15]

That God preferred Abel's offering is not simply an approval for a singular sacrifice of the dominated being by man. It signified further God's approval of the exploitation of an animal that went beyond need or survival. An animal being may be used, injured or killed for man's higher goals, for the pleasing of a higher authority, or for any "higher purpose." Here God appeared to approve and encourage shows of strength and domination, rather than care, towards the dominated. The sacrifice of the animal signified not only the essential relationship of man over animal but the essential relationship between man and all those subordinate to him. That is, it revealed a lack of duty for the dominated's care insomuch as this duty might conflict with man's reward for not exhibiting a duty of care. Can God be serious?

In our time, people still take this story seriously enough to argue over the meaning of "dominion." Some people believe, based on the Genesis story, that non-human animals have no interests that humans are morally bound to respect. Others believe that "dominion" implies a "duty of stewardship," which entails responsibility for the safety and well-being of non-human creatures. Given that at the time of the writing "dominion" over *humans* was the unlimited power of life and death, we tend to view the definition of "stewardship" as a modern conceit but without cross-examining the Deity there is no way to know His intent.

What is plain from the story is the idea of hierarchy, an idea that migrated from religion to science in the Great Chain of Being paradigm, attaining wide acceptance in the Eighteenth Century (Lovejoy 1936: Chapter VI). The hierarchy within the Great Chain was defined by distance from God, starting from nothingness to minerals to plants to non-human animals to man and onward to angels and other ethereal beings. Man was not considered to be at the top of the Great Chain but rather somewhere in the middle between God and nothingness.

Alexander Pope's *Essay on Man* described what religion had conceived and science of the time seemed to confirm:

> Vast chain of being! which from God began,
> Natures ethereal, human, angel, man,
> Beast, bird, fish, insect, what no eye can see,
> No glass can reach; from Infinite to thee,
> From thee to nothing (Pope 1891: Epistle I, VIII).

In Epistle II, Pope emphasized the length of the Great Chain and how its inherent inequality cuts both ways:

> Superior beings, when of late they saw
> A mortal man unfold all Nature's law,
> Admired such wisdom in an earthly shape
> And showed a Newton as we show an ape (*Ibid*)

This hierarchical ordering of the universe could of course be read as "subordination without subservience," (Lovejoy 1936: 207), a sort of Aristotelian world where everything gets its due and some things are inherently due more than others. More often, subordination carried plenty of entitlement to subservience. As man exists to serve God, the creatures lower in the Great Chain exist to serve man. Once some essential difference between things or creatures is established, hierarchy follows subject only to quibbles about the basis for value judgment. Fang and claw? Technology? These would serve as well as distance from God and might even be markers of distance from God since whatever hierarchy is must be right. The Great Chain exists in perpetual tension over the dominant vision of a vertical axis. The observable fact of interdependence intrudes.

> See dying vegetables life sustain,
> See life dissolving vegetate again:
> All forms that perish other forms supply
> (By turns we catch the vital breath, and die).
> Like bubbles on the sea of matter borne,
> They rise, they break, and to that sea return.
> Nothing is foreign: parts relate to whole;
> One all-extending, all-preserving soul
> Connects each being, greatest with the least;
> Made beast in aid of man, and man of beast;
> All served, all serving: nothing stands alone;
> The chain holds on, and where it ends, unknown (*Ibid*).

Pope in his poet persona comes close above to inventing the twentieth century discipline of ecology, but in the world of politics if not in the world of poetry the Great Chain remained on its vertical axis. Lovejoy's history of the Great Chain is rife with references to the "Hottentot," and the assignment of essentialized humans to separate links on the chain supported European race theory[16] and fit the vertical iteration of the Great Chain in a manner that justified slavery and imperialism. The Genesis story underpins the Great Chain of Being, "…the plan and structure of the world which, through the Middle Ages and down to the late eighteenth century, many philosophers, most men of science, and, indeed, most educated men, were to accept without question…" (Lovejoy 1936: 207).

The structuralist Claude Levi-Strauss opined that "(m)ythical stories are, or seem, arbitrary, meaningless, absurd, yet nevertheless they seem to reappear all over the world" (Levi-Strauss 1995: 11-12). Joseph Campbell has demonstrated such reappearances, literally, all over the world (1949) and argued that myths are metaphors by which we order our lives (1986). The Genesis myth, and the hierarchy that has followed in religion and science, is not the only way to envision humans and non-human animals and their relationship.[17] Cherokees believe, for example, that much disease among humans is revenge taken by animals for having been treated poorly

(Mooney 1992: 319-322).[18] A hunter who kills without need or without asking pardon of his prey takes his life in his hands. Animals are actors in Native stories, not just acted upon, and humans lack any innate superiority.

American Indian creation stories lack the ontological punch of Genesis, since they are more about coming to be *here* rather than coming to be. Humans in these stories are typically surrounded by other "peoples" (Deloria 2003: 88) who often play pivotal roles in how we came to be as we are and where we are (Hallowell 1992). How would Cherokees be without Water Beetle? (Mooney1992: 239). Navajos without Coyote? (Verney 2004: 135). Pope, speaking for Christianity, finds this dependence on mere animals rather than an omnipotent God cause for pity:

> Lo, the poor Indian! Whose untutored mind
> Sees God in clouds, or hears Him in the wind;
> His soul, proud science never taught to stray
> Far as the solar walk, or milky way;
> Yet simple Nature to his hope has given,
> Behind the cloud-topped hill, an humbler heaven;
> Some safer world in depth of woods embraced,
> Some happier island in the watery waste,
> Where slaves once more their native land behold,
> No fiends torment, no Christians thirst for gold.
> To be, contents his natural desire,
> He asks no angel's wing, no seraph's fire;
> But thinks, admitted to that equal sky,
> His faithful dog shall bear him company (Pope 1891: Epistle I).

For the "poor Indian," existence is not a vertical chain representing hierarchy but a circle representing interdependence (Lombardi and Lombardi 1993; Kidwell 2005, Noley and Tinker biblio?:48-51), or a "sacred hoop" (Neihardt 194). Deloria asserted that "(i)n all of North American Indian traditions, there is .... no sense of 'animal bondage' but rather relationships with the specific peoples of creation; hence, creation is ultimately good and humans are a part of it" (2003: 95). To be a part of it is not to be the reason for it or the master of it. American Indians could agree with Alice Walker when she said that "...animals of the world exist for their own reasons ("Preface" 1988: 10). They were not made for humans any more than black people were made for whites or women for men."

> Councils of animals that must give their blessing to human undertakings are alluded to in many ways in Indigenous mythologies. Humans rightfully and respectfully ask the permission of animals for doing things with them and to them. This is the meaning of the dances, rituals, and ceremonies Western scientists sometimes call "animism" (Cajete 2000:168).

One of the co-authors is American Indian, born and raised in Oklahoma, and has never met an Indian vegetarian, also known as "very poor hunter." Animals, human and non-human, will take what they need. Humans may dispute "need" non-human animals do not see it as a matter for dispute. They do not, as far as we know, argue over their essential nature as humans do.

The other co-author is a vegetarian, and we agree that the suffering of animals is a legitimate concern of the law and that non-human animals have interests that humans are bound to respect or face unpleasant consequences far beyond those we advocate in the law. We come to this common position from very different perspectives, one of us born to a hunting culture and one a feminist scholar. What we have in common is a philosophical rejection of the hierarchy represented by the Great Chain of Being.

Paula Gunn Allen (1986:59) described the Indian critique from a feminist perspective:

> The notion that nature is somewhere over there while humanity is over here or that a great hierarchical ladder of being exists on which ground and trees occupy a very low rung, animals a slightly higher one, and man (never woman)–especially "civilized" man–a very high one indeed is antithetical to tribal thought. The American Indian sees all creatures as relatives (and in tribal systems relationship is central), as offspring of the Great Mystery, as cocreators, as children of our mother, and as necessary parts of an ordered, balanced, and living whole (59).

Allen's position is that Indians have been stripped by European colonists of gynocentic traditions without which the sacred hoop can never be whole. For her, the status of women is not belittled by comparison to the status of animals or plants or rocks. When the sacred hoop is broken, it is broken by whatever is left out. Alice Walker's statement that animals were no more made for humans than women for men is right but it is also wrong. All is made for all. Walker's point, so eloquently made, seems self evident, but comparisons on the mutual plight and linked destiny of those who are dominated have been repeatedly surrounded by controversy.

Walker's remarks on the similar plight of lower status humans and animals were enough to get her "Banned"[19] and campaigns such as PETA's "Holocaust on your plate"[20] are considered by some people to be highly offensive. Undoubtedly, this offense is in no small part due to the perceived equation of oppressed persons (*e.g.*, Holocaust victims) with barnyard animals or in Walker's case the conclusion that she was "anti-meat." However, "Holocaust on your Plate" and similar campaigns do articulate a legitimate criticism of the hierarchy on which both the factory meat industry and genocide are premised, a critique that more gentle approaches cannot well offer. The Great Chain hierarchy is largely unthreatened by most animal welfare arguments, especially those based on the morality of pity or kindness since these arguments treat animal rights as grand benevolence from the master species rather than duty owed to fellow beings.

If the moral argument against animal abuse fails to dislodge the Great Chain's logic from our beliefs entirely, it does at least displace the time honored notion of the food chain as a proxy for the Great Chain. Meat eating is, and has been, a way that humans understand themselves as at the top of the mundane part of the Great Chain and food chain hierarchy. The moral argument, while it may problematize carnivores as superior does not particularly alter the notion of humans at the top of a hierarchy.

Peter Singer's critique of speciesism (Animal Liberation) is no less based on hierarchy than speciesism itself. Non-human animals will take the life of human animals if they perceive a need to do so based on hunger or fear, as far as we know. Humans, of course, kill without need–but they do that to each other as well as to non-humans. Leaving this moral imbalance aside, where does the duty of humans arise to treat non-humans better than non-humans would treat humans in a situation of perceived need? There is a way to avoid reciprocity of duties, and it might be called speciesist *noblesse oblige*: We have a duty because we know better than dumb animals. This is a hierarchical praxis.

Carol Adams's advice to vegetarians is also a telling evidence of hierarchic logic in vegetarian philosophy. Adams advises vegetarians to "Speak to meat eaters the way you would speak to a wild animal: softly and without any sudden movements (2001:92)". The equation of meat eating human to wild animal is clearly a demotion of said human– marking both the savageness of the carnivore and the higher moral ground of the vegetarian.[21] Vegetarianism and animal liberation philosophy alone do not reject the hierarchy upon which animal ab/use is founded when they portray the carnivore as the caveman with his broiled leg of stegosaurus in one hand and his woman by the hair[22] in the other. The idea of animal use as primitive and lesser behavior is an idea that marks most animals themselves (who are largely omnivores) as less than human–when vegetarian morality is the yardstick of advancement. The animal rights argument is particularly problematic when the duty of care arises from the moral superiority implicit in the duty of stewardship.

We return here to the story of Genesis and the sacrifice of the lower animal for the blessing of God upon man. In modern culture it appears that we observe both a right of dominance and a duty of stewardship, the extent of either depending on the animal's location in the hierarchical order[23] and balanced against human utility. Human pleasure for its own sake is devalued as the status of the harmed being increases. We seem to owe a duty of care to animals that are traditionally pets or humanlike, but not to fish, fowl and that which creepeth.

Utility is a justification for much animal cruelty when the activity harming the animal is at least hypothetically in reference to a human need. The higher an animal is on the hierarchal order the greater the utility to humans must be in order to justify harm to the animal. The justification for killing a dog must be significantly higher than for killing of a fish, which may be done for minor human pleasure or even relaxation. It is okay to kill a dog that is dangerous to humans (especially to human children) but not okay to kill a dog (especially another person's dog) because one feels it is a relaxing sport. However, at the same time, a dog deprived of pet status, and thus of utility (*e.g,.* a pound dog) is also deprived of the protection of life. Although society still would draw the line at cruelty, these animals are seen as largely disposable

or (if not seen) are at least in fact disposed of in great numbers in most cites by methods humane and not.

This understanding of utility explains why real hunting is condoned but internet hunting[24] is highly controversial and has been banned in most states–including Texas, where the "sport" originated.[25] We might also note that product and medical testing on animals follows this same rule. Cruelty to primates for cosmetics development is taboo, but not for (human) life saving medical technology. Lower rats and rabbits however find themselves subjected to torturous experiments for the benefit of minor human convenience or curiosity.

Because human life is considered sacred and thus useful in its own right, the utility to humankind must be demonstrably large in order to justify harm to another human. Alternatively, however, the human to be harmed may cease to be human. In other words, s/he may be demoted to a lower animal life form to justify harm. Patterns of cruelty to humans suggest both techniques are utilized and often simultaneously. Exacting the death penalty on a human is framed in terms of public protection but also as the justified extermination of an animal, a creature farther from God than humans on the Great Chain.

Historically, the boundaries of animal rights, and thus also the rights of those humans who we deem animal-like, have been tied to our conception of the line between human and other animal life - - the particular qualities which define and justify hu/

- "If a family dog so much as bites a person that dog is put down - this animal is worse than a ***dog*** so the death penalty is too good for him."[1]
- "Just terminate the ***cockroach***."[1]
- "Mark my word, Eric, George and Marcus have all offered rewards for whoever kills this ***dog***. He is going to be done in faster than the state could do it."[2]
- "I am sure this ***animal*** will have to live the rest of his life looking over his shoulder. I'm sure he will live in fear the rest of his life" [2]
- "Maybe God felt this man needed to suffer some before he dies. Maybe God felt that this ***animal*** needed to be raped and tortured on earth to give him a preview of what is to come after death." [2]
- "On the other hand, at least there's one less ***animal*** out in the free world. There soon will be several less." [2]
- "Powell deserves the compassion of a ***cockroach***"[3]
- "Haven't they executed the sewer ***rat*** that killed Polly Clas?" [4]
- "Drown, but first torture that animal. He should have no right whatsoever to appeal. The day that the poor boy can plead for his life is when this ***rat*** can be considered to appeal for his."[5]
- "Thank God this evil low life monster of a ***dog*** is dead. He got what he deserved."[6]
- "Our newspaper printed all the guys in our area, on death row, each and everyone of them, are the same as a ***cockroach***, or ***rat*** you would step on a cockroach or put out rat poison, if you had rats wouldn't you?"[7]

1. "Richard Ramirez Execution Date?" Uaddit.com discussion and comments section. Web. 05 September 2010.< http://uaddit.com/discussions/showthread.php?t=5627>
2. "Gary Christian: 'He deserves to die'" Knoxnews.com. Discussion and comments section. Web. 05 September 2010. <http://www.knoxnews.com/news/2009/aug/27/gary-christian-he-deserves-to-die/>.
3. "October 2002 Executions." prodeathpenalty.com. Discussion and comments section. Web. 05 September 2010. <http://www.prodeathpenalty.com/pending/02/oct02.htm>.
4. "Montebello woman receives death penalty for poisoning husband." FreeRepublic.com. Discussion and comments section. Web. 05 September 2010. http://www.freerepublic.com/focus/f-news/1056702/posts
5. "Death penalty stands for Mesa tot's killer". azfamily.com. Discussion and comments section. Web. 05 September 2010. <http://www.azfamily.com/news/local/Death-penalty-stands-for-Mesa-tots-killer-99944189.html>.
6. Amnesty International USA blog. Comment in response to the execution of Jermaine Herron. 12 July 2006. Web. 20 August 2006. <http://blogs.amnestyusa.org/weblogFeedback/1kyksfpeyk9ki>.
7. Political Hotwire Chat. Formerly posted chat board. Web. 12 July 2006. <http://www.politicalhotwire.com/the-failed-experiment-death-penalty-vt962.html>.

man's superiority. Animals have been defined as different from, and thus less than, humans on the basis of many supposed qualities. While we may largely accept our Great Chain birthright as the superior species, the need to define this superiority on the basis of scientific fact has persisted.

However, as history bears out, any observable line between the human-animal and the animal-animal[26] is arguably a matter of our ability to perceive apparent human qualities in animals, such as language (Budiansky 1998), the use of tools (Tallis 2003), rationality (Kant 2002) or abstract thought (Rieser 1943). Not only have these distinctions been largely disproved by science itself but, as many scholars have noted (Donovan 1990; Bentham 2005), arguments for the treatment of animals based on an objective division between animal and human life invariably leave questions about disabled, comatose, or very young humans.

> A full-grown horse or dog is beyond comparison a more rational, as well as more conversable animal, than an infant of a day, or a week, or even a month old (Bentham 2005: 311).

These lines between human and animal have proved not only inaccurate but rather impractical even for humans convinced of their own superiority. Those suggesting that humans are superior to cows and that this is evinced in the human employment of language and the bovine lack of language are certainly not suggesting that we eat infants or disabled humans. What they *are* suggesting is that human superiority is a rational basis for the domination of animals and quantifiable differences must exist because the human is superior. The underlying argument is circular.[27]

That human superiority rests on a problematically vague distinction between human and animal stands as testimony to the potential for rank slippage from human to animal. History reminds us of how easy slippage is even with a "scientific" line. Donovan (1990: 353) claims that all humans have not historically enjoyed the status of "sufficiently rational as to be considered 'persons'" by the framers of the U.S. Declaration of Independence and the Constitution. We are unlikely to justify the eating of human children due to their lack of reason,[28] yet we did deny women (among others) the right to vote on this basis. The lines between human and animal life which distinguish rights have also served to delineate human from human. If anything, these artificial lines serve to facilitate harms to humans rather than prevent them. The plain purpose of the human/animal dichotomy is to distinguish valued beings from unvalued beings. Singer would call concern for humans beings treated as animals "speciesism." We agree, when the concern is *only* for humans. We no more want humans harmed on the basis of their animalness than we want animals harmed on the basis of their animalness.

Animal advocates suggest that our treatment of animals should depend on *shared* qualities between human and animal, in particular that the capacity to suffer or feel might more reasonably determine those who are/are not to be treated as ends in themselves–an argument made most famously by Bentham and later by Singer and adopted by H. L. A. Hart in his dispute with Lord Patrick Devlin over the role of morality in law (Hart 1963: 34). While undoubtedly a more inclusive definition and an improvement over the less observable "rationality" criterion–particularly in terms of protecting human rights–we respectfully suggest that feeling and suffering are as subject to the limitations of human perception and definition as are rationality or complex awareness. For example, both feeling and suffering as criteria still leave open questions about those humans who we are unable or unwilling to understand as suffering or capable of suffering either because of their physical state or their social status. Furthermore, harm is neither necessarily related to observable suffering, nor have we as a species historically shown particular consideration for the suffering of other humans, thus making it unlikely for the purposes of allotting rights that suffering amounts (or can amount) to a meaningful line or offer a consequential alternative to the present hierarchal notion of difference more than any other quantitative measure of humanness.

Though Singer's Bentham-istic "suffering" criterion is problematic as a marker of intolerable interference with a sentient being it does move closer than most animal advocacy arguments to disassembling the logic of the Great Chain by arguing that similar interests must be counted equally regardless of the species of being involved. In fact, it is potentially Singer's basis for animal rights that most damages the justification for harms based on race, sex, species or other hierarchical categorization.[29] However what remains problematic is that the argument is still based on animal's perceived likeness to humans, however liberal the understanding of likeness may be. The sacred hoop, cultural feminism and ecofeminism offer alternative models to this human centered valuation of life.

Cultural feminists look to alter the "voice" of law towards a feminine "relational" voice of caring and community over a masculine rights-based ethic of justice. Carol Gilligan is critical of the masculine rights approach, suggesting instead a feminine ethic of responsibility. For Gilligan, the rights approach is based on the same formal logic that relies on separation of beings and hierarchical quantitative evaluations (1982:73). Gilligan has a point in that attempts to quantify the trait of sentience lends to hierarchical ranking. Similarity, Donovon (1990:370) argues for a cultural feminist model based on a holistic "disordered" (where hierarchical domination means "order") conception of life which finds human consciousness to be continuous with all forms of life and respects rather than reorders the environment to fit a "master paradigm." In this model, the distinctions between species may be recognized and differences are respected but no hierarchical ranking[30] occurs.

However, the cultural feminist model cannot, and is not intended to, work in a rights based system of law, where law is arguably male because of law's focus on rationality (Pruitt:1994) and justice rather than care (Gilligan:1982). Thus, a system of animal rights emerging from "women's relational culture of caring and attentive love" (Donovan 1990: 375) is oxymoronic. In other words, the *regulation of* "an emotional and spiritual conversation with non-human life-forms" (Donovan 375) just cannot exist by definition. Regardless, cultural feminism presents an important critique of the hierarchal rights based model–namely, that regardless of one's goals, the quantification of sentience is invariably subject to the whims of those doing the calculations–namely those already at the top of the chain.

We have presented two worldviews that come from very different social milieus but have in common rejection of the hierarchy enshrined in our cultural imagination by the Genesis story. We understand that it is preposterous to suppose that only American Indians or feminists can turn away from the hierarchy paradigm, but until we turn away from that paradigm humanity in general will be stuck with the instrumental view of nature. Animals, plants, and the land itself will be of concern only to the extent that human interests are obvious.

Hart's characterization of suffering by sentient beings as harm is to be applauded as a moral stance, but his objective in responding to Devlin was to take animal abuse out of the merely moral. When we reject the hierarchy, we postulate a self-interest that is the fundamental harm in animal abuse, as well as, for example, herbicide pollution, clear-cutting old growth forests, or excessive carbon dioxide emissions. What we do to the rest of the planet, we ultimately do to ourselves.

Notes

1 *e.g.* Criminals are animals. Criminals behave like animals. Criminals live like animals. Criminals are no better than animals. Criminals deserve to be treated like the animals they are.

2 What constitutes animal "abuse" is a thorny issue beyond the scope of this article, but criminologists are beginning to address the human animal-animal animal relationship (Beirne) and the categorization of abusive actions (Cazaux 1999).

3 Genesis 1: 11-13. All citations to the King James Version. This is a conscious choice by the authors because while this translation is dated in scholarly terms, it remains the coin of the realm among people inclined to literalist interpretations. For similar reasons, we ignore Christian hermeneutical theories, particularly theories of redactive or source criticism, that would "explain" the otherwise inexplicable absurdities of the King James Genesis account. The issue for us is not what Genesis is "really" supposed to say, but how it informs an archetypal understanding of how creation is ordered.

4 Genesis 1: 20-23.

5 Genesis 1: 24-31.

6 Genesis 1: 1: 26.

7 Genesis 1: 28.

8 Genesis 2: 15. This verse perhaps reveals a Divine origin for the lawnmower.

9 Genesis 2: 19. Apparently, the fish were only created once.

10 Genesis 2: 22.

11 Genesis 3: 16. Eve is not named until later. Genesis 3: 20.

12 Genesis 3: 21.

13 Genesis 4: 3

14 Genesis 4: 4.

15 Genesis 4: 5. God later becomes quite specific about that matter of meat-eating, setting out exactly what creatures were to be treated as food. Leviticus, Chapter 11. Allah, according to the Qur'an the same Divine Person as the God of Genesis, agrees that men are to be carnivorous. Surah 16: 5; Surah 23: 21.

16 Arthur de Gobineau famously developed essential difference as justifying social inequality. The Inequality of Human Races was first published in Europe in 1854 and in the United States in 1915. This stratification of human beings is a mainstay of so-called "social Darwinism" and respectable scientific racism that died out between the World Wars for lack of data supporting essential racial difference (Barkan 1992).

17 *E.g.*, Campbell, 1949, chapter 6.

18 This metaphor might have found some confirmation in modern science that suggests ill-treated animals provide inferior milk, eggs, and meat. Another Cherokee belief also has scientific support: "The river sends disease to those who insult it by such actions as throwing rubbish into it, urinating into it, etc." (Mooney 2005: 23).

19 See Alice Walker Banned (Walker 1996).

20 CNN.com/U.S. "Group blasts PETA 'Holocaust' project." http://www.cnn.com/2003/US/Northeast/02/28/peta.holocaust/ (Last retrieved September 4, 2010), February 28, 2003.

"It's shocking, it's startling, it's very hard to look at," Lange said of the exhibit. "We're attacking the mind-set" that condones the slaughter of animals.

"The very same mind-set that made the Holocaust possible – that we can do anything we want to those we decide are 'different or inferior' – is what allows us to commit atrocities against animals every single day," PETA representative Mark Prescott wrote in a statement, which added that members of Prescott's family were murdered by Nazis.

21 And of course, Adams is humorously pointing out the savagery with which carnivores attack vegetarians and vegetarianism, a point on which few vegetarians would disagree.

22 For a discussion of meat eating and masculinity see Carol Adams, The Sexual Politics of Meat: A Feminist-Vegetarian Critical Theory.

23 The modern hierarchy is less concerned with distance from God and more with anthropomorphic feelings for esteemed animals: great apes (Cavalieri , Singer), dolphins (Lilly), and companion animals. We share these feelings but question the value of redrawing hierarchical lines among animals.

24 Internet hunting was formerly available at http://www.live-shot.com/, but the site went dormant while this article was being written.

25 There has been a lot of legislative activity recently against internet hunting which, like so-called canned hunting (killing of tame exotic animals as trophies), causes unlikely allies in moral outrage: animal protectionists and hunters. See Figure 1.

Federal legislation against internet hunting is regularly proposed, but it is hard to picture the allocation of resources it would take to prosecute an individual in one country who sells the right to click a mouse in a second country that causes the death of an animal in a third country.

26 Or, in the language (Josephine Donovan), the line which objectively turns animals into "its" (1990: 353).

27 Humans are superior beings. Superior beings can be distinguished from lower ones. This distinction exists because humans are superior beings.

28 At least, we assume Jonathan Swift (*A Modest Proposal*) was not serious. And even if he was, we have presumably progressed enough to be shocked upon finding that "Soylent Green is people!" (Soylent Green).

29 This would of course apply to any new categories in the future, should we, say, discover that people with large eyebrows are more reasonable than those with small ones.

30 Donovan (1990: 375) points out that the "life boat hypotheticals" presented as arguments for hierarchical models are unrealistic and therefore unnecessary. While we know of no cases involving a man and a dog in a lifeboat, there have been many cases of humans in similar situations (Read; Simpson; Stewart). In those situations, the humans do exactly what non-human animals would do. They construct temporary and perhaps idiosyncratic hierarchies that tell us nothing about how society ought to be ordered except when we judge, or more often do not judge, the survivors.

References

Adams, Carol J. 2001. *Living Among Meat Eaters: The Vegetarian's Survival Handbook*. New York: Continuum.

______. 2000. *The Sexual Politics of Meat: A Feminist-Vegetarian Critical Theory*. Continuum.

Allen, Paula Gunn. 1986. *The Sacred Hoop: Recovering the Feminine in American Indian Traditions*. Boston: Beacon Press.

Associated Press. "Angry Chefs Sue over Chicago Fois Gras Ban." Last accessed September 4, 2010. Web. 08 Feb. 2006. <http://www.msnbc.msn.com/id/14472971/>

Barkan, Elazar. 1992. *The Retreat of Scientific Racism: Changing Concepts of Race in Britain and the United States between the World Wars*. New York: Cambridge University Press.

Bentham, Jeremy. 2005. *An Introduction to the Principles of Morals and Legislation*. Whitefish, MT: Kessinger Publishing.

Beirne, Piers. "For a Nonspeciesist Criminology: Animal Abuse as an Object of Study." *Criminology* 37 (1999): 117-147.

Budiansky, Stephen. 1998. *If a Lion Could Talk: Animal Intelligence and the Evolution of Consciousness*. Free Press.

Cajete, Gregory. 2000. *Native Science: Natural Laws of Interdependence*. Santa Fe: Clear Light Publishers.

Campbell, Joseph. 1949. *The Hero with a Thousand Faces*. Princeton: Princeton University Press.

_____. 1986. *The Inner Reaches of Outer Space: Metaphor as Myth and as Religion*. New York: Harper & Row.

_____. 1988. *The Power of Myth*. New York: Doubleday.

Cavalieri, Paola and Peter Singer (eds.) 1993. *The Great Ape Project: Equality Beyond Humanity*. New York: St. Martin's Press.

Cazaux, Geertrui. "Beauty and the Beast: Animal Abuse from a Non-Speciesist Criminological Perspective." *Crime, Law and Social Change* 31 (1999): 105-26.

de Gobineau, Arthur. 1967. *The Inequality of Human Races*. Trans. Collins, Adrian. New York: Howard Fertig.

Deloria, Vine Jr. 2003. *God Is Red: A Native View of Religion*. 3rd ed. Golden, CO: Fulcrum Publishing.

Devlin, Patrick. 1965. *The Enforcement of Morals*. London: Oxford: University Press.

Donovan, Josephine. "Animal Rights and Feminist Theory." *Signs* 15 (1990): 350-75.

Engelsman, Stephanie J. "'World Leader'-At What Price? A Look at Lagging American Animal Protection Laws." *Pace Environmental Law Review* 22 (2005): 329-69.

Gilligan, Carol. 1982. *In a Different Voice: Psychological Theory and Women's Development*. Cambridge: Harvard University Press.

Hallowell, A. Irving. 1992. "Ojibwa Ontology, Behavior, and World View." *Teachings from the American Earth: Indian Religion and Philosophy*. Eds. Tedlock, Dennis and Barbara Tedlock. New York: Liveright Publishing. 141-78.

Hart, H. L. A. 1963. *Law, Liberty, and Morality*. Stanford, CA: Stanford University Press.

Kant, Immanuel. 2002. *Critique of Practical Reason*. Indianapolis: Hackett Publishing Company.

Kidwell, Clara Sue, Homer Noley, and George E. Tinker. 2005. *A Native American Theology*. Maryknoll, NY: Orbis Books.

Levi-Strauss, Claude. 1995. *Myth and Meaning: Cracking the Code of Culture*. New York: Schocken Books.

Lilly, John C. 1961. *Man and Dolphin*. Garden City, NY: Doubleday.

Lombardi, Frances G., and Gerald Scott Lombardi. 1993. *Circle without End*. Happy Camp, CA: Naturegraph Publishers.

Lovejoy, Arthur O. 1936. *The Great Chain of Being*. Cambridge: Harvard University Press.

Mill, John Stuart. *On Liberty*. 1869. Retrieved August 4, 2010 <http://www.bartleby.com/130/>.

Mooney, James. 1992. *History, Myths, and Sacred Formulas of the Cherokees*. Asheville, NC: Bright Mountain Books.

_____. 2005. *The Swimmer Manuscript: Cherokee Sacred Formulas and Medical Prescriptions*. Oklahoma City: Noksi Press.

Neihardt, John G. 1988. *Black Elk Speaks: Being the Life Story of a Holy Man of the Oglala Sioux*. Lincoln: University of Nebraska Press.

Parmenter, Mark J. "Does Iowa's Anti-Cruelty to Animals Statute Have Enough Bite?" *Drake Law Review* 51 (2003): 817-38.

Pollard, Deana. "Banning Child Corporal Punishment." *Tulane Law Review* 77 (2003): 575-657.

Pope, Alexander. *An Essay on Man*. 1891. Web. Retrieved August 4, 2010 <http://www.gutenberg.org/ebooks/2428 >.

Pruitt, Lisa R. "A Survey of Feminist Jurisprudence." *University of Arkansas at Little Rock Law Journal* Spring 1994 (1994).

Read, Piers Paul. 2005. *Alive!* New York: Harper Collins.

Rieser, Max. "On Will or the Levels of Action." *The Journal of Philosophy* 40 (1943): 206-13. Print.

Russell, Steve. "Futility of Eloquence: Selected Texas Family Violence Legislation 1979-1991." *South Texas Law Review* 33 (1992): 353-75. Print.

Simpson, A. W. B. 1984. *Cannibalism and the Common Law*. Chicago: University of Chicago Press.

Singer, Peter. 1990. *Animal Liberation (Rev. Ed.)*. Random House, New York.

Soylent Green. 1973. Seltzer, Walter and Russel Thatcher. Film.

Stewart, George R. 1960. *Ordeal by Hunger: The Story of the Donner Party*. Boston: Houghton Mifflin.

Swift, Jonathan. *A Modest Proposal*. 1729. Web. Retrieved August 4, 2010 <http://www.gutenberg.org/etext/1080 >.

Tallis, Raymond. "Escape from Eden". 2003. www.newhumanist.org.uk. Retrieved August 4, 2010. <http://newhumanist.org.uk/661/escape-from-eden-raymond-tallis-novemberdecember-2003>.

Verney, Marilyn Notah. 2004. "On Authenticity." *American Indian Thought*. Ed. Waters, Anne. Vol. 133-139. Oxford, UK: Blackwell Publishing.

Walker, Alice. 1988. "Preface." *The Dreaded Comparison: Human and Animal Slavery*. Ed. Spiegel, M. Viborg, Denmark: Heretic Books.

_____. 1996. *Alice Walker Banned* San Francisco: Aunt Lute Books.

Chapter 12

Karl Barth and "Love for All Creatures" as a Quintessential Aspect of Christian Spirituality

☆ *William Greenway*

Love for All Creatures

In 1967 Lynn White Jr. published a famous essay in the prestigious journal *Science* in which he argued that Christianity bore a "huge burden of guilt" for the global ecological crisis (1206). When I, a Christian, relate White's accusation to Christian audiences, I stress that the first thing Christians should do is confess, for there is no doubt that White's accusation is wholly justified.

For the past few centuries, predominant Western Christian interpretation, both scholarly and popular, elevated humans above all else. Everything else was relegated to the status of "natural resources." In particular, the *Genesis* creation narrative in which humans are given dominion over all creation was taken to mean that humans should appropriate for themselves the earth and everything on it (*Genesis* 1:26-28).

Of course, people have not exploited the earth so recklessly over the past few centuries because of Christian teachings. A far larger role has been played by raw desire for power, pleasure and/or profit. But not only did mainstream Christianity fail to speak against exploitation and urge us toward reverence for, or at least stewardship of all creation, but the predominant Christian interpretation positively sanctioned the activities of those whose aim was exploitation.

Much has changed in the four decades since White issued his indictment. As Christians have awakened to the myriad ecological crises pervading most every landscape on earth, sensitivity to the biblical and traditional call to be good stewards of creation has risen among Christian conservatives and progressives alike. Indeed, it may not be too much to claim that the sense that humans are called by God to be good stewards of creation may finally be emerging as the predominant Christian understanding at both scholarly and popular levels.

But the same cannot yet be said with regard to "animals." Already in 1976, only a year after the publication of Peter Singer's influential *Animal Liberation*, Christian theologian Andrew Linzey dedicated an entire book, *Animal Rights: A Christian Assessment*, to advancing a Christian argument for animal rights. Ever since then a small but significant and steadily growing number of Christians at both the scholarly and popular levels have been working to include non-human creatures as subjects of ethical and spiritual concern. At present, however, the contention that, "love for all creatures" names a quintessential aspect of Christian spirituality would strike most Western Christians (and likely the majority of Christians across the globe, not to mention those who observe Christianity from the perspective of other faiths) as absurd.

Though in this brief essay I cannot establish that "love for all creatures" names a quintessential aspect of Christian spirituality, I can establish ample warrant for Christians and those of other faiths to consider with utmost seriousness the possible truth of this claim about Christian spirituality. Such will be the prime goal of this essay, and I will attempt to do this by making clear that this claim about Christian spirituality is affirmed by one of the most prominent Christian theologians of the 20th century, Karl Barth.

I appeal to Barth not because I am a Barthian, and certainly not because Barth is famed for his sensitivity to ecological or animal concerns. Indeed, Barth is regularly and with considerable justification accused of thinking only in terms of God and humanity. For instance, in *The Travail of Nature*, an influential study of theology and ecology, Paul Santmire accuses Barth of "theanthropocentrism," that is, of constructing a theology that forgets the world and thinks only in terms of God and humans (147-155). But precisely that is why I appeal to Barth. For if even Karl Barth, when confronted with the question of animals, clearly supports the contention that "love for all creatures" names a quintessential aspect of Christian spirituality, then there is ample warrant for taking this spiritual contention seriously, and the widespread and easy confidence that currently enables the marginalizing of those Christians who argue for "animal rights" or "love for all creatures" should be seriously shaken.

As we will see, Barth's work and perspective on animals was inspired by the life of Albert Schweitzer. In significant ways, I will suggest, Barth and Schweitzer supplement one another. Moreover, I will suggest that ambiguities in their thought quite naturally bring the work of Emmanuel Levinas into constructive play, and that it is helpful to read Barth, Schweitzer and Levinas together. Unfortunately, in the space allotted I can only sketch out the essentials of this claim, and there is no space to explain or defend my interpretation of Levinas' esoteric vocabulary. We attempt first to sketch out this claim.

Barth, Schweitzer, Levinas

It was the work of Albert Schweitzer, and in particular his doctrine of "reverence for life" that confronted Barth with the question of animals. In an uncharacteristic gesture, Barth is so moved by the spirit of Schweitzer's work that he adopts "reverence for life" as the title of a long section of his *Church Dogmatics* (a fact somewhat obscured in the English translation, which translates *Ehrfurcht*, the word used by both Schweitzer and Barth, as "respect" instead of "reverence").

As we will see, Barth rejects Schweitzer's methodology, but this rejection only makes Barth's powerful affirmation of Schweitzer's spirituality all the more stunning. Moreover, as we will see, Schweitzer would have been more consistent with the essence of his own spirituality if he had abandoned his paradigmatically modern methodological quest.

Barth's profoundly sympathetic encounter with Schweitzer also created profound and unprecedented openings at junctures where Barth's own methodology is open to significant critique. For instance, Barth is often critiqued for being a revelatory positivist (*e.g.*, for making dogmatic appeals to Christian doctrines or scripture), and for being religiously exclusive. Barth certainly provides ample warrant for these critiques. With regard to reverence for life, for instance, Barth claims first that, "Life does not itself create this respect. The command of God creates respect for it" (339). And he contends, second, that, "We may confidently say that the birth of Jesus Christ as such is the revelation of the command as that of respect for life. This reveals the eternal election and love of God" (339). Barth even goes on to claim, third, that with respect to reverence for life, "the Christian Church and Christian theology have an incomparable weight to throw into the scales. They and they alone know exactly why and in what sense respect for life is demanded from us" (340).

While the first claim is primarily misleading, the second is sheer, nonsensical assertion and the third is arrogant and offensive. Unfortunately, at this juncture in Barth's theological journey all three of these claims are typical. Fortunately, in later works Barth will significantly qualify the second and third claims. In any case, given these problematic claims, it is especially notable how, in the face of Schweitzer's testimony to reverence for life, Barth is also moved to make other claims that sound very different.

For instance, Barth says that, "Respect is man's astonishment, humility and awe at a fact in which he meets something superior – majesty, dignity, holiness, a mystery which compels him to withdraw and keep his distance, to handle it modestly, circumspectly and carefully" (339). Note that there is no question in this quote of reverence for life being mediated through knowledge of Jesus Christ or of any specifically Christian understanding. To the contrary, the astonishment, humility, and awe describe response to an immediate encounter.

Significantly, it is in this immediate context that Barth asserts, "life itself does not create this respect," for it is the "command of God" (339). That is, the "command of God" names the experience of being irresistible grasped by reverence for life. "God" names the surpassing, sacred, awe-inspiring character of the reverence. And the "command" is not external, is not found in a book or in some orthodox teaching, but names the irresistible, non-intentional character of the call of the lived encounter. This is classic Barth, for Barth stresses throughout his corpus that the command of God is always encountered as a living and present event, it is encounter with the living Word of God *in* the world, *in* history.

The "life" Barth is rejecting as the source of reverence, then, is "life" as it is known in the world, life as it is known biologically or theorized philosophically. With "command of God" Barth means to be gesturing to a source beyond history, beyond philosophy and science, beyond any understanding, reasoning or willing. Indeed, Barth stresses that the command of God is, "an incomprehensible and in relation to ourselves intangible fact ... Those who do not know *respicere* in face of it, those who are not startled and do not feel insignificant and incompetent in its presence, those who think they can understand and master and control it, do not know what obedience is" (340).

Once we have encountered the command of God that seizes us in the form of reverence for life, Barth says, we must not, "at any price ... relinquish or even to lose sight of this perception. We must be awake ... and always be wakeful...." (341). Moreover, Barth argues, if one is truly awakened to reverence for life, truly seized by the "command of God," then action necessarily follows, for "real respect for life" is "more than passive speculation" (341).

I think it is no accident that Barth stumbles here into using classic Levinasian categories and terms (*e.g.*, "awake"). On this philosophical frontier Barth lacks the rigor of Levinas, who speaks of the awakening brought about by the in-breaking of the Other within the sphere of the Same as a non-intentional consciousness finds itself taken hostage by the Face of some other. But while Barth's attack upon metaphysics is less rigorous than that of Levinas, his concern is quintessentially Levinasian, and thus I would suggest that we rightly read Barth in a Levinasian light.

Let me also note, finally, how explicitly Barth, when inspired by Schweitzer's testimony to reverence for life, and so writing in the grip of a quasi-Levinasian focus upon the "command of God," makes statements that push back against his more arrogant, offensive and exclusivist claims. For instance, later in this section Barth reflects upon the spirituality of a "good horseman," namely, one who "is so completely one with his horse that he always knows exactly to take out of it no more and no less than what it can not only give but is willing and glad to do so" (352). Somewhat stunningly, Barth without any mention of Christianity asserts that a person who enjoys such a relationship with an animal, "cannot possibly be an ungodly person" (ibid.).

Barth's Levinasian instincts are also on display when he expresses a concern that Schweitzer betrays his own reverence for life when he attempts to set it up as the "fundamental principle of ethics" (324). Schweitzer himself has a vague awareness of this problem. It may seem, he acknowledges, that "Reverence for Life" sounds "too general and too lifeless to provide the content of a living ethic," but, he continues, anyone who "comes under the influence of the ethic of Reverence for Life will very soon detect ... what fire glows in the lifeless expression" (1953: 180).

While Schweitzer continually specifies that real "thought" refers not to any thinking, but to the ethical awareness named by reverence for life, he continues to name reverence for life as a "logical consequence" of thought. For instance, Schweitzer contends that his doctrine of "Reverence for Life" is "the ethic of Jesus, now recognized as a logical consequence of thought" (ibid.). Moreover, Schweitzer never gives up on what turns into a decades-long and ultimately unsuccessful quest to ground "reverence for life" and all ethics in a first principal. From a Levinasian perspective, Schweiter's failure on this score is saving, for he has confusedly taken up the misbegotten and paradigmatically modern epistemological quest to reduce and negate the in-breaking of the Face of the Other by grasping it within reason and drowning it within the Same.

Schweitzer's confusion on this score is overt. For instance, he specifies that he wants to make an "assertion" the prime and continual object of our consciousness. Schweitzer explicitly presents this assertion, which he now puts forward as "the most immediate and given fact of [one's] own consciousness," as his own substitute for Descartes' *cogito* (*Ibid*., 125). Rejecting the *cogito*, Schwietzer maintains that, "the most immediate fact of [a human's] consciousness is the assertion: 'I am life which wills to live, in the midst of life which wills to live" (*Ibid*.).

The first and really significant problem here, one obvious from a Levinasian perspective, is that Schweitzer elides the Faces of others from view when he urges us to focus primarily and continually not upon other creatures but upon an assertion. On this score Barth's Levinasian instinct to keep us focused upon the "command of God," the living encounter which irresistibly awakens astonishment, humility, reverence and awe, is far more consistent with the spirit of Schweitzer's reverence for life than is Schweitzer's own attempt to make an assertion the most immediate focus of our consciousness.

Second, Schweitzer's epistemological project fails on its own terms because his fundamental assertion ("I am life which wills to live, in the midst of life which wills to live") is wholly compatible with a wholly individualistic, hyper-competitive and amoral vision of the world as "red in tooth in claw." Schweitzer quickly goes on to assert that, "the man who has become a thinking being feels a compulsion to give to every will-to-live the same reverence for life that he gives to his own," and he asserts that such a person "accepts as being good: to preserve life, to promote life, to raise to its highest value life which is capable of development; and as being evil: to destroy life, to injure life, to repress life which is capable of development," and that this "is the absolute, fundamental principle of the moral, and it is a necessity of thought" (ibid., 126). But Schweitzer never articulates any logic that would move one from his prime assertion (*i.e.*, "I am life which wills to live amidst...") to a compulsion to revere the lives of all creatures equally, and there is no reason to think any such logic exists.

Again, Schweitzer's theoretical ambition urges us to replace the Face of the other with an idea and so shuts us off from awakening to the Faces of others and to the realm of the Other. Neither of the two problems noted above can arise if one simply dismisses the misbegotten epistemological quest to derive all ethics from a principal. But Schweitzer never stopped affirming epistemology as first philosophy. Fortunately, Schweitzer's theoretical frustration never lessened the passion with which he acted upon and testified to a compulsion to reverence for life.

To be clear, I am criticizing Schweitzer's philosophy, not his spirituality. The phrase "reverence for life" and the naming of the "compulsion" to do good that it irresistibly inspires wonderfully captures Schweitzer's love for all creatures. But "reverence for life" never gained traction as a philosophical principle, as the lynchpin of an entire ethic, not even, despite decades of efforts, for Schweitzer himself. Sadly, Schweitzer never realized the secret wisdom in his failure. That is, he never realized, as is now clear, that the epistemological quest for a first principle was itself wholly inconsistent with the essential spiritual character of reverence for life.

As the following quote illustrates, at his best, Schweitzer sounds something like Levinas, though his need to think in terms of a basic principle continually short-circuits his ability simply to name and own the having-been-grasped by the Faces of others that brings forth a compulsion to protect, aid, and comfort all creatures. "Reverence for life," says Schweitzer, "means to be in the grasp of the infinite, inexplicable, forward-urging Will in which all Being is grounded. It raises us above all knowledge of things and lets us become like a tree which is safe against drought, because it is planted among running streams. All living piety flows from reverence for life and the compulsion towards ideals which are given in it" (*Ethics*, 214).

Precisely because Barth thought the immediate, living encounter with the "command of God" transcended all human categories he was protected from Schweitzer's quintessentially modern quest to delineate a foundational principle. And in summing up the essence of Schweitzer's spiritual stance Barth wisely cites not Schweitzer the would-be foundationalist epistemologist, but the Schweitzer whose life and testimony helped awaken multitudes, including Barth, to the Faces of all creature.

Barth cites the following long quote in its entirety, for he considers it to manifest the essence of Schweitzer's spirituality:

> A man is truly ethical only when he obeys the compulsion to help all life which he is able to assist, and shrinks from injuring anything that lives. He does not ask how far this or that life deserves one's sympathy as being valuable, nor, beyond that, whether and to what degree it is capable of feeling. Life as such is sacred to him. He tears no leaf from a tree, plucks no flower, and takes care to cruse no insect. If in summer he is working by lamplight, he prefers to keep the window shut and breathe a stuffy atmosphere rather than see one insect after another fall with singed wings upon his table. If he walks on the road after a shower and sees an earthworm which has strayed on to it, he bethinks himself that it must get dried up in the sun, if it does not return soon enough to ground into which it can burrow, so he lifts it from the deadly stone surface, and puts it on the grass. If he comes across an insect which has fallen into a puddle, he stops a moment in order to hold out a leaf or a stalk on which it can save itself. He is not afraid of being laughed at as sentimental (*Ethics*, 243).

This sort of testimony awakens people to the Faces of other creatures. And for his part Barth, whose work is justifiably criticized for being theanthropocentric, quickly and decidedly insists that his readers take Schweitzer's testimony here with utter seriousness by contending that those "who can only smile at this point are themselves subjects for tears" (349).

All of this brings us to our primary objective, which is to make it even more clear that Barth would affirm that "love for all creatures" is a quintessential aspect of Christian spirituality. Before proceeding, however, let me pause to suggest some preliminary conclusions we might draw from this brief sketch of the ways in which Barth, Schweitzer and Levinas constructively correct and supplement one another.

First, I am suggesting that what Schweitzer called the "compulsion" to reverence for all life, and what Barth called an encounter with "the command of God," and what Levinas describes as being taken hostage by the Faces of others are all attempts to gesture toward the same ethical having-been-seized. Of the three, it is Schweitzer whose spiritual sensitivities were profound and subtle enough to extend to every creature, and who bequeathed to us a phrase, "reverence for life," that beautifully testifies to that all-inclusive spirituality. Schweitzer, however, was hampered by a belief that he had to theorize "reverence for life," even though there are clear indications in his work that at some level he knew he was dealing with a form of "thought" and a "compulsion" that came to him apart from the "logical consequences" of any theory.

Levinas most rigorously describes the encounter with the Faces of others, and offers an anti-foundational, anti-theoretical and wholly reasonable argument for ethics as first philosophy. As noted, Levinas' displacement of epistemology and delineation of ethics as first philosophy offers a critical correction for Schweitzer and adds rigor to instincts extant but unevenly developed in Barth. At the same time, Levinas' rigor necessitates the creation of an esoteric vocabulary that is inaccessible to general audiences. More significant, Levinas notoriously remains even more anthropocentric than Barth. Quite inconsistently, given that the Face, according to Levinas, is not supposed to be about things like eyes and noses or, *mutatis mutandis*, fur and tails, Levinas only ever sees Faces in encounters with other humans. In this regard the all-inclusiveness of Schweitzer's ethical sensibilities offers an important correction to Levinas.

Barth is nicely supplemented by Levinas' philosophical rigor, and he evidently owes Schweitzer thanks for awakening him to a sense of reverence for the life of every creature – an awakening that apparently leads him to make claims that resist some of the more problematic aspects of his theology. While Barth's appeal to "the command of God" is indeed susceptible to dogmatic (in the negative sense) appropriation, it is now clear that Barth's understanding of the command is neither dogmatic nor positivistic. Moreover, while Scheitzer's life provides the most dramatic living witness of the three, he is marginal to major world wisdom traditions. Levinas' technical vocabulary, meanwhile, is generally inaccessible. Barth, however, speaks from and to one of the most influential wisdom traditions in the world, and while one must guard against the very real danger that his work will be appropriated towards dogmatic, positivistic and exclusive ends, it also has tremendous greatest potential for exerting substantial influence among the adherents of a major world religion. This last observation partly explains and returns us to the prime objective of this essay, which is to establish that Barth would full-heartedly affirm that "love for all creatures" is a quintessential aspect of Christian spirituality.

Barth's Affirmation of Reverence for Life

Barth is clearly moved and convinced by Schweitzer's conviction that we should reverence all life as sacred. He goes far beyond explicitly adopting Schweitzer's signature line, "reverence for life" as the title of a major section of his *Church Dogmatics*. For Barth believes that if "we are really listening in relation to the human life of ourselves and others, we cannot feign deafness with regard to animal and vegetative life outside the human sphere" and so we must give significant and passionate attention to the question of animals (349).

For fear that it could lead to a belittling of the responsibility people will assume with regard to people, Barth will not equate the responsibility we owe other humans with the responsibility we owe to "animal and vegetable life." Nonetheless, he insists that with the vegetative and animal we face a "serious secondary responsibility" (350).

Taking on the predominant and oppressive modern Western understanding of the character of human dominion on earth, Barth agrees that humans are indeed set up as "lords" on the earth, but he takes care to stress that animals and plants belong not to humans but to God. Thus, Barth argues, lordship conveys upon humans not license but serious responsibility (351). Of course, animal and vegetative life are humans' "means of life," and responsibility to humans is primary, so there are times when the appropriation of plants and animals for human ends may be necessary.

Even with regard to plants, however, Barth contends that we face serious ethical questions. There is no problem, he maintains, when the question has to do with the superfluidity of vegetable life, that is, with a harvest of seeds and fruits, since in such instances there is no destruction of a plant's life (*Ibid.*). But where the destruction of plant life is in question, Barth stresses that we face a serious ethical issue. In the first place, there must be no needless or cavalier destruction of plants: "No one should behave in wood or field 'like the boy who beheads the thistle'" (*Ibid.*). On the other hand, Barth does allow that vegetation may be destroyed if it is necessary for essential human purposes (*e.g.*, the cutting down of trees in order to provide warmth or shelter).

With regard to animal life the ethical issues become even more serious. Barth stresses that humans and animals are profoundly close as creatures. He finds the biological and behavioral similarities, especially with regard to some mammals, to be undeniable and striking (348). Indeed, Barth contends that we "may well have to reckon on an animal soul, at any rate in the biblical sense of a principle of life" (*Ibid.*). Barth notes that the proximity of humans and animals is affirmed in the *Genesis* creation narrative, which thereby signals that reverence is due "for the fellow-creature of man, created with him on the sixth day and so closely related to him" (352). Given all this, we might conclude, it is no surprise that in their encounters with animals, those who are spiritually awake encounter the command of God, which calls them to be "careful, considerate, friendly, and above all understanding" (*ibid.*).

Indeed, it is worth noting once again that Barth considers the spiritual significance of a person's reverence for other animals to be so indicative of a person's spirituality generally that he contends that "a really good horseman," that is, one who "is completely at one with his horse," "cannot possibly be an ungodly person" (ibid.). In the same spirit, Barth insists that animals that are caged or tamed "merely to provide a spectacle or pleasure will always have a doubtful element, and the revolt of a sea-lion, for example, against what is demanded from it will evoke spontaneously the sympathy of all right-thinking spectators" (*Ibid.*).

When it comes to the question of killing animals, Barth does not hesitate to assert that to kill an animal is "at least very similar to homicide" (ibid.). Barth notes the chronically overlooked fructarian ideal (*i.e.*, one only eats fruits and seeds) explicitly articulated in *Genesis* 1:28 – "Behold, I have given you every herb bearing seed, which is upon the face of the earth, and every tree, in which is the fruit of a tree yielding seeds; to you it shall be or meat" – and stresses that this expresses "the true and original creative will of God" (351-353).

Barth thinks that in our world the killing of animals is sometimes necessary for the sake of protection, food, shelter or the like. But given the original creative will of God – which, I would want to specify, we know directly through our encounter with the command of God when we are taken hostage by the Face of a creature – given the original creative will of God, even a degree of necessity that justifies a killing does not make it good. A person "must always shrink from" the possibility of killing an animal, even at the moment when necessity and the balance of goods at stake demands it.

"Across every hunting lodge, abattoir and vivisection chamber," says Barth, "there should be written in letters of fire" the words of St. Paul concerning the liberation of humans, along with all creatures and all creation, from a world in which it is necessary to kill animals (355). Indeed, Barth contends, unless a person has killed an animal in whole dependence upon the grace of God, and abandoning any and all assertion of an ethically pure right to kill, then that person "sins in the killing of animals ... murders an animal" (*Ibid.*). Thus what distinguishes the "good hunter, honourable butcher and conscientious vivisectionist" from the bad, Barth says, is the fact that even as they kill animals those who are good hear the "groaning and travailing of the creature," and thus "are summoned to an intensified, sharpened and deepened diffidence, reserve and carefulness" (*Ibid.*).

By this point it is clear that Barth would agree that "to love all creatures" is a quintessential aspect of Christian spirituality. I have left out some of Barth's more dubious formulations, and even some of the points I have lifted up would be criticized by theologians today (including myself) who advocate for reverence for all life. On the other hand, we should bear in mind that when Barth wrote this essay in 1951, talking seriously about animal souls, reverence towards vegetation and the murder of animals was truly radical. That Barth could so seriously and sincerely respond to the fullness and inclusivity of Schweitzer's testimony to reverence for life, all life, even as he vehemently disagreed with Schweitzer's methodology, gives us ample reason to believe that were he alive today, Barth himself would be the first to critique and revise aspects of positions he staked out sixty years ago. In any case, again, by this point it is abundantly clear that Karl Barth, by consensus one of the most distinguished theologians of the 20th century, and one not at all known for ecological, let alone animal concerns, nonetheless would passionately agree that love for all creatures is a quintessential aspect of Christian spirituality.

Notes

1 Barth's work on animals is familiar to those who have long labored to put non-human creatures on Christianity's ethical and spiritual agenda, and the basics of his position are not conceptually difficult. But despite the clarity of Barth's position and broad familiarity among theologians with his massive, twelve volume, *Church Dogmatics*, there is still little general awareness of his stance upon animals.

2 Barth cites the 1923 German edition of Schweitzer's *Civilization and Ethics*. I should note that in preparing this essay I only had access to the 1946, third edition of Schweitzer's work in English translation.

References

Barth, Karl. 1961. *Church Dogmatics, Volume III, The Doctrine of Creation, Part Four*. Bromiley and Torrance, eds. Edinburgh: T&T Clark.

_____. 1957. *Die Kirchliche Dogmatik, Dritter Band, Die Lehre Von Der Schopfung, Vierter Teil*. Zurich: Evangelischer Verlag A.G. Zollikon.

Linzey, Andrew. 1976. *Animal Rights: A Christian Assessment*. London: SCM Press.

Santmire, Paul. 1985. *The Travail of Nature: The Ambiguous Ecological Promise of Christian Theology*. Minneapolis: Fortress Press.

Schweitzer, Albert. 1946. *The Philosophy and Civilization II: Civilization and Ethics*, 3rd ed. London: Adam & Black.

_____. 1953. *Out of My Life and Thought: An Autobiography*, trans. Everett Skillings. New York: Mentor Books.

Singer, Peter. 1975, 1990, 2002. *Animal Liberation*. New York: Ecco, Harper Collins Publishers.

White Jr., Lynn. *The Historical Roots of Our Ecological Crisis*. *Science*, New Series, Vol. 155, No. 3767 (March 10, 1967), pp. 1203-1207.

Chapter 13

Cows and Dharmic Ecology: Perspectives from the Hindu "Bovine Activists"

☆ *Pankaj Jain*

Introduction

This is an article about the *Swadhyayis*, Swadhyaya practitioners, in the Indian states of Gujarat and Maharashtra. The Swadhyaya movement arose in the mid-twentieth century in Gujarat as a new religious movement led by its founder late Pandurang Shastri Athavale (1920-2003). In my research, I discovered that there is no category of "environmentalism" in the "way of life" of Swadhyayis living in the villages. Following Weightman and Pandey (1978), I argue that the concept of dharma can be successfully applied as an overarching term for the sustainability of the ecology, environmental ethics, and the religious lives of Swadhyayis. Dharma synthesizes their way of life with environmental ethics based on its multidimensional interpretations as I show in their work with the cows. In addition to Swadhyaya, I also mention some other key activists who played a key role in Hindu cow protection movement in India[1].

Swadhyaya's Dharmic Ecology

Having heard about the Swadhyaya, I called their office in Mumbai to visit one of such sites in summer of 2006 during my trip to India. Soon, I found myself on my way to Valsad in Gujarat. I arrived at the home of a Swadhyaya volunteer Maheshbhai[2] who took me to the water harvesting site managed by local Swadhyayis. All of them showed warmth and enthusiasm to welcome me and explain about their various activities and ideologies of the Swadhyaya movement. As the caretakers of the water harvesting site began explaining about the way they perceive the nature and the vision of their guru Athavale, I began asking questions related to environmentalism. What I present below is based on my several such interviews with Swadhyaya followers. I have also extracted relevant information from the vernacular literature of Swadhyaya that is based on the video-recorded discourses of Athavale.

Swadhyaya is one of the least known new religious movements that arose in the mid-twentieth century in the Western states of India. Although this movement now has some presence in several Western countries such as the United States, Canada, and the UK, it has not received the attention of scholars of Hinduism except few introductory articles[3]. Before I begin introducing the ecological work of Swadhyaya, it is important to mention that their environmental significance is denied by the Swadhyayis. In fact, one of the Swadhyayis was taken aback when I told him about my topic of research. In his own words,

> "You might misrepresent Swadhyaya if you choose to research it from ecological perspective. Swadhyaya and its activities are only about our devotion to almighty; ecology is not our concern. Environmental problems are due to industrialization and the solution lies beyond Swadhyaya's activities. Swadhyayis are not environmentalists!"

Based on my observations of Swadhyaya's several activities, I tend to agree with him. Athavale has repeatedly emphasized that the main goal of Swadhyaya is to transform the human society based on the Upanishadic concept of "Indwelling God". According to him, since the Almighty resides in everybody, one should develop a sense of *spiritual* self-respect[4] for oneself irrespective of *materialistic* prestige or possessions. In addition to one's own dignity, the concept of "Indwelling God" also helps transcend the divisions of class, caste, and religion and Athavale exhorted his followers to develop the Swadhyaya community based on the idea of "brotherhood of humans under the fatherhood of God". Activities of Swadhyaya are woven around this main principle, which in turn are also aimed at the Indian cultural renaissance.

Although environmentalism is neither the means nor the goal of Swadhyaya's activities, natural resources such as the earth, the water, the trees, and the cattle are revered and nurtured by Swadhyayis based on this understanding. Environmentalism does come out as an important by-product of its multi-faceted activities and this was noted by a 1992 conference in Montreal where Swadhyaya was invited to present its ecological philosophy and work[5]. I argue that a multivalent term like dharma can comprehend and describe this kaleidoscopic phenomenon and the way it relates to the ecology. Swadhyaya followers do not regard environmentalism as their main duty, their dharma. Alternatively, from the outside, one can regard their dharma, their cultural practices, as ecologically sustainable as I show below. I also want to note that my observations are based on their activities in the rural parts of India since the urban and the diaspora Swadhyayis do not have such ecological projects yet.

Pandurang Shastri Athavale and Swadhyaya's Bovine Ecology[6]

By several explanations from Indic texts[7], Athavale developed a set of preaching that I call "bovine dharma," preaching based on the qualities of cows that emerged as *gorasa,* a project to nurture the cows. Athavale cited a Sanskrit verse to show that seven forces sustain the earth, cows, Brahmins, Vedas, Satis (noble women), truthful people, charitable people, and people without lust and greed.[8] The cow gives all her belongings to humans. Her milk and other dairy products strengthen us. Bullocks are utilized in farming. Cow dung is utilized as a fertilizer. Its urine is used as an Ayurvedic medicine.[9] After its death, its bones are utilized in the sugar industry, skin is used in the leather industry, and the horns are used to make combs. According to Athavale, humans should be eternally grateful to cows. According to him, Indians did not just exploit the cows for materialistic benefits but instead regarded her as mother. Only humans drink the milk of other species such as cows. Regarding cows as mothers is to express our gratitude and respect for them. Like our biological mothers, they nourish us with their milk. According to Athavale, Indian villagers used to feed cows with love and respect before taking their own meals. Even today, many families observe the tradition of *go-grāsa,* offering of symbolic food to cows before meals. Athavale mentioned *Krsna*'s love for cows and said that *Krsna* turned into *Gopālkrsna* (*Krsna,* the cow caretaker) out of his love for cows. Krsna used to attract cows with his sweet flute in Gokul. He cited a Sanskrit verse from Padma Purāna to show the importance of cows in ancient India, "Let cows be ahead of me, back of me, inside my heart. I should reside in the midst of cows."[10] Similarly, the *Brahma Vaivarta Purāna* mentions that all the gods and pilgrimages reside in the cows[11].

Athavale explained that the cow has a quality of chewing her food thoroughly before swallowing it. He interpreted it to preach that we should also chew every new thought before accepting it. Only after carefully analyzing it, should we accept it. Such carefully accepted thoughts will not only help our own intellectual development but will also benefit greater society. Extending the bovine dharmic discourse further, Athavale cited the famous BhP verse in which all the Upanisadas are called cows, *Krsna* as the milkman supplying milk to Arjuna. We should always be nourished by the teachings of the Upanisadas and the BhG.[12]

He also cited another verse by Kālidāsa and extended the meaning of the Sanskrit word *go. Go* has several meanings such as cow, physical power, eyes, and speech. He preached cow-worship in all its different meanings, *i.e.,* we should respect cow (animals), the rulers should be powerful, women should possess long-term vision, and scholars should convey good thoughts to masses. Any kind of ritual is incomplete without *Pancāmrta,* consisting of three dairy products milk, yogurt, and ghee, in addition to honey and sugar. Milk signifies purity. Our life, character, reputation, actions, mind, and heart, all should be pure and untainted. Yogurt is the second ingredient of Pancāmrta. Describing Yogurt's qualities, Athavale explained that just a small quantity of yogurt transforms a big quantity of milk. Noble people should also develop this quality. They may be small in numbers but they can transform huge sections of societies. The third ingredient is ghee with its unique lubricant quality. Our life should also be lubricated with love for everybody. Obviously, Athavale had very skillfully utilized several metaphors, analogies, myths, and legends to inspire his followers to practice what he preached.

Athavale developed a project for cows called *Gorasa*. He established dairies in villages where people could get the milk throughout the year at a nominal price and the profit earned from such collective effort was distributed to needy local families or saved for future projects. This stopped the earlier practice of selling the milk to far-away cities. Since the intermediaries involved in selling were eliminated, the purity of milk was ensured, and the cost was minimized. This also inspired farmers to domesticate more cows and thus inspired more love and care for animals.

He sought to develop what I call a "bovine dharma", dharmic ecology for the cows inspired by the cows. Athavale describes the inherent qualities and virtues of the cows, *i.e., the dharma of the cows*. By observing and following the dharmic qualities of a cow, one can develop one's moral and ethical qualities, *the dharma of a human being*. Here again, we see several meanings of dharma interplaying with each other. The dharma, *inherent quality*, can inspire the dharma, *virtue*, and this can help develop the dharma to care for the animals and environment, *environmental ethics*. I now mention some other Hindus who played a key role in the protection of cows in India.

Radhakrishna Bajaj

Radhakrishna Bajaj was born on July 14, 1905 and since the age of 16, he was constantly with Vinoba Bhave and grew up under Bhave's tutelage and direction. He also came in contact with the leaders of the Indian freedom movement and under their inspiration took part in the non-cooperation movement and the Jaipur Satyagraha. In 1942, he took part in the Quit India movement, was charged with sabotage by the British and was sentenced to death but was later acquitted. He engaged himself in opening and running spinning and weaving centers and was responsible for spreading the village industries movement in Wardha district of Maharashtra. He also took part in the Sarvodaya movement. Late Jamnalal Bajaj made him the secretary of Goseva Sangh in 1950s. He was also the editor of *Gogras*, the monthly mouthpiece of the Goseva Sangh. He was also the President of the *All India Krishi Gram Seva Sangh*. Bajaj had also translated and published eminent Gandhian Vinoba Bhave's writings. In 2004, he was awarded the Jamnalal Bajaj award, a prestigious Gandhian award.

In his Hindi article in the 1995 special issue of Hindu magazine *Kalyāna,* Bajaj described the cow protection movement in Independent India. In 1966, All Party Cow Protection Campaign Committee started a movement for the ban on cow slaughter. Swami Karpatri (Lutgendorf 1991) was its leader. Swami Niranjan Teerth, Shankaracharya of Puri had conducted a 73 days fast in this campaign. In 1975, there was a conference held at Paunar and Vinoba Bhave decided to launch a hunger strike for this cause. Government tried to suppress this news but it spread all over India. Eventually several Indian states, such as Assam, Tamilnadu, Andhra Pradesh, and Maharashtra, made laws to ban cow slaughter but the demand for the nationwide law remained unfulfilled. Vinoba Bhave again launched a campaign on April 22nd, 1978. In 1979, *Roko Bhaai Roko* ("Stop brother stop") movement was launched to stop the cows going into different slaughter houses. First it was launched at Kosikalan from where a huge number of cows used to be sent to Calcutta and within two months there was a big drop in the number of cows to be sent for killing. Several students also participated in this movement and sometimes there was a shoot out by the police to suppress the movement. There were at least one hundred and fifty court cases registered. On January 11, 1982, Vinoba Bhave also launched a *Satyagraha* against the Devnar Slaughterhouse in Mumbai. Then there was a national convention on this issue on March 25, 1987 at Rajghat, in New Delhi in which about 1000 delegates participated (Harris 1987).

Lala Hardev Sahai

Lala Hardev Sahai was one of the foremost cow devotees in Independent India. He was also an activist in Indian freedom struggle and later worked to spread the Hindi language. He was born in 1892 in village Saatrod, district Hisar of Haryana. He had studied Hindu scripture in his adolescence and started weaving his own clothes taking inspiration from Indian freedom struggle, particularly from Madan Mohan Malaviya, Lala Lajpat Rai, and Lokmanya Tilak. He was also against the British education and propagated Hindi and Sanskrit based education by opening sixty-five schools around his native place. He actively participated in Indian freedom struggle and was arrested in 1921 and 1942 during the Quit India Movement. In 1921, he was imprisoned with Swami Shraddhananda in Nimavali, Punjab where he developed a special inclination towards the Bhagvad Gita.

In 1939, there was a severe drought in Hisar and the cattle were particularly affected. Lala joined hands with another prominent cow-devotee Maanakar to save the dying cattle. Together they collected food for the cattle and also established centers for women's' self-employment. After Indian independence, he wrote several petitions to Indian Prime Ministers and President and continued his campaign to stop the cow-slaughter. In 1954, he met Swami Karpatri and Prabhudatt Brahmachaaree. With the latter, he stopped several cows from being sent to Calcutta. On February 4, 1954, he established *Cow Slaughter Prohibition Committee*. They both also conducted a *Satyagraha* at the slaughterhouse in Mathura. When Karpatri launched his campaign to free the land of Brij from cow-slaughter, Lala actively cooperated him. Lala and Brahmachaaree also conducted a *Satyagraha* in front of the UP Assembly to protest the cow-slaughter. This movement ultimately succeeded in forcing the UP government to devise a provisional law to ban the cow-slaughter (which ultimately was cancelled). Similarly, during the 1955 Cow Protection movement in Bihar, Lala was imprisoned with Brahmachaaree in Bankipur. Subsequently, both were released when the cow-protection law was made. Lala had also taken a vow not to wear his turban and not to sleep on the cot until a nationwide ban on cow-slaughter is not declared by the government. He also wrote several books on cows. The first Indian President Dr. Rajendra Prasad wrote the foreword to Lala's Hindi book *Gaaya Hee Kyon* (why only cow?) in which he admired Lala's knowledge about the cattle.

Prabhudatt Brahmachaaree and Meharchand Paahuujaa

In 1966, Prabhudatt Brahmachaaree and Puri's Shankaracharya Swami Niranjan Teerth conducted hunger strikes at Vrindavan and Puri respectively. The former was a devotee of Swami Karpatri and the latter was a devotee of Shreekrishnabodhaashram and used to campaign very actively against cow-slaughter with slogans and flags. On November 30th, 1966, while participating in another hunger strike for ten days, he died at the *Dharmasangha Bhavan.* Similarly, Meharchand Paahujaa also died in Vrindavan on December 31st, 1966 after being on fast for about forty days.

Cow-Martyrs in Katarpur

Once in 1918, at Katarpur, a village near Haridwar, some Hindus tried to stop the cow-slaughter which led to some social unrest. Local British authorities arrested 172 Hindus and on August 8th, 1919, Mahant Brahmadas Udaaseen, Chaudhry Janakidas, Dr. Purnaprasad, and Shrimukha Chauhan were given death sentences, while 135 other Hindus were given *Kala Pani,* exiled to Andaman Nicobar islands. Shivdayal Singh, the Hindu inspector of Haridwar was also given life-imprisonment. Madanmohan Malaviya and Acharya Naradev Shastri tried very hard to fight on Hindus' behalf but did not succeed. Finally Mahant Brahmadas (age 45) and Chaudhry Janakidas (age 60) were hanged in 1920 in the Allahabad Jail. Allahabad City observed a Bandh in their honor. Dr. Purnaprasad (age 48) was hanged in Lucknow and Shrimukha Chauhan (age 32) was hanged in Banaras Jail. Mahant Brahmadas was a leader of the Udaaseen Akhaadaa and had thousands of disciples. Prominent Hindus who were exiled in Andaman-Nicobar included Sardar Jagadatt, Shrinanda, Lala Khubchand Pansari, Pt. Asaram, Shrilakshaminarayan Bhakta, Lala Daulatram, Lala Devichand, Pt Narayandatt, Chaudhary Raghuveer SIngh, Chaudhary Fateh SIngh, Pt Maakhanlaal, Lala Pyaarelaal, Shrisaadi. In Katarpur, every year these martyrs are remembered on their death anniversary.

Swami Ramachandra Vir

On August 10th, 1966, Swami Ramachandra Vir declared that he will start his hunger strike on August 20th if the Indian government does not make a law against cow-slaughter before that. At the Hindu Mahasabha Bhavan, Mandir Marg, New Delhi, on August 20th, he actually launched his fast amidst the Vedic chanting. For next 32 days, he continued his fast while several Hindu leaders and activists paid visits to his weakening health. On September 20th evening, the police came and arrested him on the charge of attempted suicide. On November 7th, 1966, he was at the Govind Ballabh Pant Hospital on his 80th day of the fast.

Conclusion

Anil Agarwal (Chapple *et al.,* 2000) had mentioned that Hindu beliefs, values, and practices, built on a "utilitarian conservationism", rather than "protectionist conservationism", could play an important role in restoring a balance between environmental conservation and economic growth. The bovine activism that I have described above, does not fall in either category. In fact, when I interviewed some of the people who were involved in such work, they vehemently denied both utilitarian and protectionist motives behind their *prayogs* and underscored the devotional motive instead, as I noted above.

The cow protection examples are Indian counterparts of what could be called "environmental activism". They are inspired by Indic religious traditions. When I asked Swadhyayis about the practical challenges or difficulties related to such work, they noted several challenges. One challenge is to be able to sustain the transformation based on the Swadhyaya's teachings. Without the dharmic perspective, the work can become "mechanical" or can take the form of another "religious ritual". If these activities fail to inspire people to develop an ethos, develop a bonding with the nature, or if Swadhyayis stop practicing this ethics in their daily life, then the work can take a "religious" shape based on the devotional faith on the words of Athavale. That would reduce the dharmic work into another religious ritual as has happened with some rituals in both India and elsewhere. Another challenge is to take these activities and replicate them at a bigger level. So far, these have remained smaller local role models at the district level rather than projects at the regional or national level. They also confessed that the number of volunteers available to work at different sites varies according to the intensity and depth of Swadhyaya's thoughts in the surrounding villages. Since the spread of Swadhyaya is not uniform across the different villages and towns of Gujarat and Maharashtra, the number of volunteers working at such *prayogs* is also not uniform. Noted environmentalist Anupam Mishra aptly remarked (1993):

> "Even without involving the environmentalists, people are bringing out miracles at the grass-root level. Upon seeing them, we should humbly accept them. Even if they may not fit our measuring scale, our measuring scale itself may be inappropriate. A work that has already reached millions belongs to the people. Media reports only political parties but it cannot represent the people."

Athavale had developed several more "prayogs" to fructify his mission for socio-spiritual transformation based on dharmic philosophy of "Indwelling God". I have described only some of them that relate to the cattle. These prayogs do not label themselves as "environmental projects" and yet they have succeeded in sustaining natural resources in thousands of Indian villages. Like any other such work, the challenge now is to maintain them and to develop new such prayogs especially now that Athavale has passed away. To my knowledge, the current leadership has not developed new ecological prayogs. However, it seems focused to strengthen the existing prayogs by inspiring more villagers to join the movement. I agree with Ramachandra Guha's remark that there was no environmentalism before industrialization; there were only the elements of an environmental sensibility. The traditional societies did not transcend their locality to offer any systematic vision of reorganizing nature (2006: 6).

The Swadhyaya followers also show similar sensibilities in their local activities in the villages. This sensibility in turn is inspired by their cosmology that is based on the texts, myths, and legends derived from the dharmic traditions. We do see a reflection of textual reverence for nature in the behavior of Swadhyayis. Whether this behavior will take a generic ecological ethos across the time – space limitation is yet to be seen. It is still a nascent movement with the charisma of the founder still fresh in the memories of its followers. Will they become environmentalists that are more active in the new century in the absence of Athavale? Swadhyaya is also emerging as a movement spread around the globe. When Swadhyayis migrate to different parts of the world, will they connect their environmental sensibility to respond to the problems of climate change? Only time will tell the answers to these questions.

Notes

1 For an excellent overview of cow shelters in India, see Lodrick (1981). Similarly, Korom (2000) and Srinivasan (1979) go in detail about the Hindu perspectives towards cows as mentioned in various texts. My attempt in this article is to provide some practical examples of these Hindu bovine perspectives.

2 In this article, I have used pseudonyms except for the well-known personalities.

3 Between 1994 and 1996, some observers and scholars had visited the Swadhyaya villages. Their observations were compiled in an edited volume by R. K. Srivastava (1998). This is a helpful introduction of the movement. In addition, I have provided some other scholarly articles in the bibliography, see Dharampal-Frick (2001), James (2005), Little (1995), Paranjape (2005), Rukmani (1999), Sharma (1999), and Unterberger et al (1990). After completing this article, I also came to know that Ananta Giri has published a monograph on self-development and social transformation brought about by Swadhyaya, see Giri (2009).

4 Since 1990s, this became the theme of celebration for the Athavale's birthday, which is celebrated as "Human Dignity Day" around the world by his followers.

5 This conference, "*Living with the Earth: Cross-cultural Perspectives on Sustainable Development, Indigenous and Alternative Practices*", took place on April 30 – May 3, 1992. It was organized by the Intercultural Institute of Montreal, Canada. A three- page report titled "Presentation by Dīdī (the current leader of the movement) on the Swadhyaya Movement" was written by Robert Vachon in the proceedings of the conference. An interview with Dīdī was subsequently broadcasted nationwide by the Canadian Broadcasting Corporation on a one-hour radio program called *Ideas*.

6 Information in this section is based on Swadhyaya's Hindi books such as *Samskriti Pūjan* and *Esa Pantha Etad Karma* (with my translations).

7 *Cow Hymn, Atharva Veda 4.21*

1. The Cow is the mother of the Rudras, daughter of the Vasus, sister of the Aditi-sons, and treasure of the ghee-nectar. I have explained to every thoughtful person never to kill an innocent cow.
2. The cows have done favor for us here at our place. May they stay comfortably at our cow-shelter and fill the atmosphere with their music sounds. May these colorful cows produce different kinds of calves and give us ritual-milk before dawn.
3. May these cows neither be destroyed, nor stolen, nor be harmed by our enemies. With the help of those cows, whose owner worships the gods and is capable of donation, may the cows live with him forever.
4. May the cows be our chief wealth, Indra provide us with this wealth, and their milk with the soma be our blessed drink for our rituals. Whoever

owns these cows, is virtually Indra himself. I want to worship Indra with my faithful heart and cow-products.

5. Cows! You strengthen a frail and enlighten a lackluster person. With your ennobling sounds, you make our homes noble. This is why, your glory is sung in the assemblies.
6. Cows! May you produce many calves, have enough food to graze upon, drink clean water in the ponds, be safe from thieves and violent animals, and may Rudra's weapon protect you from all evils.

Cow-Shelter Hymn, Atharva Veda 3.14

1. Clean, prosperous, and beautiful cow-shelters should be built for the cows. Clean water should be fed to them and their skill to produce good calves should be maintained. They should be loved so much that the best materials should be given to them.
2. Aryamaa, Pushaa, Brihaspati, and Indra should nourish the cows and the milk from these cows should nourish me.
3. Healthy cows giving good fertilizer in the form of cowdung and good food in the form of milk should come and stay in our cow-shelter.
4. The cows should come to this cow-shelter. Nourished here, they should produce good calves, love their owner, and live happily here.
5. May this cow-shelter be beneficial to the cows, may they be nourished here, produce calves, and multiply in numbers. The owner himself should manage the cow-shelter.
6. The cows should live with their owner happily. This cow-shelter is nice and the cows should be well-nourished here. Increasing their glory and health, the cows should grow here. We will obtain such cows and care for them.
7. By which cow, the three worlds (the heaven, the earth, and the space) are protected, we praise that milk-giving cow. One hundred persons are ready to milk it and one hundred caretakers are waiting for it. The gods that are nourished by it realize its importance.

8 *Gobhir vipraisca vedaiscah satibhih satyavādibhih. Alubhdhaih dānsūraisca saptabhirdhāryate mahīh.*

9 Indian government awarded a patent to Virendra Kumar Jain, a cow urine therapist, for his discovery and innovation of cow urine therapy & therapeutic Ayurvedic composition. According to him, In Mahabharata, Nakul was a cow urine therapist. Similarly, the cow urine business is booming in Hyderabad as Yoga and Ayurveda is getting popular there (*The Telegraph* May 5, 2005).

10 *Gāvo me agrath santu gāvo me santu prsthath. Gāvo me hrdaye santu gavām madhye vasāmyaham.* (From Padma Purāna, Srútī Khanda 57. 152)

11 *Sarve devā gavāmange tīrthâni tatpadesu* (Brahma Vaivarta Purāna, Sri Krsna Janma Khanda 21.91)

12 *Sarvopanisado gāvo dogdhā gopālanandanah. Pārtho vatsaha sudhīrabhokta dugdham bhagvadgitāmltam mahat*. (BhG Mahātmyam verse 6).

References

Videos

Athavale, Pandurang Shastri. n.d. *Discourses in Hindi and Gujarati* (Mumbai: Sat Vichar Darsana).

Bazaz, Abir and Meena Gaur. 2003 *Swadhyaya*. New Delhi: Public Service Broadcasting Trust, 2003.

Benegal, Shyam. 1992 *Antarnaad: Inner Voice*. India: Suhetu Films, 1992.

Books and Articles in Hindi

Athavale, Pandurang Shastri. 1999. *Esa Pantha Etat Karma*. Mumbai: Sat Vichar Darœana Trust.

_____. 1997. *Sanskrti Pūjan*, Mumbai: Sat Vichar Darœana Trust.

_____. n.d. *Kalyāna*. Gorakhpur: Gita Press.

Books and Articles in English

Daily Excelsior. Oct. 27, 2004.

The Hindu. Oct 28, 2004

The Indian Express. Nov 1, 2004

Chapple, Christopher Key and Mary Evelyn Tucker. 2000. *Hinduism and Ecology: The Intersection of Earth, Sky, and Water*. Cambridge: Harvard University Press.

Giri, Ananta Kumar. 2009. *Self-Development and Social Transformations?: The Vision and Practice of the Self-Study Mobilization of Swadhyaya*. Lanham: Lexington Books.

Guha, Ramachandra. 2006. *How Much Should a Person Consume? Environmentalism in India and the United States*. Berkeley: University of California.

Harris, Ishwar C. 1987. "Sarvodaya in Crisis: The Gandhian Movement in India Today" in *Asian Survey*, Vol. 27, No. 9 (Sept., 1987), pp. 1036-1052.

James, George A. 2005. "Athavale" and "Swadhyaya". In Bron Taylor, ed., *the Encyclopedia of Religion and Nature*. New York: Thoemmes Continuum.

Juneja, M. M. 1995 *Lala Hardev Sahai (1892-1962), A Fearless Fighter*. Hisar, Haryana: Lala Hardev Sahai Smarak Samiti.

Korom, Frank J. 2000. "Holy Cow! The Apotheosis of Zebu, or Why the Cow Is Sacred in Hinduism" in *Asian Folklore Studies*, Vol. 59, No. 2., pp. 181-203.

Little, John T. 1995. "Video Vachana, Swadhyaya and Sacred Tapes". In Laurence A. Babb and Susan S. Wadley, ed., *Media and the Transformation of Religion in South Asia*. Philadelphia: University of Pennsylvania Press. pp. 254-81.

Lodrick, Deryck O. 1981. *Sacred Cows, Sacred Places: Origins and Survivals of Animal Homes in India*. Berkeley: University of California Press.

Lutgendorf, Philip. 1991. *The Life of a Text: Performing the Rāmcaritmānas of Tulsidas* Berkeley: University of California Press. pp.384-387.

Mishra, Anupam. 1993. *Aaj Bhi Khare Hain Talab*. Delhi: Gandhi Peace Foundation.

Paranjape, Makarand. 2005. *Dharma and Development: The Future of Survival*. Delhi: Samvad India Foundation.

Rukmani, T. S. Turmoil. 1999. *Hope, and the Swadhyaya*. Montreal: CASA Conference.

Shah, V. P., N. R. Sheth, and Pravin M. Visaria. 1998. "Swadhyaya: Social Change through Spirituality" in M. L. Dantwala, Harsh Sethi, and Pravin M. Visaria. Ed. *Social Change Through Voluntary Action*. New Delhi: Sage Publications.

Sharma, Arvind. 1999. *A Metaphysical Foundation for the Swadhyaya Movement (If It Needs One)*. Unpublished paper.

Sheth, N. R. 2002. "A Spiritual Approach to Social Transformation", in Shinoda, Takashi (ed.). *The Other Gujarat*. Mumbai: Popular Prakashan.

Srinivasan, Doris. 1979. *Concept of Cow in the Rigveda*. Delhi : Motilal Banarsidass.

Srivastava, R. K. 1998. *Vital Connections: Self, Society, and God: Perspectives on Swadhyaya*. New York: Weatherhill Publications.

Unterberger, Betty Miller and Rekha R. Sharma. 1990. "Shri Pandurang Vaijnath Athavale Shastri and the Swadhyaya Movement in India", in *Journal of Third World Studies*, vol 7, no. 1, Spring 1990, pp. 116-132.

Weightman, Simon and S.M. Pandey. 1978. "The semantic fields of dharma and kartavy in modern Hindi", in Doniger, Wendy and J. Duncan M. Derrett. *The Concept of Duty in South Asia*. Columbia, Mo.: South Asia Books for the School of Oriental and Africa Studies.

Chapter 14

Compassion for All Beings: Ethical Treatment of Animals in the Indic Religions

☆ *Jeffery D. Long*

Introduction: Karma and Compassion

The ethical treatment of animals has long been a major topic of reflection in the religions of India–particularly the œramaGa traditions, Jainism and Buddhism, which, in turn, exerted considerable influence upon the dominant Vedic traditions now collectively known as Hinduism.

The thesis of this article is that these traditions' emphasis on the ethical treatment of animals is closely tied to their non-anthropocentric conceptions of the soul and their related doctrines of karma and rebirth. In contrast with Western traditions, which tend to see the soul as a rational principle distinctive to human beings, the Indic traditions see the soul as pure consciousness, which can be manifested in a wide variety of forms, of which the human form is only one. This conception of the soul serves as a metaphysical basis for an ethic of compassion for all living beings.

Jainism, Buddhism, and Hinduism affirm the reality of a universal moral principle of cause and effect called *karma*. Derived from the Sanskrit verbal root *kr*, which means 'act,' karma governs all action. Karma can be likened to Newton's Third Law of Motion: for every action there is an equal and opposite reaction. But traditional Indic worldviews do not make the sharp distinction, so typical of modern Western thought, between realms of fact and value. Karma thus manifests not only in the form of physical laws, such as gravity, but also as a moral law governing action. If one engages in violent actions that are motivated by hatred, selfishness, or egotism, the universe will respond in kind. So the one who has caused suffering to others will, in turn, suffer.

Similarly, if one engages in actions that are benevolent, pure, and kind, one will eventually have pleasant experiences. There are biblical and other Western expressions that convey much the same sensibility as that reflected in the idea of karma: You reap what you sow. What goes around comes around.

Jains, Buddhists, and Hindus deduce from the principle of karma the concept of rebirth, or reincarnation (*punarjanma*). All religions must address the issue of why bad things happen to good people and good things happen to bad people. Why, if there truly is universal justice–which is essentially what karma amounts to–does the world in which we live appear to be as unjust as it does? The Indic religions explain this phenomenon in terms of past and future lives. Today's joy or suffering may be the fruit of karma from a previous life. And the actions one takes today will inevitably bear fruit, if not in this life, then in a future one.

But are the only goods that one can anticipate the temporary fruits of one's karma, the good effects of one's actions? Is there nothing more in our destiny than constantly working for and enjoying these temporary goods, while at the same time avoiding all evil actions, with their consequent fruits? Or is there a higher, transcendent experience that is our ultimate aim? Jains, Buddhists, and Hindus see the ultimate good not as the repeated experience of the fruits of good actions–good though this certainly is, and a very widely pursued aim in all three traditions–but as escape from the cycle of rebirth. Freedom from rebirth–known as both *moksa*, or liberation, and nirvāna, or absorption–is conceived as a state of unending bliss, in contrast with the impermanent and constantly changing flux of action-generated experiences. But for most Jains, Buddhists, and Hindus, this final goal is seen as remote and difficult to attain, the more immediate goal of most religious action being merit-making: the acquisition of 'good karma.'

Each of the three traditions under consideration has its own distinctive approach to the basic model outlined here. But for all three, an important implication of this model is the interconnectedness of all life, and an injunction of compassion for all beings.

Jainism

Jainism is an ancient tradition which, according to its own accounts, has no origin point in time, but rather has been taught and re-taught repeatedly, through a beginningless succession of cosmic epochs, by a series of enlightened beings called *Tīrthankaras*. Each cosmic epoch, or *kalpa*–a period that lasts many millions of years–is characterized by a recurrent pattern of events that includes the emergence, one by one, of twenty-four such enlightened beings. The most recent *Tīrthankara* to appear on the earth is *Mahāvīra*, who probably lived from roughly 499 to 427 BCE. As the twenty-fourth *Tīrthankara* of our cosmic epoch, Mahāvīra is not, technically speaking, the founder of Jainism. There is a good deal of historical evidence that suggests he practiced and tapped into a pre-existing tradition 'proto-Jain' tradition. He could, perhaps, be described as the founder of the Jain tradition *as practiced today*–the contemporary fourfold *tīrtha*, or sacred community, that consists of Jain monks, nuns, laymen, and laywomen.[1]

One of the hallmarks of Jain teaching and practice is its emphasis on *ahimsā*, or a state of complete nonviolence in thought, word, and deed. This emphasis is closely tied to the Jain view of karma, according which the mechanism that makes our actions bear their morally corresponding fruit is an actual, material particle that adheres to the soul– literally, the 'life force,' or *jīva*. Karmic matter distorts the originally pure character of the soul, which has unlimited awareness and bliss. Karmic bondage leads the soul to have the kinds of limited experiences that constitute its journey through the cycle of death and rebirth. The ultimate goal is freedom from this bondage. This is difficult to achieve, however, because our actions are generally tinged with *rāga*, or passion, which attracts karmic particles to our soul. Whenever these karmic particles bear their fruit, we have experiences whose quality depends upon the type of karma we have attracted–which, in turn, depends upon the type of passion that has attracted that karma.

The pleasant or unpleasant character of these experiences, however, evokes more passion, which draws more karma to the soul, and the cycle begins anew. Jain tradition therefore places a very strong emphasis on cultivating a state of equanimity (sāmāyika) in all situations, responding alike to pleasure and pain, thus reducing the karma-attracting passion in the soul that keeps one in bondage.

Ahimsā is central to this process because, according to Jain thought, the strongest, most negative passions–the ones that attract the most painful and binding karmic matter to the jīva–are those associated with violence: with the desire to do harm to other living beings.

The Jain community is famous for the seriousness with which it takes its principle of ahimsā. One who has become aware of the all-pervasive presence of life forms in the world–including insects and other tiny creatures, such as microorganisms–is responsible for doing as little harm as possible even to the tiniest of beings. This results in the highly restrictive discipline observed by Jain monks and nuns, evidenced in such practices as the sweeping of the ground by these ascetics to avoid accidentally treading upon an insect, and the practice of some (though not all) Jain ascetics of wearing a *muhpattī*, or mouth shield, to avoid accidentally inhaling microorganisms.

Beyond this almost heroic commitment to utter harmlessness by Jain ascetics, the Jain lay community has taken an active role in the protection of animals, establishing the first animal shelters and animal hospitals in history in ancient India. Such institutions are widely supported by lay Jains to the present day, being one of the many distinctive Jain contributions to Indian culture.

Finally, the practice of vegetarianism is universal among observant Jains (and is widespread even among otherwise non-observant Jains, so deeply ingrained is the ethos of nonviolence in Jain culture). The Jain commitment to the compassionate treatment of animals is so strong that, despite their relatively small numbers throughout history, Jains have had a strong influence on the practices of both Hindus and Buddhists in this regard.

Buddhism

On a metaphysical level, Buddhism stands out from the other Indic religions in its conception of the cycle of karma and rebirth in not emphasizing the reality of the soul. It is not that Buddhism is a form of materialism, which identifies the self with the body and denies the reality of rebirth–though some Buddhists, especially in the West, do interpret their tradition in this way, despite the Buddha's sharp criticism of this view in the earliest Buddhist scriptures. But the Buddhist tradition sees release from the cycle of rebirth as a state requiring the relinquishment of all attachments to material circumstances–very close to the Jain idea of becoming free from the passions. The Buddha's deeply penetrating analysis of the process of consciousness reveals that at the basis of all such attachment is a constructed ego–a false self that is the one who thinks, "I desire this. I desire that," and "I am the doer of action," and "I am the one who is experiencing the fruits of my action." Eradicating attachment therefore requires one to tear such attachment up by the root, as it were, freeing oneself from false conceptions of "I" and "mine." Rather than a soul that is experiencing action and its fruits, and the consequent cycle of death and rebirth, Buddhist tradition speaks, instead, of repeated patterns of action that carry over from one lifetime to another, seeing the idea of a soul as a potential object of passionate attachment.

Like Jains, Buddhists see the cultivation of equanimity as essential to the process of achieving freedom from rebirth. And, as we shall see, like Hindus, Buddhists affirm the fundamental interconnectedness of all phenomena–and so, of all beings. What one does to another, therefore, one does to oneself. Nonviolence is therefore a central ethical principle of Buddhism, as it is for Jainism. Buddhists have not, however, been as strict in this regard, traditionally, as have Jains. Vegetarianism, therefore, although a widespread and respected Buddhist practice, is not universal among Buddhists. Historically, though, the Buddhist emphasis on nonviolence was at least as influential as that of the Jains, due to the influence of the Buddhist emperor, Asoka (304-232 BCE), who banned the killing of animals on Buddhist holy days and promoted an ethos of nonviolence in his empire.

Hinduism

The first thing that must be said about Hinduism is that there really is no singular tradition that goes by this name. The English word *Hinduism* was coined only about two hundred years ago–just yesterday, on a Hindu time scale–to denote a very large family of distinct, but also overlapping and interconnected, religious and philosophical traditions that have emerged within the Indian subcontinent over the course of the last three to five thousand years. It is therefore impossible to talk about Hinduism without generalizing so broadly that some group of Hindus or other will necessarily be excluded by most of one's claims. Bear in mind, therefore, that whatever I say here about Hindus or Hinduism is by no means true of all Hindus or of all forms of Hinduism. Indeed, the internally varied character of Hinduism–the fact that it contains as much diversity as it does, while at the same time possessing a certain measure of cohesion–is itself an ecological statement; for just as an ecosystem thrives on its biological diversity, the internal variety of Hinduism is also a catalyst for tremendous intellectual and artistic creativity.

The origins of Hinduism are a mystery. Because it is not really a single religious tradition, there is no one founding figure to which it can be traced in recorded history–no precise Hindu equivalent of a Buddha, a Jesus Christ, or a Prophet Muhammad. Indeed, Hinduism can be compared to an upside-down tree. While most religions begin with one founding figure who starts a community which, over the course of time, divides itself into many different branches–Theravada and Mahayana Buddhism, or Catholic and Protestant Christianity, or Sunni and Shia Islam–Hinduism begins with many different branches, a wide variety of diverse traditions, which gradually converge under the rubric of the Vedic tradition to form what we today call Hinduism. The beginnings of this process are lost in the mists of time. Some of the elements of Hinduism can be traced to the Indus Valley Civilization, a highly advanced, peaceful culture that was at its height from 2600 to 1900 BCE. The Indus people, according to the archaeological record, knew no war and had a very high material standard of living.

Scholars have not yet deciphered the system of writing used by the Indus people. The oldest available Hindu texts are a set of hymns and philosophical reflections known as the *Vedas*. Most scholars date these texts, in their current form, to a period somewhat later than the height of the Indus Valley Civilization, between about 1500 and 1000 BCE. But it is likely that they contain some material from a much earlier period.

Ecological wisdom can be discerned in Hinduism right from the beginning of the Vedic period. The religion of the Vedas is centered on the worship divinity manifested in the form of natural phenomena. The Vedic deities, or *devas*, are all manifestations of the divine in the form of such phenomena as fire, lightning, the sun, the moon, the wind, the dawn, the ocean, sacred rivers, and the earth itself. Agni, the sacred fire; Indra, lord of the devas, whose weapon is the thunderbolt; Sūrya, the sun; Soma, the moon; Vāyu, the wind; Usā, goddess of the dawn; Varuna, lord of the ocean; the sacred river Sarasvatī, who is also the goddess of wisdom; and Bhūdevī, mother earth: all are personifications of nature, to be given reverence and gratitude. According to ancient Vedic thought, human beings are themselves an integral part of the sacred cosmic order in which these divinities operate. The success of any human venture depends upon the proper offerings–tokens of gratitude and respect–being made to the devas. This can seen, on a symbolic level, as an expression of the deep ecological insight that humanity does not stand above or outside of nature, but is part of it. Human activity carried out in mindfulness of our dependence on nature brings with it prosperity and success. But human activity carried out without this understanding leads to disaster. That is the insight of the ancient Vedic sages.

In the later Vedic period, between 800 and 200 BCE, a set of texts are composed called the *Upanisads*, or 'Secret Doctrine.' The *Upanisads* are believed to convey the inner wisdom of the Vedas. The philosophy of the *Upanisads* is called the end or aim of the Vedas–or Vedānta in Sanskrit, the ancient language in which most Hindu sacred texts are composed. Vedānta is the philosophy that underlies much of Hinduism, from ancient times to the present. It is the conceptual superstructure shared by most Hindu traditions.

According to Vedānta, the sacred power of nature, personified by the many Vedic deities, and manifested in all natural phenomena, is an infinite consciousness that is both the basis and the original essence of all that exists. Vedānta calls this ground of all being Brahman, which means 'that which is expansive.' Some forms of Vedānta see Brahman as God–the Supreme Being whose creative power, or *māyā*, brings the entire cosmos into manifestation. Other forms of Vedānta see Brahman in more impersonal terms, as both the material and efficient cause of the natural world, which has manifested the universe out of itself. And still other forms of Vedānta see Brahman as both: the impersonal basis of our existence and the personal, supreme God. As the supreme deity, Brahman can be approached in many ways, through many forms. These forms correspond to the deities of Hinduism–personifications of the supreme reality such as Vishnu, the Preserver, or Siva, the creator, destroyer, and re-creator of the world through his cosmic dance, or Sakti, the Divine Mother and power of creation. Indeed, from a Hindu point of view, all concepts of the divine, not only from the Hindu traditions, but from all religions, are legitimate and effective ways to be nearer to God. As one personification of the divine, Lord Krsna, says in the *Bhagavad Gītā*, or 'Song of the Blessed One,' one of the most popular among the Hindu scriptures, "In whatever way that living beings approach me, thus do I receive them. All paths…lead to me." (*Bhagavad Gītā* 4:11)

All of existence, according to Vedānta, is charged with the divine presence. As Krishna, again, says in the *Bhagavad Gītā*, "There is nothing whatsoever beyond me! All of this world is strung on me like pearls on a thread." (*Bhagavad Gītā* 7:7) In keeping with deep ecology, this includes ourselves. The ability to see the divine in all of nature–as the basis of all beings–is, according to Hindu thought, the insight that brings salvation. "Whoever sees me everywhere and sees everything in me will never be separated from me, nor will I be separated from him. The spiritual aspirant who is aware of the oneness of life is devoted to me, the one who dwells in all beings. Wherever he happens to find himself, he remains within me." (*Bhagavad Gītā* 6:30-31)

The 'Secret Doctrine' that the *Upanisads* convey–though it is no longer a secret, but a basic tenet of Hinduism–is that the divine reality that is the basis of all existence is also the basis of our own existence, and is, indeed, our true self: the *ātman*. This theme of the unity of being is a fundamental insight of Hinduism. As Pravrajika Vrajaprana, a contemporary teacher of Vedānta, explains, "The unity of existence is one of the great themes of Vedānta and an essential pillar of its philosophy. Unity is the song of life, it is the grand theme underlying the rich variations that exist throughout the cosmos. Whatever we see, whatever we experience, is only a manifestation of this eternal oneness. The divinity at the core of our being is the same divinity that illumines the sun, the moon, and the stars. There is no place where we, infinite in nature, do not exist."[2]

The purpose of our existence, according to Hindu thought, is to realize our true, divine nature. This Self-realization–or God-realization–is not simply a cognitive matter. It is not just a matter of understanding Vedānta philosophy on an intellectual level. It is a *realization* in the full sense–a *making real* of this insight, such that we experience it as the fundamental basis of our existence. According to Vedānta, no being will fail to attain this realization. But clearly, many of us are very far indeed from perceiving God directly within ourselves and within all beings. Cultivating this realization takes many lifetimes. According to Hindu thought, we are not the physical forms with which our consciousness currently identifies itself. Rather, we are Brahman: we are pure consciousness. Our brief identification with a physical body over the course of a lifetime is but one stage in a very long process of trial and error in which, over the course of many lifetimes, in which we identify with many bodies and many kinds of life form, we gradually realize our true self. This process of rebirth, or reincarnation, is like a school in which we are slowly learning the great cosmic lesson of the unity of being: that we are Brahman and that Brahman is, indeed, all beings. The mechanism by which these life lessons operate is karma.

The basic idea behind karma is one that is very much in tune with deep ecology. The idea is that, because of the unity of being, because all that exists is essentially one, it follows that whatever we do to another–every action, every word, and every thought–we do to ourselves. This provides a metaphysical basis for the Golden Rule, the foundational ethical principle found in most of the world's religions–that we should treat others as we wish for ourselves to be treated. The idea of karma, connected with the idea of the unity of being, makes the Golden Rule very concrete: for whatever we do to others, we in fact do to ourselves, for we are all one. In terms of karma, this means that we will eventually experience the effects of our every thought, word, and deed just as though it were done to us. The distinction between the doer of the action and its recipient is ultimately illusory. So we actually feel the effects of our every choice because we are, ourselves, the ultimate recipients of those choices.

This of course has strong resonances with the insistence of deep ecology that all of life is interdependent, and that all actions have consequences, because every part of the ecosystem affects every other part. And we are not, ourselves, exempt from these effects, being part of the ecosystem ourselves.

In ethical terms, most reflection on karma is generally applied to human relations. We should be kind to others, helpful to others, compassionate to others, and so on, since that is what we would desire for ourselves. But according to Hindu thought, the ātman, or soul, is not human. Only human bodies are human, and the body is only the temporary abode of the soul. The injunction to treat others as we wish to be treated applies not only to other human beings, but to all beings, and indeed to the earth itself. The idea that one who is human today might reborn in the form of an animal, or even a plant, is not seen to be farfetched from a Hindu perspective. In fact, this insight–that the soul, as a being of pure consciousness, can inhabit many types of form, including multiple species–becomes the basis in Hindu thought for a wide array of practices with environmental implications, such as vegetarianism and the regarding of certain animals and plants as sacred.

Hinduism is replete with sacred animals, rivers, trees, and mountains. According to Vedānta philosophy, of course, all of existence is a manifestation of Brahman, and so is charged with the divine. Even the traditional Hindu way of saying 'hello' affirms the divine presence in all beings: one puts one's hands together in the gesture of prayer called *anjali* and says, "Namaste!" This means, "I honor the God within you!" But, for many complex historical reasons, particular animals, rivers, and other natural phenomena have taken on special significance. Perhaps one could say that Hindus have perceived the divine with a certain vividness in these phenomena, resulting in their being worshiped as special foci of the divine presence. At times, these are localized phenomena–a particular sacred tree, or a specific animal, such as the sacred turtle which was recently made an object of worship by devout Hindu villagers in Orissa.[3] But in other instances, these phenomena take on a wider significance, such as the sacred River Gangā–the Ganges–whose waters are regarded by most Hindus as purifying, with many cities along its banks, such as Haridwar and Banaras, being major centers of pilgrimage. The leaves of the tulsi tree are sacred to Vaisnavas, and bel leaves to Saivas. Respect for the cow has become emblematic of Hinduism in the minds of many, in part due to Mahātma Gandhi's efforts to promote its protection as a sacred Hindu symbol. But others animals are also sacred to Hindus, and indeed all are seen as forms of the divine, to be treated with respect and reverence. Animal symbolism is closely tied to the iconographic representation of Hindu deities. Ganesa, the son of Siva and the remover of obstacles, to whom one prays before beginning any venture, has the head of an elephant. Hanuman, the ape deity, protects one from evil, especially sickness. Several *avatars*, or incarnations, of Visnu have animal forms: Matsya, the fish; Kurma, the tortoise; Varaha, the boar; and Narasimha, the manlion. And all Hindu deities are associated with animals who serve as their vehicles.

The doctrine of the fundamental unity of being that is the heart and soul of Hindu philosophy, or Vedānta, is an intellectual foundation for compassion for all beings. The understanding that the self, as pure consciousness, is not confined to any one physical form, and the doctrines of karma and rebirth, which suggest that all of us could at some point have existed in a variety of animal and plant forms, and that we may yet do so in the future–and conversely, that animals and plants could at some point have existed, and could yet exist again, in human form–is non-anthropocentric. That is, it does not place human beings at the center of–much less standing above or outside of–nature, but sees humanity, rather, as part of the larger unified and interdependent chain of being. The saving realization that God dwells in all beings, that all of life is pervaded by divinity, grounds an ethos that sees divinity in nature, producing a corresponding respect for both plant and animal life, and for the earth itself. And the history of Hinduism is full of examples of ecologically sound practices that arise from this ethos, such as a widespread, though not universal, endorsement of vegetarianism–the lower environmental impact of such a diet in comparison with meat-eating being a well-attested fact. The protection of certain animals due to their special sanctity could be seen as an ancient Hindu precursor of the modern practice of the protection of endangered species.

Conclusion

Though actual practices vary, one can see that a metaphysical foundation exists in all three of these traditions–Jainism, Buddhism, and Hinduism–for an ethos of nonviolent and compassionate behavior toward animals. The concepts of karma and rebirth, with the de-centering of human beings from the role that they hold in many other traditions as the locus of ultimate value, to a role as one possible form among the many that the soul may take, call into question our tendency to value animals only in terms of their usefulness to us. Rather, they provide a conceptual basis for seeing animals just as we see ourselves–as centers of consciousness in the cycle of life, and so deserving of the same compassion and respect.

Notes

1 For an extended discussion of Mahāvīra and his role in regard to the historical origins of Jainism, see my Jainism: An Introduction (London: I.B. Tauris, 2009), pp. 29-56.

2 Vrajaprana, Pravrajika. Vedānta: A Simple Introduction (Hollywood, CA: Vedânta Press, 1999), p. 60.

3 "A Sacred Turtle and a Controversy," Hindu Press International, http://www.hinduismtoday.com/modules/xpress/hindu-press-international/2009/11/12/a-sacred-turtle-anda-controversy/

References

Chapple, Christopher Key. 1993. *Nonviolence to Animals, Earth, and Self in Asian Traditions*. Albany, New York: State University of New York Press.

Long, Jeffery D. 2009. *Jainism: An Introduction*. London: I.B. Tauris.

Strong, John. 2007. *The Experience of Buddhism: Sources and Interpretations* 3rd edition. Wadsworth Publishing.

Thompson, George (trans.). 2008. *The Bhagavad Gita: A New Translation*. New York: North Point Press.

Vrajaprana, Pravrajika. 1999. *Vedanta: A Simple Introduction*. Hollywood, CA: Vedānta Press.

Chapter 15

Jainism's Ethics of Non-Harm

☆ *Lawrence A. Babb*

Few visitors to Old Delhi's *Chandni Chowk* fail to notice the Bird Hospital, for it is very conspicuously located, directly facing the Red Fort. The hospital is a charitable institution supported by local Digambar Jains, and it stands as a testament to Jainism's deep commitment to animal welfare. This commitment is but one facet of an ethic of non-harm (*ahimsa*, also often translated as non-violence) that Jainism has supported and promoted more than any other major religion. That Jainism posits a special kinship between the human and animal worlds is vividly symbolized by the image of the *samavasarana*, central to Jainism's concept of its teaching mission. The word refers to the great congregations that assemble to hear the discourses of Jainism are founding teachers, the Tirthankaras. In attendance are not only human beings and deities, but also animals. And while there has been some scholarly debate about the antiquity of Jainism's commitment to vegetarianism (see Alsdorf 6-15; also Dundas, *The Jains* 177), there is no question that vegetarianism is deeply intertwined with Jain doctrine and the Jain way of life. But more, the Jains are famous for extending their ethic of non-harm not only to the animal kingdom, but to the humblest forms of life, including plants and microscopic organisms. Indeed, the very organization of Jain communities is arguably in part a response to the desire to extend the protective umbrella of non-harm to the very limits of what human beings can achieve. Although Jainism is fissured by major and minor sectarian divisions, the commitment to an ethic of non-harm is shared by all Jains.[1] For the Jains, writes Padmanabh Jaini, non-harm "is not only the first among virtues but *the* virtue; all other restraints are simply elaborations of this central one" (167).

This paper provides a short sketch of the Jain ethic of non-harm and its doctrinal foundations. It also explores its implications for actual behavior. We shall see that these implications are many, and that the ethic of non-harm shapes Jain life in profoundly important ways. It can also be seen as a boundary marker reflecting the Jains' position in a wider social landscape of culturally similar and dissimilar communities. The paper concludes with a brief reflection on possible areas of common ground between Jainism and environmentalist and animal rights movements in the West.

Doctrinal Background

As will be seen later in this paper, the Jains possess a *religious culture* of non-harm that cannot be reduced to religious doctrine as such. Still, this culture is intertwined with the doctrines developed in classical Jain texts, and the Jain ethic of non-harm cannot be understood at any level without taking into account Jain teachings about the nature of the soul, the cosmos, and the causes of the soul's worldly bondage.[2]

Fundamental to the Jain world view is a division of the cosmos, itself uncreated and eternal, into living and non-living constituents. On the living side, the cosmos is as warm with indestructible, immaterial entities known as *jivas*, a word conveniently rendered as "soul" in English. The cosmos is vast, but even so there are souls everywhere within it. They possess the qualities of consciousness, energy, and bliss, and they have existed from beginningless time and will continue to exist eternally. Each soul is, in its essence, identical to all the others, but they are embodied in a vast range of different forms. Embodiment covers the soul in layers of lifeless matter, and because of this the true nature of the soul is occluded; that is, the soul cannot know itself as it truly is. Embodiments are temporary, and after the body's death the soul instantaneously migrates to a new body. The soul is thus trapped in a cycle of birth, death, and rebirth that will never end unless the soul achieves liberation. Indeed, given the infinitude of cosmic time, each soul has already inhabited every kind of body that exists and has found itself in every possible location within the cosmos, and has already done so since ages. And although the soul may experience episodes of relative enjoyment, its overall world career is dominated by pain and suffering that stretch forward into the future as a prospect of infinite pain and suffering to come. The issue, however–and this brings us to the heart of the matter–is that this cycle of birth, suffering, and death can be terminated for the individual, but to understand how such liberation is possible requires an understanding of the mechanics of the soul's bondage.

Jainism is unique among Indic religions in its understanding of the nature of *karma*. While other Indic traditions place the emphasis on action (the literal meaning of the word *karma*) and its effects, the Jains see *karma* as an actual material substance, a form of lifeless matter omnipresent in the cosmos that adheres to the soul and is responsible for its embodiment. The relationship between karmic matter and the soul presents a puzzle, because it is difficult to see how they could interact at all, given the fact that they are utterly different in essence. One image is that karmic matter floats free in the cosmos as a kind of fine dust that is somehow attracted to the soul and then sticks to it as dust does to a damp surface. The attraction is caused by the soul's embodied actions, and the adhesion to the soul is a result of the desires and passions underlying the actions. The soul experiences good or bad rebirths as a consequence of its past actions, good or bad, as manifested in the nature of its karmic accumulations.

But Jain teachings insist that no worldly existence, even the most seemingly pleasant, can be anything but bondage. "Good karma," Jains often say, is a chain of gold, and "bad karma" a chain of iron, but *all* karma is a chain that binds the soul to the unending cycle of birth, death, and rebirth.

Because karmic accumulations are the cause of the soul's bondage, liberation (*moksha*) requires two things: the influx of new karmic matter to the soul must be attenuated and stopped, and the karmic deposits of the past must be shed. Then, and only then, can the soul realize its true nature; it then rises to a special zone at the top of the cosmos where liberated souls exist without rebirth in a state of omniscient bliss, totally uncontaminated by the material world below. An account of the complex series of measures and stages of progress leading to the liberated state would be beyond the scope of this paper. Let it suffice to say that, speaking very roughly, non-harm is crucial to the control of karmic influx. The accumulation of karmic matter is a direct consequence of the actions the soul performs in its embodied form, and the karmic potency of actions is a consequence of the degree to which they engage one's desires and aversions, which is especially true of violent and harmful actions. Avoiding harm to other beings is therefore a necessary condition for liberation.

It is not a sufficient condition, however, and the equation is completed by the ascetic practices for which Jainism is justly famous. Asceticism is often said to "burn away" the accumulations of karmic matter and is therefore the royal road to liberation. While asceticism is not central to our subject, it is related to it through the idea of restraint, fundamental to Jainism, which can be seen as an area of common ground between ascetic practice and the ethic of non-harm. Ascetic practice is a factor in all Jain life, but is central to the lifestyles of initiated Jain mendicants. The mendicants are a small spiritual elite totally dependent on support from the lay community for the most basic necessities of life. Vowed to non-harm, truthfulness, non-stealing, celibacy, and propertlylessness, they are possessionless wanderers who lead lives of extreme austerity. In supporting the mendicants, lay Jains earn religious merit (*punya*) that will result in favorable rebirth and ultimately advance along the long road to liberation.

Perhaps the most striking difference between lay and mendicant life, however, is the extent to which the ethic of non-harm is pursued. Although lay Jains are certainly expected to adhere to the main tenets of this ethic, mendicants must exercise restraint and self-control at a level that requires an extraordinary extension of it. But to understand this ethical division of labor fully will require a detour into the realm of Jain biology, after which we shall return to the lay/mendicant divide.

Biology

The Jains possess a complex system of biological knowledge that can be seen, in part, as a response to their ethic of non-harm. The need for a biological system arises from the ethic's universality, by which I mean its applicability to *all* living things. The problem is that an umbrella of non-harm protecting all living things *in equal degree* would be irreconcilable with human subsistence needs for the simple reason that the consumption of food must always involve some form of harm inflicted on living things. If there is to be human life at all, therefore, some non-human forms of life will have to be more protected than others, and this requires a system of taxonomic distinctions making such discriminations possible. That is, a system of biological classification will have to exist that can enable the entire range of forms of life to be protected from harm *to the furthest extent possible*, given the exigencies of human life.

The Jains maintain that living things of all types belong to an inclusive category of "sentients" (Jaini 97). These entities are the *jivas*, or "souls," that we have already discussed–each (except for the liberated) endlessly circulating through the cosmos. In their wanderings, and as a consequence of their karmic burdens, souls find embodiment in a fantastically diverse array of life forms. Some are submicroscopic and very simple; others are very large indeed and quite complex. Jain biology is concerned with the systematic classification of these forms of life. It differs, however, from the Linnaean system in that it utilizes moral and soteriological criteria in addition to structural criteria as principles of classification. To employ McKim Marriott's usage, it is a system of "biomoral" (112) knowledge. What follows is a simplified sketch of this system.

In common with other Indic religious traditions, Jainism teaches that 8.4 million (right) actual forms of life (*yonis*) exist.[3] This is probably best seen as a default concept, a generic South Asian vision of the web of life. Specific to Jainism, however, is a taxonomy that distributes all forms of life into four great categories called *gatis*, or "destinies." They are: 1) animals and plants, 2) dwellers in hell, 3) human beings, and 4) gods and goddesses.

The lowest of the classes, and also the most internally complex, is the animal and plant category. Vast in extent, it includes the most microscopic of beings together with highly complex forms of animal life. We shall return to this category for closer look shortly. Not as low as the plants and animals, but probably the most miserable of beings, are those who dwell in multiple layers of the Jain hell; they endure punishments eons long as they exhaust the karmic deposits of their past misdeeds. Their opposites are the gods and goddesses who inhabit the several levels of the Jain heaven. They are reaping the rewards of meritorious actions in former lives, and they live for vast amount of time in states of total enjoyment. As in the case of the dwellers in hell, however, the karmic residues that account for their status will be used up in the end, and they will find themselves once again in the round of birth, death, and rebirth. Soteriology enters the picture in the case of the category of human beings. According to Jain teachings, liberation is possible only in a human body. This is because the kinds of ascetic practices that lead to liberation are not possible for other forms of life. Thus, a human birth is the most valuable of all.

Most relevant to the ethic of non-harm, however, is not the four-*gati* system but another taxonomic scheme that internally differentiates the animal and plant class. The issue of harming does not arise with respect to deities or denizens hell for the obvious reason that these beings are beyond the reach of human behavior, good or bad. What remains are other human beings and animals and plants. Humans are of course automatically covered by the basic injunctions against inflicting pain and killing, a matter that requires more reiteration and emphasis than elaboration. In the case of the large category of animals and plants, however, the sheer variety of contexts in which we encounter these beings–including and especially dietary contexts–necessitates complex distinctions. Significantly, these distinctions are largely based on the number of senses these forms of life possess. This fact points directly to the capacity to sense, and thus the capacity to suffer, as a key differentiating criterion; at its heart, that is, this is a scheme based on the idea susceptibility to violence-inflicted pain.

At the very bottom of the scale are submicroscopic beings known as *nigodas*. They possess only one sense, that of touch, and they are found in colonies in every cranny of the cosmos where they wink in and out of their brief spans of life. So tiny and ubiquitous are they that they even inhabit the bodies of human beings and other creatures. Somewhat higher on the scale are souls embodied as the four basic elements of earth, air, fire, and water. They, too, are one-sensed, but they differ from the *nigodas* in that they are not omnipresent; rather, they are embodied as the four elements, and they can live only where these elements exist. Their presence in nature means that earth, air, fire, and water must be regarded as alive and capable of feeling pain and being injured or killed. Moving upward beyond these simplest forms of life, we encounter the divergence between the animal and plant categories. Plants are more complex than the simplest forms of life, but they are similar in possessing the single sense of touch. Some plants exist only as individual entities–that is, a single body for each soul. Others accommodate vast swarms of souls (said to be infinite) per body. This distinction is quite important, for it bears directly on certain dietary rules.

Animals stand apart from the one-sensed forms of life in two respects. First, they (along with human, dwellers in hell, and deities) are mobile. Second, they have senses other than the sense of touch. Their mobility can be seen as a corollary of their additional senses; it enables them to escape sources of pain to which they are more vulnerable than the simple one-sensed entities. The internal ordering of the animal category is a hierarchy of genera based on number of senses. The lowest forms of animal life possess two senses; the highest, five. The two-sensed animals–which are very simple creatures such as worms, certain mollusks, etc.–have only the senses of touch and taste. Three-sensed creatures–mainly various sorts of simple insects–have the senses of touch, taste, and smell. Four-sensed animals–flying insects such as flies and bees are prominent in this category–add sight to the mix. The highest forms of animal life–fish, birds, mammals, etc.–add hearing. All life forms belonging to the other three *gatis* also possess five senses.

The foundation of the Jain ethic of non-harm is the assumption that each and every living thing, of whatever category or level, has the capacity to suffer. This is true even of the one-sensed beings; in their primitive way, they feel pleasure and pain through their single sense of touch and can be harmed by the actions of humans. This is the ethic's universality. And although these beings may seem inconsequential, inflicting harm on them creates negative karmic residues that give rise to worldly troubles and inhibit progress on the path to liberation. Nonetheless, the karmic consequence of harming one-senses beings is less than that of harming more complex forms of animal life; the simple rule is that the more the senses of the victim, the greater the karmic harm to the inflictor of harm. This is extremely important, because it makes possible the application of a universal ethic of non-harm in degrees: the strictest observance will require avoiding harming those entities with only one sense (along with all the rest); minimal observance will require avoiding harm to beings with five senses. It seems likely that this scheme evolved precisely in order to make such gradations possible.

But to integrate such a scheme into a workable set of socially transmitted norms concerning permissible (or at least tolerable) levels of harm inflicted on others also imposes requirements on the organization of society. If some members of society are to live in a manner protective of *all* forms of life, other members of society will have to inflict harm on *some* non-human forms of life, and this will require a distinction between social as well as biological genera.

The Social Organization of Non-harm

Jain society is idealized as the *chaturvidh sangh*, the "fourfold order" of society, consisting of monks and nuns and laymen and laywomen. This points to the fact that the division between the mendicant elite and the lay majority is seen as being as fundamental as the gender divide; indeed, it is even more basic in some ways. As noted already, the social break between mendicants and laity has two main features. The first is that mendicants lead lives at a level of austerity and self-denial not required of laity. The second and more relevant to this paper is that mendicants must adhere to the ethic of non-harm at a much higher level of inclusiveness than the laity. Let us look at this latter circumstance in greater detail, beginning with the laity.

Lay Jains must shape their lifestyle in a way that avoids harm to living things, especially multi-sensed forms of life. One implication of this is occupational: Jains must avoid occupations that involve inflicting harm on human and non-human living things. For this reason, trade and banking–fields that involve little physical violence–have been the occupations most favored by Jains (at least in northern India). It is by no means the case that all Jains are in trade; they are found in many other occupations, and apparently most of the Jains of Karnataka are actually farmers (Sangave 259), but Jains are very prominent in Indian commerce. There have indeed been Jain military men, but the martial life is obviously not highly esteemed by Jains.

But the area of Jain life most highly regulated by the ethic of non-harm is diet. At a bare minimum, lay Jains are vegetarians, and it is fair to say that no Jain who abandons this requirement can be regarded as truly Jain. But there is much more. Jains, even lay Jains, must not only avoid eating meat, but must also avoid consuming certain vegetables because of the extra harm to non-human life involved in their preparation and consumption. These restrictions are less strict than the basic requirement of vegetarianism, but the orthoprax adhere to them. For example, vegetables that grow underground should not be eaten because they contain multiple souls and/or because their harvesting requires injuring and killing creatures that live in the soil. This means that Jains are not only supposed to eschew onions and garlic (also avoided by some non-Jain groups), but also such widely consumed items as potatoes and carrots. Figs and other seedy fruits and vegetables should be avoided because each seed contains a separate soul. Honey is avoided because of the multitudes of microscopic beings that swarm as its sweetness ferments and also because of the violence required to obtain it (honey is like the "blood of bees," some say). Alcohol is forbidden because of the fermentation, and hence the death of countless microscopic creatures, that its manufacture requires. Leafy green vegetables are seen as suspect because of the possibility that small creatures may lie hidden in crevices in the leaves. Eating after sunset is forbidden to Jains because small insects might fall into the food undetected. This latter restriction is very difficult to observe in today's India, especially by shopkeepers who must stay open into the darkness of night, but it remains a significant ideal and is often mentioned by mendicants in their sermons.

Water is a source of special concern because of the multitude of small living things that it contains. Water spigots in Jain households are typically equipped with cloth filters to prevent the inadvertent consumption of small creatures. Jains tend not to allow pots and other containers of water to stand unemptied overnight lest microscopic organisms multiply within. Jains sometimes drink the washings of their plate at the end of a meal; were this not done, the careless diner would be responsible for the deaths of innumerable small creatures that would otherwise arise in the dirty water with which the plates were washed. Jains also favor boiled water, an option for laity but required of mendicants. It is true that boiling kills living things, but the harm done by boiling (which is done only by laity) is mitigated by the fact that *not* boiling the water would allow additional living things to arise in it, and the deaths of these would then be added to the total harm done when the water is consumed.

But if the ethic of non-harm shapes lay life in important ways, this is even more profoundly true of mendicant life. Although laypersons may undertake a formal vow (made only by some) to avoid harm to all creatures with two or more senses, the mendicant must vow to extend the umbrella of his or her restraint from harm to beings of one sense. In this perspective, it is highly significant that of the five vows that mendicants must take upon initiation, the first is that of non-harm.

Attention to the welfare of the most microscopic of beings is responsible for some of the most noteworthy and even dramatic features of mendicant life among the Jains. To protect earth-bodied beings, mendicants cannot dig in the ground; to protect water-bodied beings, they cannot bathe or even walk in the rain. This latter prohibition means that although mendicants are normally peripatetic, they must not travel during the rainy season. In order to protect fire-bodied beings, mendicants can neither extinguish nor start fires (by starting a fire one is responsible for the deaths of the fire beings one has created). The prohibition on starting or using fires extends (at least in some sects) to a prohibition on turning on or off any electric device on the theory that electricity consists of living fire-bodied beings. To protect air-bodied beings, mendicants cannot fan themselves or even move suddenly, and mendicants of some Shvetambara denominations wear cloth shields permanently over their mouths in order to prevent their hot breath from harming airborne forms of life. Mendicants are given brooms at the time of their initiation with which to push aside small insects before sitting or lying on any surface. Even the vow of celibacy can be interpreted as an aspect of the ethic of non-harm; innumerable microscopic beings swarm in sexual fluids for whose demise one engaged in the sex act would be responsible.

The extreme severity of the monastic disciplines associated with non-harm plays a central role in lay-mendicant relationships. Among the many things that mendicants cannot do is prepare food, which necessarily entails the destruction of myriads of living things. Thus, they depend completely on laity for nourishment, the most basic of life's needs, along with much else. The result is a religio-social system founded on a basic transaction uniting the lay and mendicant communities. The mendicants are visualized as being on a more direct road to liberation than the laity. It is appropriate, therefore, that they adhere to the dictates of the ethic of non-harm to the fullest extent humanly possible. Laymen and laywomen likewise pursue liberation as a long-term goal, but their path is less direct. They must make their way in the world in a manner that requires some compromise of the ethic of non-harm (and also less ascetic self-denial). So when, for example, a housewife feeds a mendicant, the karmic residues of the violence occurring in the cooking of the food fall on her, not on the mendicant recipient. In recompense, however, she receives religious merit (*punya*) that will generate good fortune in her present lifetime or in subsequent births and assist her on the long road to final liberation. More generally, the lay community benefits from the teaching of mendicants in return for their support. For his or her part, the mendicant is protected from the karmic residues of food preparation by the requirement that the food not be prepared on his or her behalf. This is a very serious requirement, and to ensure its fulfillment the mendicant is expected to move about randomly while seeking alms from the laity. The mendicant on alms-rounds should move "like a cow grazing," Jains say–at random, a little here, a little there.

These customs and usages suggest that Jainism can be regarded as a limiting case for an ethic of non-harm. By pushing it as far as it can logically go–that is, by extending its reach even to the microscopic world–Jains have pushed it beyond the realm of the socially realizable for all members of a given society. You cannot combine a universalist ethic of non-harm (*i.e.*, an ethic protecting all forms of life whatsoever) with the satisfaction of human subsistence needs without making a sinner out of somebody. Thus, in order that some may avoid the sin of harm altogether (or nearly so), the religio-social order must be structured in such a way that others take it upon themselves.

Compassion

At one level, the Jain ethic of non-harm can be seen as primarily a soteriological strategy. Thus, Padmanabh Jaini characterizes Jain non-harm as rooted in the idea that violence is *unwise;* that is, knowing what the Jains know about our situation in the cosmos, only a fool would behave violently because harming others is self-harm in the end. Here is the way Jaini puts it:

> *Himsa* has ordinarily been understood in India as Harm done to others; for Jainas, however, it refers primarily to injuring *oneself*–to behavior which inhibits the soul's ability to attain moksa [liberation]. Thus the killing of animals, for example, is reprehensible not only for the suffering produced in the victims, but even more so because it involves intense passions on the part of the killer, passions which bind him more firmly in the grip of samsara (167).

It must not be supposed that this interpretation of Jain non-harm actually precludes a sense of remorse about the effects of one's violent acts on others as a motivating factor, for it does not, but its main emphasis is on the effect harming others has on the actor's own liberation prospects because of the desires and aversions–the passions–with which harm and violence have a mutually reinforcing relationship.

Even so, while there is no question that the ethic of non-harm has part of its foundation in Jain soteriology, this does not exhaust the matter. This missing ingredient is compassion. Far from being monopolar, the Jain ethic of non-harm is deeply intertwined with the concept of compassion (*karuna*), which is central to the Jain view of the world and is based on the idea of fellow feeling between self and others. As Kristi Wiley has noted, compassion has tended to take a back seat to other concepts in Western scholarship on the Jains (438-9), but in Jain writings it is closely associated with the ethic of non-harm, and this is true both in contemporary writing and in classical texts.

Wiley's excellent discussion shows that the fit between the concept of compassion and Jain soteriology is far from seamless. The difficulty is that compassion is an affective mental state involving pity or empathy of some kind. This being so, it would seem to be in tension with Jainism's principal goal, which is to achieve an affectless state beyond all desire and aversion. The result is what seems to be a tendency to emphasize less the image of compassion as feelings of pity for concrete others and more the notion of compassion as generalized wish to assist all beings on the path to liberation. There is, moreover, disagreement within Jainism on the spiritual value of actual positive acts to alleviate the sufferings of others. Acharya Bhikshu, (1726-1803), founder of the Terapanthi sect of the Shvetambara branch of Jainism, famously argued that mitigating the physical sufferings of others cannot be, in itself, spiritually beneficial; spiritual benefits can only arise, he insisted, from imparting religious instruction to or alleviating the fears of others. This doctrine of non-assistance, however, is a minority view within the Jain world.

However, leaving aside convoluted (and sometimes casuistic) doctrinal debates about how compassion should be understood and acted upon, when we turn to Jain life as lived we find a very different picture. It is here that religious *doctrine* gives way to religious *culture*. On the basis of my personal contact with Jain communities, I can say that the ethic of non-harm is not usually seen in primarily soteriological terms, nor are ordinary Jains much preoccupied with abstruse debates about how compassion might fit or not fit with soteriological theory. Rather, the concept of compassion is part of a cultural complex that is rooted in a genuine sense of empathy for the sufferings of others and that encourages–in many, at least–deep revulsion at the spectacle of cruelty or violence, whether perpetrated against humans or non-human living things. Far from being a monopolar ethic, the Jain ethic of non-harm considers the feelings of others to matter very much. The basic idea, often reiterated, is that other souls, whatever their embodiment, suffer pain and wish to live, just as oneself. And although this in an ethic that manifests itself principally as restraint, it is not merely negative. Rather, it posits a vast kinship that encompasses every form of life, an idea that encourages goodwill and friendliness toward all living things.

Social Boundaries

Leaving Jainism aside for the moment, it is widely believed that non-harm is also a teaching of that somewhat amorphous religion called "Hinduism." This is only partly true. Hinduism is best seen as a bundle of religious traditions that share certain family resemblances but are also quite unlike in other respects. It is indeed true that many Hindu religious traditions promote non-harm, but it is also the case that many groups considered to be Hindu do not share this ethic. Perhaps most notable among these are groups seen as belonging to the *Kshatriya* category. Classical Hindu law books divided society into four great classes: *Brahmans* (priests and teachers), *Kshatriyas* (warriors and rulers), *Vaishyas* (originally farmers, but later coming to refer to the merchant class), and *Shudras* (whose function was to serve the other classes). Groups such as the Rajputs–who were once the ruling aristocracy of much of northern India–belong to the *Kshatriya* category. Although they are Hindus, they adhere to a warrior ethic, bear a martial culture, are non-vegetarian, place great emphasis on martial valor, pursue military careers, enjoy blood sports, and practice animal sacrifice. Certain other Hindu communities share some or all of these traits to varying degrees. Non-harm is thus normative only among some groups in the Hindu world.

Of the groups that do adhere to the ethic of non-harm, the most prominent are the traders and bankers, those traditionally assigned to the Vaishya category. In northern India, the traditional trading castes belong to a socially recognized cluster that carries the label "Baniya." Some of these castes are Jain, some are Hindu, and some have both Jain and Hindu subdivisions. The collective identity of these groups as a cluster is partly based on their common occupation of trade and banking, but it also reflects the fact that they are culturally quite similar, irrespective of whether they are Jain or Hindu. At the core of their cultural similarity is the ethic of non-harm. While there are of course wide variations in the actual temperament of individuals belonging to these groups, their value systems stress the desirability of the vegetarian, non-martial, non-animal-sacrificing, non-blood-sport-loving life of the merchant. We see, therefore, that non-harm is not a monopoly of the Jains. Rather, it is arguably a generic feature of trading-caste culture, at least in India's north. From this standpoint, the Jain non-violent lifestyle has the appearance of a variant form of this culture. It is arguably an extreme variant, but it is a peninsula, not an isolated island.

So strong is this association between the ethic of non-harm and trade that the non-violent, vegetarian lifestyle can be seen as a social boundary marker for the trading communities as a cluster. By this I do not mean that these groups are vegetarian in order to mark themselves off from other groups. The sources of trader non-harm are religious, not sociological. That said, it is also true that an accentuation of the ethic of non-harm in trading-caste culture does indeed establish a cultural boundary between traders, as a group of similar groups, and the old martial aristocracies of such regions as Rajasthan (on this point see Carstairs). Indeed, the contrast is virtually a polar opposition, with Rajputs and traders each constituting the cultural "other" of the other. Elsewhere (Babb *Alchemies of Violence*) I have argued that a culture of non-harm enabled trading groups to coexist with martial groups in the same social space without challenging their legitimacy as rulers, thus making it easier for traders to serve, as they so often did, as bankers and advisors in the indigenous states of the subcontinent. But whether this is true or not, there can be little doubt that trader non-harm is part of an identifiable cultural complex that has its own distinctive niche in the variegated quiltwork of Indian cultures.

From this perspective, what is most distinctive about the Jain ethic is not non-harm as such, but its extraordinary extension to the entire kingdom of life. If other vegetarian groups prohibit the eating of meat, Jain vegetarianism (for the orthoprax, at least) takes matters much further: they avoid vegetables that grow underground, distrust green leafy vegetables, prefer boiled water, forbid eating after sundown, and so on. And of course the lifestyle of initiated mendicants presses matters even further, extending the ethic of non-harm to the protection of the most microscopic forms of one-sensed life. These practices have doctrinal foundations that we have already surveyed. Still, it is also the case that these severe restrictions on diet, movement, and so on can also be read as another boundary-marking symbolism, appended as an addition to non-violent vegetarianism. If non-harm marks off traders as a class from martial communities, the intensity of Jain non-harm establishes a boundary between them–as vegetarians who are even more vegetarian than other vegetarians–and other trading communities that are otherwise culturally quite similar.

Translatability

It is tempting to regard the Jain view of the natural world and accompanying ethic of non-harm as an Indic version of two movements that have acquired significant followings in the contemporary West: the environmental and animal rights movements. A closer look, however, reveals significant differences in respect to both movements.

It is hard to argue that there is a Jain environmental ethic as such.[4] To say this is not to criticize anything but merely to highlight the fact that environmentalism, as we now know it, has a specific cultural and historical context, and that context is Western and modern. John Cort has put the matter succinctly in pointing out that environmentalism is a new "episteme" that has achieved prominence only recently and has no clear analogue in Jain tradition ("Green Jainism?" 65-66). To this he adds the observation that Jain soteriology is a poor foundation for such an ethic: "At its heart is the goal of permanent separation of the soul from all matter. In such a dualist ideology positive environmental results are largely incidental" (70). Paul Dundas goes even further, noting that in the Jain world view, nature is "perceived not as a source of comfort or succor, but as a potentially menacing force that by its mere existence has the capacity to threaten the embodied soul at all times " ("The Limits" 98), and to this he adds a warning against "the attempt, however well-intended it may be, to compel a traditional soteriological path originally mapped out for world-renouncing ascetics to fit the requirements of modern, ultimately secular, Western-derived agenda" (111).

We find that similar caveats might be raised about attempts to press Jainism into the mold of the Western animal rights movement. This movement draws partly from western traditions of political theory based on the concept of natural law and the derivative concept of the natural rights of the individual. The basic idea is that human rights, as envisaged in natural rights theory (and as enshrined in such documents as the United Nations Universal Declaration of Human Rights), should be extended to the non-human animal kingdom. There are other approaches to the issue of animal rights within the movement, most notably Peter Singer's utilitarian approach, but the entire movement is firmly anchored in western philosophical traditions. However, to the best of my understanding, the Jain ethic of non-harm does not employ a discourse of "rights" at all. Rather, its foundational concept is compassion. Furthermore, as Laidlaw (162 n. 11) points out, it is difficult to imagine that Western animal rights advocates could ever agree with the universalism of the Jain ethic of non-harm. This would entail imputing human rights to bacteria and including viruses in the "greatest number" (the greatest happiness of which is the utilitarians' goal). This is not a small point, because the universalism of the Jain ethic of non-harm is not a disposable adjunct of Jainism; rather, it is interwoven with the most basic Jain ideas about the nature of life and the cosmos.

But I would like to conclude by suggesting that these debates about whether Jainism is or is not the same thing as western environmentalism or animal rights advocacy are beside the point. In fact, Jains are natural allies of these movements; indeed, Jains have been conspicuous supporters of the environmentalist and animal rights movements. If Jainism is not in itself a form of environmentalism, there is surely plenty of material in the Jain understanding of life and its place in the cosmos that can be taken in that direction if Jains wish to do so. As Chapple points out, for example, there is the potential for agreement between the Gaia hypothesis and the Jain view of the ubiquity of life in the cosmos (*Nonviolence to Animals* 68-71). Readers who wish to see what the outlines of such an environmental ethic might look like should investigate L. M. Singhvi's *The Jain Declaration on Nature* (reprinted as an appendix to Chapple Jainism and Ecology and elsewhere). And if Jain soteriology, strictly speaking, has little affinity with animal rights advocacy, Jain religious culture is another matter. As we have seen, this is a religious culture permeated by the concept of compassion. Respect for and kindliness to other forms of life come easily when you know that you have been in the position of other creatures before and will probably be in that position again. One is therefore compelled to say that although the Jains and Western defenders of the environment and non-human species come from different starting places, and although the roads they travel are different in very significant ways, their destination is the same.

Notes

1 The two main branches of Jainism are divided on a point of monastic discipline, namely whether Jain monks should wear clothing Monks and nuns belonging to the Shvetambara ("white clad") sect wear white garments; monks (not nuns) belonging to the sect known as Digambara ("space clad") are nude.

2 The doctrinal foundations of Jainism can be explored in a number of well known works on the subject. These include Dundas, Jain, and Jaini. For studies emphasizing the ethnography of Jainism, see Babb *Absent Lord*, Cort *Jains in the World*, Laidlaw, and Vallely.

3 In the interest of brevity, the sketch given here omits various distinctions (such as that between womb-born and spontaneously-generated animals and humans) that are actually quite important within the scheme as a whole. A detailed account of Jain biology can be found in Javeri and Kumar.

4 On these debates, see the essays in Chapple's *Jainism and Ecology*.

References

Alsdorf, Ludwig. 2010. *The History of Vegetarianism and Cow-Veneration in India*. Trans. Bal Patil and Revised by Nichola Hayton. Ed. with additional notes, bibliography, and appendices by Willem Bollée. London and New York: Routledge.

Babb, Lawrence A. 1996. *Absent Lord: Ascetics and Kings in a Jain Ritual Culture*. Berkeley: University of California Press.

_____. 2004. *Alchemies of Violence: Myths of Identity and the Life of Trade in Western India*. New Delhi: Sage.

Chapple, Christopher Key, ed. 2002. *Jainism and Ecology: Nonviolence in the Web of Life*. Cambridge: Harvard University Press.

_____. 1993. *Nonviolence to Animals, Earth, and Self in Asian Traditions*. Albany (NY): State University of New York Press.

Cort, John E. 2002. "Green Jainism? Notes and Queries toward a Possible Jain Environmental Ethic." In *Jainism and Ecology: Nonviolence in the Web of Life*. Ed. Christopher Chapple. Cambridge: Harvard University Press. 63-94.

_____. 2001. *Jains in the World: Religious Values and Ideology in India*. Oxford: Oxford University Press.

Dundas, Paul. 2002. *The Jains*. Second Edition. London and New York: Routledge.

_____. 2002. "The Limits of a Jain Environmental Ethic." In *Jainism and Ecology: Nonviolence in the Web of Life*. Ed. Christopher Chapple. Cambridge: Harvard University Press. 95-117.

Javeri, Jethalal S. and Muni Mahendra Kumar. 2008. *Jain Biology*. Ladnun: Jain Vishva Bharati University.

Jain, Jyoti Prasad. 1983. *Religion and Culture of the Jains*. New Delhi: Bharatiya Gyanpith.

Jaini, Padmanabh S. 1979. *The Jaina Path of Purification*. Berkeley: University of California Press.

Laidlaw, James. 1995. *Riches and Renunciation: Religion, Economy, and Society among the Jains*. Oxford: Clarendon Press.

Marriott, McKim. 1976. "Hindu Transactions: Diversity without Dualism." In *Transaction and Meaning: Directions in the Anthropology of Exchange and Symbolic Behavior*. Ed. B. Kapferer. Philadelphia: Institute for the Study of Human Issues. 109-42.

Sangave, Vilas A. 1980 *Jaina Community: A Social Survey*. 2nd rev. ed. Bombay: Popular Prakashan.

Singhvi, Satish Kumar. 2002 *The Jain Declaration on Nature*. Appendix to *Jainism and Ecology: Nonviolence in the Web of Life*. Ed. Christopher Chapple. Cambridge: Harvard University Press. 217-24.

Vallely, Anne. 2002. *Guardians of the Transcendent: An Ethnography of a Jain Ascetic Community*. Toronto: University of Toronto Press.

Wiley, Kristi L. 2006. "Ahimsa and Compassion in Jainism." In *Studies in Jaina History and Culture: Disputes and Dialogues*. Ed. Peter Flügel. London and New York: Routledge. 438-455.

Chapter 16

The Qur'ân, Shari'a Law and the Exclusion of Nonhuman Animals

☆ *Rod Bennison*

Stewardship and Care, and Islamic Law and Traditions

Like the Bible, the *Qur'ân* provides an equally judicious foundation for the human interaction with nonhuman animals based on principles of stewardship and care, but unlike Christian precepts, the distinction between the religious and the secular in Islam is less clear. Several writers have recently examined how Islam provides an ethic of environmental concern and nonhuman animal protection, they include Al-Masri (1986, 1987 and 1992), Callicott (1994), Foltz (2000 and 2006), Foltz *et al* (2003), Hamed (1993), Izzi Deen (1991 and 1992), Khalid and O'Brien (1992), Wersal (1995) and Zaidi (1981). Yet, such considerations as paid by these authors are not limited to the recent past, as the works of the Ikhwān al-Safā' from the 10th century (1978), and ibn Rushd (Averroes, 1978) and Farid al-din 'Attar (1984) from the 12th century, amongst others, show.

Islam, like Christianity, is a revealed religion, where it is claimed that the Prophet Muhammad received a revelation from God. This revelation is laid down in the *Qur'ân*, and is supplemented by a set of traditions about Muhammad's words and deeds, including the hadiths, which are authenticated sayings ascribed to Muhammad by his companions. The Shi'a, Sunni and other doctrines each have their own particular traditions, which are in turn supplemented by rules and

---

This paper is a revised version of the one appeared in *The Journal of the Commonwealth Veterinary Association.*

analogies drawn from the Muslim community by consensus and precedent, as noted by Wescoat (1995). Importantly, and unlike Christian and Jewish precepts, these sources dictate that what is legal and what is ethical in Islam are seen as inseparable.

Muslims believe that Islamic law (called Shari'a, literally the 'source of water') was laid down by God and that human actions should reflect what that law says. Muslims hold that all facets of human life and action exhibit an interconnectedness or unity, known as At-Tawhid (Zaidi, 1981), and that any attempts to break the law must be prevented (Izzi Deen, 1991). This is unlike Western legal concepts and ethical principles which have developed through processes that are more humanistic (Izzi Deen, 1991), and where Christians and Jews have, in the main, relied much less on religious teachings and more on secular moral and ethical principles developed over the centuries, especially from the time of the Scientific Revolution in the 17th century. Izzi Deen notes the division of the Shari'a relevant to human action into five categories; namely, the obligatory (wajib), the recommended (mandub), the permissible (mubah), the offensive (makruh) and the forbidden (harām) (1991). As al-Masri points out, human actions relating to nonhuman animals should therefore be conceivably categorised as to their effects (1987).

Like the *Bible*, the *Qur'ân* and the hadiths, which form the basis from which the Shari'a can be applied, are explicit as much as they are vague. Much has been open to interpretation and, just like in Christianity, Islam too has suffered from a broad interpretation and from cultural underpinnings from a remnant and historic past. For example, the *Qur'ân* first developed within Arab culture and most adherents are no longer Arabic. Local traditions like nonhuman animal sacrifice and other practices involving nonhuman animals have remained, but have since been greatly regulated by religious edict or have even been declared as harām (Foltz, 2006).

Despite this, the *Qur'ân* is in some places most specific. For example, it says that the "creation of the heavens and earth is greater than the creation of men" (*The Believers 40:57*), and although "God has created every beast of (the common substance) water" (*Light 24:45*), "it is he (God) who appointed you (people) as viceroys in the earth" (*The Angels 35:39*). The *Qur'ân* explicitly states human dominion over other life forms, but notes that vicegerency or trust was only made after it was first offered to other 'beings' within nature, which refused it; as the *Qur'ân* says:

> *we offered the trust to the heavens and the earth and the mountains, but they refused to carry it and were afraid of it; and man carried it* (*The Confederates 33:72*).

Some assert that such a stewardship tradition is in itself a test of how humans should act as caretakers (Zaidi, 1981), while others believe that, this vicegerency is actually a test of "how responsibly humans will act as trustees" and maintain environmental integrity (Wersal, 1995, p. 452). Al-Masri (1987) further notes that respecting nature is obligatory.

Undoubtedly, the *Qur'ân* emphasises human dominion over nonhuman animals as much as it is asserted in the *Bible*. It also stipulates that nonhuman animals were created for human use; for example, it states that:

> *And the cattle – He created them for you; in them is warmth, and uses various, and of them you eat, and there is beauty in them for you, when you bring them home to rest and when you drive them forth abroad to pasture; and they bear you loads unto a land that you never would reach, except with a great distress ... And horses, and mules, and asses, for you to ride, and as an adornment; and He creates what you know not* (*The Bee 16:4-9*).

The *Qur'ân* states that "God created that not save with the truth" (*Jonah 10:5*), for it is "he (who) created everything and then ordained it very exactly (with due potential)" (*Salvation 25:2*) not to allow any "change to God's creation" (*The Greeks 30:30*). Zaidi notes that the *Qur'ân* actually forbids human actions which may lead to harm (1981); the *Qur'ân* states:

> *transgress not in the balance, and weigh with justice, and skimp not in the balance .... earth, He set it down for all beings* (*The All-Merciful 55:8-10*).

Humans are not only entrusted with due care and should actually conserve the environment, but should show a reverence for nature, as is noted in the *Musnad of Ahmad hadith* that "on doomsday, if any one of you has a palm-shoot in hand, he should plant it" (*5:440* and *3:114*, as cited by Regenstein, 1991, p.257). Similarly, the *Bukhārī hadith* notes:

> *there is none amongst the Muslims, who plants a tree or sows seeds, and then a bird, or a person or an animals eats from it, but is regarded as a charitable gift for him* (*Bukhārī hadith 10:3764-70*, as cited in Foltz, 2006, p.19).

The *Qur'ân* provides for a caring stewardship at least equal to the *Old Testament*. In the case of nonhuman animals, the *Qur'ân* is often more specific than the *Bible*. For example, the *Qur'ân* states that there is:

> *No creature crawling on the earth, no bird flying with its wings, but they are nations like unto yourselves* (*Cattle 6:38*).

The passage indicates that all nonhuman animals are in fact 'peoples' or 'communities', not unlike humankind (Al-Masri, 1986); hence, nonhuman animals should be afforded due respect and protection, and are due as much respect as other human communities. At this point, it should be indicated that the *Bible* does nevertheless contains similar references, albeit not as inclusive, such as:

> *for that which befalleth the sons of men befalleth beasts; even one thing befalleth them: as the one dieth, so dieth the other; yea, they have all one breath; so that a man hath no pre-eminence above a beast; for all is vanity* (*Ecclesiastes 3:19*).

The Islamic tradition teaches an interconnectedness and orderliness of nature, indicative of a perceived design and purpose where all of 'creation' is in fact a "divine work of art" (Callicott, 1994, p.32). Like Christianity, but unlike Aristotelian philosophy, this order is supposedly sustained by God; its complexity a sign of God's omnipotence. 'Creation' is by implication to be treated with the utmost care and respect. Humans, therefore, are enjoined to act responsibly toward the environment; further, that such actions should include the care and protection of nonhuman animals (Wersal, 1995). For example, humans are told to protect she-camels:

> *Leave her that she may eat in God's earth, and do not touch her with evil, lest you be seized by a painful chastisement* (*Battlements 7:73*).

Al-Masri (1986 and 1987), amongst others, notes the specific Islamic teachings that would be relevant for instilling an ecological ethic inclusive of nonhuman animals, teachings that prescribe actions that Muslims would specifically describe as obligatory. These teachings can be gleaned from the *Qur'ân* and hadiths and range from describing actions that may be described as being wajib, to those that are harām.

Significantly, it is implied in the *Qur'ân* (in *Cattle 6:38*, which is discussed above) that human and nonhuman animals must share resources (Al-Masri, 1987). In fact, the *Qur'ân* is quite specific in stating that food and other resources should be shared equitably between species when it says:

> *let man consider his nourishment .... an enjoyment for you and your flocks (He Frowned 80:24-32)*; and, *no creature is there crawling on the earth, but its provision rests on God (Hood 11:6).*

In other words, Islamic teachings note that it is God, and not humans, who controls the division of resources. Further, the *Qur'ân* and hadiths speak explicitly about the "sanctity of life as a whole, and declare animals as possessing souls, and place animals physically on a par with human beings" (Al-Masri, 1987).

Foltz also notes that, within Islam, nonhuman animals can receive divine revelation (2006, p.17), as when it states in the *Qur'ân*:

> *And the Lord revealed unto the bees, saying: 'Take unto yourselves, of the mountains, houses, and of the trees, and of what they are building. Then eat of all manner of fruit, and follow the ways of the Lord easy to go upon* (*The Bee 16:70*).

In relation to the sanctity of life, the *Qur'ân* states that:

> *God has created every beast of water, and some of them go upon their bellies, and some of them go upon two feet, and some of them go upon four* (*Light 24:44-45*).

So, all life forms therefore have an implied right to care and protection, as do humans; also borne out in the *Mishkāt al-Masābīh hadith* in a similar way to the Pythagorean tradition, which states:

> *a good deed done to a beast is as good as doing good to a human; while an act of cruelty to a beast is as bad as an act of cruelty to a human* (*Book 6, ch.7, 8:178*, as cited by Al-Masri, 1992, p.18).

Islamic teachings dictate the application of the same legal and ethical codes to both humans and nonhuman animals alike. Most specifically, Wescoat examines the right of thirst for nonhuman animals in Islam, and notes that human and nonhuman animals are equally dependent upon God for water and life (1995), noting that the *Qur'ân* states:

> *We loose the winds fertilising, and We send down out of heaven water, then We give it to you to drink, and you are not its treasurers* (*El-Hijr 15:22-23*, as cited by Wescoat, 1995, p.263).

Thus, for Muslims true to their faith, human actions should encapsulate, at the very least, the rights of nonhuman animals to water (Wescoat, 1995); ironic given the long history in Muslim countries of diverting water for irrigation, by damming rivers and flooding or drying wetlands. The point can be extended when discussing the slaughter of meat by the killing of domesticated nonhuman animals.

Islamic law is most prescriptive in its insistence on humane treatment. The killing of nonhuman animals for meat and hides by halāl (that is, permissible based on set ethical and religious standards) methods is obligatory, with meat considered forbidden if the nonhuman animal has in any way been subjected to inhumane treatment (Al-Masri, 1987). Regenstein cites several sources that stipulate the need to ensure humane and efficient practices are fulfilled, including edicts from the second and fourth Caliphs (1991, p.254).

The *Muslim hadith* stresses humane slaughter when it says that:

> *Verily Allah has enjoined goodness to everything; so when you kill, kill in a good way and when you slaughter, slaughter in a good way. So everyone of you should sharpen his knife, and let the slaughtered animal die comfortably* (*Muslim Hadith 21:4810* as cited by Foltz, 2006, p.26).

If nonhuman animals are killed humanely and efficiently, as prescribed by Islamic law, then this should not by implication preclude stunning the nonhuman animal before slaughter, as long as the law is followed. Al-Masri notes that stunning has been declared as acceptable by a fatwa (unanimous verdict) of the Al-Azhar University in Cairo (1987). Furthermore, the Muslim World League declared in 1986 that pre-slaughter stunning is lawful when the weakest electric current renders a nonhuman animal unconscious before slaughter (Al Masri, 1987).

The *Qur'ân* gives confusing signals to adherents suggesting eating meat is acceptable, even obligatory, albeit within strict humane treatment standards, whilst also suggesting that eating flesh is wrong. It offers an alternative to the slaughtering of meat when it states:

> *the beasts of sacrifice ... We have subjected them to you; haply you will be thankful. The flesh of them shall not reach God, neither their blood, but goodliness shall reach Him* (*The Pilgrimage 22:37-8*).

Foltz has recently considered the vexed issue of vegetarianism in Islam, noting that vegetarians have encountered much resistance, if not outright hostility (2006). Opposition has also been from environmentalists insistent on their own meat eating, even in light of evidence which shows the direct links between the meat industry, the deleterious effects of industrial meat production on the environment, social injustice and the subjugation of women (Foltz, 2006). In regards to the latter point, Adams has been particularly damning (1999 and 2004).

Dutton describes human interaction with wildlife (1992, pp.63-4). He notes that it is permissible to hunt herbivores, which should be killed in the same way as a domestic nonhuman animal, or killed by a trained nonhuman animal such as a dog or bird of prey, but it is harām to kill a carnivore except when it poses a threat to human life or property, or when on pilgrimage. The major issue here is one of the protection and preservation of species, besides the biological problems associated with bio-accumulated vitamin A in meat products that are to be consumed.

Generally, the killing of wildlife for any other reasons than food is always prohibited, as is the caging of birds, sports hunting and animal baiting (Foltz, 2006). Reality within Muslim countries has been most different. As in the Christian West, wildlife has been decimated. Although in certain countries, like Iran, where wildlife was hunted to extinction or where numbers are now critically endangered, some areas are now protected (Foltz, 2006).

Given the increase in human numbers and unrestrained environmental exploitation, and if Muslims are allowed to continue to kill wildlife without little if any state-based regulation, it will eventually become imperative to stop that killing if the survival of that particular species that is being exploited becomes threatened, let alone protect the biodiversity of the local area.

The answer lies, in part, in the *Muslim hadith*, which relates a story where God reproached one of the previous prophets for the killing of an entire ant colony after he was bitten by a single ant. It says that:

> *have you destroyed a whole community that glorifies Me because of one ant that bit you?* (*Muslim hadith* as cited by Dutton, 1992, p.64).

By implication, therefore, the care and protection of nonhuman animals by Muslims is obligatory, even if it were in the name of God. However, much work is needed to make wildlife protection a part of everyday Muslim life.

Lynn White Jnr. argued that Abrahamic stewardship of the Judaeo-Christian tradition has led to environmental destruction and an instrumentalist view of nonhuman animals (1967). Yet, on this key point, by equating Islamic stewardship or dominion with those other religions, Islam should conceivably share a burden of responsibility with Judaism and Christianity for justifying the utilisation or exploitation of the environment, and nonhuman animals more specifically. Given the number of Muslims in the world today consuming as much as those in the West, while at the same time practicing their faith, this burden of responsibility seems well-founded.

Islamic Literature and Philosophy

Discussion of the relationships between human and nonhuman animals in Islam is neither a recent phenomenon nor one of the distant past. The care and protection of nonhuman animals in Islam was espoused in the 10th century in one of the most important books of Islam, *The Case of the Animals versus Man Before the King of the Jinn*

(Ikhwān al-Safā', 1978). Compiled in Basra by scholars known as the Rasa'il Ikhwān al-Safā' wa khillan al-wafa ('Brethren of Purity'), the fable examines the relationships between human and nonhuman animals in great detail and asserts the need for human obedience to the faith. Foltz calls it the "most extensive critique of mainstream human attitudes towards animals in the entire vast corpus of Muslim literature" (2006, p.50). Callicott believes that it shows Islam to be equal to the task, as is the Christian tradition, of providing a "direct biocentric, stewardship environmental ethic" (1994, p.35).

Another important book of Islam is by the 12th century Sufi mystic, Farid al-din 'Attar (c.1130-c.1229), entitled *The Conference of the Birds* (Attar, 1984). In this allegory, 'Attar uses several parables to portray the way that humans should seek unity with the divine through pilgrimage. Although not dealing directly with the interaction between human and nonhuman animals *per se*, the text is particularly profound providing insight into the way in which Islamic mysticism seeks closer connections between human and nonhuman animals, at least through parable.

Sufism in fact provides many examples of how humans can otherwise interact with nonhuman animals. Many Sufis are vegetarian and base their strict living on Qur'ânic teachings. The 13th century founder of the Mehlevi Sufi Order of Whirling Dervishes, Mawlānā Jalāl al-Dīn Mohammad Rūmī (1207-73), used extensive nonhuman animal imagery in his poetry. Inspired by 'Attar, Rūmī forcefully upheld vegetarianism when he relates the story of a vengeful mother elephant who slew a group of hungry travellers who ate her calf, except for the one individual who counselled the others not to eat the calf and then refused to accept the eating of its meat (Rūmī, 1993). Abu Abdallah Muhammad ibn Battuta (c.1304-c.1377) also told a similar famous story of elephants in Islam (Bartel, 2007). The 15th century Sufi mystic Kabir (c.1440-1518), also unequivocally condemned meat eating, and labelled the act of eating meat as the "ultimate failure of compassion, deserving of eternal punishment" (as cited by Seidal, 1999).

Although Sufism has ebbed and waned, the late 20th century saw its small but significant resurgence with Sufi teachers and academics such as Seyyed Nasr, M.R. Bawa Muhaiyaddeen (d.1986) and Chisti Inayat Khan expressing deep concern for nature and compassion to nonhuman animals in both diet and deed. Current followers of Sufism, such as Ahmet Yüksel, go as far as pressing the need for outright 'love' of nonhuman animals. Yüksel states that "if someone does not have any love for animals, it is impossible for them to carry other kinds of love" (2000). In the 12th century, Abu al-Walīd Muhammad ibn-Ahmad ibn Rushd (latinised to Averroes) (1126-98) advocated an Aristotelian approach to nonhuman animals (1978), particularly in relation to the concept of plenitude. Recognised as the last of the great classical Muslim thinkers, ibn Rushd's exegesis subsequently, and perhaps like no other, influenced Islamic thought as well as the Christian scholastic school of philosophy, including that of Aquinas (Glacken, 1967).

The school of thought advocated by ibn Rushd held sway at the expense of the more holistic worldviews espoused by the Ikhwān al-Safā' and the Sufis. The various Islamic exegeses and scientific works of the Middle Ages, to which time ibn Rushd belongs, greatly influenced the development of scientific philosophy that was to emerge in the Scientific Revolution and the Age of Discovery. In fact, science as we know it today not only emerged from the Middle Ages, it is in fact and in many ways "the heir to all the sciences of the past, especially perhaps to the work of the great Islamic scientists of the Middle Ages, who so often outdid the ancient Greeks in skill and perspicacity" (White, 1967, p.1204).

Islamic Science

During the time of monastic communion with nature and the European introspection of the High Middle Ages in the 11th and 12th centuries, Islamic science and philosophy were proceeding apace. Importantly, several scientists and philosophers influenced the interrelationship that was developing between human and nonhuman animals. Islamic exegetical writings were more voluminous than Christian ones and, in fact, reinterpreted Greek philosophy and extended those Christian exegeses that had already been written prior to the life of Muhammad. The most voluminous was that of ibn Rushd (Glacken, 1967). Islamic scholars were largely influenced by Aristotelian logic and reason, then considered to be the most adequate system with which to study nature (Marangudakis, 2001).

Islamic philosophers began to theorise that individual nonhuman animals have not only imagination and reason, but an instinctive estimative faculty, enabling the individual to distinguish or deduce significant intentions in mental images. This position was expounded by both Abu Ali al-Husayn ibn Abdollah ibn Sina (italicised as Avicenna) (980-1037), later by Abu Bakr Muhammad ibn Yahya ibn as-Say'igh (or ibn Bajja, italicised as Avempace) (c.1095-1138) (Sorabji, 1993), and eventually by Saint Thomas Aquinas (1225-74) in the 13th century (1947, First part, Q.78, Art.4). Further, ibn Rushd denied that nonhuman animals have abstract thought (Sorabji, 1993), and that "divine law makes it an obligation to apply rational speculation to one's reflections on the universe" (Glacken, 1967, p.221)[1].

There are many scientific works from the early Muslim world that focus on nonhuman animals. The most important was the seven volume encyclopaedic Kitāb al-hayawān (the *Book of Animals*) by Abu 'Uthman 'Amr ibn Bahr al-Fuqaymi al-Basri al-Jahiz (776-869) (2001). Al-Jahiz studied Aristotle's work in great detail, and was writing largely in praise of the work of God's creation. As Foltz notes, the *Book of Animals* is largely a compendium of zoological knowledge, but he did anticipate evolution, the effects of the environment on animal physiology, and provided a taxonomy of nonhuman animals, including kangaroos (2006, pp.55-59).

Other famous scientific works from the Middle Ages include *Benefits of Animals* by Jabril ibn Bakhtyshu (c.late 8th–early 9th century) and *Natures of Animals* by Ahmad ibn 'Abdallah Habash al-Hasib al-Marwazi (c.770-c.874). Later works include those by Muhammad ibn Musa ibn 'Isa Kamal al-Din al-Damari (1341-1405) entitled *Hayāt al-hayawān al-kubra* (The Great Book on the Life of Animals), and *The Wonders of Creation* by Abu Yahya Zakariyya 'ibn Muhammad al-Qazwini (1203-83) (Bartel, 2007; and, Foltz, 2006).

Conclusion

The history of the interrelationships that have developed between Muslims and nonhuman animals is indeed a rich one. And just as in the *Bible* and for Christians, an equally strong foundation based on principles of stewardship and care can be found within the *Qur'ân*, the early Islamic texts and exegeses, and within Islamic literature and scientific texts of the Middle Ages. However, if Muslims, as much as Christians, are to become much more responsible towards nature and more inclusive of nonhuman animals, a much deeper commitment to what is prescribed in those writings and teachings, and by this is not meant orthodoxy, is urgently needed.

Notes

1 It was Muhammad ibn Muhammad ibn Tarkhan ibn Uzalagh al-Farabi (italicised as Alfarabius) (c.873-950) who was the first to claim that revelation is subordinate to reason, but it was this view, extended by others like ibn Sina and then ibn Rushd, that was eventually to lead to the decline of Islamic philosophy and science, in large measure brought about by the rejection of such rationalism and neo-Platonic theories by more orthodox Muslims. Chief amongst those leading the rejection of such philosophy as against Islam was Abu Hamid Muhammad ibn Muhammad at-Tusi al-Ghazali (italicised as Algazel) (1058-1111).

References

Adams, C. 1999. *The Sexual Politics of Meat: a feminist-vegetarian critical theory*. Tenth Anniversary Edition. New York: Continuum International Publishing Group.

_____. 2004. *The Pornography of Meat*. New York: Continuum International Publishing Group.

Al-Jahiz (Abu 'Uthman 'Amr ibn Bahr al-Fuqaymi al-Basri al-Jahiz). 2001. *Kitāb al-hayawān (Book of Animals)*. Published as *Le cadi et la mouche: Le Livre des Animaux*. Paris: Albin Michel.

Al-Masri, H.B.A., (1986). "Animal Experimentation: the Muslim viewpoint". In: *Animal Sacrifices: religious perspectives on the use of animals in science*. Edited by: Regan, T.. (Series: Ethics and Action. Series editor by: Regan, T.) Temple University Press, Philadelphia, United States of America. 0877224110. pp.:171-97.

_____. 1987. *Animals in Islam*. Sections reprinted online by The Athene Trust, Petersfield, United Kingdom. http://www.chai-online/islam.htm. Accessed: 4 October, 2001.

_____. 1992. "Islam and Ecology". In: *Islam and Ecology*. Edited by: Khalid, F., and O'Brien, J.. (Series: World Religions and Ecology). Published for the World Wild Fund for Nature by Cassell Publishers Limited, London, United Kingdom. 0304323772. pp.:1-23.

Aquinas, St. Thomas, 1947. *The Summa Theologica*. Translated by the Fathers of the English Dominican Province. Originally published by Benziger Bros., New York, United States of America. Reprinted online by Christian Classics Ethereal Library of Calvin College, Grand Rapids, United States of America. http://www.ccel.org/a/aquinas/summa. Accessed: 13 August, 2005.

Attar (Farid al-din 'Attar), 1984. *The Conference of the Birds*. Penguin Classics, London, United Kingdom. 0140444343.

Averroes (Abu al-Walīd Muhammad ibn-Ahmad ibn Rushd), 1978. *Averroe's Tahafut Al-Tahafut: The Incoherence of the Incoherence*. UNESCO Collection of Great Works under the auspices of The E.J.W. Gibb Memorial Trust and the International Commission for the Translation of Great Works, Warminster, United Kingdom. 0906094046.

Bible, 1986. *The Holy Bible*. King James Version - Red Letter Edition. P.S.I. & Associates, Inc., Miami, United States of America. 0938261371.

Callicott, J.B. 1994. *Earth's Insights: a survey of ecological ethics from the Mediterranean Basin to the Australian Outback*. University of California Press, Berkeley, United States of America. 0520085590.

Dutton, Y. 1992. "Natural Resources in Islam". In: *Islam and Ecology*. Edited by: Khalid, F., and O'Brien, J.. (Series: World Religions and Ecology. Published for the World Wide Fund for Nature by Cassell Publishers Limited, London, United Kingdom. 0304323772. pp.:51-67.

Engel, J.R., and Engel, J.G. 1991. *Ethics of Environment and Development: global challenge, international response*. Reprint Edition. University of Arizona Press, Tucson, United States of America. 0816412639.

Foltz, R.C., 2000. "Is There An Islamic Environmentalism?" *Environmental Ethics* 22(1):63-72.

_____. 2006. *Animals in Islamic Tradition and Muslim Cultures*. Oneworld Publications, Oxford, United Kingdom. 1851683984.

Foltz, R., Denny, F.M., and Baharuddin, A., (eds.), 2003. *Islam and Ecology: a bestowed trust*. (Series: Religions of the World and Ecology. Series editors: Tucker, M.E., and Grim, J.. Harvard University Press for the Center for the Study of World Religions Harvard Divinity School, Cambridge, United States of America. 0945454406.

Glacken, C.J., 1967. *Traces on the Rhodian Shore: nature and culture in Western thought from Ancient times to the end of the Eighteenth Century*. University of California Press, Berkeley, United States of America. 0520032160.

Hamed, S.E. 1993. "Seeing the Environment Through Islamic Eyes: application of Shariah to natural resources planning and management". *Journal of Agricultural and Environmental Ethics* 6(2):145-64.

Ikhwān al-Safā' 1978. *The Case of the Animals versus Man Before the King of the Jinn: a tenth-century ecological fable of the Pure Brethren of Basra*. Translated by: Goodman, L.E.. Library of Classic Arabic Literature, Vol.3. Twayne Publishers, a Division of G.K. Hall and Co., Boston, United States of America. 0805781617.

Izzi Deen, M.Y. 1991. "Islamic environmental ethics: Law and Society". In: *Ethics of Environment and Development: global challenge, international response*. Edited by: Engel, J.R., and Engel, J.G.. Reprint Edition. University of Arizona Press, Tucson, United States of America. 0816412639. pp.:189-200.

_____. 1992. "Islamic Ethics and the Environment". In: *Islam and Ecology*. Edited by: Khalid, F., and O'Brien, J.. (Series: World Religions and Ecology. Published for the World Wide Fund for Nature by Cassell Publishers Limited, London, United Kingdom. 0304323772. pp.:25-35.

Khalid, F., and O'Brien, J. 1992. *Islam and Ecology*. (Series: World Religions and Ecology. Published for the World Wide Fund for Nature by Cassell Publishers Limited, London, United Kingdom. 0304323772.

Marangudakis, M., (2001. "The Medieval Roots of Our Ecological Crisis". *Environmental Ethics* 23(3):243-260.

Qur'ân, 1983. *The Holy Koran*. (Translated by: Arberry, A.J.. The World's Classics Paperback Edition. Oxford University Press, Oxford, United Kingdom. 0192816284.

Regan, T., (ed.), 1986. *Animal Sacrifices: religious perspectives on the use of animals in science*. (Series: Ethics and Action. Series editor: Regan, T.. Temple University Press, Philadelphia, United States of America. 0877224110.

Regenstein, L.G. 1991. *Replenish the Earth: a history of organized religion's treatment of animals and nature - including the Bible's message of conservation and kindness to animals*. The Crossroad Publishing Company, New York, United States of America. 0824510755.

Rūmī (Mawlānā Jalāl al-Dīn Mohammad Rūmī), 1993. *Tales from the Masnavi*. (Translated by: Arberry, A.J.. Curzon Paperbacks. Routledge, London, United Kingdom. 0700702733.

Seidel, K. 1999. *Serving the Guest: a Sufi Cookbook*. Extract printed online by the author, Peterborough, United States of America. http://www.superluminal.com/cookbook/essay_vegetarianism.html. Accessed: 30 December, 2001.

Sorabji, R. 1993. *Animal Minds and Human Morals: the origins of the Western debate*. Gerald Duckworth & Co. Ltd., London, United Kingdom. 0715624709.

Wersal, L. 1995. "Islam and Environmental Ethics: tradition responds to contemporary challenges". *Zygon* 30(3):451-9.

Wescoat, J.L., Jnr. 1995. "The 'Right of Thirst' for Animals in Islamic Law: a comparative approach." *Environment and Planning D: Society and Space* 13(6):637-54.

White, L., Jnr., 1967. "The Historical Roots of Our Ecologic Crisis". *Science* 155(3767):1203-7.

Yüksel, A.F., 2000. *The Love For Animals*. First published in afyuksul.com and reprinted online by A.F. Yüksel, Istanbul, Turkey. http://www.sufizmveinsan.com/sohbet/animal.html. Accessed: 9 December, 2001.

Zaidi, I.H., 1981. "On the Ethics of Man's Interaction with the Environment: an Islamic approach". *Environmental Ethics* 3(1):35-47.

Chapter 17

Animals in Buddhism: In Defence of Hierarchical Evaluation

☆ *Colette Sciberras*

Recent years have seen copious publications that bring together Buddhism and environmental ethics, and the question regarding the status of animals within the tradition has formed an important part of the inquiry into its green credentials. Often, Buddhism is charged with being anthropocentric, or with involving some other, similar bias towards human interests, and these accusations generally arise due to the hierarchical organisation of living beings within its traditional cosmology. In this paper, I review these charges in light of some basic Buddhist doctrines, and argue that the objections stem from a misapplication of two philosophical assumptions to certain extracts from the Pāli texts. My general claim is that if we consider Buddha's teachings as a whole, there is little to suggest that he recommended discriminatory behaviour towards nonhuman animals.

I start by defining anthropocentrism, speciesism, and human chauvinism, and go on to provide a few reasons why such charges have been brought against Buddhism. It will emerge that the objections rest on two key ideas; there is, first, the assumption of a fundamental divide between humans and other animals, and second, a belief in a restrictive moral circle, which, by including certain beings, automatically excludes others. Buddhism, however, does not endorse these ideas, and the charges against it only arise when commentators misappropriate these, or similar, beliefs into their account of the Buddha's teachings. The teachings emphasize *continuity* between living beings, and not difference, and I will argue that despite the fact that Buddhism's hierarchical view of life places humans at the summit, there are no prejudicial implications, since it does not discriminate between forms of life in terms of moral worth. Although there are some passages from the Pāli canon that are generally taken as evidence of anthropocentrism, I conclude that these are too infrequent to support the charge, especially when considered against the far more common references to continuity between, and concern for, all sentient beings.

Anthropocentrism, Speciesism, and Human Chauvinism

The basic idea behind the concept of anthropocentrism, as suggested by the term itself, is a view that regards human beings as the centre of all significance, and human needs and wants as foremost among the interests of all beings. Simon Blackburn defines it as "any view magnifying the importance of humans in the cosmos, for example, by seeing it as created for our benefit" (Blackburn 19). 'Anthropocentric' and 'egocentric' are analogous, therefore, in that both terms suggest that whatever is *not* at the centre–nonhuman animals in the former case, and all other people in the latter–are there *for the sake of* the person described as anthropocentric or egocentric, and can be used in whatever way he or she sees fit.

Gary Steiner reveals the depth at which human centeredness is ingrained in the West. He argues that even in *theo*centric societies, where the world is believed to be "centred on god," still, deities are generally depicted as human beings, and, one might add, in pre-modern societies, they were often believed to display qualities and to act in ways that were all too human indeed. This, Steiner says, shows the influence of anthropocentric thinking, which, even when it posits a 'Higher Being,' conceives of it as similar to us, and always regards animals as occupying an inferior place in the "cosmic order" (Steiner 1–2) (1). Many environmentalist thinkers and supporters of animal rights have made it their purpose to overturn this kind of worldview, which they regard as being based on an unfounded prejudice that favours the human.

A related term is that of 'speciesism,' originally conceived as an extension of the anti-racist and anti-sexist critiques. Speciesism is defined as "the improper stance of refusing respect to the lives, dignity, rights, or needs of animals other than the human species" (Blackburn 358). Paul Waldau characterizes it as the inclusion of all humans within the moral circle and the exclusion of all other animals from it. If a being is included within the moral circle, this means that its life, its freedom from captivity, and from instrumental use are protected, while direct, intentional infliction of harm, pain, or unnecessary suffering are seen as morally unacceptable (Waldau 38–39). Just as racism can be represented as the belief that it is wrong to keep white people captive, but permissible to enslave blacks, similarly, speciesism can be characterized as the claim that, say, conducting experiments on human babies is unacceptable, but unproblematic if performed on chimps (Singer 38–39). As the example shows, if this kind of critique is valid, and such discrimination between animals is truly based on an unfair bias, then, much of the world is, in fact, speciesist. The discussion, therefore, centres on the reasons that are provided for differential treatment of humans and other animals. If such criteria can be found and shown to be valid, then the charge of speciesism can be rejected.

Consideration of the criteria for discriminatory treatment of animals brings us to the third, and least known of our cluster of concepts, that of human chauvinism. This occurs whenever the grounds for inclusion within the moral circle are defined from the outset in such a way that only humans are included. Often, they are altered every time new information is obtained, which would otherwise build a case for the inclusion of some other animals (Hayward 53–54). A human chauvinist, for instance, will consider only intelligence to render a being worthy of moral consideration, and, when it is proved that some other animals are also intelligent, she will redefine the criteria, so that, say, only those that can use language will be included in the moral circle.

The charges of anthropocentrism and speciesism have each been brought against Buddhism, for diverse reasons. In the next section, I shall review this criticism, and examine certain assumptions on which it relies.

The Charges against Buddhism

The charge of anthropocentrism arises against Buddhism, at times, because the latter is seen as occupied predominantly, if not exclusively, with human matters. As Malcolm David Eckel puts it, it is concerned with the "human achievement of human goals," that is, ultimately, the attainment of *nibbana* (Eckel 64; Holder, 119) (2). Although, as we shall see below, neither Holder nor Eckel endorse such as simplistic view, one version of the objection seems to be that the teachings, and the path they set out, are intended for human beings alone. Famously, the Buddha emphasized that he taught nothing but suffering (*dukkha*), the causes, of suffering and the means to extinguish suffering (*M* i 140; *D* i 189). He set up an order of monks and eventually–and somewhat reluctantly–allowed women to be ordained as nuns. He accepted lay followers too; yet, the Pāli canon specifically states that he refused to ordain nonhuman animals, or even to allow them to listen to the teachings.

In itself, this does not seem to be such a serious offence; indeed, it stands to reason that animals, being unable to grasp the Buddhist teachings, cannot perform the role of monastics, and any attempt to teach them to meditate would immediately strike us as a colossal waste of time. Elsewhere, I have argued that a position that differentiates between animals in terms of their qualities alone–as in this case, where the difference has to do with their level of intelligence–does not always imply a distinction in moral worth, or that animals ought to be treated differently. The charges of anthropocentrism, and especially speciesism, will only arise if, from the differences that we acknowledge to exist between us and other animals, we wrongly infer that it is appropriate to treat other animals in ways that we would not accept for human beings (Sciberras 228).

The prohibition against ordaining animals originates from a canonical story about a *nāga* (3) who, having grown weary of life, pretends to be a human being and joins the order as a novice. When the Buddha discovers his true form, he advises the *nāga* to practice some basic acts of discipline, so that, in his next life, he will be born as a real human being, and be able to follow the Buddhist path. The implication, then, is that animals, "are not capable of (spiritual) growth in this doctrine and discipline" (Rhys Davids and Oldenberg 219) (4) and they too, like all other forms of existence, such as worldly gods and ghosts, must aspire to be reborn as humans first. Thus, it appears that, in regarding spirituality as an exclusively human domain, the Buddha ended up elevating the status of humans above that of animals. This was further consolidated through the hierarchical view of life portrayed in the traditional image of the *Wheel of Sansāra* (*bhavacakka*). The worlds of humans and of the gods are depicted in the upper part of the wheel, forming the three "happy destinations" or "higher realms" (*sugati*), while other animals, hell beings, and "hungry ghosts" occupy the three "lower realms" (*apāya*) (Nyanatiloka 119). Eventually, humans would be raised above the gods too; the Tibetans, for instance, speak of the "precious human life" (Tib. *mi lus rin po che*).

The question, therefore, is whether this consists of the same sort of "magnification of human importance" which amounts to anthropocentrism, and whether it makes any implications about the way we ought to treat other animals, which could justifiably be called speciesist. Before I turn to these questions, however, I shall consider some other reasons why such charges have been made.

Several authors have suggested that Buddhism's hierarchical view of life brings with it a deprecatory outlook towards other animals, which reveals a serious lack of respect for the nonhuman inhabitants of this earth. Ian Harris points out that in the traditional Buddhist cosmic order, animals are usually classified alongside "matricides, patricides, thieves, hermaphrodites and slayers of the Buddha" (105). Lambert Schmithausen makes a similar compliant, when he argues that the violent aspect of nature, where animals kill and eat the weaker, is "deeply abhorred" in Buddhism, and in its "ultimate analysis and evaluation of existence" there is no motivation for preserving animal life (10–11).

Paul Waldau infers from this allegedly derogatory view that Buddhism might be speciesist. In brief, his claim is that there is a "constant disparagement" of all nonhuman animals, which are lumped together in one group, and negatively evaluated. Nonhuman animals, he says, are seen as culpable, "products of bad conduct," that endure a fundamentally unhappy existence. "The tradition," he argues, "has never emphasized seeing animals in terms of *their* realities" (154) Worse yet, there is also an acceptance of harmful instrumental use of animals, which, in his view, is revealed through the fact that the canonical texts often contain references to such uses, for instance, they include several stories about elephants that are held in captivity. All this, Waldau believes, is "suggestive of…the existence of pervasive species based exclusions," that is, speciesism (155).

There is no space to discuss fully the merits of Waldau's arguments here (5). Instead, I would like to examine certain assumptions that such arguments rely on. First, it appears that such critiques *start from* the position that there is a fundamental difference between humans and other animals. While it is true that Buddhist cosmology divides existence into separate realms containing different forms of life, I shall argue in the following section that the sense of continuity between living beings is far more pronounced. To begin from the presupposition that it is the divide, rather than the connection, that is more crucial, is to read Buddhism from a European perspective, in other words, to apply certain Western categories that do not necessarily belong to the doctrine. That is to say, to argue that Buddhism views humans as occupying a higher status than other animals assumes from the outset that the tradition endorses the dualistic view that contrasts humans with all other natural forms of life, and this seems to beg the question that is being addressed.

Second, there seems to be the assumption that any differences that are recognized among beings imply a difference in our moral obligations to them. Again, this belongs properly to Western ethics, where creatures are assumed to have different grades of moral worth, and where the very concept of a moral circle implies a border between those that are and those that are not worthy of moral consideration. Once again, it seems unlikely that such a conception exists in Buddhism, and below, I shall argue that the tendency is far more inclusive than Western ethics has generally been. That is, to assume that there is a divide between beings that are morally significant, and those that are not–an assumption on which the charge of speciesism depends–is, once again, to carry over Western ideas into one's critique of Buddhism.

Continuity versus Hierarchy in Buddhism

The emphasis on continuity in Buddhism can be appreciated from various doctrines, including rebirth (*jāti, utapatti-bhava*), not-self (*anattā*), and the view of all beings as one's mother. Taken together, these portray a striking vision of universal connectedness, which arises from our shared predicament of having participated in cyclic existence for countless rebirths, and from 'beginningless time,' as conveyed by the image of the *Wheel of Sansāra*. The Buddha conceived of rebirth as an endless trajectory through the different realms, with their respective highs and lows, and maintained that we have all been all born, at some time or other, in every single domain–as gods, demons, ghosts and hell beings, as well as animals, and people. For this reason, he says,

> A being who has not been your mother at one time in the past is not easy to find... A being who has not been your father... your brother... your sister... your son... your daughter at one time in the past is not easy to find (*S* ii 189; Thanissaro "Mother").

Stereotypical accounts of Buddhist peoples often portray them as anxious to avoid harming even the tiniest creature (cf. Norbu) and this kind of attitude is normally held to emerge from a belief that that creature might have been one's mother in a past life. I propose that this is somewhat inaccurate; it is not the mere *possibility* of there having been a relation that inhibits Buddhists from harming creatures. Rather, the concept of 'mother sentient beings' is based on the secure knowledge that, given that we have all been reborn an infinite number of times, each creature we encounter *has actually been*, at some time, our mother, our father, teacher, lover, enemy, and so forth.

It is commonly accepted that there is a "close correspondence" between the hierarchical realms of Buddhist cosmology and our inner states of mind, including higher levels of meditative consciousness (Gethin 202). The various strata of *sansāra* and the range of beings that inhabit them, that is, have always had both a literal sense and a psychological one, so that as well as having autonomous existence and external reality (6), hell beings, ghosts, gods, and animals can be understood as representing anger, lust, sensual pleasure, and ignorance respectively (7). The two interpretations are tightly coupled, and sometimes even conflated. According to Rupert Gethin, for example, during the later, scholastic phase of Buddhism, the difference between them was regarded as a mere "shift in time scales" (195) On the psychological interpretation, therefore, an individual moves from an experience of hatred to one of attachment–and in this way, 'travels' from the hell realm to that of the hungry ghosts–in a matter of moments. To be physically reborn in another realm is a continuation of the same process, although one that spans a much longer period of that individual's existence (Gethin 195).

At least one author believes that the literal interpretation of Buddhist cosmology is not appropriate. Relying on Richard Gombrich's claim that the Buddha was "not really interested in what existed 'out there,'" (cited in Hamilton 73) Sue Hamilton argues that the Buddha's talk of the realms of *sansāra* was intended to serve as a metaphor for the way the mind works, and that the literal interpretation was a later addition (73). Even the term *loka*, which is generally translated as 'world' or 'realm,' is most commonly used, she says, as "a metaphor for the entire life of the individual human being." (81) She goes on, "such metaphors serve to emphasize the extent to which the Buddha [is] consistently drawing attention away from the external world to the subjective world of experience" (82).

Familiarity with the Buddhist doctrines of not-self and emptiness (*sansāra*) will further reinforce this view. These can be characterized as the claim that there are no sharply delineated individuals, whether we consider persons or things, nothing is independent or irreducible, and nothing has a fixed, abiding nature. Rather, what we generally conceive of as an individual, when analyzed carefully, turns out to be a collection of causally related but ephemeral phenomena; the traditional account of the constituents of a human being, for instance, lists the body, feelings, perceptions, volitions, and consciousness, and these too can be broken down into moments of consciousness and particles of matter. Our tendency, however, is to reify complex phenomena, that is, we view ourselves, other beings, and the furniture of the world as somehow 'concrete,' or as truly existing 'from their own side,' separate from all the other factors that participate in their being, and independently of our experience of them. We conceive of individual 'things' as simple and existentially or logically independent, and as having an essential nature, and this causes us to desire some and to fear or be repulsed by others. This reification, therefore, is the ignorance that the Buddha speaks of, and which needs to be eradicated in order to attain *nibbana.* Only when we realize the emptiness of all phenomena, and the lack of self of all beings, will this reification, and our consequent suffering end.

In short, when we consider Buddhist doctrine as a whole, it appears that the assumption that was identified above, that posits a fundamental divide between humans and all other creatures, does not fit the tradition. No being is essentially either a human or a nonhuman animal, but on the literal interpretation at least, we all can and do take each form at different times of our existence. To think of a being as a clearly defined and separate self is misguided, and therefore, so is the assumption that there is a fundamental difference between humans and other animals, rather, as we have seen, there is an intimate relation between all beings, and all share the same characteristics. It seems that, in positing a hierarchy of life, the Buddha was not referring primarily to the external world, or to actual natural beings, but setting up a distinction in various inner states. Therefore, what appear to be derogatory remarks about animals were probably intended as a commentary on certain forms of human behaviour, and the Buddhist emphasis on continuity, as I hope has become clear, suggests that human and nonhuman animals hold much more in common than is generally supposed by those that start out with dualistic presuppositions. To internalize fully the belief that all other creatures have, at some time, been one's mother, and to understand that we all have nonhuman elements within our own psyche–whether these are what are generally known as 'animal instincts,' or perhaps, rather more elevated 'divine attributes'–cannot but lessen the opposition some perceive between ourselves and other animals.

Buddhism and the Moral Circle

The concept of a restricted moral circle is the second assumption on which the charges of speciesism and the like were said to rely. As I have argued elsewhere against attributing this idea to Buddhism, I will only briefly outline my claims here (8). Buddhist morality is encapsulated in the form of the Five Precepts (*paxca-sīla*), the first of which can be construed as a prohibition against killing, or else, in its stricter interpretation, involves an injunction against all forms of intentional harm (Schmithausen 11). The scripture containing rules for monks and nuns (*Vinaya Pitaka*) sets out the meticulous care that must be taken to prevent harming creatures; for instance, water must be strained before being drunk, in order to avoid ingesting small organisms, and even picking flowers, or digging soil, is considered an offence (Ariyesako n. pag). Although there is less detail when the Buddha addresses his lay followers, still, the proscription against killing living beings is made very clear.

Waldau has picked out a few paragraphs from the *Vinaya*, which imply that the gravity of killing a human being is far greater than that of killing another animal. One of the most conspicuous among these is the fact that a monk who accidentally kills a human being will be expelled from the order, whereas killing an animal intentionally "only requires expiation" (123–124). While the existence of such passages is undeniable, their frequency appears negligible, especially when compared to the vast number of canonical texts that encourage the development of loving-kindness (*mettā*) and compassion (*karunā*), and which promote an attitude of non-harm (*ahimsā*). In each case, these benevolent attitudes and acts are directed towards "all living beings," and the sheer number of repetitions of this phrase, contrasted with the paucity of extracts that suggest discrimination, militates against Waldau's charge of speciesism.

Schmithausen points out that the class of 'living beings' was originally conceived as including plants, soil, water, air, and seeds. Eventually, it was narrowed to include only *sentient* beings, and certain forms of life began to be conceived as lacking any form of consciousness. Consequently, the restriction against damage caused to minerals and vegetation was loosened to some extent, although, Schmithausen adds, it was never abolished entirely. It is likely, he says, that the rules were relaxed a little in order to free monks and nuns from what might have seemed like "unnecessary inconvenience and qualms" (Schmithausen 6–8). Later still, when Buddhism had spread to China and Japan, the idea that sentient beings *alone* were the proper recipients of benevolence appears to have been considered something of a limitation. A long debate ensued about whether plants, trees, and even nonliving objects were capable of attaining enlightenment, that is, whether they could be considered to possess the seed of Buddhahood. It seems that the trend was to argue in favour of widening the class of sentient beings (La Fleur 95).

In short, Buddhism does not share the predilection of Western ethics to conceive of a restrictive moral circle; rather, the tendency has always been towards inclusion. In the West, Aristotle, Kant, and indeed, most prominent ethicists, have started from the assumption that there is a characteristic, or else a set of characteristics, that render certain beings morally considerable. Until modern times, it was generally believed that while all people possess this characteristic, no other animal does, and favourite candidates included, of course, rationality and the use of language. Then, Jeremy Bentham argued that neither of these was relevant to moral worth, and instead he proposed sentience, since the capacity to feel pleasure and pain is presupposed by any moral theory that, like utilitarianism, derives its principles from the universal pursuit of happiness and well-being. The tradition of seeking to account for moral considerability through the possession of some characteristic continues to our times. Critics of speciesism, such as Waldau, argue that since many other animals possess such characteristics–sentience, primarily, but also a degree of intelligence, communication, tool-making and social organisation–then these animals ought to be included within the moral circle (Waldau 63–75).

Yet, Buddhism does not take this approach at all. Instead, as we have seen, it begins from as wide a moral circle as possible–that is, all living beings–and this is occasionally narrowed down to include only sentient ones, merely as a matter of expediency. Nowhere does there appear to be the suggestion that it is the possession of any particular characteristic that renders a being worthy of love and compassion, or else, that it deserves to have its life and well-being protected, because it is sentient, or rational, or whatever. In fact, the *KaraGīya Mettā Sutta* suggests precisely the opposite:

> Whatever beings there may be,
> without exception, weak or strong,
> long, huge, or middle-sized
> Or short, minute, or bulky.
>
> Whether visible or invisible,
> And those living far or near,
> Those born and those seeking birth:
> May all beings be happy! (*Sn* 146–7; Buddharakkhita)

Refuting the Charges and Defending Hierarchy

It is time to return to the charges of anthropocentrism and speciesism, and to ask whether these are valid in light of what was said above. As both Eckel and Holder acknowledge, such critiques are not altogether straightforward, because although Buddhism is mostly concerned with human goals, these do not exclude the good of other beings. The path to *nibbana*, as we have seen, inherently includes concern for the "wider network of life" (Eckel 64). Therefore, the fact that human beings are regarded as the sole candidates for spiritual fulfilment is not the sort of 'magnification' that constitutes anthropocentrism. This is because, first, no being is either human or nonhuman essentially; rather, there is a great degree of fluidity in the way beings progress towards, and regress from, the state of human being. Second, the elevation in status of humans has no moral implications, or at least, there are too few to overturn the general prohibition against harming beings.

If we consider Steiner's account of anthropocentrism, we can see the complete lack of such thinking in Buddhist doctrine in that, although the gods are conceived as human-like, this is also true for animals and other nonhuman creatures. There are plenty of stories where animals are anthropomorphized, especially in the *Jataka Tales*, and examples of the reverse, where humans are described as displaying animal behaviour. As we have seen, the psychological interpretation implies that we all have elements of every form of life within us, and since the teachings are not primarily about objective reality, it seems that the worst charge that can be brought against Buddhism is that it uses animals to symbolize everything that is shallow or depraved in human beings. While there seems to be a valid argument here that points to a deep lack of respect, it is far less serious than the claim that Buddhism endorses or encourages prejudicial discrimination against nonhuman animals.

The absence, in Buddhism, of a dualistic conception of humans and other animals, and of a restrictive moral circle, provide strong grounds for an egalitarian outlook that even those with the best intentions cannot reach, if they begin with such assumptions. To return to Waldau's critique, a large part of his argument about Buddhist speciesism relies on his having identified certain characteristics that other animals share with human beings; in brief, these include intelligence, communication, social-organization, and tool-making (69-86). He argues that since Buddhism fails to recognize these qualities in elephants, apes and other "key animals," then it must be speciesist (117-36; 154-5). From another perspective, however, one can see that Waldau's argument contains elements of human chauvinism, in the sense that he defines moral considerability in a way that relies on characteristics that are, primarily, human. One might ask, what is so special about these qualities that only they should be valued? Compared to the Buddhist view, Waldau's account, which only deems certain animals to be morally considerable, seems very partial indeed.

Finally, a few words must be said in defence of the Buddhist hierarchical organisation of beings. It has become common, as Sponberg has shown, to assume that any trace of hierarchy leads inevitably to domination and exploitation, and therefore, he goes on, "we have even banished that dreaded h-word from all forms of polite conversation" (n. pag). Yet, Buddhism would be seriously diminished without the belief in higher and lower forms of life, that is, if it were read as implying continuity alone. Without having something to aspire to, whether worldly enjoyment in a god-like rebirth, or something much higher, like *nibbana,* there would be no sense in our following any path at all. In other words, if our environmental- or animal-friendly tendencies lead us to reject altogether the deprecatory view of nonhumans that is suggested by the Buddhist hierarchy, it is important that we do not end up being unable to discern the differences between beings that motivate us take up one path rather than another.

Summary

I have defended Buddhism's hierarchical view of life from the charges of anthropocentrism and speciesism, by claiming that it does not imply a fundamental divide between humans and other beings, either in nature or in moral worth. Instead, I have suggested that the sense of continuity between beings is far more pronounced. Continuity emerged from the way that beings move up and down the hierarchy through the course of their lives, and that humans display animal qualities and vice-versa. Buddhism's basic doctrines support the idea that there are no hard-edged beings with essential natures, and therefore, they contradict the dualistic assumption that opposes humans to all other animals.

The second assumption that I argued against was that of a restrictive moral circle. I claimed that Buddhism generally tends towards inclusion, as seen in the First Precept, which is an injunction against harm to all living beings. Buddhism does not share the tendency of Western ethics to regard moral considerability as being dependent upon the possession of any particular characteristic.

In short, the charges of anthropocentrism do not appear to succeed, for Buddhism undeniably displays a broad concern for all forms of life. While it is not entirely free of problematic attitudes towards animals, these are rather minor, and can easily be redressed. One suspects, however, that this could be true of all traditions, that is, there is no reason why any religion should be inherently inimical towards animals. To that extent, therefore, those critics, like Harris and Waldau, who enable us to see the less savoury aspects of our traditions are doing crucial work. It is important that we do not allow our assumptions to cause us to take either an overly-critical or a rose-tinted view of our beliefs and those of others.

Abbreviations

D *Digha Nikāyā*
M *Majjhima Nikāyā*
S *Samyutta Nikāyā*
Sn *Sutta Nipata*

Notes

1 Clearly, the examples of gods that are a cross between humans and animals, such as Anubis, in Egypt, and Ganesha in India, militate against this argument to a certain extent.

2 Although some Sanskrit terms, like *nirvana*, are better known, I use Pāli here, and throughout, for consistency, and because the discussion is mostly about early Buddhism.

3 A snake-like creature, capable of appearing as a human. These mythological creatures have both divine and demonic aspects, and therefore, *nāgas* cannot be

taken as a straightforward representative of the class of all nonhuman animals. This renders problematic the entire rule against ordaining animals, since the moral of the story is later interpreted as applying to all animals (see note 3).

4 Actually, in this passage the Buddha says only that *nāgas* are not capable of spiritual growth (Rhys Davids and Oldenberg 219), while the later commentary specifies that it applies to "all nonhuman beings" (Thanissaro "Monastic Code" 190). Whether or not this extrapolation is justified is an interesting question, which, unfortunately, there is no space to go into here.

5 For a more extensive review, see Sciberras 2008.

6 Below, it will be seen that not all authors accept this literal interpretation, and many argue, in fact, that the Buddha had little to say about objective reality.

7 This account is somewhat over-simplified. It is not the case that hungry ghosts, for instance, do not experience anything apart from attachment, rather, it is their dominant emotion, or mental 'poison.' All beings are said to contain a mixture of the negative emotions.

8 For further discussion, see Sciberras 2008.

References

Ariyesako, Bhikkhu. "The Bhikkhu's Rules: A Guide for Laypeople." *Access to Insight*, 5 June 2010. Web. 21 Aug 2010.

Bentham, Jeremy. 2005. *Introduction to the Principles of Morals and Legislation*. Eds. J.H. Burns and H.L.A. Hart. Oxford: Oxford University Press.

Blackburn, Simon. 1996. *Dictionary of Philosophy*. Oxford: Oxford University Press.

Buddharakkhita, Acharya. "Karaniya Metta Sutta: The Hymn of Universal Love." *Acces to Insight*. 14 June 2010. Web. 21 Aug. 2010.

Eckel, Malcolm David. 1998. "Is There A Buddhist Philosophy of Nature?" *Philosophies of Nature: The Human Dimension*. Eds. Robert S. Cohen and Alfred I. Tauber. Dordrecht, The Netherlands: Kluwer Academic Press. 53–69.

Gethin, Rupert. "Cosmology and Meditation: From the Agganna-Sutta to the Mahayana." *History of Religions* 36.3 (1997): 183–217.

Gombrich, Richard F. 1996. *How Buddhism Began. The Conditioned Genesis of the Early Teachings*. London: Athlone Press.

Hamilton, Sue. "The External World': its Status and Relevance in the Pali Nikāyas." *Religion* 29 (1999): 73–90.

Harris, Ian. "How Environmentalist is Buddhism?" *Religion* 21 (1991): 101–14.

Hayward, Tim. "Anthropocentrism: A Misunderstood Problem." *Environmental Values* 6 (1997): 49–63.

Holder, John J. "A suffering (but not irreparable) nature: environmental ethics from the perspective of early Buddhism." *Buddhism and the Environment*. Spec. issue of *Contemporary Buddhism* 8.2 (2007): 113–30.

La Fleur, William R. "Saigyō and the Buddhist Value of Nature: Part 1." *History of Religions* 13.2 (1973): 93–128.

Norbu, Jamyang. "Wildlife and Nature Conservancy in Old Tibet." *Phayul.com*. N.p., 8 Dec. 2009. Web. 21 Aug. 2010.

Nyanatiloka, Mahathera. 1980. *Buddhist Dictionary: Manual of Buddhist Terms and Doctrine*. Kandy, Sri Lanka: Buddhist Publication Society.

Rhys Davids, Thomas William, and Hermann Oldenberg. 1881. *Vinaya Texts: Part I; The Pâtimokkha and The Mahâvagga, I–IV*. Oxford: Clarendon Press.

Schmithausen, Lambert. 1991. *Buddhism and Nature; the lecture delivered on the occasion of the EXPO 1990 – an enlarged version with notes*. Tokyo: International Institute for Buddhist studies.

Sciberras, Colette. "Buddhism and Speciesism: on the Misapplication of Western Concepts to Buddhist Beliefs." *Journal of Buddhist Ethics* 15 (2008): 215–40. Web. 21 Aug 2010

Singer, Peter. 1998. "All Animals Are Equal." *Environmental Philosophy: From Animal Rights to Radical Ecology*. Eds. Michael E. Zimmerman *et al*. Upper Saddle River, N.J: Prentice Hall. 26–40.

Sponberg, Alan. "Green Buddhism and the Hierarchy of Compassion." *Western Buddhist Review* 1. n. d. Web. 21 Aug. 2010.

Steiner, Gary. 2005. *Anthropocentrism and its Discontents: animals and their moral status in the history of Western philosophy*. Pittsburgh: University of Pittsburgh Press.

Thanissaro, Bhikku. "Buddhist Monastic Code II." *Access to Insight*, May 29, 2010. Web. 21 Aug 2010.

_____. "Mata Sutta: Mother." *Access to Insight*, 18 June 2010. Web. 21 Aug. 2010.

Waldau, Paul. 2002. *The Specter of Speciesism: Buddhist and Christian Views of Animals*. New York: Oxford University Press.

Literary and Media Positions

Chapter 18

The Problem with Fern: Leaving the Childhood World of Animals for Heteronormative Adulthood

☆ *Alyce Miller*

"Where's Papa going with that ax?" (White, 2006:1) constitutes the startling opening of E. B. White's classic children's novel, *Charlotte's Web*, immediately establishing `the life-and-death stakes conflict driving the story. The question emanates from tender-hearted, eight-year-old Fern Arable, who has just learned from her practical mother that her father plans to "do away" with the runt of a litter of pigs born the night before in the hoghouse.

Fern chases, sobbing, after her father. In the Garth Williams illustration accompanying the narrative, a pig-tailed Fern, wearing a red gingham dress and brown sneakers, fiercely yanks at the ax her father grips in both hands.

Mr. Arable admonishes her to control herself. To this Fern responds, "Control myself? . . . This is a matter of life and death, and you talk about *controlling* myself" (*Ibid*, 2). Put another way, Fern feels personally responsible. She is taking a moral stand.

Insisting that a weak pig will only make trouble, her father unsuccessfully tries to send her on her way. Instead, Fern stands her ground.: "But it's unfair . . . The pig couldn't help being born small, could it? If *I* had been very small at birth, would you have killed *me*?" (Here Fern attempts to appeal to her father's potential empathy for another living being by having him imagine a direct substitution of herself for the pig). And though her father counters the analogy by distinguishing the two: "A little girl is one thing, a runty pig is another," but to no avail. Instinctively. not only does

Fern refuse to accept his speciesist justification, she has already made the imaginative leap past "self" and "other," blurring the boundaries between herself and the little pig. This ability to imagine her way into the skin of another being allows her to acknowledge that what is about to happen is an injustice.[1] Eventually, Mr. Arable relents, somewhat condescendingly, to her pleas, not because he thinks Wilbur has a right to moral consideration as Fern seems to believe, but simply to appease his daughter. He agrees to bring her the pig, but with the caveat that the pig will be a lot of "trouble."

Though Fern is a farm girl, and presumably familiar with the cycle of breeding and killing of animals raised for food, her heart has been touched by this particular pig, small and helpless, though of course White himself never explores why Wilbur's plight should be distinguishable from that of his siblings, except for the one detail that he's a "runt" and therefore perhaps more easily perceived as vulnerable. And though, as the book progresses, Wilbur does indeed turn out to be, thanks to the clever spider, Charlotte, "Some Pig," noticeably absent is any mention of the others in his litter whose brief lives will already of course have ended at the slaughterhouse.[2]

White describes Fern's role toward Wilbur as instantly maternal, and her love, unconditional. Though White's narrator doesn't explicitly state this, the implication implies that girls (as represented by Fern) possess a natural empathy toward animals:

> "Fern loved Wilbur more than anything. She loved to stroke him, to feed him, to put him to bed. Every morning, as soon as she got up, she warmed his milk, tied his bib on, and held the bottle for him" (*Ibid*, 8).

Wilbur, the size of a human infant, is small and helpless and dependent, and functions as a surrogate child. Fern is learning to care for another creature, and her parents approve, perhaps implicitly in preparation for what they might expect to be her own future motherhood. In this way, her tenderness is both assumed and naturalized.

> "If she took her doll for a walk in the doll carriage, Wilbur would get tired, and Fern would pick him up and put him in the carriage alongside the foll. He liked this. And if he was *very* tired, he would close his eyes and go to sleep under the doll's blanket... . Fern would wheel the carriage very slowly and smoothly so as not to wake her infants" (*Ibid*, 10).

For a period of time, with Wilbur tagging along behind her everywhere she goes, "Every day was a happy day, and every night was peaceful" (*Ibid*, 10). But when Wilbur reaches five weeks of age, their idyll is interrupted. Mr. Arable announces the pig needs to be sold (he has already sold Wilbur's ten brothers and sisters): "You have had your fun raising a baby pig, but Wilbur is not a baby any longer and he has got to be sold," Mr. Arable instructs the distraught Fern. [3] Fern's mother intervenes with a more humane alternative: that Wilbur be moved from living under the apple tree to a manure pile in Uncle Homer Zuckerman's barn so that Fern can walk down every day to visit. (In an interesting twist, Fern actually sells Wilbur to her uncle for six dollars, a fact which is never fully explained.) With Wilbur now safely ensconced at the Zuckerman farm, Fern visits almost daily, and begins to develop an even deeper connection to "the world of animals":

> She found an old milking stool that had been discarded, and she placed the stool in the sheepfold next to Wilbur's pen. Here she sat quietly during the long afternoons, thinking and listening and watching Wilbur. The sheep soon got to know and trust her . . . All the animals trusted her, she was so quiet and friendly (*Ibid*, 15).

The Zuckermans' barn is described as a magical place: it "smelled of hay and it smelled of manure. It smelled of the perspiration of tired horses and the wonderful sweet breath of patient cows. It often had a sort of peaceful smell–as though nothing bad could happen ever again in the world" (*Ibid*, 15).

Fern's uncle expressly forbids her from taking Wilbur out of the pen, or going into the pen with him (a significant physical separation point). Perhaps White felt it necessary to keep the human body and the animal body separate, particularly since Wilbur is a "farm animal." But Fern spends hours in the barn close at hand, a careful observer who watches and listens to everything that goes on. And though the setting of the book is grounded in "realism," the device of talking animals who express feelings and hopes and ideas manages to feel organic. One reason may be that the reader is squarely positioned with Fern on that milking stool listening to the animals talk with one another, so that we never doubt what Fern knows. In addition, the animals never speak to Fern or with other human characters. In this way we can more easily suspend disbelief that they are actually communicating in human language, specifically, English. In addition, the conversations feel unmediated; we actually "hear" their voices. . Initially Fern is our conduit, but eventually, when Fern is absent at various moments throughout, the reader is left alone with the animals.

White's representation of animal cognition understandably suffers from some of the problems of anthropomorphism, but fortunately the animals are in no way meant to be figurative or simply instructive. The sheep, the geese, the rat Templeton, and the spider Charlotte, emerge as distinctive characters, and the interactions in the barn create a complex stew. The strategy of a partially-omniscient and alternating point of view that shifts among the human characters and animal characters grants reader access to Wilbur's interiority as well as that of Fern's.

I have taught this book a number of times, in three different courses, Children's Lit, Animals and Ethics, and The Literary and Legal Animal, and each time I discover the rich opportunities it offers for discussion. When I was a child and early on a self-described lover of animals, I adored this book, and never doubted the possibilities for such communion between the human and nonhuman animal worlds. Different from books like *Wind in the Willows* in which animal characters behave exactly as people would (sweeping their doorsteps, driving cars, rowing boats, etc.), *Charlotte's Web* offers an interesting hybrid world, and one of its delights is the very Elizabeth Costello-esque suggestion that if we just listen, we, too, can "become." And though White makes it clear that Fern is not touching the animals in the barn, the relationship with Wilbur remains fully embodied until the moment when she begins to pull away.

Even as a child reader, I understood the exchanges between the animals to be a quasi-translation to which the attentive, empathic, and open-hearted could surely be privy. Fern's ability to see and hear "the animal world," a world I deeply imagined and "heard" myself, is part of the book's charm. For the first half of the book, she is wiser than the human adults around her, and her intuitive observations function as a kind of layperson's ethology. Though founded on an assimilative model (imagining how like us animals might be), *Charlotte's Web* offers, like many children's books featuring animal characters, the potential to teach empathy, even if through a fictionalized world in which animals are consistently lovable and entertaining, and whose worldview, desires, and behaviors are not so different from our own. Each time I teach this book, I find my relationship to it shifting. I teach it for a number of different reasons, including what it contributes to discuss about representations of animals in literature, and what it ultimately has to say about human/nonhuman animal interactions.

On the one hand, the fundamental argument that lies at the heart of *Charlotte's Web*, and what distinguishes it from many coming-of-age novels which rely on the sacrificed animal, is not only that Wilbur is not forfeited, but that he expressly states his desire to live, a desire repeated and echoed throughout the book. Briefly lulled into thinking he's safe, Wilbur is confronted by the truth of his impending death in that terrible moment in the chapter titled, "Bad News," when the sheep explains to Wilbur that the Zuckermans are fattening him up for Christmas dinner. Naive to the cruelties of the world (not unlike Fern), Wilbur reacts with utter disbelief and horror.[4] And while Fern, too, leaps up in distress, she is quickly brought up short against the limits of her own ability to protect him any further. It is only with the device of the intervention of the clever spider Charlotte that Wilbur's life can once again be spared.

But in those cries of despair, Wilbur expresses what many animal rights theorists would describe as the most fundamental reason we should not kill animals: that he has "not only an interest in being free of inflicted pain and suffering but that he has a positive interest in living. Of course, however, his status as legal property of the Zuckermans strips him of any autonomy and makes his future welfare completely dependent on their choices.

The two other main characters in the book, first Fern and then Charlotte, share in common an investment in Wilbur's interest in life.[5] "You shall not die," said Charlotte, briskly.

Fern is the first to see that his life is spared in the beginning, but Wilbur is still in danger as he matures, and Charlotte continues in Fern's spirit to see that Wilbur remains alive. It is not an accident that both characters are female, and in varying ways, maternal, and both resort to ploys to keep him safe. The "gendered" aspect of this novel is intriguing and not as dated as one might think. Fern is the child "mother-in-training" and Charlotte is the fiercely protective who, together, manage to preserve Wilbur's life. The feminine becomes a repository for protection. It is first Mr. Arable, then Mr. Zuckerman, who are threats to Wilbur's life. And of course, Fern's compassion stands in direct contrast to the rambunctious cruelties of her ten year old brother Avery whose recklessness and cruel treatment of animals is often played for laughs.

First introduced to the reader as "heavily armed–an air rifle in one hand, a wooden dagger in the other" (*Ibid,* 4), Avery has little interest in Wilbur except to ask if he could have a baby pig, too (he likes to collect animals). Later in the novel, when the children return home from playing, Fern is described as having made a crown of daisies for her hair, while Avery, we're told, carries a frog he's caught and which he places on the kitchen sink. The frog jumps into the dishwater, splashing soap all over the blueberry pie his mother has made. The narrative tone toward Avery's cruelty is one of amusement, and in keeping with a "boys will be boys" ethos the book seems to promote. We're told that Avery keeps the frog in his pocket all morning, and when the frog gets hot while Avery swings in the hayloft, he places the frog in an empty candy box. White writes, "The frog seemed tired after his morning in the swing" (*Ibid,* 70). The fate of the frog is never clarified, but it's not difficult to assume that Avery's complete indifference to the frog's wellbeing could end in tragedy.

In fact, despite the light-hearted tone of the novel, Avery, like other male figures in the book, poses a danger to animals, and his active presence on the page often signals a threat. When Avery follows Fern out to the barn with mischief on his mind, for example, she again steps in to protect against male cruelty. Spotting Charlotte the spider overhead in her web, Avery announces he's going to capture her ("It's tremendous") (*Ibid,* 72), and place her in the candy box alongside the frog (it's unclear whether the frog has managed to survive or not). Again, Fern intervenes, shouting at Avery, "Leave it alone! . . . You've got a frog–isn't it enough?" (*Ibid,* 72).

While Avery's conquests of animals is viewed by his parents throughout the book as "normal" (and, I might add, normal for a boy), it is Fern, who chooses an opposite approach. Instead of imposing her will on animals, or viewing them as objects of conquest, she chooses to listen to and observe them quietly. And yet, as the novel evolves, it is she who becomes "the problem" for her parents, the child Mrs. Arable worries about.

After learning from Fern that she can hear the animals speak, Mrs. Arable becomes convinced that her daughter spends too much time in Uncle Homer's barn. After Fern sweetly describes to her mother how Charlotte is "the best storyteller I ever heard" and proceeds to recount the tale Charlotte has told about the fish caught in her cousin's web" (*Ibid,* 105), Mrs. Arable accuses her daughter of inventing. Without hesitation, Fern replies, "I'm not inventing . . . I'm just telling you the facts" (*Ibid,* 105). Concerned, Mrs. Arable suggests to Fern she should find playmates and "do something nice outdoors. . . . it isn't good for you to be alone so much" (*Ibid,* 107).

By contrast, Avery who also seems to spend a fair amount of time alone does not seem to concern Mrs. Arable. In addition, though Avery is also interested in animals in a destructive way, Mrs. Arable believes "it didn't seem natural for a little girl to be so interested in animals" (*Ibid,* 107). Put another way, Avery's interest, even as dangerous as it is to animals, is acceptable. Fern's is not. First, Fern has, in Mrs. Arable's estimation, crossed a line, and moved eyond harmless fantasy and invention. In fact, she seems to feel her daughter's social development, as well as her own (maternal) responsibility for it, are at stake here.

As a result, Mrs. Arable decides to consult a medical expert, known ambiguously only as "old Dr. Dorian" and described as having a "thick beard" (*Ibid*, 107). In the Garth Williams illustration of their appointment, Dr. Dorian wears glasses and reclines in a chair behind his desk, in a sort of central-casting version of a therapist. He patiently listens as Mrs. Arable's recounts her worry that Fern's behavior isn't "normal" and offers her description of the way Fern "sits on a milk stool in a corner of the barn cellar, near the pigpen, and watches animals, hour after hour. She just sits and listens . . ." (*Ibid*, 107).

Of course by this time, the reader has also heard the animals talk, and we know Fern is telling the truth. In fact, Fern's experiences in the barn have become so naturalized to the reader that Mrs. Arable's worries strike us as foolish and wrong. Therefore, Dr. Dorian's initial response comes as a relief to the reader: "If Fern says that the animals in Zuckerman's barn talk, I'm quite ready to believe her," he says. "Perhaps if people talked less, animals would talk more. . . " (*Ibid*, 110).

The discussion continues with Dr. Dorian inquiring about Fern's appetite and sleep patterns, and then he asks her age, which is eight. "Well," said Dr. Dorian, "I think she will always love animals. But I doubt that she spends her entire life in Homer Zuckerman's barn cellar. How about boys–does she know any boys?" (*Ibid*, 111)

Mrs. Arable mentions a boy named Henry Fussy. Dr. Dorian assures her then that it's fine to let "Fern associate with her friends in the barn if she wants to . . ." and adds that "I predict that the day will come when even Henry will drop some chance remark that catches Fern's attention" (*Ibid*, 111).

This somewhat condescending prediction comes true later in the novel when Fern does find Henry Fussy of more interest than the doings in the barn.

Dr. Dorian then inquires about Avery.

"Oh, Avery," chuckled Mrs. Arable. "Avery is always fine. Of course, he gets into poison ivy and gets stung by wasps and bees and brings frogs and snakes home and breaks everything he lays his hands on. He's fine" (*Ibid*, 111-112).

"Good!" says the doctor (*Ibid*, 112). And there is no more discussion of Avery. He has been approval-stamped as "all boy," while Fern, we are reassured along with Mrs. Arable, will grow out of this stage with Wilbur and the other barn animals, as soon as she finds a boy (*Ibid*, 130). Now convinced that Fern will be "normal": read "sexually normal" (meet a boy and begin to develop romantic interests that will eventually lead to adult heteronormativity), Mrs. Arable allows her to continue her visits to the barn. [6]

The county fair becomes a turning point in the novel for two reasons: It is the event that triggers Fern's break from Wilbur, and it is also the climax of the various preparations for Wilbur who is under pressure to win first place, or else. The idea that Wilbur has to keep "earning" his right to life pervades the book. And the way he earns it is by making himself appealing to human beings. Context is everything. Wilbur must continue to distinguish himself (with help from Charlotte) from pig-as-dinner. Commerce is also clearly at the center of farming life and rules Wilbur's life.

Wilbur as the runt is worthless. Wilbur as Fern's companion and friend has intrinsic value only for Fern and Wilbur. Wilbur's role at the fair is to bring fame and limited fortune to the Zuckermans. Before Wilbur is transported to the fair, with everyone's hopes placed on his continued notoriety, Charlotte must work feverishly to create the dazzling spectacles of descriptive words she weaves into her web. It is again female labor that saves and protects Wilbur. And though at one point Mrs. Zuckerman very humorously points out that maybe it's not the pig who's remarkable, but the spider, Wilbur's value still is determined by the human beings for whom the words on the web signal his uniqueness.

Yet despite the enormous stakes for Wilbur at the fair, Fern proves herself to be fickle. She is more interested in winning a doll at one of the booths and riding the Ferris wheel with Henry Fussy.

> "The children grabbed each other by the hand and danced off in the direction of the merry-go-round, toward the wonderful music and the wonderful adventure and the wonderful excitement, into the wonderful midway where there would be no parents to guard them and guide them, and where they could be happy and free and do as they pleased" (*Ibid*, 131).

Mrs. Arable fights off tears watching them go and asks her husband if it's all right. "Well, they have to grow up some time . . . And a fair is a good place to start, I guess" (*Ibid*, 133)

Growing up, an important theme of the book, is equated with a new-found autonomy from parents (though Fern is safely in the care of Avery) and, in Fern's case, a move into the world of humans.

In "The Cool of the Evening" chapter, Mrs. Arable takes special delight in observing Fern and Henry Fussy riding on the Ferris wheel (he has bought the tickets, another detail of traditional gender roles). Fern looks happy as she ascends into the starry sky next to Henry, and from below, her mother murmurs approvingly, "My, my! . . . Henry Fussy. Think of that!" (*Ibid*, 139). The chapter ends with Mrs. Arable tucking Fern into bed and asking if she had a good time at the Fair.

"I had," Fern says, "the best time I have ever had anywhere or any time in all of my whole life" (*Ibid*, 143). Without almost no explanation or critique, Wilbur is fading from Fern's sphere of interest, and Henry Fussy is, as Dr. Dorian predicted, replacing him.

After this, Fern further retreats from Wilbur's life. The next day at the Fair after noting that the web above Wilbur now has the word "HUMBLE" stitched into it, Fern has little interest in hanging around and asks her mother for money to go to the midway. When her mother refuses, Fern cries (*Ibid*, 180). And even though Wilbur's fame is about to be celebrated at the judges' booth in front of the grandstand, Fern again pesters Mrs. Arable for money so she can presumably go in search of Henry Fussy. And though she is still denied the money, and made to join the family in the truck to when they take Wilbur to the judges' booth, Wilbur is the farthest thing from Fern's mind: "As they passed the Ferris wheel, Fern gazed up at it and wished she were in the topmost car with Henry Fussy at her side" (*Ibid*, 54).

In "The Hour of Triumph," as the crowd gathers around the truck bearing the crated Wilbur and the family, Fern is distracted again by the sight of Henry Fussy in the distance. This time she succeeds in wheedling Ferris wheel ticket money from her beleaguered mother, and racing off "... ducking and dodging through the crowd, in search of Henry" (*Ibid*, 156).

When Wilbur faints with all the heat and attention, Fern is absent, and it is the rascally rat Templeton who saves the day by biting down on Wilbur's tail to revive him. Fern is mentioned again only briefly at the beginning of Chapter XX1, "Last Day," when the reader is simply told, "The families had gone to look for Fern" (*Ibid*, 163).

Wilbur's return home follows on the heels of the death of Charlotte. Left to his own devices, he is forced to mature by accepting responsibility for her five hundred and fourteen unborn spider children he has placed, at her request, in a safe corner. Fall comes and goes, and Christmas arrives. Out sledding with Avery, Fern claims "The most fun there is ... is when the Ferris wheel stops and Henry and I are in the top car and Henry makes the car swing and we can see everything for miles and miles and miles" (*Ibid*, 173).

Bored by her new obsession with a boy, Avery reminds her the Fair was weeks ago, and Fern replies, "I think about it all the time" (*Ibid*, 174).

Thus, her transition from childhood to adulthood is represented, in part, as Dr. Dorian predicted. In the last chapter, there is one final mention of Fern. "As time went on, and the months and years came and went, he [Wilbur] was never without friends. Fern did not come regularly to the barn any more. She was growing up, and was careful to avoid childish things, like sitting on a milk stool near a pigpen."[7] White's language here is interesting in its gentle condemnation of "childish things." Spending time in the company of animals becomes one of the most pronounced judgments in the book. After this, Fern disappears entirely from the book.

That Fern initially occupies the moral center of the book (perceiving that Wilbur has intrinsic value and is worth saving, protecting, and allowing him to flourish and become "his own pig") makes her rejection of not just Wilbur, but the animal world that she once thrived in, and her vanishing at the end of the novel in search of Henry Fussy (the first of what will presumably be the men who will follow him in her life) even more acutely problematic. It is as if White ultimately concludes, albeit gently, that while young girls with tender hearts (not boys like Avery) can appropriately care for and about animals, that same level of intimacy with animals is incompatible for the move into heteronormative womanhood.

The views implied in *Charlotte's Web* still seem very much alive, starting with the idea that women have an innate ability to nurture though that nurture is mostly appropriately directed toward other human beings (children? husband?). A recurring theme in many children's and young adult books involving human-animal bonds seems to suggest that the adoption of appropriate relationships with human beings and a kind of "toughening up" "involves either the abandonment or death of an animal. In many cultures, adults who maintain close relationships to animals (even traditional companion animals like cats, dogs, and horses), particularly in the case of women[8], are often pathologized, specially those women who deviate from the traditional roles like marriage and motherhood, and live instead with animals.[9]

In that slippery, catch-all category known crudely as "the animal rights movement," women have historically played disproportionate roles, starting with early advocacy groups like the Battersea Dogs and Cats Home, [10] and animal protection organizations like the National Anti-Vivisection Society,[11] and continuing into this last century with Christine Stevens' Animal Welfare Institute, founded in 1951, Joyce Tischler's pro bono group, The Animal Legal Defense Fund (ALDF), founded in 1978, and Ingrid Newkirk's establishment of People for the Ethical Treatment of Animals (PETA) in 1980. Connections between women and animals make much more sense when viewed through the lens of the rhetoric, mechanisms, and structures of power supporting various forms oppression, and the suffering inflicted.[12]

The problem with Fern, then, is not the problem Mrs. Arable describes to Dr. Dorian–that her daughter has spent too much time alone and believes she hears animals talking—but instead is the problem that the book never really tackles in full–the undoing of compassion as a part of the human maturation process. The relationship Fern develops with Wilbur, starting with her moral outrage over his physical welfare, and her recognition that his life has intrinsic value, is ultimately consigned to little more than a childish pastime, one that must be abandoned if she is to successfully the world of "adulthood." A beautiful, eloquent, and eminently teachable book, *Charlotte's Web* does distinguishes itself from many others in the genre by not relegating Wilbur solely to symbolic evidence Fern's evolution or sacrificing him through death in order for her to move on with her life. In fact, in an odd twist, at the end, after Fern has vanished, he is briefly the sole point-of-view character, and is given some agency. I would therefore argue that a different sort of sacrifice of Wilbur is made; that by yoking animals and "childish things," White makes it more palatable for the reader to conclude that Fern's abandonment of Wilbur is an appropriate step for her evolution into heternormative womanhood. In addition, the naturalizing of Wilbur as a happy barnyard animal is made complete with the final, rosy image of him living out his life, the property of the Zuckermans, in the barn with Charlotte's descendents. So though as a child Fern advocates for Wilbur's interest in life, by the end she has abandoned this concern, and taken her proper place in the human hierarchy, having, it appears, altogether lost interest altogether in him and the other animals. In addition, the implied triumph of heteronormativity, as predicted by Dr. Dorian, that Fern will learn to take a healthy interest in boys and eventually grow up to marry and have her own family, seems to suggest, in part, that intimacy with animals and adulthood are incompatible, and that inhabiting the world of animals keeps one connected to the infantile, and the adult world, proscribed by heteronorms, demands that only by giving up the one do you fully gain entrance to the other.

Notes

1 Though I'm quite certain it is not even close to White's intent, even as he invites us to feel sympathy for the rightness of Fern's position in this brief moment, Fern's touching assertion–her recognition that she, like Wilbur, could be disposed

of— could be an excellent entry point into ecofeminism and a look at the ways in which categories of "woman" and "animal" function both materially and theoretically as "other," as well as the vulnerabilities of both within male power structures.

2 A common schema in children's literature and films like Babe is the focus on the plight and safety of particular animals, thereby, avoiding a larger-scale commentary on our inconsistent treatment of animals, connected to the context of our relationships to them (Wilbur becomes a "pet"; his siblings will be bacon).

3 This is an interesting moment in which White makes the claim that Wilbur is clearly still not disposable. His purpose is not simply to entertain Fern or to make the family some money: he is apparently entitled to have a life, an important idea around which the rest of the book is built.

4 "I don't want to die! Save me, somebody, save me!" "I can't be quiet. I don't want to be killed. I don't want to die." (P 50) "I want to stay alive, right here in my comfortable manure pile with all my friends. I want to breathe the beautiful air and lie in the beautiful sun." 51

5 "You shall not die," said Charlotte, briskly.

"What? Really?" cried Wilbur. "Who's going to save me?"

"I am," said Charlotte.

"How? asked Wilbur.

"That remains to be seen. But I am going to save you, and I want you to quiet down immediately. You're carrying on in a childish way. Stop your crying! I can't stand hysterics." White, 51

6 In a terrific article, "The Love Whose Name Cannot Be Spoken: Queering the Human-Animal Bond," Journal for Critical Animal Studies, Volume VIII, 2010, Carmen Dell Aversano writes:

"Human love for animals is ridiculed, marginalized, despised and repressed . . .

An individual belonging to the human species is *assumed*, by the way language works, to identify primarily with the human species, to feel emotions and loyalties coherent with this identification, and to act accordingly."

Though it's close to impossible to imagine that White's idea of "norm" for Fern had anything to do with anxieties about actual homosexuality, the idea that a young girl might spend time alone with animals she loves and not gravitate toward boys lies at the very center of her mother's fears.

7 White, 183; cf. *Flush*, Virginia Woolf. Flush, the canine protagonist of this eponymous novel, has a good, intimate life with the poet Elizabeth Barrett until his life is turned upside down when she marries and becomes pregnant. "It was not merely that she was calling herself Mrs. Browning now; that she flashed the gold ring on her hand in the sun; she was changed, as much as Flush was changed. Flush heard her say, 'Robert,' 'my husband,' fifty times a day, and always with a ring of pride that made his hackles rise and his heart jump. But it

was not her language only that had changed. She was a different person altogether." p. 114 (*Flush*, Harcourt, Inc., 1983.)

8 There is no male corollary in the West for "the crazy cat lady," a possible product of earlier times when women and cats were considered consorts in witchcraft.

9 In her wonderful memoir, *Pack of Two*, Caroline Knapp devotes a fair amount of time to the question of the role of animals in the lives of people. "Did I up and leave my boyfriend for a dog?" she asks/ In one poignant example, she describes an exchange between a woman and her daughter: "'Mom, . . . sometimes I think that dog is your best friend. . . .' The mother replies, 'Honey, some days that dog is my *only* friend,'" 163. In a chapter called "Surrogate Dog," Knapp analzyes what she feels are gender-based accusations that women who are close to their dogs are sublimating desires for children and mates. A friend reacts by asking, "'Do men with dogs get accused of this? Can you imagine seeing a guy walk across a field with his German shepherd and having your first association be, Oh, he's obviously sublimating his biological need to procreate into the dog?'," 187.

10 Founded in England in 1860 by Mary Tealby, and initially known as "the Temporary Home for Lost and Starving Dogs"

11 Founded in England in 1871 by Frances Cobb

12 Feminist scholars like Carol Adams, for example, do an excellent job of demonstrating specific links between oppression of women and animals. Josephine Donovan challenges traditional male philosophical rationality and proposes an ethics of care based on reciprocity instead of domination. Emily Gaarder, in a paper titled, "The Gender Question of Animal Rights: Why Are Women the Majority?", thoughtfully examines ideas some of the reasons historically that more women than men have been involved in animal welfare and protectionist movements and why today, for example, women constitute somewhere between 68-80 percent of those involved in animal rights.

References

White, E.B. 2006. *Charlotte's Web*. New York: Harper Collins.

Chapter 19

In Ancestral Times When Birds were Still Humans: Kinship and Dialogue in Amerindian Ornithological Narratives in Southern South America

☆ *Ricardo Rozzi*

Introduction

Indigenous narratives correspond to oral traditions whose transcriptions are open to multiple interpretations. The narratives possess many layers of significance whose decodification and interpretation vary depending on the analyst's questions and perspectives (Turner 1967, Geertz, 1974). The present analysis of Yahgan and Mapuche stories about the birds of the southern South American forests aims to allow for comparisons of traditional ecological knowledge (TEK) with notions of ecological and evolutionary sciences, and of environmental ethics. For this purpose, the analysis centers on two metaphors utilized extensively in environmental and evolutionary sciences, both of which involve an ethical dimension: the "tree of life" and the "web of life".

These two metaphors were used by Charles Darwin in the *Origin of Species*, and today they synthesize central concepts of evolutionary-ecological sciences as well as of environmental ethics (Rozzi 1999). In the first metaphor, the common trunk of the evolutionary tree offers a visual representation of Darwinian evolutionary theory that emphasizes the common biological origin that humans share with all living beings. This image of a common evolutionary genealogy stimulates a sense of kinship which defies the traditional western ethics by demanding an extension of ethical respect beyond the limits of our own species. The sense of evolutionary kinship provides one of the most appealing scientific foundations to support the notion of the *intrinsic value* of non-human life, and appeal to an ethical treatment of animals.

The second metaphor, the web of life, provides a complementary image that evokes the multiple ecological interactions that happen between nutrients, plants, seaweed, animal, fungi, bacteria, nutrients, water, solar energy and soil that form biotic communities and ecosystems. This understanding of ecological interrelations provides one of the most appealing scientific foundations to support the notion of *instrumental value* of biodiversity, because it illustrates how biodiversity delivers ecosystem services for human societies, a concept that converges with TEK hosted by many Amerindian communities, including the Yahgan and Mapuche in southwestern South America (Figure 1).

Storytelling formed an integral part of Yahgan and Mapuche life. According to the Austrian ethnographer Martin Gusinde (1961) in Yahgan communities there were men who narrated stories around the fires at night in this austral archipelago. However, the stories recorded in this work, as well as the majority of those recorded by ethnographers in the early 20th century, were told by women. At the beginning of the 21st century, sisters Úrsula and Cristina Calderón were the only persons that fluently spoke the Yahgan language, and they were the major holders of Yahgan traditional ecological knowledge. The stories were recorded *in situ* between the years 2000 and 2003, and they provide the basis for the analysis of Yahgan traditional ornithological knowledge presented in this chapter.

Regarding the Mapuche narratives, these originate principally from stories recorded with the Lafkenche poet Lorenzo Aillapan. I focus on a particular aspect of the intimate tie that the Mapuche people and their culture maintain with both their land, which today includes the south of Chile and Argentina, and the animals that inhabit that land: their traditional ecological knowledge about the birds from the point of view of a current poetic expression (Aillapan & Rozzi 2001). Both the name "Mapuche" that means "People of the Land" (*mapu* = land, *che* = people), and their language "Mapudungun" that means "Language (=*dungu*) of the Land" express the ancestral identification that these people maintain with their land (Cona & Moesbach 1930), and their capacity to communicate with other animals, especially birds (Rozzi *et al.*, 2010).

The communication of the Mapuche culture with the diversity of living beings in the region of South American temperate forests is expressed by and captured through very diverse cultural practices (Ovalle 1646, Gusinde 1917a, 1917b; Domeyko 1971, Bulnes 1980, Massardo & Rozzi 1996). Among these, poetry constitutes an expression that allows us to get particularly close to the birds, thanks to the rich onomatopoeic nature of Mapudungun and to the close attentive bird watching of Mapuche poets (Rozzi 2001). Mapuche poetry today acquires a bilingual form of Mapudungun and Spanish that permits a poet like Lorenzo Aillapan to express himself, and through him, the habitats and the birds speak. Aillapan lives in the countrysides and forests

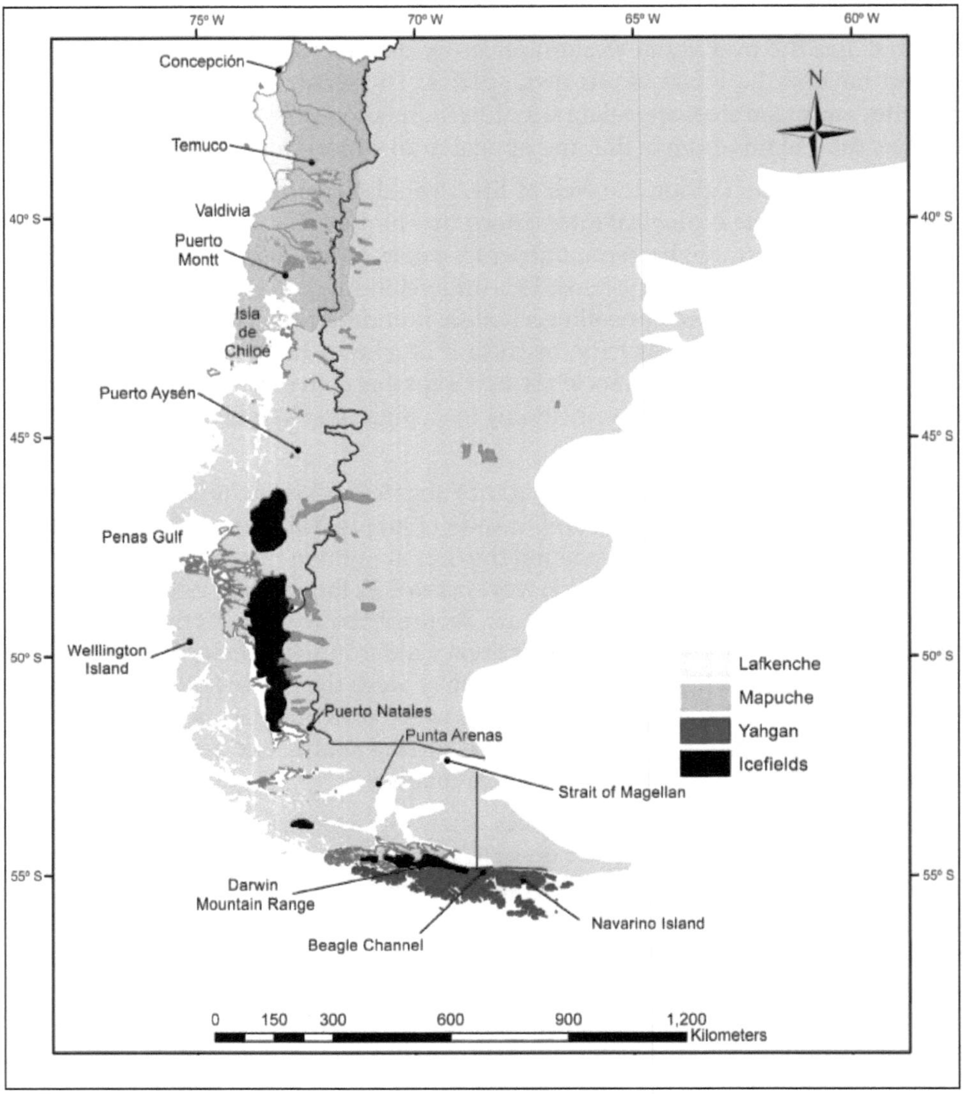

Figure 1. Ethnographic map of the southern South America illustrating the location of the Yahgan and Mapuche (including the *Lafkenche*) territories (Modified from Rozzi *et al.*, 2010).

of the Lafkenche region inhabited by Mapuche communities associated to the coast and ocean (= *lafken*) in southern Chile. Between the ages of 8 and 9, Aillapan was initiated as the Birdman through a dream, initiation which was guided by renowned Mapuche masters. Today, Aillapan is a Mapuche Birdman or *Ünümche* (*ünüm* = bird, *che* = human being), and as such he communicates with the birds, especially in his ancestral native language: Mapudungun (Aillapan 2003). Before doing this, on each occasion Aillapan asks permission of the birds and Mother Nature to speak about them and for them. The capacity of the Mapuche people to listen to the birds and "the land that speaks to us" is expressed in poems based on refined observations in the field and the cultural tradition of Ünümche. They allow us to understand some key features about the relations established by the Mapuche communities with their natural environment and avifauna. Our ethno-ornithological analysis shows how interrelated biological and cultural diversities are in the Mapuche natural medicine, social and environmental ethics, language, traditional ecological knowledge and worldview. Interestingly, the biocultural approach discloses not only differences, but also fundamental similarities between the ornithological knowledge rooted in the ancestral Amerindian cultures and that rooted in contemporary biological sciences.

A Sense of Kinship Between Birds and Humans

Yahgan Ornithological Narratives Associated with the Tree of Life Metaphor

From an environmental ethics perspective, the central concept of the tree of life metaphor is the notion "of evolutionary kinship" (*sensu* Leopold 1949). We consider the Yahgan story of the Hummingbird (*Sephanoides sephaniodes*) or *Omora* as told by Cristina Calderón (Appendix 1). Processes of transformation from human beings into birds–like those evoked in the story about the Hummingbird or *omora*–are encountered in numerous Yahgan stories, and constitute a primary foundation for the notion of a relationship between birds and humans. Grandmother Úrsula Calderón related stories like that of the beautiful young women transforming into little Black-Chinned Siskin birds (*Carduelis barbatus*) or *twin*, or of two Yahgan siblings who fell in love and changed into a pair of Magellanic Woodpeckers (*Campephilus magellanicus*), or *lana*, while collecting fruits from the *chaura* (*Pernettya mucronata*). Such transformations imply a common ancestral lineage (Figure 2A).

Transformation of humans into animals, or animals into humans, was conceived by numerous Amerindian cultures. Mexican anthropologist Eva Hunt referred to this type of cosmogony as a "pre-Hispanic theory of evolutionary biology" (Hunt 1977, 52). In scientific evolutionary theory as well as pre-Hispanic evolutionary perspectives, the humans, birds, and other living creatures share a common evolutionary origin and, as a consequence, are linked through the notion of a genealogical relationship.

In the story of the Yahgan hummingbird *omora*, we also encounter a notion of "interspecies relationship" between the birds. The White Owl (*Tyto alba*) or *sirra* is the grandmother of another species, the Hummingbird. In other Yahgan narratives, this spectrum of kinship and transformation is projected onto the physical world: rocks are transformed into human beings, and humans transformed into stars, the moon or the sun (Gusinde 1961).

Interspecies kinship relationships, as well as transformations from one species to another, support a notion of common nature that presents a parallel to the biological theory of the unity of life. Such a notion is further affirmed when the story speaks of "the time of the ancestors, when birds were still human". This affirmation begins the *omora* narrative and many other Yahgan stories. For example, the grandmother Úrsula

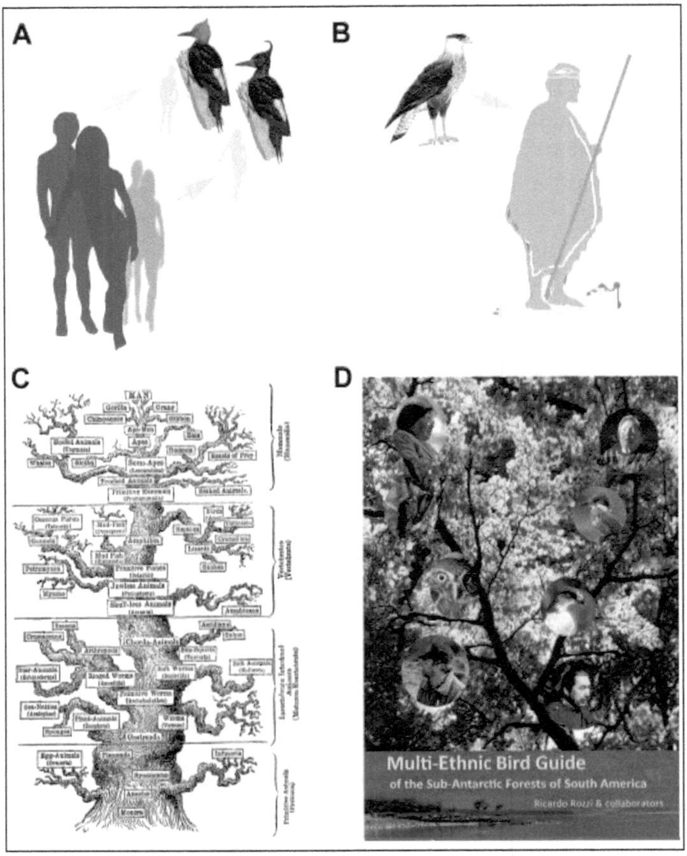

Figure 2. South American Amerindian and scientific genealogies of birds and humans. In a broad sense, the Yahgan and Mapuche worldviews share with contemporary biological sciences concepts about common genealogies implying a sense of kinship between birds and humans.

A) Yahgan ornithological narratives frequently begin with the sentence "in ancestral times when birds were humans," and one of these narratives tells that a pair of Yahgan siblings who felt in love transformed themselves into woodpeckers, or *lana*, (*Campephilus magellanicus*). B) According to the Mapuche worldview, several human lineages descend from birds, and one of their bird stories tells that *Lautaro* or *Leftraru*, the greatest of Mapuche warriors, descended from the lineage of caracaras or *traru* (*Caracara plancus*); indeed, the name *Leftraru* means fast (=*lef*) *traru*. C) The tree of life depicted by German biologist Ernst Haeckel illustrates the scientific-evolutionary theory of Charles Darwin, which like Amerindian worldviews implies a common genealogy shared by humans and all animal species. However, in contrast to the Yahgan and Mapuche worldviews that do not imply a hierarchy between humans and birds, in Haeckel's tree "Man" is portrayed at the top of the trunk. D) In order to illustrate equitable kinship relationships among birds and humans, on the cover of the contemporary Multi-Ethnic Bird Guide of the Sub-Antarctic Forests of South America each of the main participants is portrayed on branches of a native tree near a bird with which he or she has close affinities (Figure modified from Rozzi *et al.*, 2010).

Calderón relates that, in ancestral times, an old man, the inventor of canoes or *anann* was the widower, *Tuto*. He then transformed into the little Magellanic Tapaculo (*Scytalopus magellanicus*). Today this bird collects and weaves vegetable fibers of roots, branches, lichens, and mosses to carefully build his nest between roots or in tree-trunk cavities, resembling the way the old man gathered and prepared these materials to build the Yahgan canoes (in Rozzi et al. 2010: 52-5).

In summary, Yahgan ornithological stories present three elements that in an extensive sense imply a notion of kinship between human beings and birds: 1) Yahgans note that human beings have given origin to various species of birds through diverse processes of transformations; 2) Yahgans recognized kinship relationships between diverse species of birds, such as those between the White Owl and the Hummingbird; 3) Yahgan cosmogony refers to an ancestral time when birds were still humans.

Mapuche Ornithological Narratives Associated with the Tree of Life Metaphor

In order to analyze Mapuche ornithological stories from the point of view of the "tree of life" metaphor, we primarily consider the Andean Condor (*Vultur gryphus*) or *manke* who is the king of birds in traditional Mapuche ornithology, as well as in other Andean cultures (Appendix 2).

With a wingspan up to three meters, the Andean Condor is the largest flying bird in the world. For the Mapuche culture, *manke* symbolizes the Andean mountain range through its grand size and by its colors: white like the snow and black like the rocks; it also flies to grand heights, embodying fundamental Mapuche virtues. Today, Aillapan emphasizes that both manke and the Mapuche virtues – wisdom (*kim*), justice (*nor*), kindness (*küm*) and discipline (*newen*) – are threatened with extinction. Just like the Andean Condor is hard to see because it soars so high, so the essential exalted Mapuche values are difficult to appreciate and cultivate.

The Mapuche have a special admiration for predatory birds. Of those, the Crested Caracara (*Caracara plancus*) or *traru* has great historical significance because its lineage gave origin to the greatest of Mapuche warriors, *Leftraru*. This valiant warrior, known by the Spanish as *Lautaro* (1535-1557), was aided in battle by the topography of his land, his capacity to predict climatic phenomena, and the speed to attack with his people from different flanks (Barella 1971). Poet Aillapan asserts that the Crested Caracara is a thousand-year-old warrior who belongs to the lineage of eagles and gave origin to *Leftraru*, whose name signifies fast (= *lef*), Caracara (*traru*) indicating that "*traru* transformed itself into *Leftraru* " (Figure 2B).

The story of *Leftraru* shows how certain essential attributes such as quickness and wisdom are conserved through a common lineage of birds and human beings. In the Mapuche worldview, morphological and behavioral attributes are shared by humans with many bird species. For example, the Black-throated Huet-huet (*Pteroptochos tarnii*), or *wed-wed*, is thought of like a buffoon or a crazy man that, with his vocalizations, narrates stories while accompanying travelers through the forests along ravines. The Southern Lawping (*Vanellus chilensis*) or *tregul* is a bird that acquires various human or animal forms: "it is a man that comes with his sex dangling, it is a woman that walks another way, it has four legs, it can be a dog or a fox" (in Rozzi et al. 2010:118-9). The flight and song of the *tregul* courtship are mimicked during the Mapuche ceremony, *ngillatun*. Participants dance in circles as the *tregul* does, with arms open toward the north, the west, the south and the east, in order to invoke the Grand Spirit of Universal Fertility (Foerster & Gundermann 1996, Aillapan & Rozzi 2001). Identification between birds and humans is expressed, then, in multiple aspects of Mapuche culture, generating a living feeling of familiarity and relationship. Furthermore, representation of birds as humans and vice-versa implies that these beings possess common characteristics, an assertion consistent with the scientific "unity of life" paradigm. Analogous to scientific evidence based on similarities of cellular structure or embryological and anatomical patterns between the vertebrates, the Mapuche worldview establishes similarities between human beings and birds based on observation of certain patterns of conduct. In this way, the genealogical histories of the birds are intermingled with those of human beings. Both, Mapuche cosmogony and scientific evolutionary biology converge by affirming kinship and a common origin for all live beings (Rozzi 2001).

A Dialogue Between Humans and Birds

Yahgan Ornithological Narratives Associated with the Web of Life Metaphor

The main ecological concept of the "web of life" metaphor refers to the numerous and complex interactions between biological species, and also between those species and the physical environment they inhabit (Carson 1962). Humans and other biological species form part of this web of interactions. From a biocultural perspective, the "web of life" metaphor allows one to compare the "ecological order" and the "social order". To this end, we consider the story of the Black-necked Ibis (*Theristicus melanopis*) or *lewuja* (Appendix 3).

This story evokes the association between migratory conduct and the seasons of the year. For the Yahgans, arrival of the Black- necked Ibis signifies the beginning of spring, and its departure involves snowstorms and cold weather. This bird has a brilliant size and color, and emits powerful metallic vocalizations. Thus, at the end of autumn upon its departure it announces very clearly the changes of the seasons and marked climatic changes. This kind of relation between the Yahgans and the birds involves a type of code (behavior) coordination, with birds acting as signs to predict the climate.

The Buff-necked Ibis story also expresses a social rule prohibiting the disruption of these birds. Such a mandate is encountered in numerous Yahgan bird stories. For example, the grandmother Úrsula observes that the Fire-eyed Diucon (*Xolmis pyrope*) or *hashpúl* is a powerful shaman who should be treated well and never have stones launched at him. If this rule is broken, *hashpúl* gets irritated and provokes storms and tempests with south winds or *ilan*. This Yahgan prohibition of casting stones at certain birds has been documented since the first registered ethnographies (Lothrop 1928). Grandmother Cristina says that no one is permitted to imitate the hum of the

South American Snipe (*Gallinago paraguaiae*) or *shakóa*. Whoever imitates or bothers *shakóa* will awaken with their toes cut by the fibers of *ushkulampi*, or the rushes (*Marsippospermum grandiflorum*). This plant grows in the southern wetlands inhabited by the Snipe, wetlands where Yahgan women collect rush fibers for weaving baskets (Massardo & Rozzi 2006). The narratives thus express a conscience about ecological interactions with the birds that they should not be bothered. Through the stories, rules and norms are established for interacting with the birds. Whoever disobeys these rules will suffer the consequences.

Stories like those of the Buff-necked Ibis, the Fire-eyed Diucon, and the South American Snipe are narrated by Yahgan grandmothers and grandfathers. Thus the bird stories also express respect toward elders. For example, grandmothers Úrsula and Cristina tell how when the Falkland Thrush (*Turdus falcklandii*), or *hakasir*, has chicks, the mother first educates the oldest, and then together they educate and feed the other sibling chicks in the trees and rivers of the forests. *Hakasir* moves from branch to branch talking, whistling, and educating her child. Úrsula remembers one day when "I was laid out watching, and the mother *hakasir* said to her oldest 'when you have a little brother you should educate him, enliven him to work, to bathe himself'. And then the little bird went singing, flying toward the river where together they soaked and washed themselves, then shook and returned to the same branch. This is how the *hakasir* grow (Rozzi et al. 2010: 148)." Úrsula adds that "if the son Thrush does not listen to its mother, if it does not obey, it is abandoned. The little thrush remains alone and has nothing to eat. It is fed bad *dihuenes* (mushrooms of the genera *Cyttaria*), spoiled ones, that make him sick and he dies. On the other hand, if he obeys his mother he grows healthy, feeding himself on good fruit, with *amai* or prickly heath (*Gaultheria mucronata*)" (Rozzi *et al.*, 2010: 148).

Yahgan social relations are projected onto the narratives about the birds. At the same time, observation of the birds' conduct inspires Yahgan habits and social norms. In order to survive, both the thrush and the small Yahgans should obey their parents and grandparents, and learn to distinguish good foods from bad foods. Respect toward elders is also encountered in the story of *omora*, since only with the wisdom of its grandmother, the White Owl, is the Hummingbird able to achieve success in maintaining: (1) the Yahgan social order, through punishing greediness; and (2) the ecological order, through the conservation of the community of birds and the reestablishment and creation of waterways. These stories express a close parallel between the social order of the Yahgans and the ecological order observed between the birds.

Mapuche Ornithological Narratives Associated with the Metaphor of the Web of Life

Among the birds of southern South America, the Buff-necked Ibis is also central to the Mapuche worldview. The poem that Aillapan dedicates to this bird, *Raki*, is organized around the number four: "The *raki*-bird is the one whose ability is to count best. From one to four they go along like wise counselors," "Flying overhead, in pairs of two and pairs of four (Aillapan & Rozzi 2001: 42))." The number four is the main number in the Mapuche worldview. There are four cardinal points, four guardians of the Earth, and there are four spaces in the vertical axis of the world: 1)*Wenumapu* (the upper (= *wenu*) land, that is like a reflection of the lower land); 2) *Ankawenu* (the sidereal space where the Buff-necked Ibis and other birds fly and where wind, air, clouds, rain, and thunder all dwell; it is placed between (= *anka*) the upper land and the *mapu*); 3) *Mapu* (the land "of here" where human beings live, suffer, pay, and love; it is also called *nukemapu* (= mother earth) and is delicate and cannot be sold or perforated, instead it must be loved and farmed with care. Aillapan emphasizes that in the *nukemapu* the land is worked freely and fraternally, like *peni* (= brothers), and birds and trees should be treated as equals); 4) *Minchemapu* (the land of below (= *minche*), she encompasses all that is underneath the rivers and the soil, it is like the subsoil).

With their metallic voice and "song- language", the Buff-necked Ibis or *Raki* cross the sidereal space communicating with the divine, the human, and the natural. Their sounds are heard in the worlds of *wenumapu*, *ankawenu*, and *mapu*, and over *mapu*. In the Mapuche community of *Ruka Raki* (home of the Buff-necked Ibis), today exists a school also called *Ruka Raki* to express the profound identification of Mapuche culture with the birds, at the same time than stripping the human ego by identifying the site of teaching with a bird. Such a gesture emphasizes the feeling of belonging that the Mapuche culture has with birds, animals and the whole biotic community. Still today, many places in southern South America preserve their original Mapudungun names, which refer to the Buff-necked Ibis, the Pygmy Owl, the Fire-crown hummingbird, and other birds (Massardo & Rozzi 2004). The Mapuche ornitho-toponomy helps us keep ever-present in our minds how interwoven our human lives are with those of the birds, and the habitats we co-inhabit with them (Rozzi et al. 2010).

For the Mapuche, the Long-tailed Meadowlark or *lloyka* is also a powerful bird, capable of connecting the human with the divine. Mapuche cosmogony distinguishes four cardinal points that involve four central divinities, the "Four Wind-Spirits of the Earth" who come from the west with the guardian of the sea called *Dumpall*, from the southeast the guardian of the volcanoes or *Pillan*, from the northeast the princess of the sun or *Anchümallen*, and from the north the great visitor who had the riches or *Witranalwe* (Faron 1968). In the Mapuche culture, the natural, human, and divine worlds live together, a unity that is expressed in the healing practices.

According to Aillapan, the Long-tailed Meadowlark's call says "I have always healed my people with pure herbs. For this, my name is 'full-time healer", alluding to the refined Mapuche natural medicine based in medicinal plants (San Martín 1983, Aldunate 1996, Massardo 1997, Villagrán 1998). Early in the 18th century, the Jesuit naturalist Juan Ignacio Molina admired the Mapuche herbalist, noting that "vegetables, especially herbs, form the capital of the Mapuche pharmacy...and their doctors, called *machi* and *ampife*, are expert herbalists that possess through tradition the secret of a large number of simple [plants], adaptable to every kind of illness, with which they make marvelous cures every time; and although...they try to hide what they know in this matter, nevertheless, moved by friendship, they have demonstrated up through today the medical virtues of many trees and more than 200 healing herbs that they use with much success..." (Molina 1978, 45). Still today, just as the Mapuche shaman (*machi*), the sister bird *lloyka* is a full-time healer who brings his medicine made of the forest plants. Regarding her Mapudungun name *lloyka*, Aillapan highlights that this name stems from the terms *llako* (= healing) and *lawen* (= remedy). The Mapuche poet and the *lloyca* affirm that one should not be afraid for one's health when medicinal plants exist in the region; the plants are the source of life and health for both the birds and the humans.

Among the ecological interactions expressed through the Mapuche stories, the trophic relations stand out. Mapuche narratives recurrently refer to the flows of matter and energy generated by all animals, more broadly said by the "beings of air, water, and earth." For example, for the Mapuche worldview, "the hypnosis that the Ringed Kingfisher (*Megaceryle torquata*) or *challwafeünüm* exercises over the fishes unchains the flow between beings of the water and those of the air through the food chain" (Rozzi *et al.*, 2010: 96). For Aillapan, the woodpeckers are seen as "birds that pass the day pecking at a tree, nibbling at and removing larva that, at the same time, are fed by the sap and tissues of the trees" (Rozzi et al. 2010: 135). Thus, a "cord of life" is established that flows from the minerals in the mountain soil through the sap of the trees, the insects and the birds. Finally, the Mapuche worldview pays attention to all the members of the food chain, and carrion birds like vultures or *kanin* are admired as "full time cleaners of the cities, beach and open countryside" (Aillapan & Rozzi 2001: 38). Predatory birds and men are notable hunters.

The ecological network of interactions among birds and humans is finely interwoven and has a rhythm that imposes a requirement: the rhythm of the birds must be heard and respected. In his poem dedicated to the Magellanic Tapaculo (*Scytalopus magellanicus*), *Churrin* or Pocket-Watch Bird, Aillapan first highlights that this "millenary bird" "begins giving the hour from the daybreak, throughout the morning, to the noon-time and into the afternoon... symbol of life, work, results and production, family, planting, harvest and animals...pocket-watch bird that indicates the human pair always in action and harmony with nature" (Aillapan & Rozzi 2001:24-5). Then, at the end of his poem, the Mapuche poet questions the notion of progress: "it is said that the native people are behind the times, so I will set my watch forward" Thus, with his song, the Pocket-watch bird – which has inhabited the forests of southern Chile since its beginnings, just as the Mapuche have – ridicules the notion of progress that is so sacred to the conquistadors. The hurry and acceleration of the rhythm of life is as easy and mechanical as accelerating the "tick-tock" of a clock, but why do such a thing? Further, what sense does it make to abandon the forests' own rhythm, in order to accelerate it in a career that dissociates from the forests and drives to suicide all its beings? The notion of the "web of life" thus acquires in the Mapuche worldview a normative character: one must listen and respect the rhythm of nature in order to maintain the health of humans and animals alike.

Implications for Environmental Ethics and Biocultural Conversion

During the last decades, traditional ecological knowledge [TEK] (Inglis 1993) has increasingly captured the attention not only of anthropologists, but also of scientists in many areas of biology (Berkes 1993), conservation (Cunningham 1993, Balick et al. 1994), and environmental ethics (Callicott 1997). The Convention of Biological Diversity and Agenda 21, both of which resulted from the UN Earth Summit in Rio de Janeiro (1992), underline the importance of traditional indigenous knowledge and conservation of both cultural and biological diversity (Rozzi & Feinsinger 2001).

At the beginning of the 21st century, the Yahgan and Mapuche narratives from southern South America help global society to recover the awareness that birds form part of the community of beings with whom we humans spend our existence. Birds and human beings can sing together, enjoying, taking care of, and loving one another. This identification and communication with birds encourages our living together with them and with nature. Birds inhabit in their own languages and songs, as much as they inhabit in the multiplicity of languages of our human species. The poetic dialogue that arises through perception of our kinship with the birds recovers a sensibility toward the profound relations and interdependencies that exist between human life and nature.

This dialogue helps not only conservation, but also current intercultural educational programs that aim to recover an integrated vision of the dimensions of knowledge and values, based on interdisciplinary and transversal curricula, integrating ancient Amerindian and novel scientific forms of ecological knowledge. Such integration is urgently necessary to reconstruct a sense of history and the social and environmental responsibility in the austral and other Latin American countries (Rozzi & Feinsinger 2001). The intensity of current economic globalization, as well as biotic and cultural homogenization, surpasses our capacity to understand and assess the consequences that these processes have for a fulfilled life in the continent inhabited by the world's highest bird diversity. It is the task of students, teachers, researchers, and lovers of biocultural diversity to decipher the metaphorical language of the Mapuche-Bird-Man and the Yahgan narratives and to project them in this search for meaningful lives of humans, co-inhabiting with birds, other animals, and the diversity of living beings in which we have co-evolved. The Mapuche and Yahgan traditional ecological knowledge can offer a communication bridge among persons from different cultures, and among these persons and the birds.

Metaphors can also assist the dialogue between scientific and Amerindian forms of knowledge. As we stated at the beginning of the chapter, the metaphors of the "tree of life" and the "web of life" synthesize central concepts for evolutionary and ecological sciences, and for environmental ethics (Rozzi 1999). Darwin utilized the first metaphor ("tree of life") at the end of his chapter on natural selection in "The Origin of the Species" in order to communicate that: "The affinities of all the beings of the same class have sometimes been represented by a great tree. I believe this simile largely speaks the truth. The green and budding twigs may represent existing species; and those produced during each former year may represent the long succession of extinct species... As buds give rise by growth to fresh buds, and these, if vigorous, branch and overtop on all sides many a feebler branch, so by generation I believe it has been with the great Tree of Life, which fills with its dead and broken branches the crust of the earth, and covers the surface with its ever branching and beautiful ramifications" (Darwin 1859: 129-130).

These metaphorical lines communicate synthetically that: 1) all living creatures share a common origin, and 2) we are part of a course of evolution of diversification where some groups of organisms are born, others are transformed and others disappear. The ethical implications of this notion were lucidly captured by Aldo Leopold who in the mid 20th century wrote that "it is a century now since Darwin gave us the first glimpse of the origin of species. We know now what was unknown to all the preceding caravan of generations: That men are only fellow-voyagers with other creatures in the odyssey of evolution. This knowledge should have given us, by this time, a sense of kinship with fellow creatures; a wish to live and let live; a sense of wonder over the magnitude and duration of the biotic enterprise" (Leopold 1949: 116-7).

This Leopoldian ethical imperative, derived from a notion of evolutionary kinship based on modern sciences, resembles the ethical implications of the indigenous Yahgan and Mapuche bird narratives. From the point of view of environmental ethics, the three cultural perspectives –Yahgan, Mapuche, and scientific– highlight a notion of kinship that provides one of the most appealing foundations to support the notion of the intrinsic value of animal life. Birds and other animals are perceived as our distant evolutionary relatives. This implies that, to some degree, their existence can be subject to moral considerations based on ontological and ethical judgments on par with those we use to judge the value of human life. This conclusion is very important for effectively implementing arguments and treatises for biological conservation that are based in intrinsic value of humans, birds, other animals and the life of the diverse biological species that inhabit the planet, such as the Convention of Biological Diversity (CBD) that begins with the statement: "Conscious of the intrinsic value of biological diversity and of the ecological, genetic, social, economic, scientific, educational, cultural, recreational and aesthetic values of biological diversity and its components; Conscious also of the importance of biological diversity for evolution and for maintaining life sustaining systems of the biosphere" (Shiva 2000).

The "web of life" metaphor complements the evolutionary kinship arguments with ecological notions. In *The Origin of Species,* Darwin utilized the metaphor to illustrate the "complete web of [ecological] relationships," describing how: "humble-bees alone visit the red clover (*Trifolium pratense*) as other bees cannot reach the nectar. Hence I have very little doubt, that if the whole genus of humble-bees became extinct or very rare in England, the heartsease and red clover would become very rare, or wholly disappear. The number of humble-bees in any district depends on the number of field-mice, which destroy their combs and nests...Now the number of mice is largely dependent, as every one knows on the number of cats" (Darwin 1859: 73-74).

In this web of biotic interactions, the pink cloverleaf plants depend directly on the bumblebee for their pollination, and indirectly on predators of mice, like cats, to hunt these rodents that destroy the nests of bumblebee pollinators. Comprehension of the interdependency between species currently represents a key notion for conservation biology's approach to the levels of community and ecosystem (Thompson 1997). More recently, conservation sciences have emphasized the role of human participation in this type of web of biotic interactions (McDonell & Pickett 1993).

Poetically, Rachel Carson expressed in *Silent Spring* that: "the earth's vegetation is part of a web of life ... [and] although the modern man seldom remembers the fact, he could not exist without plants that harness the sun's energy and manufacture the basic foodstuff he depends upon for life"(Carson 1962: 63-4).

This evocative image of the "web of life" is analyzed today in quantitative terms by ecological economics (Daily 1997). More extensive applications of this metaphor establish that biological species as well as physical components of ecosystems (such as the water, the soil, and the atmosphere) must be protected in order to allow the continuation of the flow of "goods and services" that emanate from ecosystems, and that are essential for the well-being of current and future generations of humans (Norton 1991).

Considering again the Yahgan story of the Hummingbird or *omora*, we can now appreciate clearly how this relation teaches us that: 1) social well-being and conservation of biological species go hand-in-hand, and 2) solidarity ends up being better than selfishness and we create a better life for everyone. This story of the Hummingbird reveals substantial coincidences between concepts from the ecological sciences and Yahgan traditional ecological knowledge. From the perspective of contemporary science, conservation of the community of birds contributes to ecological processes such as pollination, seed dispersal, insect population control, predation of rodent populations, or nutrient cycles through the guano of the birds that transport mineral salts between marine and mountainous ecosystems. At the same time, maintenance of these ecosystemic processes contributes to the conservation of vascular and non-vascular plants in the hydrographic basins. This vegetation is a key for regulation of hydrological flows, preventing floods and avoiding droughts. Therefore, in order to maintain the long-term water supply, both the Yahgan story of *omora* and the scientific ecosystem analysis recommend the protection of birds, other animals, and biodiversity in general.

From the perspective of contemporary environmental ethics, the ethical imperatives implicit in the Yahgan narrative (to conserve the community of birds in order to ensure the supply of water) are consistent with the notion of instrumental value. The Yahgan story of *Omora* mandates to protect the birds and other animals because they are instrumental for human survival. At the same time, in the Yahgan narratives Omora is seen as a bird, yet at the same time it is seen as a small person, a spirit who maintains social as well as ecological order. In summary, the ancient traditional ecological knowledge contained in the story of the hummingbird, as well as in other Yahgan and Mapuche tales of birds, as well as the novel knowledge of ecological and evolutionary sciences, nourish and inform the notions of instrumental and intrinsic values of birds, and biological diversity. Both types of ethical values appeal today to establish respectful and alert relationships with the birds, other animals, and the habitats where we co-inhabit at the southern end of the Americas (Rozzi et al. 2006).

Acknowledgements

I deeply thank the teaching of grandmothers Cristina and Úrsula Calderón, and from the poet Lorenzo Aillapan, as well as the support from the Academia Mapuche

PÜLLÜMAPUKIMUNWEFTUY from Puerto Saavedra. Francisca Massardo did valuable comments to the manuscript. Kelli Moses and Charlee Tidrick helped with the English translation and editions. Research and biocultural conservation work in southern Chile has been primarily funded by the Millennium Scientific Initiative (ICM, Chile, P05-002 ICM-MIDEPLAN) and Basal Financing Program of CONICYT (Chile) PFB-23 (Basal-CONICYT).

References

Aillapan, L. 2003. Breve biografía del Hombre Pájaro Mapuche o Ünümche. P. 33 in Rozzi, R., F. Massardo, C. Anderson, S. McGehee, G. Clark, G. Egli, E. Ramilo, U. Calderón, C. Calderón, L. Aillapan, & C. Zárraga (eds.) *Guía multiétnica de aves de los bosques templados de Sudamérica austral*. Fantástico Sur - Univ. de Magallanes, Punta Arenas, Chile.

_____ & R. Rozzi. 2001. *Veinte poemas alados de los bosques nativos del sur de Chile*. Ciudad de México, México: Editorial Plaza y Valdés.

Aldunate C. 1996. Mapuche: gente de la Tierra. Pp. 111-134 in Hidalgo, J., V. Schiappacasse, H. Niemeyer, C. Aldunate, & P. Mege (eds.) *Etnografía. Sociedades indígenas contemporáneas y su ideología*. Santiago, Chile: Editorial Andrés Bello.

Balick, M., R. Arvigo, & L. Romero. 1994. "The Development of an Ethnobotanical Forest Reserve in Belize: Its Role in the Preservation of Biological and Cultural Diversity. *Conservation Biology*. 8: 316-317.

Barella, C. 1971. *Lautaro Guerrillero*. Santiago, Chile: Nueva Universidad.

Berkes, F. 1993. "Traditional ecological knowledge in perspective." Pp. 1-9 in Inglis, J. (ed.). *Traditional ecological knowledge: Concepts and Cases*. International Program on Traditional Ecological Knowledge, International Development Research Centre, Ottawa, Ontario.

Bulnes, G. 1980. *Los Mapuches y la Tierra*. Rotterdam, Holanda: Pequenas ediciones INC..

Callicott, J.B. 1997. *Earth's Insights: A Multicultural Survey of Ecological Ethics from the Mediterranean Basin to the Australian Outback*. Berkeley and Los Angeles, California: University of California Press.

Carson, R. 1962. *Silent Spring*. Boston, Massachussetts: Houghthon Mifflin.

Cona, P., & Ernesto Wilhelm de Moesbach. 1930. *Vida y costumbres de los indígenas araucanos en la segunda mitad del siglo XIX*. Santiago, Chile: Imprenta Cervantes.

Cunnigham, A. 1993. *Ethics, Ethnobiological Research, and Biodiversity*. Gland, Switzerland: World Wildlife Fund International Publication.

Daily G. 1997. *Nature's Services: Societal Dependence on Natural Ecosystems*. Washington D.C.: Island Press.

Darwin, C. 1859. *The Origin of Species by Means of Natural Selection*. (First Edition, London: Murray). Reedited by E. Mayr, Cambridge, Massachusetts: Harvard University Press (1964).

Domeyko, I. 1971. *Araucanía y sus habitantes*. Buenos Aires, Argentina: Editorial Francisco de Aguirre.

Faron, L. C. 1968. *The Mapuche Indians of Chile: An Ethnography*. New York: Waveland Press.

Foerster, R., & H. Gundermann. 1996. Religiosidad mapuche contemporánea: elementos introductorios. Pp. 189-239 in Hidalgo, J., V. Schiappacasse, H. Niemeyer, C. Aldunate, & P. Mege (eds.) *Etnografía. Sociedades indígenas contemporáneas y su ideología*. Editorial Andrés Bello, Santiago, Chile.

Geertz, G (ed.) 1974. *Myth, Symbol and Culture*. New York: Norton.

Gusinde, M. 1917a. Medicina e higiene de los antiguos araucanos I. *Revista Chilena de Historia y Geografía* 22: 376-379.

_____. 1917b. Medicina e higiene de los antiguos araucanos *II*. *Revista Chilena de Historia y Geografía* 23: 137-145.

_____. 1961. *The Yamana: The Life and Thought of the Water Nomads of Cape Horn*. Volumes I-V, translated by F. Schutze. New Haven: New Haven Press.

Hunt, E. 1977. *The Transformation of the Hummingbird*. Ithaca: Cornell University Press.

Inglis, J. 1993. *Traditional Ecological Knowledge: Concepts and Cases*. International Program on Traditional Ecological Knowledge, International Development Research Centre, Ottawa, Ontario.

Leopold, A. 1949. *A Sand County Almanac*. New York: Oxford University Press.

Lothrop, S. 1928. *The Indians of Tierra del Fuego*. New York: Museum of The American Indians.

Massardo F. 1997. Plantas medicinales y medio ambiente. Pp. 189-211 in *Actas 2° Congreso de Plantas Medicinales*. Santiago, Chile: Ministerio de Salud.

_____ & R. Rozzi. 2004. Etno-ornitología Yagán y Lafkenche en los Bosques Templados de Sudamérica Austral. *Ornitología Neotropical* 15: 395-407.

_____. 2006. *The World's Southernmost Ethnoecology: Yahgan Craftmanship and Traditional Ecological Knowledge*. Punta Arenas, Chile: Ediciones Universidad de Magallanes.

_____. 1996. Plantas medicinales chilenas: un recurso subvalorado. *Ambiente y Desarrollo* 12: 76-81.

McDonnell, M.J., & S.T.A. Pickett (eds.). 1993. *Humans as Components of Ecosystems: The Ecology of Subtle Human Effects and Populated Areas*. New York: Springer-Verlag.

Molina, J. I. 1978. *Historia natural y civil de Chile*. Santiago, Chile: Editorial Universitaria.

Norton B. 1991. *Toward Unity Among Environmentalists*. New York: Oxford University Press.

Ovalle, A. 1646. *Histórica relación del reyno de Chile*. Santiago, Chile: Editorial Universitaria .

Primack, R., R. Rozzi, P. Feinsinger, R. Dirzo, & F. Massardo. 2001. *Fundamentos de Conservación Biológica: Perspectivas Latinoamericanas*. Ciudad de México, México: Fondo de Cultura Económica, México. Rozzi, R. 1999. The reciprocal links between evolutionary-ecological sciences and environmental ethics. *BioScience* 49: 911-921.

_____. 2001. Éticas ambientales latinoamericanas: raíces y ramas. Pp. 311-362 in Primack, R., R. Rozzi, P. Feinsinger, R. Dirzo, & F. Massardo (eds). *Fundamentos de Conservación Biológica: Perspectivas Latinoamericanas*. Ciudad de México, México: Fondo de Cultura Económica, México. —, & P. Feinsinger. 2001. Desafíos para la conservación en Latinoamérica. Pp. 619-688 in Primack, R., R. Rozzi, P. Feinsinger, R. Dirzo, & F. Massardo (eds.). *Fundamentos de Conservación Biológica: Perspectivas Latinoamericanas*. Ciudad de México, México: Fondo de Cultura Económica, México. —., F. Massardo, C. Anderson, K. Heidinger & J. Silander Jr. 2006. "Ten Principles for Biocultural Conservation at the Southern Tip of the Americas: The Approach of the Omora Ethnobotanical Park. *Ecology & Society* 11(1): 43. [online]: http://www.ecologyandsociety.org/vol11/iss1/art43/

Primack, R., F. Massardo, C. Anderson, S. McGehee, G. Clark, G. Egli, E. Ramilo, U. Calderón, C. Calderón, L. Aillapan, & C. Zárraga. 2010. *Multi-Ethnic Bird Guide of the Sub-Antarctic Forests of South America. Denton TX and Punta Arenas*. Chile: UNT Press – Ediciones Universidad de Magallanes.

San Martín J. 1983. Medicinal Plants in Central Chile. *Economic Botany*, 37: 217-27.

Shiva, V. 2000. *Monocultures of the Mind: Biodiversity, Biotechnology and Agriculture*. Fourth Printing. New Delhi: Zed Press.

Thompson, J.N. 1997. "Conserving Interaction Biodiversity." Pp. 171-190 in Pickett, S., R. Ostfeld, M. Shachak, & G. Likens (eds.). *The Ecological Basis of Conservation*. New York: Chapman and Hall.

Turner, V. 1967. *The Forest of Symbols*. Ithaca: Cornell University Press.

Villagrán, C. 1998. Botánica indígena de los bosques de Chile: Sistema de clasificación de un recurso de uso múltiple. *Revista Chilena de Historia Natural* 71: 245-268.

Appendix 1. Yahgan Story of the Fire-Crown Hummingbird Omora

In ancestral times, when birds were still humans, a great drought occurred in the region of Cape Horn and the people were dying of thirst. Only the wily fox, known as *Cilawáia*, found a lake. He told no one and built, instead, a solid fence around it so that no one could enter. Hidden as such, he drank lots of water, concerned only with himself. However, after some time the other people discovered this lake and went as a group to ask the fox for a little bit of water. *Cilawáia* did not even listen to their supplications and expelled them with rude words. The condition of the people got increasingly worse and when they were at the point of death, they decided in their desperation to send a message to the hummingbird or little *Omora*, an occasional visitor who had saved them before in times of crisis. *Omora* was always prepared to help others and arrived quickly. Although diminutive, this little tiny bird-man-spirit is braver and bolder than any giant. When he arrived, the downcast people told him about their great sufferings, and, indignant, *Omora* raised himself and rapidly undertook the flight towards where the egotist *Cilawáia* was, and he confronted him directly saying: "Listen! What is it that the people are telling me? Have you had access to a lake full of water, but you don't want to share it with other people who soon will die of thirst!" The fox replied, "Why should I worry myself about the others? This lake has only a little water, and it barely is enough for me and my closest family. I cannot give anything to other people because soon I'll suffer thirst myself."

Upon hearing this, Omora became furious and without replying to the fox, returned to the settlement. He reflected briefly and rapidly rose again, took his sling and flew back to the fox. Along the way he collected several sharp stones, and when he was within sight of the fox and sufficiently near, he shouted to him: "Are you going to share your water for now and for always with the other people? Come on, don't be so selfish. If you don't spare water, they surely will die of thirst!"

The fox answered indifferently, "They can all die of thirst. Why should I worry? I can't give water to each one of the people who live here, or else my family and I will also die of thirst." Omora was so offended that he could not control himself any longer and furiously fired a sharp stone with his sling, killing the fox with his first throw..The rest of the community had been watching and became happy, running to the place. Rapidly they broke down the fence, went to the lake and drank, satiating their thirst. They drank so much that the lake became completely empty and a few birds that arrived too late barely found a few drops with which they moistened their throats. Then, the wise owl Sirra or Sita, the grandmother of Omora, said to the birds that arrived late, "collect mud from the bottom of the lake and fly to the peaks of the mountains, and fling the mud over them." The little birds flew, and their balls of mud gave birth to springs that became water courses that spouted from the mountains, forming small streams and large rivers that flowed through the ravines. When all the people saw this, they were extremely happy and drank great quantities of fresh, pure water, that was much better than the water of the lake, and now all find themselves safe again. Today all of those water courses still flow down from the mountains and provide exquisite water. Since that time no one has died of thirst.

Appendix 2. Mapuche Narrative of the Andean Condor

The Andean Condor or the *manke* is for the Mapuche culture the King of the Birds. It is a symbol of the mountains, not only due to its great size and impressive wing-length, but also its black and white colors, which signify the snow-capped peaks and the black mountain rocks. For the Mapuche, the condor unites the virtues of being *Kimche*, a wise person, *Norche*, a person who loves justice, *Kümeche*, a good-natured person, and *Newenche*, a powerful, lordly person. This King of Birds is the national bird of Colombia, Ecuador, Bolivia and Chile. *Manke* was once abundant on *Manquehue* Mountain, which means place (*hue*) of the condors (*manke*), in the center of Santiago, the capital of Chile. It is still found today throughout territory of the austral forests, but paradoxically, this king of birds, this national symbol of four countries, finds itself now threatened with extinction.

Appendix 3. Yahgan story of the Buff-Necked Ibis Lejuwa

In the times of the ancestors, one day when spring arrived, a *yamana* leaned out of his *akar* or hut and saw a Buff-Necked Ibis or *lejuwa* flying in the sky. The *yamana* was so happy that he yelled to the other members of the community, "An ibis is flying over our *akar*. Look!" Immediately the others came out of their huts, shouting loudly, "The spring has come, the ibises are already in their return flight." They jumped for joy, and they jumped hard.

Lejuwa is, however, a very delicate and sensitive woman and likes to be treated with special deference. Upon hearing the noise of the shouts, she became furious, and profoundly offended, causing a strong snow storm. It snowed incessantly; it was freezing cold; all the land was covered with ice; and all the waters froze. Many people died because they could not navigate in their canoes to search for food. Neither could they leave from their *akars* to look for firewood because all was covered with snow. More and more people continued to die. Only after a long time did it stop snowing, and the sun begin to shine.

The sun warmed so much that begin to melt the ice and snow that had completely covered the land, and consequently floods began to flow towards the channels and the oceans. The ice that covered the wide and narrow channels broke apart and melted, and the Yahgans could finally go to the shores and navigate in their canoes to collect food. On the big mountainsides and deep valleys, however, the ice was so thick that not even this sun could melt it. Even today, glaciers can be observed coming down to the sea, that remind us of the severe frost and snowstorm that *lejuwa*, who is a very delicate and sensitive woman, provoked in those times. Since then, the Yahgans treat the Buff-Necked Ibis or *lejuwa* with much respect, and when they come near the houses or *akars*, the people keep quiet, particularly the little children who are not allowed to imitate them or to cry.

Chapter 20

Animal-Human Nexus in Gerald Durrell's Fictional Writings

☆ *Dhishna. P*

The relation between animals and human beings are devoid of any boundaries as human beings themselves could be considered as social animals. Human beings on the other hand, lack many specific traits that animals on the larger extent have in their physical framework though they consider themselves as the masters of the world. All animals to a larger extent differ and become common based on the physical and mental attributes that they have when compared to that of human beings. When bringing human beings and animals together in the same framework, the unique nature of the two, needs to be further explored.

The word 'animal' as the *Longman: Dictionary of Contemporary English* coins, has multiple meanings: "1) A living creature such as a dog or cat, that is not an insect, plant, bird, fish, or person. 2) Any living creature that is not a plant or a person. 3) Any living creature, including people. 4) Someone who behaves in a cruel, violent or very rude way." (49) Durrell's animal world is understood based on the second definition that *Longman Dictionary* provides, as that of 'any living creature that is not a plant or a person' (*Ibid*, 49). Though there are many subdivisions based on their body structure and distinct characteristics the animal world remains as the only group naturally untouched by modernity. The animal world is vast and widely seen around us. But we seldom look at their peculiarities or uniqueness. Human beings are more self-centered and nature often had remained the 'other' that remains much less acquainted with than with the working space that we generally tend to occupy. Among human beings, the tribes could be considered as the only group of people who live in rapport with the nature and its products. The world of tribes has much to do with the nature and the natural environment around them. This happens to be a reason for their oral literature to focus on the land and its inhabitants though much less is documented by them. In oral literature, myths and folksongs of the tribes life style, their relation with nature and with the animal world is being presented.

Nature remains to be a well explored place for those interested in the subject especially ecologists, ornithologists. anthropologists, geologists, and archeologists. From Charles Darwin's *On the Origin of Species* (1859) there had been a remarkable increase in the documentation of the animal world. Another group of enthusiasts on nature includes the writers. They were the only group who were able to sketch the nature through their creative writings. The world of animals and human beings are presented through various literatures of the East and the West through fictional and poetic representation.

There are certain characteristics involved in literary representation of the animal world. One major feature is that animals are presented as closely related to human beings in narration. Animals are presented as taking up major and minor roles as characters in literary production. Writers employ the animal representations as tools to convey their message to the readers. Animals are often compared with human beings in literary productions. This kind of comparison makes readers realize what they are and what they need to be. Animals, as seen in literary productions, occupy only a marginal or nominal space in real life when compared to human beings. They remain tamed, subjugated, used as messengers and overworked in real life. They are the marginalized owners of the earth. An exploration of the animal world is a way of giving voice to the voiceless or to give space to the space less. In fact animal literature is an area that remains less focused in literary studies and criticism.

Animals could be seen occupying space in the Western and Eastern myths, scriptures and religious texts of the ancient times. Its inception could be seen from the sixth century B.C., Aesop's *Fables.* Greek and Egyptian mythical tradition had given an important space for the animals. The Griffin and the Sphinx are very relevant examples of the dominant animal existence in their literature. It is evident in *Bible* where the best representation is the Noah's Ark and Indian scriptures including *Ramayana* and the *Mahabharata*. The land and the forest are inevitable aspects of Indian literary tradition with the diversity of animal life that is being presented. Many western writers have used animals in their narratives. Poetry is an area where much is said of the animals. Nature and animals got the required attention from the Romantic age to the present. The major poets who had focused on the animal world are William Wordsworth, Percy Bysshe Shelley, Lord Byron, and John Keats. The nightingale, albatross, tiger, lamb, pigs, horses are all familiarly used in literature. Noteworthy among the prose writers who represented the animal world and their related issues are Jonathan Swift, Rudyard Kipling, Jim Corbett, Yann Martel, George Orwell, J. M. Coetzee, Gerald Durrell, Joyce, Upton Sinclair and Franz Kafka among others.

Among the major writers who had brought in the image of the animals in literature, Gerald Durrell (1925-1995), deserves special attention. He was a naturalist, zookeeper, conservationist, more over an enthusiast in ecology. He was born in India of English and Irish descent parents and had undergone extended expeditions and worked for the welfare of animals in various places. He is the founder of the Jersey

Zoological Park and had extensively written on animals and their connection with human beings. His major works remain mostly focused on the animal life and its relation to humanity. They include *The Overloaded Ark, The Bafut Beagles, Encounters with Animals, The Drunken Forest, Three singles to Adventure, A zoo in my Luggage, The Whispering Land, My Family and other Animals, Menagerie Manor, Rosy is My Relative* and *The Donkey Rustlers*. Durrell was a writer with a purpose. The purpose was mainly to make his readers aware of the voice of the animals and to generate an interest in their mind to conserve the extinct species of animals.

Durrell had provided a vivid portrayal of animal world through his fictional narrative. All his fictional works lay as a testimonial for his literary excellence and his interest in the animal world. The world of animals is very vast and beyond any systematic documentation, so lies the plethora of animals in his works. Gerald Durrell or Gerry's fascination for the animal world began from his early childhood onwards. His interest in the subject was often received by his family, during those days with coldness and envy. His obsessive interest in natural history later developed into an interest for the preservation of animals leading to his growth as a natural scientist. *My Family and other Animals* (1956), stands as a testimony to his development as a natural scientist. But the question is: what brought in Durrell's mind such an interest towards animals? What where the factors that contributed to such an interest? What does Durrell try to bring to the reader's attention through the vivid portrayal of the animal world?

Durrell is a writer whose creative skills excel in giving a clear portrayal of the animal world. He lays more emphasis on visual presentation of his experiences to the readers. An important aspect of his writing skill is that the animals speak a language that is quite familiar to human beings. The animals are often presented as having close acquaintance with human beings. *My Family and other Animals* is a work which would give a complete idea of Durrell's interest in the subject and the evolution of the conservationist in him.

Durrell was a writer who had given lot of significance to the landscape and its tenants in his writings. Land is the basis of his interest in the animal world. As evident from *My Family and other Animals,* his interest in the animal life begins from his visit and stay in the Corfu islands. Corfu Island is located in Greece and is thirty miles long and seventeen miles broad. It is separated by the Ionian Sea. This is the place that had brought in the inquisitive spirit in Durrell's mind. His creative skills excel in the realistic description of the island and its landscape. It is evident from his comment: "The sky was pale and stained with yellow on the eastern horizon. Ahead lay a chocolate-brown smudge of land, huddled in mist, with a frill of foam at its base. This was Corfu" (Durrell 1956:18) The writer's interest in giving a visual description of the place that he had visited is evident from these lines. Another instance to show the importance that he had given for the presentation of the landscape in the eighteenth chapter of the book where he gives a description of the Chessboard fields, where he mentions: "Each neat little patch of earth, framed with canals, was richly cultivated and green with crops of maize, potatoes, figs and grapes. These fields, small coloured squares edged with shining waters, lay like a sprawling, multi-coloured chess board on which the peasant's coloured figures moved from place to place." (261) The element of giving 'colour' to the representation is what gives Durrell the credit of being a realistic writer. He was a writer who was able to give life to his landscape and animals through the magic wand of his writing skills. It is a method of preserving history and also a way of making the readers understand nature and also to make them aware of the need to conserve it.

Durrell was not only a writer and an enthusiast on nature and the animal world but also a traveller. His major interest was in wandering through unknown landscapes and exploring the flora and fauna. His interest in nature was mainly rid by his interest in the animal world. In order to understand the animal world he often kept the company of the natives or the aborigines of a particular place. He learned many difficult and foreign languages for this purpose which often his family members found difficult to understand. This shows the keen interest of the environmentalist in him. He was not afraid to keep company of the strange people whom he had met, to make his mission a success. The difficulty that his family members had in such 'foreign' living conditions is mentioned by him when he said: "Mother had, often considerable mental effort, managed to commit to memory two or three Greek words. This lack of vocabulary had a restrictive effect on her conversation at the best of times, but when she was faced with the ordeal of exchanging small talk with a murderer, she promptly forgot all the Greek she knew" (*Ibid*, 276). Durrell remained acquainted with new people that he had met in order to get more information from them regarding the land and the animals.

Indigenous people were an important source for Durrell to know about a place and the animals that it contained. They provided information to Durrell regarding the vitality of the land. This was a way for him to know the unknown or unfamiliar animals. He was often provided with animals by them. This was a rare habit that he had cultivated in the long run. We could sense it in the long hours of silent communication through gestures that he made with the Rose- Beetle man in "The Rose-Beetle Man" (*Ibid*, 49), to fix up the price for Achilles the tortoise. Though Durrell was not acquainted with the language of that indigenous person he fixed up the price using various numerical gestures of his fingers and owned the tortoise for a lesser price.

Durrell's interest in the world of animals had nurtured in him a desire to own them as evident from the dealings that he had with the indigenous people. This inturn made him to have companions during his further adventures and exploration. Young Gerry's interest in the ecology outside Corfu was kindled by Dr. Theodore Stephanide's gift of a microscope. Though the microscope did not have a higher magnification, it enhanced the interest in Gerry for outdoor investigation and inquiry of the animal world. Dr. Theodore was not just a physician of the family he had close acquaintance with Durrell's family members as Lawrence Durrell, Gerry's brother, who was an eminent writer and traveller comments in *Lawrence Durrell: Spirit of Place*: "For Durrell's, Theodore became much more than a physician; for one would be hard pressed to find a more erudite, civilized and charming man; a mine of every kind of knowledge relating to the island, he became the perfect friend and mentor of their new world" (*Ibid*, 27). Dr. Theodore was a constant companion to Gerry as he knew the landscape and the people well. He was also sure of the versatile animal world around.

Gerry's interests in pets began with owning the family pet dog, Roger. He describes Roger as a perfect pet and mentions: "He was a perfect companion for an adventure, affectionate, without exuberance, brave without being belligerent, intelligent, and full of good-humoured tolerance for my eccentricities" (*Ibid*, 42). The presentation of Roger was that of a 'real' person. For Durrell, his pets where not mere animals but often shared a living space along with him. So was his interest in owning the other animals when he mentions the bargain for Achilles with the Rose-Beetle Man , thus: "Excited by owning this new pet, I wanted to get back home as quickly as possible in order to show it to everyone, so I said good-bye, thanked him, and hurried off along the road" (*Ibid*, 49). The possession of a pet was seen by Durrell as a precious gift so his mannerisms were often eccentric on such occasions.

The naming of Durrell's pets, beginning with Roger to the other animals, shows the human-animal nexus in the author. The animals in his narratives often had human traits and human beings had animal traits in them. This was clearly brought out by the author through his works. The big list of pets that Durrell had from his early childhood till his death cannot be listed completely in this paper. But one could have a wilder guess of the animal world that he had possessed and of his obsessions even from the title of the works that he had written and from the logo of the Jersey Zoological Park in later years.

My Family and other Animals is a work that had presented the wide description of Durrell's pets in Corfu Island. His pets were evoked with colour, scent, sound, beauty, peculiar habits and physique. Even in his literary productions or documentation he showed peculiar interest and exactness in the presentation of the animal world. The right words were used to give the clear picture of the animals that he had come across during his stay at Corfu though he used to provide exaggerations in his documentation of an often viewed animal. The extra-ordinary presentation of a normal animal life made his works peculiar and varied from that of other environmentalists.

Durrell presented through his works the liberated animal world and the animals in confines. To list a few of his pets as evident from *My Family and other Animals*, are Roger- the dog, Achilles-the tortoise, Quasimodo-the pigeon, Alecko- the seagull, and so on. His interest was also drawn towards insects and their world, which is seen in Chap. 3, "The Strawberry Pink-Villa" where he describes the plethora of insects in the garden. It is evident again in his description of earwig (Durrell 1956:38). His pets were the contributions of the wilderness of the island and its surroundings. He had scorpions, Mantids, Mantis Cicely-his adopted pet and so on.

It is difficult for human beings to have an interest in reptiles. Owning a reptile at home itself is a difficult thing to think about. But for Durrell, reptiles were another area of ecological interest. He often had the habit of collecting tortoises in the spring season as evident in Chap.6. He had to his collection Madam Cyclops, Giant toads, Water Snakes, Gecko, Geronimo and so on. The animal world of Durrell stands incomplete without the mention of his interest in birds as he often felt that birds are easier pets than reptiles. Magpies and seagulls where often looked with greater interest and curiosity by the writer. Durrell's pets could be seen further in *Birds, Beasts and Relatives* (1981), where he discussed about the Tarantulas, dung beetles, sea horses, water spiders, cuttlefish, owls and house lizards. Rosy, the elephant in *Rosy is My Relative* (1968) also stands in testimonial to his special love for the animal world. Durrell's sense of naming the animals with familiar names could be seen in the animals that he had discussed in various works. His interest to provide names of historical figures for animals could be seen in *Catch Me a Colobus* (1981), where he mentions the Claudette squirrel/Claudius and its Caesar baby. Sheena the Chimpanzee also shows his peculiar naming skills for animals.

Most of the animals that he had come across were looked at by Durrell scientifically and anthropomorphically. The need for a scientific laboratory for the animals and a properly maintained zoo were the issues and appeal that Durrell had put forth with the readers. Durrell's interest in the animal world could be evident in the Methuen *Notes on Gerald Durrell's My Family and other Animals* where it is commented that: "As long as Mr. Durrell has to do with animals, he will never lack subjects which arouse in him the zest and affection so evident in his prolific writings" (Methuen 1972:2).

Animals were seen by Durrell as presenting certain characteristics as that of human beings. For him, human beings also presented the nature of animals. The comparison that he brings in is human in terms of animals and animals in terms of humans. It could be evident in *My Family and other Animals* where he presents the tortoise with human qualities. He says:

> His legs were spread out from his shell, his neck extended as far as it could, his head resting on the ground; with eyes closed, the creature seemed to be absorbing sunshine through every bit of his body and shell (Durrell 1956: 113).

He also assigns masculinity and femininity to animals. Achilles is referred as 'he' (*Ibid,* 50) and Rosy the elephant as 'she' (Durrell 1968: 27) which is again a human construct. Assigning human element in animals could be seen in the references to the dogs that were teased by the Magenpies as: "They would rush up to the wire of the cage, barging and squawking, treading on each other's feet, pecking at each other, and then stand in a disorderly, panting crowd looking up into the cage where the Magenpies, sleek and elegant in their black and white suits, would stare down at them and chuckle, like a pair of city slickers that have successfully duped a crowd of bumbling and earnest villagers (Durrell 1956: 246) The narrative style of bringing in human qualities in animal world could be seen further in the elephant Rosy and also in the Marigolds. The gecko life is presented through Geronimo with a human touch when he says: "Geronimo took up his usual stand in one corner while the new comer chose a spot in the centre of the ceiling. They set about the task of insect-hunting with immense concentration, completely ignoring each other" (*Ibid,* 208) He exaggerates the human elements in Marigold in *My Family and other Animals* when he comments: "Marigolds like broods of shaggy suns stood watching their parent's progress through the sky" (*Ibid,* 30).

Durrell often used poetic imagery and human descriptions with the landscape and the animal world in Corfu. Human beings where measured for the animal traits in them as evident from the description of Mr. Kralefsky who was seen: "bobbing like a courting sparrow." (215) Agathi's voice, whom he met on the Rose-Beetle Man's place is presented as "childish and sweet as a flute." (*Ibid*, 43) This comment shows that landscape and the animal world had an important role in the psyche of the writer blending perfectly in his writings. The use of metaphoric style of visualizing animals even in the presentation of landscape could be evident in the description of the sea from the boat. He states: "the sea lifted smooth blue muscles of waves as it stirred in the dawn-light, and the foam of our wake spread gently behind us like a white peacock's tail, glinting with bubbles" (*Ibid,* 18) From the reading of Durrell's works, it is evident that there is a close relation between the animal world and the human world. This is made possible by the writer mainly by bringing in the animal and human binaries. Though they have lots of differences in biological and physical factors, human beings and animals have equal rights to share their abode *i.e.*, the mother earth. It is at this juncture that the need for conservation of environment gets its emphasis.

Durrell had his own precise ideas on conserving the animal life and had often focused on the need to change our attitude towards the animal world. As Durrell mentions in *Catch Me a Colobus* thus:

> Conservation means preserving the life of the whole world, be it trees or plants, be it even man himself. It is to be remembered that some tribes have been exterminated very successfully in the last few hundred years and that others are being harried to extinction today-the Patagonian Indians, the Eskimoes, and so on. By our thoughtlessness, our greed and our stupidity we will have created, within the next fifty years or perhaps even less, a biological situation whereby we will find it difficult to live in the world at all. We are breeding like rats and this population explosion must be halted in some way. All religious factions, all political factions, the governments of the world, must face facts, for if we persist in ignoring them then, breeding like rats, we will have to die like them also. (Durrell 1981: 211)

Durrell's works are an appeal to the readers for the need to see the animal world as equal to us. They do need a better life and are not to be subjugated and kept in captivation. Making animals as pets or building conservation sites in the form of zoos and zoological parks are a means to sustain animal life and to give them the best facilities that could be provided. The zoo hence needs to be seen as the best place to support and sustain animal life and not as a place to watch the extinct species and to derive a pleasure out of it. The Jersey Zoological Park is a means to make the people aware of the significance of such conservation steps. Durrell in his *The Stationary Ark* (1976) mentions the aim of zoo and animal keeping thus:

> My plan, then, is that all of us, zoo opponents and zoo lovers alike, should endeavour to make them perfect; should make sure that they are a help to animal species and not an additional

> burden on creatures already too hard-pressed by our unbeatable competition. This can be done by being much more critical of zoos and other animal collections, thus making them more critical of themselves, so that even the few good ones will strive to be better (Durrell 1976: 147).

Durrell's concerns to the world are deeper and of much significance in the current world. The concern towards the environment and related issues are increasing though the awareness has already been generated among the people. This lack of concern might be due to the feeling of animals as the 'other' or outside the human world, though they have so much of similarity when compared to that of human beings. Why human beings lack this knowledge? This may be answered as due to the inability of animals to communicate in the human tongue. Durrell had to often face queries on his inquisitive nature from people. He often addressed such queries as evident from *Catch Me a Colobus* thus:

> When asked, as I frequently am, why I should concern myself so deeply, I reply that I think the reason is that I have been a very lucky man and throughout my life the world has given me the most enormous pleasure. I feel indebted for it, and I would like to try and do something to repay the debt. People always look at you in a rather embarrassed sort of way when you talk like this, as though you had said something obscene, but I only wish that more people felt that they owed the world a debt and were prepared to do something about it (Durrell 1981: 241).

Durrell through his narratives was able to make the animals speak. Durrell's works could not be placed as mere stories or novels due to the rich animal world that he presents in relation to human beings. His works could also not be seen as travelogues or travel narratives as they form a part of his documentation of the places and the animals that he had seen during his travels. They also do not fit into scientific literature but could be classified into a new genre of literature called eco-centric literature as they focus on bringing in a detailed world of both the animals and human beings in relation to nature or could be generally classified as animal literature.

The animals in Durrell's works spoke of what they are and what they need to be provided. The world outside us is colourful and pleasing only due to the animal life that it contains. So human beings are much indebted to the animals for the pleasing environment that they provided around us. Much yet need to be learned by human beings. This knowledge is often provided by nature. Human beings need to have an inquisitive mind to understand it. This is the message that evolves out of Durrell works.

References

Alan, Thomas. G. 1969. *Lawrence Durrell: Spirit of Place; Letters and Essays on Travel.* London: Faber and Faber Limited.

Cormack, Mac Carol and Marlyn Stranthern. 1989. *Nature, Culture and Gender.* Cambridge: Cambridge University Press.

Durrell, Gerald. 1956. *My Family and Other Animals.* Great Britain: Penguin Books.

_____. 1968. *Rosy is My Relative.* Glasgow: William Collins Sons.

_____. 1970. *The Drunken Forest*. London: Heinemann.

_____. 1976. *The Stationary Ark*. Glasgow: William Collins.

_____. 1981. *Birds, Beasts and Relatives*. Glasgow: William Collins.

_____. 1981. *Catch Me a Colobus*. Glasgow: William Collins.

Quirk, Randolph (ed.) 2005. *Longman: Dictionary of Contemporary English.* England: Pearson Education.

Study Aid Series. 1972. *Notes on Gerald Durrell's My Family and other Animals.* London: Methuen.

Chapter 21

Hottentot, Hottentot,
I am not a Hottentot.
- Jacobus Coetzee, *Dusklands*

Subject to Sacrifice: The Repression of the "Animal" in J.M. Coetzee's *Dusklands*

☆ *Brian Deyo*

In the past several years, theoretical discussion in the humanities has been enriched by an interdisciplinary convergence of formerly isolated fields of scholarship. The move towards interdisciplinarity demonstrates a refreshing openness to the risks – and excitement – of venturing outside the safe parameters of received paradigms of knowledge-production.[1] In addition, the popularity of interdisciplinary studies evinces a growing need amongst scholars to balance the rigor demanded by professionalization into insular fields of scholarship with a surging interest in addressing real-world ethical questions and problems often circumscribed by institutional politics. One interdisciplinary field that stands out as a salutary corrective to the constraints of specialization is animal studies.

Cary Wolfe situates the recent upsurge of interest in the philosophical, ethical, and political problematic(s) of historical and contemporary relations with non-human animals within the context of the ascendance of theory in the academy (Medoro 2003). As he notes, though critical theorists have been relatively swift to assimilate the theoretical apparatus of civil rights activists, feminists, gay and lesbian activists, and third world liberation movements into legitimate academic discourse, the animal rights and welfare movements' respective ambitions for nonhuman justice have proved relatively intractable to theorization. Thus the late arrival of animal studies onto the academic scene begs a host of questions as to why the question of nonhuman justice – or, more broadly, "the question of the animal" – has only recently emerged.[2] Wolfe contends that animal studies had to wait on the theorization of the formerly mentioned movements to make headway. Whereas postcolonial studies, queer theory, and critical race and gender studies could fruitfully draw from liberal humanist discourse to point up the contradictions of ostensibly democratic policies and laws inflected with racist and sexist ideologies, animal rights theorists and proponents of animal welfare, who, as Wolfe mentions, tend to draw upon the very same discourse, have had significant trouble on the score of incorporating animal subjectivities into a theoretical framework based in the humanist tradition. This, of course, should not only invite skepticism about the ethical adequacy of the kind of subject invoked to reach a political accommodation with patriarchal, imperialist, and authoritarian powers on behalf of human beings; it also rightly raises questions about the practicality, not to mention the ethical logic, of corralling diverse forms of animal subjectivity under the heading of a discursive creation that seeks out recognizably "human" qualities to discern the viability of what is ultimately a quite restricted form of ethical relation.[3]

One method of dealing with this problematic is to interrogate the common philosophical, legal, political, and ethical assumptions behind the manifold articulations of the humanist subject. As a variety of historians and critical theorists have pointed out, the subject of rights is intricately interwoven with Enlightenment models of subjectivity founded on rationality, a power historically arrogated by European males.[4] In addition, this discursive criterion for establishing "humanity" – not to mention political prerogative – is commonly constructed in stark opposition with a putative "animality" throughout the history of Western philosophical and political thought. As the philosopher and historian of ideas Richard Sorabji has demonstrated, many hegemonic European constructions of the human generally assume an essential ontological break between adult males and women, children, non-European others, and animals (Sorabji 1993).[5] Cary Wolfe emphasizes the deadly ramifications of the logic of species difference by commenting on Jacques Derrida's critical assessment of philosophical humanism. He astutely coordinates humanism with speciesism, which, as he mentions, is itself imbricated with cultural logics of racial difference (Wolfe 2003: 43). So long as the discourse and institution of speciesism exists, its symbolic economy remains ready at hand as a potential justification for violence against politically vulnerable human communities, which for Wolfe is sufficient grounds for making what he calls "the ethical priority of the question of the animal" the basis of serious debate in the humanities (Wolfe 2003: 43).

Nevertheless, the force of this question has met with considerable resistance in the humanities. As Philip Armstrong explains, some of the apprehension generated by posthumanist ethical critique may owe much to the fact that it aims to subvert the very models of subjectivity upon which legal doctrines of human rights are animated; and, as he notes, postcolonial critics and theorists are particularly, if understandably, cautious (Armstrong 2002: 413). However, a host of postcolonial critics, including Bill Ashcroft, Gareth Griffiths, and Helen Tiffin, and Graham Huggan, seem receptive to the possibility of developing a posthumanist ethics within the context of postcolonial studies.[6] Huggan and Tiffin's identification of the cultural logics of human *and* animal oppression within imperial discourse – enmeshed, as it is, with philosophical humanism – demonstrates a willingness to critically interrogate the terms of the species/race boundary, with due sensitivity to the conflicting interests and ideological conflicts posthumanist ethical critique entails (Huggan and Tiffin 2010: 134-140).

The urgency of the ethical priority of the question of the animal is made all the more clear if we consider the historical permutations of structures of feeling and attitude toward *human* animality from the Enlightenment through the rise of European imperial power in the nineteenth century, up to the contemporary moment. The historical ascendance of Cartesian rationality, with its attendant ontological differentiation of the "human" from "non-human" forms of life – not to mention its implicit, sometimes patent contempt for putatively arational or irrational forms of life – played a decisive role in terms of shaping imperial attitudes toward non-human animals, the environment, and indigenous peoples during colonial history. The practical consequences of such attitudes were, and still are, far-reaching in scope and magnitude.[7]

On a more positive note, in the contemporary moment, the politics of representation seems to have reversed, even if slightly, in tendency – mainly in response to destructive consequences of anthropocentrism. As Tiffin and Huggan explain:

> While the European trajectory of humanistic essentialism demanded the repression of the animal and the animalistic in all its latent and recrudescent forms, it is not until our own century, in the urgent contexts of eco-catastrophe and the extinction of many non-human species, that a radical re-drawing of this foundational relationship has occurred. Contemporary humanity, having materially destroyed vast areas of wilderness – and many other animals – is now routinely configured as spiritually hollow, as lacking the essence of the human through repression, withdrawal, destruction or absence, rather than latent threat, of the 'inner wild.' The repression is expressed, in both literal and spiritually refractive terms, as a result of the all too successful extermination of that earlier Heart of Darkness; and so it is that what had initially been banished by the Enlightenment in order to constitute human civility – the animal and the animalistic – is now paradoxically being returned as its essence, its inner core.
>
> (Huggan and Tiffin 2010: 134-35)

On the score of understanding in depth the "paradoxical" shift of current reconfigurations of human and nonhuman, the rational and the irrational, self and other, postcolonial studies, as Tiffin and Huggan argue, is advantageously positioned to offer insight.[8] And, as Philip Armstrong suggests, literature offers an especially advantageous point of entry to probe for it. He explains, "In documenting the potent but sometimes evanescent dispositions that link humans and animals over time, the study of fiction has a special role to play.... Literary texts testify to the shared emotions, moods, and thoughts of people in specific historical moments and places, as they are influenced by – and as they influence – the surrounding sociocultural forces and systems" (Armstrong 2008: 4).[9] Literature not only reflects the social and material processes that condition subjectivity, but also offers insight into the affective dimensions of the lived experience of human beings embedded in such processes. In so doing it provides a rich site of study to inquire about contemporary relations between humans and non-humans, not to mention the way they shape and reshape the reconfigurations Huggan and Tiffin describe – reconfigurations that, ethically productive though may be, are always in danger of atavistic reversals that sacrifice the interests and welfare of "animalized" humans and non-humans alike.

In the sphere of contemporary world literature, one would be hard pressed to imagine an author who more adequately registers collective anxieties about our relations with non-human animals than J.M. Coetzee. In his most recent fiction, he has imagined, with virtuosic precision, the psychological complexity of inhabiting a world in which collective feelings about our species' role and status with respect to our animal cousins and the environment is undergoing momentous, indeed bewildering changes. Coetzee's fiction imagines the affective dynamics and difficulties of forging a meaningful sense of humanity alongside the knowledge of what we do to non-human others. By elucidating the psychological challenges of our embattled historical moment, characterized as it is by human-induced climate change, the rise of industrialized factory farming, the ever-accelerating destruction of natural habitats, and the concomitant loss of biodiversity upon which both humans and non-humans depend, Coetzee's fiction, while it does not pretend to *prescribe* novel models of ethical relation with non-human others, nevertheless intimates the grounds for thinking about the unexamined biases and historical inadequacies of anthropocentric ethical traditions.

In his critical writings, Coetzee has expressed skepticism as to whether a rational basis for ethics is advisable, much less practical. In a series of written exchanges on moral perfectionism, *The Death of the Animal: A Dialogue*, Coetzee remarks, "Regarding the project of a rational ethics, it is worth saying that there are people (among whom I number myself) who believe that our ethical impulses are prerational, and that all a rational ethics can achieve is to articulate and give form to ethical impulses" (Cavalieri 2009: 121). Such statements, considered alongside his fictional character Elizabeth Costello's impassioned pronouncements against the formal dictates of reason-centered ethics, indicate a sensibility deeply informed by Romanticism's hostility toward Enlightenment valorizations of the virtue and power of reason over and against the passions. As Costello contends, philosophies that discount the centrality of the passions to the formation of ethical judgment arbitrarily limit the range of beings to which it might extend consideration. She takes serious issue with a formal syllogism that has congealed into what she takes to be a dogmatic cornerstone of Enlightenment humanism:

> '*Cogito ergo sum*,' he famously said. It is a formula I have always been uncomfortable with. It implies that a living being that does not do what we call thinking is somehow second-class. To

> thinking, cogitation, I oppose fullness, embodiedness, the sensation of being ... a heavily affective sensation – of being a body with limbs that have extension in space, of being alive to the world. This fullness contrasts starkly with Descartes' key state, which has an empty feel to it: the feel of a pea rattling around in a shell (Coetzee 2003: 78).

Costello hones her wits against what she takes to be the excessively *formal* aspect of ethical humanism. As she complains, arguments for ethical relation based on *thinking* alone have a tendency to foreclose awareness of means of relation based on intuition, imagination, and our instinctual resources as beings who inhabit bodies through which reason plays an important, albeit limited role in relations between self and other. She also submits, on numerous occasions, that humanist ethical discourse, inasmuch as it exalts rationality over and against sensation, sentience, and emotion, provides the ideological basis for "animalizing" others that "fail" the litmus test of reason. Costello insinuates, on various occasions, that racial ideology is cognate with the institution and discourse of speciesism. In addition to locating the origins of the ideological machinery of marginalization, oppression, and abuse in Enlightenment theories of subjectivity, she alludes to "classic" works of eighteenth century fiction that document the way colonial encounters with animalized humans and nonhumans attempt to negotiate the ontological confusion of boundaries that were enacted during colonization in the long eighteenth century.

Costello's critical preoccupation with Cartesian rationalism, coupled with her pithy commentaries on works like Daniel Defoe's *Robinson Crusoe* (1719) and Jonathan Swift's *Gulliver's Travels* (1726), arguably reflect Coetzee's interest in the intersection between philosophical constructions of subjectivity, the emergence of the novel as a popular genre, ideological discourse, and the gradual consolidation of imperial theories of political rule during the Enlightenment.[10] However, as Coetzee suggests, during the course of European colonial expansion, theorists of civil society in England and throughout Europe, in addition to according reason special status as a guarantor against the abuses of political authoritarianism and religious bigotry, also actively promoted "sympathy" in their writings as a weapon against the evils of political inequality and religious intolerance.[11] Asked to clarify what he means by the term, "sympathy," in an interview, he states:

> I was talking about empathy. But empathy is a recently coined word.... Its earliest occurrence, according to the OED, is in 1912. Sympathy used to mean what empathy means now, until sympathy began to be used in the sense of feeling sad because someone else is feeling sad.... I use sympathy where I can because there is a whole body of writing about sympathy (=empathy), particularly in the 18th century, which it would be a pity to lose touch with (Coetzee and Kurtz 2010: 48).

By sympathy, one infers he is referring to an affective tendency or process that is, by and large, correlated with the "pre-rational" ethical impulses he mentions above–impulses that would seem to be blunted, reshaped, and redirected during the course of one's interpellation into normative forms of political, social, and ethical rationality specific to a particular time and place.

In light of Coetzee's expressed interest the power of sympathy, taken in consideration with his early fiction, which is largely concerned with the political and psychological dynamics of racial oppression, alongside his most recent fiction, preoccupied as it is with questions of nonhuman justice, the remainder of this essay will attempt to provide grounds for plotting out connections between representations of relations between colonizer and colonized in his first work of fiction, *Dusklands* (1974), and his most recent work on relations between humans and non-humans in *Disgrace* (1999), *The Lives of Animals* (1999), and *Elizabeth Costello* (2003). Though the scope of this essay does not permit extensive commentary on his later work, it aims to facilitate a dialogue that allows for a more comprehensive approach to the "question of the animal" as it resonates throughout his writings.[12] Arguably, Coetzee's entire corpus navigates the vexed, unsettled (and unsettling) liminal space where "human" and "non-human" worlds collide. Though a great deal of critical discussion has been paid to the ethical implications of Costello's anti-humanism, comparatively little discussion has focused on what I take to be a concerted interrogation of the human-animal divide that sustains itself throughout his oeuvre. In the process of analyzing and commenting on *Dusklands*, I hope to convey that, for Coetzee, the project of post-colonial justice is ineluctably embroiled with the by no means settled question of our duties and responsibilities with respect to non-human others.

Dusklands consists of a pair of twinned novellas whose main characters mirror each other in critically important ways. The first narrates the experiences of Eugene Dawn, an expert in psychological warfare, whose otherwise stoical personality succumbs to the stresses of rationalizing feelings of guilt and complicity wrought by his involvement in U.S. efforts to win the Vietnam War. The second takes place in the eighteenth century, and imaginatively reconstructs the experiences of one of the author's ancestors, Jacobus Coetzee, a brutally callous frontiersman whose efforts at self-deception are similarly satirized. Both are characterized as hyper-rational and disturbingly impervious to even the slightest irruptions of feeling. Interestingly, both narratives are preoccupied with the psychological and ethical costs of repression and the rational mastery of affective life. I take the narrative of Jacobus Coetzee to be a dramatic representation of a moment in the history of the West whereby the culture of reason begins to consolidate itself and supplant an emerging culture of sensibility.[13]

In *Doubling the Point*, David Attwell asks about the composition of *Dusklands*. Coetzee explains that he had been reading up on the history of the settlement of Southern Africa as early as 1962. He says that before he decided to write the novel he had read up on William Burchell, an English explorer, naturalist, and author, whose memoir, *Travels in the Interior of Southern Africa*, a two-volume work, was published in 1822 and 1824.[14] Coetzee explains that after reading and taking voluminous notes on Burchell, he "[knew] that they would go into some such book as *Dusklands* turned out to be" (Coetzee 1992: 19).

One infers that Coetzee's fascination with the 18th century explorer's obsession with the South African environment deeply informs the creation of Jacobus Coetzee, a character whose pious devotion to the task of classifying the environment and domesticating its inhabitants in accordance with his vision of civilization reflects a similar, if exaggerated zeal. It deserves mentioning that Burchell, after he returned to England after conducting a survey of Southern Africa, was questioned by a select committee of the British House of Commons about the suitability of Southern Africa for emigration in 1819. Burchell enthusiastically remarked on the economic advantages of colonization (Keegan 1996: 62-63). The settlers arrived shortly afterwards in 1820. Thus it seems hardly accidental that Coetzee's literary re-enactment of the settlement of South Africa, told from the imagined perspective of one of his ancestors, expresses a desire to understand and critique a set of historically specific cultural attitudes toward the natural world that may shed light upon the history of colonial settlement.

In *Dusklands,* Jacobus Coetzee is represented as an explorer, adventurer, naturalist, and a "tamer of the wild" (Coetzee 2004 [1974]: 77). The narration, more or less exclusively filtered through the first person point of view, is modeled in accordance with mainstream Enlightenment philosophical discourse (Attwell 1998: 31). The discourse of Cartesian rationality upon which Jacobus's character is generically constructed is, however, far from objective or what one might call scientific – *i.e.,* the rationalist discourse Jacobus projects onto his environs arguably reveals a pathological dimension of the settler mentality. For Jacobus, whatever fails to measure up to the standard of rationality elicits frustration and aggression, and the violence he metes out to the "object-world" is cannily represented as proportionate to the extent to which "things" frustrate mental categorization. Graham Huggan notes:

> Coetzee – in all his work – is not only concerned to challenge the certitudes of reason, but also to undermine those 'discourses of power' that claim reason for their own. Anthropocentrism emerges as one of these discourses; colonialism is a second; 'phallogocentrism' is third (Huggan 2004: 711).

I would add that all three of these discourses intertwine and mutually reinforce each other in subtle and complicated ways throughout Coetzee's entire oeuvre. *Dusklands* can be said to set the literary stage in which Coetzee challenges the discourses of rationality, anthropocentrism, and what Derrida refers to as "carnophallogocentrism."[15]

Significantly, the indigenous peoples of South Africa in *Dusklands* are represented by Jacobus as animals. In his study of the links between the oppression of animals and human genocide, *Eternal Treblinka: Our Treatment of Animals and the Holocaust,* Charles Patterson cites several historical documents that indicate congruence between the racial attitudes of western European explorers and naturalists that facilitated the settlement of South Africa and the rationalized bigotry Coetzee strategically represents in Jacobus to critique the settler mentality. Patterson notes that the natives were described through an arsenal of unabashedly contemptuous animal tropes: "rude and beastlie," "like to bruite beasts," "ill-look'd stinking, nasty People," "heardes," "the very Reverse of Human Kind," "filthy creatures that hardly deserve the name of rational creatures," and so on (Patterson 2002: 28). The Reverend John Ovington, an

English chaplain who sailed to Africa in 1696 came to the following conclusion about the Hottentot people: "If there is any medium between a Rational animal and a Beast, the Hotantot lays fairest claim to the Species" (quoted in Patterson: 28). In Jacobus Coetzee, we discern a similar attitude toward the Hottentots. Jacobus characterizes the Hottentots as savage, primitive, irrational, and incapable of transcending the demands of a putatively essential subservience to biological instinct through the faculty of reason, of which he mistakenly believes himself to be sole proprietor.

Jacobus is a character that obstinately and violently refuses to be humbled by the natural world. He is represented as a person who has become unwittingly subject to Western metaphysical traditions of thought. Like David Lurie at the beginning of Coetzee's *Disgrace*, Jacobus is immured in a solipsistic and markedly abstract worldview that prevents his arid sensibility from being affected by the presence of otherness. For Jacobus, otherness hinges on the logic of species difference. And, interestingly, the novella is literally littered with animal references. Animals figure in Jacobus's imagination in ways that trouble – for him and for us – the ideational status of the human with respect to the "extra-human," living, material world.

As Jacobus encounters the sublimity of the Southern African landscape, one gleans a vivid sense of the manner in which his mentality has a tendency to subsume the variegated complexity of the environment under the heading of an abstract idea:

> In the wild I lose my sense of boundaries. This is a consequence of space and solitude. The operation of space is thus: the *five senses* stretch out from the body they inhabit, but *four* stretch out into a vacuum. The ear cannot hear, the nose cannot smell, the tongue cannot taste, the skin cannot feel: the sun bears down on the belly, flesh and skin move in a pocket of heat, the skin stretches vainly around, everything is sun. *Only the eyes have power*. The eyes are free, they reach out to the horizon all around. Nothing is hidden from the eyes. As the other senses go numb or dumb the eyes flex and extend themselves. *I become a spherical reflecting eye* moving through the wilderness and ingesting it. Destroyer of wilderness I move through the land cutting a devouring path from horizon to horizon. There is nothing from which my eye turns, *I am all that I see*. Such loneliness! Not a stone, not a bush, not a wretched provident ant that is not comprehended in this traveling sphere. What is there that is not me? (79)

Jacobus is represented as an extreme illustration of what is finally a degraded sensibility. The loss of boundaries he experiences is directly proportional to a loss of *sense* in general. His pictorial schematization of the world also implies that the material world is lost to him. It is as if his sensible nature is usurped upon by a tyrannically spiritual rationality that can "see" only by virtue of its power to subtract or negate the irreducibly *material* reality of his environs. This negative power functions by means of abstraction: actuality is hollowed out exclusively in accordance with what his consciousness demands; anything that does not bend to his will is annihilated.

Reality, for him, is conferred upon things only so much as they confirm his power to destroy. Thus, for him, the "wilderness" is a hall of mirrors: "I am all that I see" (79).

Jacobus's is an ego-centered, anthropocentric mentality to be sure; in fact, one is tempted to infer he is a representation of *pure* mentality. The pun that equates the organ of sight with the seat of consciousness is ironic. By virtue of seeing only what he wishes, he sees nothing. His wish to see things in metaphysical terms precludes objectivity and, more importantly, sympathy. When the living, material world reflects back its recalcitrant alterity – or jars against his ideals – Jacobus confers a verdict tinctured with a nihilistic quality. What we witness in Jacobus's "sense of things" is a dramatic demonstration of the way philosophical idealism can take a dangerous nihilistic turn. But what is especially interesting is the way *Dusklands* intimates a psychological relationship between the metaphysical drive and the ritual of sacrifice, not to mention the violence it entails – whether physical or symbolic.

As the narrative persistently indicates, for Jacobus, sacrificial violence – or, violence without practical utility – is a response to a world that fails to align itself with metaphysical fancy. As he states:

> The death of the hare is the logic of salvation. For either he was living out there and is dying into a world of objects, and I am content; or he was living within me and would not die within me, for we know that no man ever yet hated his own flesh, that flesh will not kill itself, that every suicide is a declaration of the otherness of killer from victim. The death of the hare is my metaphysical meat, just as the flesh of the hare is the meat of my dogs. The hare dies to keep my soul from merging with the world. All honour to the hare (79-80).

The living, material universe, shorn of spiritual import, has a savage beauty and completeness that makes Jacobus appear puny and insignificant.[16] As he states, "My despair was despair at the undifferentiated plenum, which is after all nothing but void dressed up as being" (101). Like Melville's Ahab, Jacobus cannot penetrate alterity; thus his compulsion to annihilate it. The extent to which his imagination and intelligence is affronted by alterity is directly proportional to his need to destroy: "I was undergoing nothing less than a failure of the imagination before the void. I was sick at heart" (102).

Consider, for instance, Jacobus's narrative classification of a Hottentot camp he alights upon in his travels throughout the African wilderness:

> The Hottentot camp was laid out on the bank of one of the streams that feed the Leeuwen River. It consisted of perhaps forty huts arranged in a rough circle with outliers, plus five set quite apart from the stream. These would be huts for menstruating women, who during their flux are permitted congress with neither husbands nor cattle. The huts were of uniform construction: bark mats and animal skins spread over hemispheres of supple branches that had been thrust into the ground and lashed together at the apex. The apex is open, allowing the Hottentot

> abed a barred view of the night sky. It has led to neither a special relationship with the sky-gods nor a Hottentot astrology. It is *nothing but* a smoke hole (77, italics my emphasis).

Jacobus expresses a dark fascination with the Hottentot's apparent unreflective innocence, which he adduces from their "failure" to conceive of metaphysics or organized religion.[17] The tone of the passage is infused with chauvinistic contempt. As he notes earlier, "The one gulf that divides us from the Hottentots is our Christianity. We, however, are Christians, a folk with a destiny" (57). The Hottentots are conceived to be *nothing but animals*, and for him this is proved by the fact that they appear to be "locked into the present" (57). In other words, the Hottentot "does not care where he comes from or where he is going" (57-58). Thus, according to Jacobus, the "human" is indexed by the capacity to imagine oneself – or *a* self – within the framework of historical time. This particular conception of the human depends on a belief in divine Providence, which, of course, lends an illusion of legitimacy to his evangelical mission in the African wilderness.

If the long passage cited above is read as critical illustration of the failure of the sympathetic imagination, its irony is rather transparent. The fact that the apex is described, neither as window onto revelation, nor as a spur to metaphysical flights of fancy, intimates Coetzee's impatience or skeptical reserve with regard to his creation's judgments. The phrase "nothing but" – a phrase charged with affective intensity – quite subtly indicates Coetzee's understanding of the pathological dimension of settler culture, a culture that would appear to demand from the world explicit confirmation of the unique value and sanctity of what goes by the name of human experience: to be more specific, the experience of white males of European origin in the particular time and place *Dusklands* represents. Succinctly put, the attitude of contempt for the Hottentot's apparent lack of interiority – and everything this would seem to imply; for instance, a narrative sense of the relationship between past, present, and future – is ironized and transvalued for the reader. By doing so, the narrative manages to inculcate an attitude of benign indifference to pretensions that would elevate an ethically sterile version of the "human" over and against the so-called natural world.

Jacobus's attitude of contempt towards the "non-human" is made manifest in ritual, which represents an attempt to extinguish the presence of animality, *both within and without*. In *Dusklands*, ritual is made to perform a function similar to ideological discourse and, as the work is at pains to articulate, it inevitably involves the *sacrifice* of the animal or the animalistic; just as the fictions Jacobus tells himself about his identity disavow animality within, his zealous involvement in rituals of sacrifice function to formally reinforce and enact such fictions. The practice of sacrifice supplements his subjectivity, and the sacrificial object plays a crucial role in its animation: an opposing term that would grant subjectivity the appearance of substantive life. But sacrificial rites that sustain subjectivity, to the extent that they sacrifice or repress animality, can only sustain life as fiction or appearance – not as it is in its embodied immediacy. Moreover, sacrifice, conceived in the most general sense, tends to involve the destruction or negation of life: symbolized as the absolute Other, the not-I, the not-us.

A critical assumption of some versions of posthumanist theory posits that the exemplary figure of otherness in humanist discourse – the term to which subjectivity is opposed – is "the animal."[18] Cary Wolfe sees the figure of the animal as an indispensable prop to humanist versions of subjectivity. He equates humanism with what he calls the "discourse of species," which, he maintains, is a sacrificial symbolic structure of human self-reference:

> The figure of the 'animal' in the West is part of a cultural and literary history stretching back to at least Plato and the Old Testament, reminding us that the animal has always been especially, frightfully nearby, always lying in wait at the very heart of the constitutive disavowals and self-constructing narratives enacted by that fantasy figure called 'the human.'
>
> It is the pervasiveness of the discourse of species that has made the *institution* of speciesism fundamental (as Georges Bataille, Jacques Derrida, Rene Girard, and others have reminded us) to the formation of Western subjectivity and sociality as such, an institution that relies on the tacit agreement that the full transcendence of the 'human' requires the sacrifice of the 'animal…' (Wolfe 2003: 6).

In light of Wolfe's remarks, it is very remarkable to note that, in *Dusklands*, it is precisely the *absence* of the ritual of sacrifice in Hottentot culture that leads Jacobus to mark them as animal. As he scornfully reflects to himself, "The Hottentots have no feeling for ceremony" (72).

At one point in the novella, Jacobus finds himself in a state of physical, moral, and spiritual destitution. He pines for death. In a dream-state he yearns for a satisfyingly *aesthetic* conclusion to his travails. He himself hopes to become an object of sacrifice, a possibility to which he clings but must ultimately abandon:

> I did not mind dying but I did not wish to die of a putrefying backside. I would gladly have expired in battle, stabbed to the heart, surrounded my mounds of falling foes…. I might even have consented to die at the sacrificial stake: if the Hottentots had been a greater people, a people of ritual, if I had been held until moonrise and then led through rows of silent watchers to a stake where, bound by stone-faced priests, I underwent the Arcadian ordeal of losing toenails, toes, fingernails, fingers, nose, ears, eyes, tongue, and privates, the whole performance accompanied by howls of the purest anguish and climax in a *formal* disemboweling, I might, yes I might have enjoyed it, I might have entered into the spirit of the thing, given myself to the ritual, become the sacrifice, and died with having a feeling of having belonged to a satisfying aesthetic whole, if feelings are any longer possible at the end of such aesthetic wholes as these. But while it was conceivable that in a fit of boredom the Hottentots might club my brains out, it was unlikely that, lacking all religion and

> a fortiori all ritual, they would subject me to ritual sacrifice (82-83, italics my emphasis).

Jacobus seems willing to endure almost any kind of suffering, *so long as it has form*. Naked, unmediated, "meaningless" suffering is so abominable to him that he feels compelled to extirpate it at all costs.

Suffering exists, for him, to be redeemed by sacrifice, which, as it happens, is instrumental to the production of history. After brutally executing two of his servants and several Hottentots for what amounts, in his mind, to the sin of insubordination, he rationalizes murder in accordance with the "logic" of Providence:

> What did the deaths of these people achieve? Through their death I, who after they had expelled me from the desert like a pallid symbol, again asserted my reality. No more than any other man do I enjoy killing; but I have taken it upon myself to be the one to pull the trigger, performing this sacrifice for myself and my countryman, who exist, and committing upon all the dark folk the murders we have all wished. All are guilty, without exception. I include the Hottentots.... God's judgment is just, irreprehensible, and incomprehensible. His mercy pays no heed to merit. I am a tool in the hands of history (106).

Jacobus's person has become identical with an ideological project that utilizes theology to justify, in symbolic terms, an evangelical mission that necessitates "human" sacrifice. Of course, for Jacobus, the human is an *aesthetic* category, so that beings incapable of divining the necessity of History are "cleared out of its way" (106). "History," conceived as a divine agency operating in lockstep with his will, mercilessly annihilates "nonentities" caught within the gaze of reason and appear, at least within the narrow parameters of its ken, to be an impediment to the implementation of order and progress. Before Jacobus trains his predatory will upon the Hottentots, he searches for "evidence" that might save them from damnation:

> What evidence was there, indeed, that they had led a life of any coherence? I had lived in their midst and I had seen no government, no laws, no religion, no arts beyond the singing of lewd songs and dancing of lewd dances. Aside from their greed for the trash in my wagon, had they exhibited any consistent attributes but sloth and an appetite for meat? To these people to whom life was *nothing but* a sequence of accidents had I not been simply another accident? (97-98, italics my emphasis)

Beings incapable of divination, without laws that would bind them into a formal, rule-governed community, without a mutually agreed upon sense of the beautiful, without a social code to curb their natural proclivities and instincts, are so much a part of *natural history* the arrival of the likes of Jacobus, he thinks, must seem to them as accidental as a drought or a famine. Hence, for Jacobus, total immersion in the natural world equates to "nothing but" in his evaluative calculus. "Heartless as baboons they are," he states, "and the only way to treat them is like beasts" (58). "It is only when you hunt them as you hunt jackals that you can really clear a stretch of country" (59). His subsumption of the Hottentots under the heading of natural history vividly suggests that sacrificial violence of an abstractly posited natural order is a necessary precursor to the reproduction of Western European political society in South Africa.[19]

Their exclusion from a historical narrative that equates reason with progress and order, of course, is conditioned by Jacobus's *psychological* exclusion of the taint of animality within. His revulsion from anything that stands in the way of reason and the extension of his will is, as suggested above, symptomatic of the repression of his own animality, which makes it impossible for him to relate to, much less sympathize with beings, human or otherwise, who appear to be immured within sentient life – as opposed to life that is disciplined by the iron dictates of reason. The repression of animality within, of course, as *Dusklands* so clearly demonstrates, was devastating to the indigenous peoples of South Africa. Thus Jacobus' maintenance of the human-animal divide – or what Kelli Oliver refers to as the hallowing of the human "against abjected others who serve as depositories of unwanted animality, embodiment and affect, those reminders of our earthy and finite existence" – is tantamount to the production of colonial history, which itself is a product of the Freudian process of sublimation, whereby the "timelessness" of human sexual and aggressive instincts are discharged "into time, thereby denaturing ('animal') instincts and fomenting ('human') drives and temporality" (Oliver 2005: 89, 90).

At the novella's conclusion, Jacobus recollects the atrocities he has committed and finds himself compelled to wonder about the legitimacy of sacrifice and whether he has acted in vain. For a moment, he is riddled with uncertainty about the justice of his acts. He asks himself:

> How do I *know* that Johannes Platje, or even Adonis, not to speak of the Hottentot dead, was not a world of delight closed off to the senses? May I have not killed something of inestimable value? (106, italics my emphasis)

The lines allude to William Blake's *The Marriage of Heaven and Hell* where the poet asks, "How do you know but ev'ry Bird that cuts the airy way, / Is an immense world of delight, clos'd by your senses five?" (Blake 1988 [1790]: 35). Jacobus, if only for a short spell, agonizes over the question as to whether he has indeed destroyed a world of sensation and delight that his heartlessly narrow subjective world seems so disturbingly unable to allow.[20] It's a powerful and bracing moment, for his act of violence engenders a degeneration from a confident position of stoic rationality to one of overwhelming self-doubt and skepticism that merely *verges* on sympathy: so that the conscious "I" implodes under the weight of the sheer terror of the implications of what his violence reveals to him about himself.

However, as one might expect, Jacobus then proceeds to regain his composure by invoking the first person to defend himself from the horror, guilt, and dismay he feels: as if the invocation of the "I" could finally allay the irruption of affect brought on by the rational self's dissolution into skepticism. He states, "I am an explorer. My essence is to open what is closed, to bring light to what is dark. If the Hottentots comprise an immense world of delight, it is an impenetrable world, impenetrable to men like me, who must either skirt it, which is to evade our mission, or clear it out of the way" (106).

Nevertheless, the moment of skepticism forces one to wonder what it is that has caused Jacobus to question his ethical judgment, which for the entire novella appears risibly confident. This moment, brief as it is, suggests the alterity of Platje, Adonis, and the Hottentot dead has impinged upon and thus *unsettled* his field of vision, so that he cannot see himself and his actions as clearly as he might wish – or maybe it's the case that he sees himself *too* clearly. It also suggests that they fail to figure as abstractions in his consciousness, as if they pose a considerable if not overwhelming resistance to cognition. Whatever the case may be, the allusion to Blake suggests that the horror Jacobus is made *sensible* to does not appear to depend on whether he thinks that what he has destroyed is capable of embodying qualities he would identify as recognizably "human" – say, for instance, consciousness or rationality. What he fixes on, if only for a moment, is the capacity for delight or joy. Nevertheless, once he has regained his balance, he rationalizes away his doubts and fears: "*I know with certainty* that their life held nothing but anxiety, resentment, and debauch. They died in a storm of terror, understanding nothing. They were people of limited intellect and people of limited being" (106, italics my emphasis). Jacobus rationalizes the execution of his servants and the Hottentots by reasoning that their suffering and extinction is of no ethical significance, namely because of their lack of "understanding." He in effect claims "suffering" without consciousness is therefore ethically negligible.

Jacobus represents carnophallogocentrism to a tee. Interestingly, however, Coetzee's narrative manages to show that what is subsumed under Jacobus's *idea* of the animal *resists* the agency of mere thought. In the end of the novella, Jacobus has intimations that his worldview is but a pale and bloodless travesty of the world itself. As it turns out, "things," "animals" and so-called primitive humans are revealed, even if provisionally, to be in possession of an autonomous agency and power that refuses to be represented as idea. The fact that Jacobus's consciousness and identity is momentarily disabled by this intimation is suggestive of a kind of power within himself that he is at pains to disavow – the power of a potentially shared sensation of the joy of mere "being." Tragically, however, his fear of being drawn into the sphere of passionate feeling – a realm of affective life that bears the ideological stigma of the "feminine" or the "animal" in patriarchal, capitalist, and colonial societies – binds him to an alienated form of masculinity that radically compromises his capacity for sympathy, fellowship, and meaningful community.

Notes

1 The sheer volume of work produced by theorists devoted to understanding the affective basis of knowledge is considerable. Theorists from a variety of disciplines, including Daniel Heller-Roazen, Michel Serres, and Cary Wolfe, in their respective critiques of rationalist assumptions pertaining to knowledge-acquisition, have opened a rich terrain for reflecting upon the way the senses and the passions inform aesthetic and ethical cognition. See Heller-Roazen (2007), Serres (1995, 2008), and Wolfe (2003, 2010). For an excellent collection of essays on the "affective turn" in the social sciences, see Clough (2007).

2 For Jacques Derrida's seminal essay on "the question of the animal" and a collection of essays that speak to problematics of identity and difference, particularly as they relate to animal ethics, see Atterton and Calarco (2006).

3 For a trenchant analysis and critique of ethical universalism, see Spivak (1991; 1999: 1-37).

4 See Latour (1991:13-26; 130-45). For a brief discussion of the ideological separation of reason from trhe sensory/sensuous world of "Nature" in imperial discourse, see Seamus Deane's "Imperialism/Nationalism" in Lentricchia & MacLaughlin (1995:354-68).

5 See also Etienne Balibar's discussion of the role of "anthropological universals" in the "animalization" of humans in colonial societies (Balibar & Wallerstein, 1991: 54-64).

6 See Ashcroft, Griffiths, and Tiffin (2002: 215). See also Huggan and Tiffin (2010: 134-40).

7 For a representative textual instantiation of the idealization of the powers of reason with respect to "Nature," see Descartes (1990 [1637]: 84).

8 For a more developed account of the similarities between the political and ethical commitments of postcolonial critics, eco-critics, and practitioners of animal studies, see Huggan and Tiffin (2010: 11-16; 134-39).

9 Armstrong makes ample use of Raymond Williams' *Marxism and Literature* to argue the importance of discerning the material conditions that determine hegemonic attitudes toward non-human others as they are reflected in literary texts. For a detailed explanation of the phrase, "structure of feeling," see Williams (1977: 128-35).

10 Carol Houlihan Flynn comments on how the fictions of Defoe and Swift "represent an imperialistic society that compromises their most compassionate instincts" (1990: 423-32). She also offers a telling analysis of the failure of sympathy in their representations of relations between colonizer and colonized, with particular attention to the way the natural impetus of sympathy is foreclosed by the disavowal of animality in Crusoe and Gulliver.

11 Jonathan Lamb explores the evolution of the discourse of sympathy in the long eighteenth century, with brief interludes on its inflections in Coetzee's *Disgrace* and *The Lives of Animals* (Lamb 2009: 105-28).

12 Columbia University Press has recently published two works devoted to the ethical implications of Coetzee's novels. For a discussion of animal ethics in *The Lives of Animals* and *Elizabeth Costello*, see Cavell, Diamond, Hacking, McDowell and Wolfe (2008). For a broader discussion of the connections between philosophy, literature, and ethics, see Leist and Singer (2010)

13 David Attwell observes that the construction of Jacobus's character reflects a "period of scientific ascendancy" coincident with "the beginnings of colonial expansion" (Attwell 1998: 31).

14 See *White Writing: On the Culture of Letters in South Africa* for an in-depth scholarly assessment of the writings of Burchell and other naturalist and explorers of Southern Africa prior to, during, and after European settlement of the region (Coetzee 1988).

15 See "Eating Well, or the Calculation of the Subject" (Derrida 1992).

16 Jacobus's avid identification of reason with spirituality, coupled with his derogatory identification of the senses, embodiment, and affect with animality/materiality, dovetails precisely with Val Plumwood's discussion of the othering of nature. See Plumwood (2002: 97-122).

17 William Burchell describes Hottentot life as though it was limited to a state of prelapsarian innocence; see Coetzee (1988: 32-33).

18 For an extended discussion of the ethical consequences of the continued maintenance and reproduction of the human-animal binary that insightfully delves into its existential underpinnings, see Wolfe (2010: 61-78).

19 Kelli Oliver comments on Giorgio Agamben's work on the political consequences of symbolic economies that derogate animality. She states, "The human-animal divide ... is not only political but sets up the possibility of politics. Who is included in human society and who is not is a consequence of the politics of 'humanity,' which engenders the polis itself (Oliver 2007: 1).

20 Coetzee's *Elizabeth Costello* critiques ontological assumptions about the centrality of reason to ethical judgment. Against reason, she opposes our shared embodiment with non-human animals as a criterion for ethical consideration. Thus the question as to whether a being is capable of delight or joy, coupled with Bentham's question as to whether a being can "suffer," becomes central to what may amount to a more embodied approach to ethics (Coetzee 2003).

References

Attwell, David. 1998 . "'The Labyrinthe of My History': J.M. Coetzee's *Dusklands*." *Critical Essays on J.M. Coetzee*, ed. Sue Kossew. New York: G.K. Hall & Co.

Armstrong, Philip. 2008. *What Animals Mean in the Fiction of Modernity*. New York: Routledge.

_____. "The Postcolonial Animal." *Journal of Human-Animal Studies* 10.4 (2002): 413-419.

Ashcroft, Bill, Gareth Griffiths and Helen Tiffin. 2002. *The Empire Writes Back: Theory and Practice in Postcolonial Literatures*. New York: Routledge.

Atterton, Peter and Matthew Calarco. 2006. *Animal Philosophy: Ethics and Identity*. New York: Continuum.

Balibar, Ettienne and Immanuel Wallerstein. 1991. *Race, Nation, Class: Ambiguous Identities*. New York: Verso.

Blake, William. 1988. *The Complete Poetry and Prose of William Blake*, ed. David V. Erdman. New York: Anchor.

Cavalieri, Paola. 2009. *The Death of the Animal: A Dialogue*. New York: Columbia University Press.

Cavell, Stanley, Cora Diamond, John McDowell, Ian Hacking, and Cary Wolfe. 2008. *Philosophy and Animal Life*. New York: Columbia University Press.

Clough, Patricia, ed. 2007. *The Affective Turn: Theorizing the Social*. Durham: Duke University Press.

Coetzee, J.M. 1988. *White Writing: On the Culture of Letters in South Africa*. New Haven: Yale University Press.

Coetzee, J.M. and Arabella Kurtz. "Nevertheless, My Sympathies are with the Karamazovs": An Email Correspondence: May – December 2008." *Salmagundi* 166-167 (2010): 39-72.

_____. 1999. *Disgrace*. New York: Penguin.

_____. 1992. *Doubling the Point: Essays and Interviews*, ed. David Attwell. Cambridge: Harvard University Press.

_____. 2004. *Dusklands*. New York: Vintage.

_____. 2003. *Elizabeth Costello*. New York: Penguin.

_____. 2006. "The Animal that Therfore I Am (More to Follow)," in *Animal Philosophy: Ethics and Identity*, ed. Matthew Calarco and Peter Atterton. New York: Continuum.

Derrida, Jacques. 1995. "Eating Well, or the Calculation of the Subject," collected in *Points ... Interviews, 1974-1994*, ed. Elisabeth Weber, tr. Peggy Kamuf. Stanford: Stanford University Press.

Descartes, René. 1990. *Discourse on Method*, collected in *The Rationalists*. New York: Anchor.

Flynn, Carol Houlihan. 1990. *The Body in Swift and Defoe*. Cambridge: Cambridge University Press.

Heller-Roazen, Daniel. 2007. *The Inner Touch: Archaeology of a Sensation*. Brooklyn: Zone.

Huggan, Graham. "'Greening' Postcolonialism: Ecocritical Perspectives." *Modern Fiction Studies* 50.3 3 (2004): 701-733.

Huggan, Graham and Helen Tiffin. 2010. *Postcolonial Ecocriticism: Literature, Animals, Environment*. New York: Routledge.

Keegan, Timothy. 1996. *Colonial South Africa and the Origins of the Racial Order*. Charlottesville: University Press of Virginia.

Lamb, Jonathan. 2009. *The Evolution of Sympathy in the Long Eighteenth Century*. London: Pickering & Chatto.

Latour, Bruno. 1991. *We Have Never Been Human*, tr. Catherine Porter. Cambridge: Harvard University Press.

Leist, Anton and Peter Singer. 2010. *J.M. Coetzee and Ethics: Philosophical Perspectives on Literature*. New York: Columbia University Press.

Lentricchia, Frank and Thomas MacLaughlin. 1995. *Critical Terms for Literary Study*. Chicago: University of Chicago Press.

Medoro, Dana and Allison Calder, "Ethics, Activism, and the Rise of Interdisciplinary Animal Studies: An Interview with Cary Wolfe," *Topia: Canadian Journal of Cultural Studies* 10 (2003): 39-52.

Oliver, Kelly. "Stopping the Anthropological Machine: Agamben with Heidegger and Merleau-Ponty," *Phaenex* 2.2 (2007): 1-23.

_____. "The Good Infection." *Parallax* 11.3 (2005): 87-98.

Patterson, Charles. 2002. *Eternal Treblinka: Our Treatment of Animals and the Holocaust*. New York: Lantern.

Plumwood, Val. 2002. *Environmental Culture: The ecological crisis of reason*. New York: Routledge.

Serres, Michel. 2008. *The Five Senses: A Philosophy of Mingled Bodies*. Trans. Margaret Sankey & Peter Cowley. New York: Continuum.

_____. *The Natural Contract*. Trans. Elizabeth MacArthur & William Paulson. Ann Arbor: University of Michigan Press.

Sorabji, Richard. 1993. *Animal Minds and Human Morals: The Origin of the Western Debate*. Ithaca: Cornell University Press.

Spivak, Gayatri. 1999. *A Critique of Postcolonial Reason: Toward a History of the Vanishing Present*. Cambridge: Harvard University Press.

_____. 1991. "Remembering the Limits: Difference, Identity, and Practice." *Socialism and the Limits of Liberalism*, ed. Peter Osborne. London: Verso.

Wolfe, Cary. 2010. *What is Posthumanism?*. Minneapolis: University of Minnesota Press.

_____. 2003. *Animal Rites: American Culture, the Discourse of Species, and Posthumanist Theory*. Chicago: University of Chicago Press.

Chapter 22

No Higher Life: Ethical Theory and Practice in J.M. Coetzee's *Disgrace*

☆ *Sally Borrell*

In her response to J.M. Coetzee's *The Lives of Animals*, Barbara Smuts observes that 'in a story that is, ostensibly, about our relations with members of other species, none of the characters ever mentions a personal encounter with an animal' (107). The protagonist, Elizabeth Costello, lectures on and debates the subject of animal rights without a word about the cats she lives with. To Smuts, 'the lack of reference to real-life relations with animals is a striking gap in the discourse on animal rights contained in Coetzee's text' (108). Such omissions of the personal in theorising nonhuman animals result from various, often important considerations, including fears of the stigma of sentimentalism, the avoidance of generalising from the specific, or conversely, the inapplicability of a theory in practice. However, difficulties like these can also be confronted directly, both by theorists and by writers of fiction.

Julie Ann Smith (2003), discussing her relation to rescued rabbits, provides a helpful example of an attempt to negotiate between theory and practice. Despite extensive modifications and sacrifices to allow them to share her house, Smith writes, 'Because I expected to live with rabbits on a free and equal basis, I was dissatisfied with my arrangements' (192). Finding that the theory of egalitarianism was not working in practice, she sought an alternative, and came up with what she calls 'performance ethics'. She writes, '"Ethics" suggests a code of moral values toward others based on reliable knowledges about their needs and desires, while "performance" relaxes these strictures by gesturing toward imperfect acts based on uncertain understandings'(2003, 192). Smith explains that seeing animal actions as performance rather than 'behaviour' helps her to minimise the imposition of her own human interpretations. She concludes that performance ethics is 'a more practical, more positive framework with which to think about my relationships to rabbits than dominance or freedom' (2003, 196). Thus, Smith reconciles theory with practice by developing theory *from* practice.

J. M. Coetzee's *Disgrace* arguably promotes a similar endeavour. The medium of fiction of course means that *Disgrace* does not deal in 'real life' experiences of animals, but in other respects, the process gone through by the protagonist, David Lurie, leads to something resembling Smith's practice-based ethics. Although he is indifferent to animals at first, proximity to them leads him to develop compassion and even a spiritual regard for them. Coetzee also adds to this a postcolonial perspective. Lurie's changing relation to animals is partly facilitated by his changing relation to South Africa, as he comes to terms with his position in the new political climate. The problems of species and postcolonialism are also related to the problem of creativity, as Lurie tries to write a chamber opera. Derek Attridge and Eluned Summers-Bremner, in their discussions of *Disgrace*, both perceive connections in the novel between the treatments of species and art. Attridge argues that the novel conveys 'an experience ... of the fundamental value' of both art and responsibility to animals (177), while Summers-Bremner sees Lurie's developing 'obedience to a principle of senseless, but necessary, ethical action' as 'carried through into his work' (153). However, neither critic extensively examines the interrelationship between these themes. Here, I explore how Lurie's ideas about both culture and species develop to the point where he can start work on what becomes a very hybrid opera.

The intimate narrative style that Coetzee employs means that although *Disgrace* is written in the third person, the reader is apparently offered Lurie's views throughout. At the beginning of the novel, he is in the position diametrically opposed to Elizabeth Costello. In *The Lives of Animals*, the question of animals is, as Marjorie Garber puts it, 'clearly staged as a debate between poetry and philosophy' (79). Costello is an accomplished writer, who, when pressed to state exactly how she thinks humans should approach animals, says, "open your heart and listen to what your heart says" (37). Lurie, an academic, is struggling even to begin as a poet: he wants to write an opera about the relationship between Byron and Teresa Guiccioli, but he effectively has writer's block. Meanwhile, he does not appear to approach the question of animals at all, with or without his heart. An argument made by Steve Baker in relation to visual art can be employed here to offer a diagnosis of Lurie's problem. Following Lyotard and Derrida, Baker observes that in postmodern thought, 'Expertise is not considered creative' (41), whereas dissent is. Baker writes that 'Despite postmodernism's supposed mistrust of dualistic thought', one can find 'all manner of broadly contemporary attempts to distinguish creative practice from a narrower and more defensive 'expert' mentality' (41). He argues that animals figure in postmodern art as part of a creative resistance to 'expert-thinking':

> [T]he postmodern animal is there in the gallery not *as a meaning* or a symbol but in all its pressing thingness... It does not so much set itself against meaning as operate independently of it. Humans

> have typically wanted things of animals, wanting them to be meaningful, and wanting to control and to be consoled by those meanings. Postmodernism mistrusts this comfortably centred self's desire (82).

The application of Baker's ideas about the postmodern animal to *Disgrace* provides a possible explanation of the difficulties that Lurie has in writing *Byron in Italy*, and of the remedy that his eventual engagement with animals seems to provide.

The expert-thinking nature of Lurie's initial position is apparent in what Mike Kissack and Michael Titlestad call 'the intellectual and rational hubris that characterizes his life as an academic' (137). In planning his opera, Lurie wants to produce 'high art', 'a chamber-play about love and death ... an action with a complex, restless music behind it, sung in an English that tugs continually toward an imagined Italian' (180). This first conceptualisation of the project is problematic not just in postmodern but in postcolonial terms. The genre of the opera, its setting of nineteenth-century Italy and its focus on Byron all distance it from its South African context. Lurie's project contrasts in this respect with a play being rehearsed by his student, Melanie, which is 'a comedy of the new South Africa' in which 'Catharsis seems to be the presiding principle: all the coarse old prejudices ... washed away in gales of laughter' (23). Although the practical realisation of this play is very different from the realisation of Smith's performance ethics, it is another attempt to use performance to get beyond preconceptions. The opera Lurie imagines is still constrained *within* preconceptions. Not just any expertise, then, but a culturally dislocated expertise impedes his creativity.

Animals, meanwhile, do not yet work against this uncreative expert-thinking in the way that Baker suggests they can, because Lurie's ideas about animals, like his ideas about opera, are limited to the conventional at this point. Indeed, animals appear in primarily figurative terms at first, to characterise his increasingly inappropriate relationships with women. When he telephones his preferred escort, Soraya, at her home, her defensive response makes Lurie think, 'what should a predator expect when he intrudes into the vixen's nest, into the home of her cubs?' (10). During his subsequent affair with Melanie, Lurie becomes the fox: he imagines her passive response to him as 'like a rabbit when the jaws of the fox close on its neck' (25). Unlike the postmodern artists discussed by Baker, Lurie is keeping the meanings of animals well within his control here: they obviously symbolise aspects of his relationships with Soraya and Melanie. However, as the exchangeability of species demonstrates, his ideas have little to do with real animals. The rabbit and the two foxes are what Carol Adams describes as 'absent referents', images used 'without acknowledging the originating oppression ... that generates the power of the metaphor' (4). Rabbits as Julie Smith knows them are nowhere in sight.

Soon, however, animals become more significant in the novel. Unlike *The Lives of Animals*, *Disgrace* does include personal encounters with 'real' animals. Escaping the aftermath of his affair, Lurie goes to stay with his daughter Lucy on her farm. Here, animals are not absent; Lucy and her friends live in physical and emotional proximity to other species. Indeed, animals are suddenly too close for Lurie's comfort. When

Lucy takes him to visit her friends Bev and Bill Shaw, 'he is repelled by the odours of cat urine and dog mange'. Indoors are 'birds in cages, cats everywhere underfoot', and outside, 'chickens scratch around and what looks uncommonly like a duiker snoozes in a corner' (73). As they drive away, Lurie says, "I don't mean to be rude. It's a subculture of its own, I'm sure" (73). Lucy accuses him of disapproving of the Shaws "because they are not going to lead me to a higher life" (74). She says "there is no higher life. This is the only life there is. Which we share with animals" (74). This view of a shared life is of course what Smith seeks to realise as fully as possible through her idea of performance ethics. Lurie, however, tries to maintain distinctions between humans and other animals. He says, "As for animals, by all means let us be kind to them. But let us not lose perspective. We are of a different order of creation from the animals. Not higher, necessarily, but different" (74). He borrows dated theories from Europe as a way of reinforcing this argument. He tells Lucy, "The Church Fathers ... decided they don't have proper souls... Their souls are tied to their bodies and die with them" (78). This appears to be a reference to Thomas Aquinas' argument that 'brute animals' do not have immortal souls (Aquinas 267-272). Later, when Bev says, "we eat up a lot of animals in this country... I'm not sure how we will justify it to them", Lurie is astonished, thinking, 'Justify it? When? At the Great Reckoning?' (82). Rather than opening his heart like the 'poet' Elizabeth Costello, Lurie is still deploying 'philosophy' in the form of received ideas. This is one of the clearest instances in *Disgrace* of the conflict between expert-thinking and creativity, and of its postcolonial dimension: Lurie's acceptance of an expert view put forward in Europe means that he has not formulated any alternative ideas on the subject, and is not inclined to listen to Lucy and Bev's opinions.

Nevertheless, Lucy functions as a catalyst to Lurie's development away from expert-thinking. It is evident to her that his stubbornness is an obstacle that needs to be overcome. She tells him: "You shouldn't be so unbending, David. It isn't heroic to be unbending" (66). However, if Lurie is unbending, then arguing with him is unlikely to have much effect. Lucy negotiates this by confronting him with practical realities instead. She suggests that, in order to avoid becoming bored with country life, he could help her African neighbour, Petrus, to work on his new land, and Bev with her animal shelter. It is as a result of these and other practical experiences that Lurie abandons much of his expert-thinking.

In terms of cultural relations, Lurie slowly comes to acknowledge that his position has changed in the new South Africa. He realises that Petrus, who is nominally Lucy's assistant, in fact has considerable self-defining power, while Lucy is becoming dependent upon him. When Lurie jokingly asks Lucy whether Petrus will pay him for his labour, she replies quite seriously: "Ask him to pay you. He can afford it. I'm not sure I can afford him any more" (77). Lurie also starts to consider the relative power of different South African languages. In *White Writing*, a critical examination of South African literature, Coetzee asks 'Is there a language in which people ... of a highly problematical South African-colonial identity, can speak to Africa and be spoken to by Africa?' (7-8). Spending time with Petrus leads Lurie to ask himself a similar question. 'He would not mind hearing Petrus's story one day. But preferably not reduced to English. More and more he is convinced that English is an unfit medium for the truth of South Africa' (117).

The limits to Lurie's cultural power are also vividly underlined in an apparently racially-motivated attack on Lucy's farm. He is locked in the toilet while she is robbed and raped. Some critics, understandably, have seen this episode as criminalising black Africans; however, Attridge points out that it is important to examine 'its place in the novel and the responses to it that seem to be endorsed' (170-71). Coetzee in fact explores opposing responses to the attack through Lurie and Lucy. Lurie is angry, but the experience still promotes his cultural humility. He cannot understand them as 'they stand beneath the window, inspecting their prisoner, discussing his fate... Italian and French will not save him' (95). Thus, he is forced to acknowledge the local irrelevance of his expertise. Lucy's reaction to the encounter, meanwhile, confronts Lurie with her much stronger sense of cultural humility. Although she seems to agree that the attack was an assertion of racial re-empowerment, she does not agree about how she should respond. Lurie says, "If you fail to stand up for yourself at this moment, you will never be able to hold your head up again" (133). However, Lucy is not trying to 'hold her head up' or even, as she says, "to save my skin" (112). On the contrary, she argues that insecurity may be a condition of her remaining on the farm. She says, "what if *that* is the price one has to pay for staying on? Perhaps that is how they look at it; perhaps that is how I should look at it too" (158). Ultimately, she not only refuses to report the rape, she also proceeds with the pregnancy resulting from it and enters into an arrangement with Petrus. She agrees to be known as one of his wives, and to give him all her land except the house, "in return for which I am allowed to creep in under his wing" (203). At the conclusion of their debate on this subject, Lurie stops trying to influence Lucy's response to local cultural politics, even though he disagrees with the extent of her humility. Lucy speaks first:

> "Yes, I agree, it is humiliating. But perhaps that is a good point to start from again. Perhaps that is what I must learn to accept. To start at ground level. With nothing... No cards, no weapons, no property, no rights, no dignity."
>
> "Like a dog."
>
> "Yes, like a dog" (205).

This reference to a dog also illustrates Lurie's progress in terms of species relations, because instead of being figurative, it derives from a new understanding of animal realities. Practical experiences in Bev's clinic and on Lucy's farm lead him to accept and indeed to share their respect for nonhuman life. He reaches a decisive turning point when Petrus brings home two sheep to be slaughtered for a party, and leaves them on bare ground with no water. Lurie's response is one of compassion. He asks, 'even if the sheep are for the party, don't you think they could graze?' (123). When Petrus does not move them, Lurie does. He discovers,

> A bond seems to have come into existence between himself and the two Persians, he does not know how. The bond is not one of affection. It is not even a bond with these two in particular, whom he could not pick out from a mob in a field. Nevertheless, suddenly and without reason, their lot has become important to him (126).

Contemplating their fate, Lurie reflects that their bodies will be used up, almost entirely, with the possible exception of the gall bladder. Then he thinks, 'Descartes should have thought of that. The soul, suspended in the dark, bitter gall, hiding' (124). Thus, having once unquestioningly subscribed to the Church Fathers' opinion that animals have no proper souls, Lurie now toys with Descartes' arguments on the subject. In fact, the theories with which he is familiar no longer correspond to what he feels. He says to Lucy, "I still don't believe that animals have properly individual lives. Which among them get to live, which get to die, is not, as far as I am concerned, worth agonizing over. Nevertheless ... in this case I am disturbed" (127). Lurie's words show that, in his own mind, he is struggling to negotiate between theory and practice: he is still trying to govern emotion with reason, but he is failing. When he is given a plate of mutton at Petrus's party, Lurie is forced to choose and the shift in his attitudes solidifies. The idea that humans could be answerable to animals – the same idea earlier expressed by Bev – is encapsulated in his own reaction to the meat: 'I am going to eat this and ask forgiveness afterwards' (131).

From here on, this problem of how to respond to an animal's dead body becomes an increasingly important one in *Disgrace*. Julie Smith's ideas are again pertinent to this. In '"Viewing" the Body', she describes trying to establish whether rabbits understand death. Not wanting to impose interpretations, her observations were again based on the idea that rabbits' actions are performances:

> Rabbits demand responses from each other and have a variety of ways to impose them... Therefore, a rabbit who tries over time to get a response and fails, and who eventually gives up trying and withdraws, most likely understands that something drastic has occurred (190).

Lurie, helping Bev Shaw in her animal clinic, finds the experience more and more intense as he reaches the similar conclusion that the dogs have an understanding of death; in this case, their own. 'His whole being is gripped by what happens in the theatre. He is convinced the dogs know their time has come... They flatten their ears, they droop their tails, as if they too feel the disgrace of dying' (143). Once they are dead, Lurie tries to avoid subjecting their bodies to further disgrace: he feeds them into the hospital incinerator himself so that the workers will not beat them into 'a more convenient shape for processing'. He can find no rational explanation for his behaviour: 'He saves the honour of corpses because there is no one else stupid enough to do it' (146). Though he does not feel Smith's comfort with it, he is now experiencing and acknowledging an emotional response to animals and their mortality. Lurie's reactions are thus beginning to accord with Elizabeth Costello's exhortation to listen to one's heart in response to animals. Thus, he has acquired humility in relation to species too. Indeed, Kissack and Titlestad go so far as to claim that ultimately, his 'ethical disposition, completely divested of any transcendent or optimistic expectation, one that in fact embraces the perversity of life, becomes a clear example of secular humility' (137).

In a shift that appears to correspond to the expert/artist opposition that Steve Baker describes, it is once Lurie has thus abandoned his expert-thinking theories and begun consciously to engage with ideas about animals that he is able to proceed with his opera. As he does so, his creative attitudes start to seem more like the postmodern ones which Baker discusses. Baker writes that postmodern knowledge is 'a kind of partisan knowledge ... which refuses to conform to rules and which may even embrace the apparently fallacious, *not knowing what will follow from that*' (41). In returning to the opera, Lurie adopts this kind of dissenting and experimental approach by revising his first, expert-thinking ideas of what it ought to be like. Although 'Formally speaking, the conception is not a bad one' (180), he recognises that 'the project has failed to engage the core of him. There is something misconceived about it, something that does not come from the heart' (181). Now, having acknowledged that his linguistic expertise will not let him access 'the truth of South Africa', he must find a different mode of expression. The problem of localised creation is of course a self-reflexive feature of the novel; the process in question is not just that of writing, but of white writing in South Africa, which, as Coetzee has suggested, seems to call for a medium that is neither African nor European. Lurie achieves this in his opera through his choice of musical instrument. Deciding that his piano sounds 'too rounded, too physical, too rich' (184), he experiments with Lucy's old toy banjo instead. Its sound, described as a 'silly plink-plonk' and later a 'flat, tinny slap', obviously breaks the rules of opera, making it literally discordant: 'the voice ... strains to soar away from the ludicrous instrument but is continually reined back' (184-85). Lurie is experimenting to the point that he rejects the conventions of operatic sound. Furthermore, because banjos are African in origin, the decision turns his previously Anglo-Italian opera into an instance of locally relevant hybridity.

Lurie's new willingness to engage with animals further contributes to the destabilisation of expert-thinking by raising the possibility of species hybridity in the opera. Baker writes, 'Neither the aesthetics of modernism nor the philosophical values of humanism, it is believed, can cope easily with hybrid forms which unsettle boundaries, most especially the boundaries of the human and the non-human' (99). The idea of combining human and canine voices is suggested to Lurie by a dog at the animal clinic. As he hums along with the banjo,

> The dog smacks its lips and seems on the point of singing too, or howling. Would he dare to do that: bring a dog into the piece, allow it to loose its own lament to the heavens between the strophes of lovelorn Teresa's? Why not? Surely, in a work that will never be performed, all things are permitted? (215)

At the level of the opera's composition, this question is for Lurie to decide. If he does include a dog, then his piece further dissents from the expert-thinking conventions of opera. A live dog would not only unsettle species boundaries, it would introduce an element of unpredictability, and Baker suggests that 'non-manipulation of the animal can perhaps be seen as one postmodern ambition or ideal' (82), even though 'it is impossible for the artist wholly to stand back from its interpretation or manipulation' (81). On the other hand, if Lurie's opera is never to be staged, then this experimentation at the animal shelter is already as close to performance as it will come. At that level, the dog's participation is controlled only by Coetzee. From Lurie's perspective, it is the dog's idea, and it is for the dog to decide whether to *introduce himself* into the music. Lurie simply continues, open, just as Julie Smith recommends, to the possible performance of a nonhuman animal. To this extent, despite and paradoxically because of his belief that the opera will never be performed, it is already a cultural and species hybrid performance.

Disgrace thus presents a gradual negotiation between theory and practice in relation to animals, within a postcolonial context. Personal encounters with animals influence Lurie's attitudes to them, until his newfound respect for the lives of non-human others becomes so strong that it redefines his own. He is forced to abandon received ideas in favour of a form of humility. Ultimately, far from seeing his artistic pursuits as 'the higher life' and animals as soulless, Lurie is tempted to combine the two in his opera. *Disgrace* thus suggests that practical experiences can and perhaps should shape our theories about human-animal relations. It still may not be possible to act upon the results; Julie Smith can put her new theories into practice, whereas Coetzee can only posit ideas in the context of fiction. In Lurie's perception, however, theory and practice have become blurred together: he no longer has a sense of a higher life than the one that he shares with animals.

References

Adams, Carol J. 2000. *The Sexual Politics of Meat*. New York: Continuum.

Aquinas, Thomas. 1975 . 'That the Souls of Brute Animals Are Not Immortal.' *Summa Contra Gentiles*. Trans. Anton C. Pegis. Vol. II, chap. 82. Notre Dame and London: University of Notre Dame Press.

Attridge, Derek. 2004. *J. M. Coetzee and the Ethics of Reading: Literature in the Event*. Chicago and London: University of Chicago Press.

Baker, Steve. 2000. *The Postmodern Animal*. London: Reaktion Books.

Coetzee, J. M. 1999. *Disgrace*. London: Random House.

_____. 1999. *The Lives of Animals*. Ed. Amy Gutmann. Princeton, NJ: Princeton University Press.

_____. 1988. *White Writing: On the Culture of Letters in South Africa*. New Haven and London: Yale University Press.

Garber, Marjorie. 1999. Reflections on *The Lives of Animals*, by J.M. Coetzee. Ed. Amy Gutmann. Princeton, NJ: Princeton University Press.

Kissack, Mike, and Michael Titlestad. 'Humility in a Godless World: Shame, Defiance and Dignity in Coetzee's *Disgrace*.' *The Journal of Commonwealth Literature* 38.3 (2003): 135-47. Web. 5 Feb. 2007. <http://jcl.sagepub.com>.

Smith, Julie A. 'Beyond Dominance and Affection: Living with Rabbits in Post-Humanist Households.' *Society and Animals* 11.2 (2003):181-97. Web. 7 Dec. 2007. <http://www.animalsandsociety.org/assets/library/506_s1124.pdf>.

_____. '"Viewing" the Body: Toward a Discourse of Rabbit Death'. *Worldviews: Environment, Culture, Religion* 9.2 (2005):184-202..

Smuts, Barbara. 1999. Reflections on *The Lives of Animals*, by J.M. Coetzee. Ed. Amy Gutmann. Princeton, NJ: Princeton University Press.

Summers-Bremner, Eluned. '"Poor Creatures": Ishiguro's and Coetzee's Imaginary Animals.' *Mosaic: A Journal for the Interdisciplinary Study of Literature* 39.4 (2006): 145-60.

Chapter 23

Ethics in/and Poetry: Them and Us

☆ *Indira Nityanandam*

Ethics is the branch of philosophy that deals with the values of human life in a coherent, systematic and scientific manner. It is concerned with the type of conduct or character that is approved or disapproved of in terms of right and wrong or good and bad. Ethics is also called moral science or moral philosophy. It is generally agreed that ethics seeks a critical grasp of the principles and standards that guide a man in making morally right choices in his daily activities. The scope of conduct that can be judged morally covers the whole range of human experience that is subject to intelligent judgement and voluntary action. Ethics is the integral science of valuation. Ethics as a branch of study is probably as old as the earliest period of formal study and thought. Different ages have attempted to answer the problem of the highest good in various ways. From theories of antiquity through the ideas of Hedonism, we reach the evolutionary theory of ethics. Evolutionary ethics reflects the spreading influence of the biological theory of evolution on all humanistic studies. One effect of this development was that the evaluation of human conduct was done by the same principle as evolutionary biology–that is, survival value, or survival of the fittest. Some other evolutionists sought a restatement of the ethical standard itself in terms of survival value: not only that what is good contributes in the long run to individual or social survival, but that what thus contributes to survival is good, and the chief good for that reason. Some others have emphasized the advance from conflict to co-operation. The process of mutual adjustment and co-operation points towards the greatest happiness of the greatest number. Human intelligence not only finds itself in a certain environment; it transforms or may even produce an environment to realize its purposes. In his purposes and choices and conduct, man is guided by more than merely animal equipment and by more than merely animal reactions. Beyond his choice of survival, he is moved by the values to which he commits his life and by which he judges it. Our moral career is concerned with our highest aims and ideals; it engages our highest capacities and stands in judgement on our lower nature even though it may continually yield to it. Thus peaceful co-existence has been of paramount interest to mankind. As literature is a reflection of the society in which the writer lives and writes, one can see this in literature. This paper looks at the various ways in which poets have looked at the animal world, vis-à-vis human experience. From a feeling of awe to one of camaraderie, from love to hate, praise to pity, threat to dependence, poets have experienced and expressed the entire gamut of human feelings and sentiments.

In Indian mythology, Gods and Goddesses have as their vahanas (vehicles) the small rat, the powerful tiger, the beautiful swan, the nocturnal owl, the strong garuda etc. Literature, both ancient and medieval , is replete with descriptions of the relationship between the animal kingdom and the world of man. Man killed animals for food (a natural necessity) but probably not for sport. In Kalidasa's Abhignyana Shakuntalam, when the king Dushyanta arrives in the forest of sage Kanva , he is told : *Na khalu, na khalu baanaha sannipathyo yamasmin/mriduni mrigasharire pushparashaha iva agnihi/Kva bata harinakaani jeevitam yaati lolam/kvacha nishita nipaathaaha vajrasaraaha sharaha te/* (Act 1, *shloka* 10). It means: "Like fire on the tender flowers is your sharp arrow on the delicate body of the deer, do not aim your arrow on it. What is the comparison between living animal and your cruel arrow which is harder than vajra?"

The role of animals in the day-to-day life of man gained importance with the domestication of animals. This led to a closer association and dependence between the two. Poets also sought to express similarities and differences between animals and human beings. Though man is on the highest rung on the ladder of evolution, man has to learn a great deal from the animal kingdom. Living together and sharing the same bounty of nature, man realizes that he is a part of the same animal kingdom. A feeling of peaceful co-existence is expressed by the Bard of Avon when he sings in Ariel's Song ,"Where the bee sucks, there suck I ;/ In a cowslip's bell I lie;/ There I couch when owls do cry,/ On the bat's back I do fly/ After summer merrily./Merrily, merrily shall I live now/ Under the blossom that hangs on the bough". (English Lyrical Types, 74).

In direct contrast is Daruwalla's distinction between easy and difficult animals. His poem of the same title concludes with, "You divided in two/ this animal delirium we call 'life'/ into 'easy animals', 'difficult animals'./ All that moved on legs/ flew on wings/ crawled on the belly/ inhaled through fins/ hedgehog and weasel and polecat/ all that went to the taxidermist/ marmot and buzzard and bat/ you lumped together as 'easy animals'./And pitched against this menagerie/ one solitary cry/ the 'one difficult animal'/ that was I". (40-41) The comparison between the two may often prove to be more in favour of the animal, as is the case in this poem.

Looking at the animals around him, man begins to wonder about the creator, who made both man and beast! Can both the lamb and the tiger be created with the same raw materials and by the same hand? A poet like William Blake wonders along these lines: "Little Lamb, who made thee?/Dost thou know who made thee?/" (*Pageant Poems*, 25) and soon after we have a poem where he questions the tiger, "Did he who made the Lamb make thee?"(*Pageant Poems*, 26). Both tigers and lambs, all kinds of people inhabit this world. Innocence and experience combine to make the world, including man and beast, what it is today. Clearly each species has its own place and its own role in the scheme of things. The lamb as a symbol of innocence and the tiger as a symbol of authority both have their individual importance. Man has to balance the relative importance of both in order to make his life possible and meaningful. The experience of this difference in the animal kingdom highlights the importance and significance of difference in the world of human beings too. Man is able to accept difference without thinking of difference in black and white. Ted Hughes speaks of animals as powerful symbols. In his poem, The Jaguar, he says that though caged in a zoo the jaguar is a symbol or apotheosis or idealization of a free condition. In caging these wild animals, man probably imagines his supremacy but that is merely illusory or ephemeral: "He spins from the bars, but there's no cage to him/ More than to the visionary his cell:/ His stride is wilderness of freedom:/ The world rolls under the long thrust of his heel./ Over the cage floor the horizons come/". (*Pageant Poems*, 193)

Life on this earth is one mad rush for many of us. Caught in the hustle and bustle of everyday life, he has no time to stand and stare. Man loves the same kind of peace that the kingfisher loves, "I also love a quiet place/That's green, away from all mankind;/ A lonely pool, and let a tree/ Sigh with her bosom over me/" (Pageant Poems, 136). To the poet, W.H. Davies, the bird becomes a pretext for expressing himself. He attempts to interpret the bird and give it some human significance. The character of the kingfisher seems to match his own as both have a similar bent of mind. However, these lines suggest a touch of indulgent self-pity as the poet yearns for what seems to be easily available to the bird. Man realizes that the peace that he craves for can be found only in close communion with nature and all her creatures. The Romantic poets have many poems about birds like the nightingale, the cuckoo, the skylark etc. In each of these poems, the poet realizes that the world these inhabit is far superior to that we inhabit. Whether it is in the desire to fly away from the sorrows of this world into the land of the nightingale or to sing as the skylark does, these poets could sensitively relate to the world of animals and birds. "Teach me half the gladness/ That thy brain must know,/ Such harmonious madness/ From my lips would flow/ The world should listen then, as I am listening now" pleads Shelley with the skylark. (English Lyrical Types, 26). Keats, on the other hand, wants to fly away with the nightingale so that he can forget, "What thou amongst the leaves hast never known,/ The weariness, the fever, and the fret/" (Pageant poems, 71).

The need to live together is paramount in man's scheme of things, but the fear of man in the mind of even a small bird is portrayed by Emily Dickinson when she writes, "A Bird Came Down the Walk", where the bird seems to be comfortable with all other living creatures except man. Not willing to accept the crumb that the poet offers, he flies away. His flight is indicative of man's inability to convince animals of his peaceful intentions, as they seem rare expressions of noble intentions. Galaway

Kindell describes how he attempted duck-chasing: "Little duck!/ I cried. It paddled away,/ I paddled after it." (Poems to Remember,96)

Tagore sums it up beautifully, when he says, in his essay The Religion of the Forest: "For us the highest purpose of this world is not merely living in it, knowing it and making use of it, but realizing our own selves in it through expansion of sympathy; not alienating us from it, but comprehending and uniting it with ourselves in perfect union" (49).

References

Daruwalla, Keki N. 1991. *Under Orion*. New Delhi: Harper Collins.

Pendlebury, B.J. 1975. *English Lyrical Types*. Bombay: Blackie and Sons.

Petit, Dorothy. 1975. *Poems to Remember*. London: Macmillan.

Sheppard, C.A. 1977. *A Pageant of Poems*. Bombay: Orient Longman.

Tagore, Rabindranath. 1922. *Creative Unity*. London: Macmillan.

Chapter 24

A Study of Endangered Species and *The Hungry Tide*

☆ ***Anju Bala Agrawal***

Amitav Ghosh's *The Hungry Tide* has been touted as both a prophetic and poetic novel. Its description of the natural beauty as well as the perils of the Sundarbans–the tide country lying to the south of West Bengal and Bangladesh–is in many respects poetic. While it does not rival the quoted poetry from Rainer Maria Rilke's Duino Elegies which concludes many of the chapters, the book lends new meaning to the lines "Tiger, tiger, burning bright...." For indeed there are fearsome man-eating tigers prowling in the mangroves of the Sundarbans–a romantic, haunting trope that the author uses to bring the implicit debate between environmentalists and survivalists. Sundarbans area is one of the principal remaining domains of the endangered Bengal tiger; although the tiger is officially protected, illegal poaching is known to occur. He summons up the aura and ambience of this singular fascinating place from its history and myth. There is an archipelago of islands in this area. Some are inhabited by human beings but the existence of some islands totally depends on tides. They appear and disappear according to the tides. Some islands merge in water forever and some new sandbanks come into existence. This region is called tide country or 'bhatir desh' inhabited by people who eke out a precarious existence from mud and water. The settlers are evicted forcibly by the police and the forest guards for the alleged reason that they are intruders from Bangladesh. This heartless action is justified in the name of conservation of forest and wild life. There are manifest threats posed by human settlement to the unique diversity of aquatic and terrestrial life in the mangrove swamps of the Sundarbans. Though aquatic species are constantly depleting by fishing and trawling, there are equal dangers for the human settlers. Not only does the forest take its toll; the tides, too, exact their revenge. The constant erosion of dykes and embankments, the silting up of channels, the flooding by storm-waters, makes human life on the islands an accident or miracle. The writer seems to convey the message that people living in these areas are more eco-friendly than officials sitting in offices claiming to save environment. If the government takes proper care of the people inhabiting the Sundarbans, they may be more helpful in saving environment than people sitting in government and making policies. The apparent cause is that these people are generally very simple and naïve. They regard these animal species very pious as Fokir calls dolphins 'the messenger of Bon Bibi' (235)

In *The Hungry Tide,* a cetologist of Bengali parentage, Piyali Roy, comes to the Sundarbans or 'tide country' region of West Bengal to study the trail of a breed of freshwater dolphin, the *orcaella brevirostris*. Grown up in Seattle, she is said to be 'out of place' on a Kolkata station platform. Her positioning in relation to the novel's setting is complemented by that of the novel's other main localizer, a character who comes to the Sundarbans at the same time as she does: Kanai, a Delhi-based linguist who runs a translation agency and crucially, unlike Piya, speaks Bengali. So, if we view her as an outsider/insider, we might describe him the other way round: as an insider/outsider. He knows the region and speaks the language, but he has lived away from West Bengal.

Piya is a skilled Western professional who comes to one of the world's most delicately balanced eco-systems on a research trip that has clear ecological implications. Her equipment includes binoculars and a GPS (Global Positioning System) and these can be seen as a metonym for The Hungry Tide's concern with cross-cultural geographies and cosmopolitan perspectives on the local: as a work concerned with global issues set in a very specific South Asian location, a region with a uniquely varied biodiversity. The novel, too, is centrally engaged with how one sees and gives voice to the local. *The Hungry Tide* dramatizes competing visions of place doing so, by using angles of vision that are in consonant with its own positioning in relation to its subject-matter.

The Hungry Tide provides multiple optics on the Sundarbans' eco-system of mangrove-forested islands and mudflats, representing the constant transformations it undergoes, because of daily tidal flows, with sections of land being temporarily submerged, and with seawater and freshwater intermingling. It also demonstrates the extent to which the tide country has been a contested site, as a consequence of the differing visions of the many stakeholders in the region – state politicians, local inhabitants, conservationists and tigers among them – and at bipolar extremes have variously been seen as uninhabitable and as fertile territory for utopian projects. In the past such projects have included a cooperative founded by an idealistic Scottish colonist who attempted to create a community that would transcend the caste and regional divisions of Indian society and a post-independence settlement by Bengali refugees, who established squatters' rights on a previously uninhabited Sundarbans island named Morichjhapi that had been designated as a wildlife conservation site. In the present these projects are paralleled by a more pragmatic hospital-centred Development Trust, which Kanai's aunt, Nilima, runs on an island called Lusibari and which is seen as a model for NGOs working in rural India. The novel, then, offers various visions of the possible ways in which a place can be shaped, viewing a supposedly remote region as a product of both its ever-changing physical geography and of human agency. Place, it suggests, is never static, but rather the manufactured product of varying perspectives, which are often laden with particular political agendas.

The Morichjhapi episode particularly points up the contested nature of land usage, since the refugees who established themselves on the island did so illegally; criminalized by poverty, they settled in an area designated for the preservation of the region's wildlife, particularly its trees and endangered species such as the Royal Bengal Tiger and they were forcibly evicted by the local authorities, with many of them being killed. Ghosh's representation of this episode may seem to suggest that dispossessed subalterns should be favoured over animals, but bearing in mind that Piya is investigating the behaviour of another unique animal species, the river dolphins of the region, and that there are numerous other references to animals that play their part in sustaining the delicately balanced ecology of the tide country, the issue is never this simple.

In the present novel, the debate is mainly seen from Piya's perspective and in terms of her responses to the various people, subaltern and middle class, that she meets in the Sundarbans. In her attempt to track the movements of the river dolphins, which appear to be behaving in a strange manner, she is assisted by a fisherman, Fokir, with whom she develops a close non-verbal relationship that seems to bridge the gap between the mobile cosmopolitan elite and the more locally rooted poor. When at one point she discovers that her work, taking soundings to ascertain the position of the group of dolphins she is tracking, is complemented by his work, catching crabs, she feels that a 'seamless intertwining' (*Tide*, 141), which transcends the need for voice, is operating. Specifically, she notes that his laying a line 'acts like a guide rail' (141), a task for which she would have had to use her GPS. This, then, suggests the possibility of a symbiotic wordless collaboration across class and culture that erodes differences between the cosmopolitan elite and the globalized subaltern and the novel seems to be offering this as an attractive possibility. However, Piya's relationship with Fokir is contrasted with her response to Kanai, whose work and temperament make him acutely aware of 'words', their etymologies and, as he sees it, their vital role in human communication. When Kanai later joins Piya and Fokir on the Sundarbans waterways, he disagrees to her view that it does not matter that she is unable to communicate with Fokir in words, telling her, '"Piya, there's nothing in common between you at all. You're from different worlds, different planets. If you were about to be struck by a bolt of lightning, he'd have no way of letting you know"' (268). And, as if to order, the proverbial bolt of lightning as it were appears a moment later and vindicates him. It manifests itself in the form of a crocodile, which Piya has not seen and when Fokir utters the word kumir ('crocodile') she has to ask Kanai to translate it for her.

For the ecological balance, proper number of animals is also a must. Marine mammals are a great source of study. A proper and scientific study of their movement can help scientists learn the arrival of cyclones or other storms. Piya as a scientist is well aware of the fact. When talking to Kanai, she comments, "When marine mammals begin to disappear from an established habitat, it means something is gone very, very wrong."(266-67) Her experience proves true. During her trip to the region, she herself sees the changed behaviour of dolphins.

> Logging the dolphins' appearances, she saw that they were surfacing with unusual frequency, with barely a minute or two separating their exhalations. And more than once, along with the breathing, she heard a sound not unlike a squeal. There was something odd here, she decided; this was not the way these animals normally behaved. (366)

While observing the behaviour of these mammals, she ruminates about an article written by Profesor G.Pilleri on the environmental adaptation of river dolphins, suddenly her eyes rest on the horizon 'with a peculiar, steel-grey glow' (367) which is a sign of storm. The storm comes with a lot of devastations and destructions. This signifies a need of thorough research of these aquatic animals which can save human race from the destruction of cyclones. But the most important aspect is the need of saving these aquatic animals which are on the verge of extinction due to several environmental factors. Politicians, by taking bribe, permit the traders to supply nylon nets. The nets when they are meant to catch chingeer meen that is the 'spawn of tiger prawns', they catch the eggs of all the other fish also and thus prevent the breeding of other fish. The result will be that "in the fifteen years the fish will all be gone, what with the new nets and all. And after that what?"(134) Nilima tried her best to ban such nets but in vain. Prawns are a large source of money for traders but poor people will have to suffer due to this policy. Though Moyna speaks these words to show the scarcity of fish in future which will destroy their livelihood, it also indicates the environmental imbalance caused by this act.

This past event takes a new turn in present. The things that left Kusum aghast, comes in a new way before Kanai. During his journey to search dolphins with Piya, a horrible event takes place. During one night when Kanai, Piya, Horen and his grandson are taking rest on their (Bhobhoti) Megha, a strange noise wakes Kanai and Piya up. Thought there was mist and the outlines of shore were obscure but at a distance, 'glowing points of orange flame could be seen moving quickly'(289). It appeared that something was going on in the nearby village. Though Horen and his grandson were thoroughly familiar with the region, they were unable to understand the cause of uproar in the village. Piya asks Kanai to ask Fokir who himself is unaware of the situation. Fokir decides to find out the cause of the situation and invites Kanai and Piya to accompany him. Horen also joins them. After a few minutes journey, they reach near the village. Fokir steers the boat. Then they all move forward. After a few steps, Horen sees the footsteps of a tiger. As the earth was neither dry nor wet the footprints were clearly visible. The tiger, after passing a long distance, must have passed alongside their boat also and then must have entered the village. The tiger was still in the village. Hundreds of villagers were standing in a group. They, with a mixed feeling of fear and rage, were demanding to kill the tiger. The villagers set the animal on fire as it had killed two men and several animals. For Horen, Fokir and Kanai, it is not a disturbing factor, but Piya has a breakdown after seeing the tiger being attacked and set on fire by the villagers. When Fokir asks her not to be upset,

Piya bursts out: "How can I not be upset? That's the most horrifying thing I've ever seen – a tiger set on fire"(295). Like the more extended account of the eviction of the Morichjhapi settlers, the tiger episode can be seen as a metonym for the larger ethical debates raised by the novel and it is also a litmus test for readers' responses. Later on, with his conversation with Piya, Kanai tells her that Fokir was not some kind of grassroots ecologist but a fisherman who kills animals for a living. Again, in their discussion on the prospect of killing a tiger, Kanai's view point is somewhat similar to Kusum: "Isn't that a horror too – that we can feel the suffering of an animal, but not of human being?" (300-1). Piya, who is down to earth ecologist, does not agree with Kanai. She does not like the action of killing rare species in their habitat. She says that we should neither interrupt nature's working nor indulge in believing anthropocentric ideology. If we kill all other species, it will be human beings next, specially people who are poor and unnoticed. Though the conversation ends unresolved, it raises an issue. Like the more extended account of the eviction of the Morichjhapi settlers, the tiger episode can be seen as a metonym for the larger ethical debates raised by the novel and it is also a litmus test for readers' responses.

How wildlife has been destroyed all over the world is an important issue. Piya, while relating her story of her love to orcaella brevirostris, tells Kanai that these animals are being killed either to make money or in the name of civilization. She describes the situation of Mekong area. She tells Kanai that the Orcaella population of the Mekong is declining rapidly and is expected to fall below sustainable levels. In 1970, the animal had to share Cambodia's misfortune as America ravaged this area by bombing. Besides, Khmer Rouge cadres massacred dolphins with the idea of using dolphin oil to supplement their dwindling supplies of petroleum. Not only that, to make the river Mekong navigable to China, some projects have been made which will definitely result in destruction of several species along with Orcaella. This fact has been hinted at but it is historically true that it has affected the environment seriously. To quote:

> The basin of the Mekong is an important agricultural area, with rice as the main crop. Without irrigation, rice cultivation is impossible during the long dry season. The United Nations (UN) started the Mekong River Development Project in 1957 to improve flood control, navigation and irrigation, and to develop hydroelectric power plants along the river. However, the project's progress was impeded by the Vietnam War (1959-1975) and political instability in Cambodia and other countries in the region. In the 1990s interest was renewed in developing the hydroelectric power potential of the Mekong River. Officials from four of the six nations that share the Mekong–Cambodia, Laos, Thailand, and Vietnam–met to make plans regarding this development; China and Myanmar did not attend the talks. By 1996, 54 dams were scheduled to be built on the Mekong. Manwan Dam in China's Yunnan Province was the first to be completed; it began producing power in 1993. Thailand completed Pak Mun Dam in 1994. Some of the nations involved

> have voiced concerns about the dam construction. The nations downriver worry that China's dam plans will interfere with the flow of the river, either flooding land downriver or changing the nature of the river so that hydroelectric projects downstream will lose some of their power potential. Additionally the two existing dams have had a negative effect on the environment, flooding certain areas and destroying fish habitats, which affects the fishing industry of native villagers (Ghosh, 2004: n.p.).

Man is a part of environment, yet he does not act to maintain this ecological balance. The misuse and overuse of natural resources by man has resulted in disturbing the balance. Animals are being killed indiscriminately and it has caused ecological imbalance. Now many species are on the verge of extinction.

Plant and animal species are dying out at an unprecedented rate. The leading cause of extinction is habitat destruction, particularly of the world's richest ecosystems. If the world's rain forests continue to be cut down at the current rate, they may completely disappear by the year 2030. In addition, if the world's population continues to grow at its present rate and puts even more pressure on these habitats, they might well be destroyed sooner. The tall and upright trees act as breakers of destructive power and velocity of winds and water. It is necessary that there should not be any cutting of mangroves and other trees in Sunderban region. In the process of this novel, we see how Piya and Fokir become the prey of cyclone. It was only a tree which could save Piya though Fokir could not save himself as the tree was not so strong.

Piya's scientific quest constitutes another pole of perception. Not that she is lacking a background: she has a family history, which she recalls in moments of reflection, a present involvement with Fokir and Kanai, and a future, which she calmly claims at the novel's close. But it seems to me that her function in the novel is to represent the life she studies in the mingled waters, salt and sweet, of the tidal pool at Garjontala; the always-threatened, always-precious material of scientific enquiry which is also, in the fragile and immensely overburdened ecosystem of the Sundarbans, nature itself. She is by no means the novel's only conduit for reflections on the unique environment that affords her material for study. Nirmal, thinking of the necessity that compels the refugees to clear the forest and till the land on Morichjhapi, as of the honey-collectors and woodcutters who go into the forest and are eaten by tigers and crocodiles, also asks a question central to the novel's concerns: whose is this land, nature's or man's? It is a question that cannot be answered, not even by the idealizing solution of co-existence. Piya is ultimately too practical and obsessive to bother with large answers; Nirmal is overcome by the very effort of posing the question.

Thus, Pia discovers interesting facts about the dolphin species that inhabits the Sunderban waters. She is aided by Fokir, a fisherman with deep intuitive knowledge of the surrounding ecology. Pia, Fokir and Kanai undertake a monumental field trip, one that ends with Fokir's death and transforms the lives of the other two. In the backdrop of folklore and recent political history, Ghosh carves out a tale that is vivid in the narrative details of characters and environs but restrained in the aspects of political opinion. His handling of facets from marine biology to regional myth is masterly. The crying need of the hour is to maintain balance between the moral responsibility of helping the poor and the conservation of wild life and vegetation of this region. Indeed the novel, in its aura and ambience is unique among the fictional corpus of Ghosh. Here, Ghosh has preferred to move beyond the limited confines of the 'poetics of memory' and to listen to "that other form of address, that intimacy which writing creates. The form of communion which one used to discover in novels" (Ghosh, 2003: 221).

References

Ghosh, Amitav. 2003. "An Interview with Amitav Ghosh.' *Amitav Ghosh: Critical Perspectives*. Ed. Brinda Bose, Delhi: Pencraft International.

_____. 2004. *The Hungry Tide.* New Delhi: Harper Collins.

Microsoft Encarta Encyclopedia 2004. © 1993-2003 Microsoft Corporation.

Chapter 25

"Hush," Said His Mama, "You're Gonna Be a Pig:"
A Pragmatist Philosophical Consideration of the
Pig in Literature, Song and Film,
With Special Emphasis on the Ones That Got Away

A Pragmatist Philosophical Consideration of the Pig in Literature, Song and Film, With Special Emphasis on the Ones That Got Away

☆ *Lisa Heldke*

Introduction

In life, people eat pigs. In art, they befriend them. There exists a collection of stories, songs and movies in which ordinarily-omnivorous humans lose their comfort with eating animals, and choose instead to cast their loyalties in with them–sometimes with the very same animals they had been prepared to eat just a few stanzas or chapters or scenes earlier. Fern befriends Wilbur, Farmer Hoggett puts Babe to work as his right-trotter man, and E.B. and his dachshund Fred sit at his unnamed sick pig's deathbed for days.

There's nothing very surprising about this discrepancy; after all, art doesn't have to imitate life. So if, in art, sometimes the pig gets to live happily ever after, it nevertheless remains true that, in life, it usually gets slaughtered and eaten.

But art also often teaches us something useful about life–which is why I'm interested in those pigs who lived to tell the tale. I want to know how they got away. To put it precisely, I want to know what transformation took place in the relationships between these pigs and the humans who would eat them. What happened to make those humans choose to befriend the very animals whose "meat potentiality" they'd earlier been so eager to actualize?

My interest in these mostly-fictional pigs arises from a real-life interest in the question of how we make decisions about what to eat. What can we learn about choosing to eat animals, by examining the transformations of these very particular human-animal relationships–transformations that turned food into friend?[1] It may appear that, by asking this question, what I am really doing is tacitly or implicitly or covertly arguing against eating animals. Appearances to the contrary, that is not my aim. In the interest of full disclosure, I will admit that I do not, for the most part, eat land animals. But my project in this paper is *not* to argue for some form of vegetarianism. Many theorists have already presented the cases for and against vegetarianism, and their arguments have already sparked enough dinner table arguments. I'd like to see if it might be possible to begin this discussion somewhere other than in the middle of the food fight scene from "Animal House."

The feuds over animal or meat eating arise, I believe, because discussions on the topic so often get mired in a slough of competing claims about whether animals possess various inherent capacities (intelligence, ability to feel pain) and, in turn, whether having those properties does or doesn't remove a being from the category of "food."[2] What I've noticed is that combatants on all sides of the debate tend to come to the dinner table sharing the beliefs that there *are* criteria–necessary and sufficient conditions, if you will–to which one can appeal in order to make this decision, and that all one must do to win the debate is to persuade opponents that their particular animal-capacity factoid is the relevant one that will rule animals in or out of the dinner category.[3]

I don't think the question of animals' status as food is the sort of question that could be settled–not even by a miraculous discovery of some crucial fact.[4] Instead, I think the question of how we ought to decide what we eat is itself in need of some examination. So, the question I ask is not "should we eat animals, if it can be proven that they possess quality A or B?" but "how should we go about making our choices about what sorts of things we want to eat? What sorts of considerations do we find relevant to such choices? What sorts of processes ought we undertake?"[5]

It bears noting, at this point, that I've just executed a standard John Dewey[6] move; namely, I have shown that adherents of two supposedly-opposing positions actually share some fundamental assumption, and then urged that this assumption be dug up and removed, or tunneled under, before anyone attempts to make any further progress in the matter at hand. And indeed, in the remainder of this essay, I propose to undertake a Deweyan excavation, by examining three cases in which people (albeit mostly fictional people) decide not to eat particular animals as a result of a change in their relationship to that animal.

Enter the pigs.[7] What can we learn from these pig-selves? How have their relations–to other animals, including humans; to objects–constituted them? How did the particularities of their various pig-selves leave them either vulnerable to, or immune to, being eaten? What can we learn from these relationships about how we might go about deciding for ourselves whether or not to eat particular animals, or particular kinds of animals, or animals-in-general?

Enter the theoretical scaffolding. In order to explore these questions, I'll employ three theoretical commitments, which I borrow from Dewey, and which I elucidate here briefly.

Pragmatist Underpinnings

Ontology: The Transactional Self

Beings–both animate and inanimate–are not independent, self-contained substances that secondarily have relations with other beings. Rather, they *are*–they come to *be* their particular selves–as a result of their interactions (transactions) with all sorts of other things in the world.[8] Being is relational–which means that the relations in which one engages have the power literally to constitute the self.

Dewey gets at this point in various ways. Here is an example about the human mind, which he defines as an "emergent property" of animal being, not some utterly unique, independent substance that only humans possess, and that is utterly separate from the body:

> That the use [in this case, advanced thought] reshapes the prior materials [in this example, the human nervous system] so as to adapt them more efficiently and freely to the uses to which they are put, is not a problem to be solved: it is an expression of the common fact that anything changes according to the interacting field it enters. (*Experience and Nature*, 217)

Consider this more general statement of the transactional nature of existence: "whatever else organic life is or is not, it is a process of activity that involves an environment. It is a transaction extending beyond the spatial limits of the organism" (*Logic*, 32). Elsewhere Dewey observes that even our capacities may be understood in transactional terms: "potentialities are not fixed and intrinsic, but are a matter of an indefinite range of interactions in which an individual may engage" ("Time and Individuality," 110).

We come to be the particular beings that we are *in* and *through* our interactions in environment. Knowing is one such interaction–a peculiarly important one, from humans' point of view. Dewey actually argues that knowing changes the nature of both the knower and the known–even if "the known" is a star hundreds of light years away.[9]) Beings are not infinitely malleable, but neither is our nature fixed and predetermined. The pigs my neighbor is raising in a field are quite literally different *kinds* of beings than are the pigs being raised at the confinement system further down the road. They become different, Dewey would say, because they have markedly different interactions with their environment–particularly with the humans in their environment. Similarly, the human beings in question also become different from each other, in part because of the differences in their interactions with those pigs.

Ethics: Ontology Begets Moral Theory

This ontological fact about the transactional nature of our being leads also to an ethical/social feature about how we are in the world–how we are warranted in relating to others. For Dewey, "is" *does* lead to "ought."[10] Recognizing that interactions shape beings on both ends requires us to make ethical decisions about the character those interactions should assume. If I'm right that, like the hogs in their care, my free-range neighbor and the confinement system farmer are being rendered different sorts of people as a result of the kinds of interactions they have with those hogs, then it behooves them–and us–to ask whether we respect, admire, value, or can at least can *tolerate* the kinds of people each is becoming. (This is my answer to the flippant question, "well, if we can transform the nature of animals through our relationships with them, then let's just make them all unambiguously into meat!" We can perhaps do so–but our doing so will also transform *us* in the bargain. And it may well do so in ways that we are not so eager to embrace.) Using Dewey's preferred term, we might ask, "do both sorts of farming relationships allow for those farmers to *grow?*"–where "growth" is a process with ethical dimensions.[11] We can ask this same question of ourselves as eaters, as literal consumers: do we value the kinds of persons we become as a result of the interactions we have with our food and its producers?[12] Dewey puts it this way: "We have to make up our minds, when we want two conflicting things, which of them we really want. That is choice. We prefer spontaneously, we choose deliberately, knowingly" (*Ethics*, 286). Further,

> Now every such choice sustains a double relation to the self. It reveals the existing self and it forms the future self. That which is chosen is that which is found congenial to the desires and habits of the self as it already exists. ... Not all [choices] are as momentous as the choice of a calling in life, or of a life-partner. But every choice is at the forking of the roads, and the path chosen shuts off certain opportunities and opens others. In committing oneself to a particular course, a person gives a lasting set to his own being. Consequently, it is proper to say that in choosing this object rather than that, one is in reality choosing what kind of person or self one is going to be. Superficially, the deliberation which terminates in choice is concerned with weighing the values of particular ends. Below the surface, it is a process of discovering what sort of being a person most wants to become. (287-8)

I continually *make myself* through the food choices I make. Each choice–that is, *each relationship chosen or refused*–may not, in and of itself, be momentous. But altogether, they go a considerable distance toward making me–or failing to make me–the "sort of being [I] most want to become."

Ethics: The Experimental Nature of Moral Choices

Dewey argues that ethics is an experimental science. Even the aims of our ethical system are not fixed, absolute ends; rather, they develop through our experimental investigation, and they remain permanently open to reevaluation and transformation, using new insights, relationships and experiences. Here is how Dewey states it:

> There are still those who think they are in possession of codes and principles which settle finally and automatically the right and wrong of, say . . . , the extent to which legislation should go in deciding what individuals shall eat, drink, wear, etc. But there are also many other persons, an increasing number, who see that such questions as these cannot be settled by deduction from fixed premises, and that the attempt to decide them in that fashion is the road to the intolerant fanaticism, dogmatism, class strife, of the closed mind. (*Ethics*, 317)

For Dewey, the nature of good is genuinely a consequent, not an antecendent fact; it quite literally emerges, comes into existence, through our investigation. It is not a given, waiting there for our discovery.

The Upshot

These three presuppositions lead me to assert that, in a non-metaphorical and nontrivial sense, we humans *make* animals into food or "not-food" through our interactions with them.[13] There isn't a fact of the matter or a set of criteria to which we can appeal to settle this question once and for all; it is not the case that, say, steers simply *are* food and that dolphins are *not*. A line does divide "food" from "not food," but the location of that line moves and develops gaps in it–sometimes during the life of a particular animal, sometimes over the course of a human individual's life, and sometimes over the course of the life of a community or of a society. It moves as a result of the kinds of interactions we have with animals, and the kinds of values, aims and desires that develop for us as a result of those interactions. It moves (or stays the same) because of the kinds of beings we become–and its movement also makes us into those persons.

What sorts of interactions matter? Given Dewey's experimentalist understanding of ethics, it follows that the answer to this question will necessarily vary across time and culture–even across the life of any given individual. In the context of current animal-rights-based conversations about meat eating, any interactions that give the animal a chance to be sociable with us, to show how much like us they are (*e.g.* with respect to intelligence or capacity for pain), or what good company they make, are the ones that seem to matter the most. On the other hand, interactions that, *e.g.*, show them mostly to be made of the same chemical elements as humans are do *not* seem to have much relevance.

We cannot make a list of all possible features of animal-human interactions that might be relevant in making decisions about whether or not they are food for us–cannot, because the list must in principle remain open to addition and deletion. However, in searching for general principles, we might appeal, as Dewey does, to the notion of "growth"; if we wish our relationships to animals to be ethical, we might always ask ourselves, "do these particular relationships promote our growth as humans?"[14] For we must remember that, just as pigs come to be the particular animals they are as a result of our relationships with them, so too do we become the particular animals *we* are as a result of their relationships to us. *Pigs* make *us* also.

Babe

Babe, the movie, could have been marketed as a pragmatist fairy tale about the transactional nature of selves–although admittedly it wouldn't have been such a box office success if it had. In this film, the received wisdom that beings are the way they are and cannot be changed (delivered, variously, by a chorus of mice, a cat, two sheepdogs, a cow, and a flock of sheep) is constantly challenged and undermined by beings who don't behave at all the way they should. Indeed, one gets the very distinct sense that the "wisdom" is delivered as often and as heavy-handedly as it is, precisely because every one of the animals in the barnyard knows, or is in the process of realizing, that their natures are anything but fixed. It starts with the duck who is performing the job of the rooster–and is beating the rooster out of bed every morning in order to do it. But soon, it spreads to the pig who is performing the job of a sheepdog–much to the vexation of Rex, one of two actual sheepdogs on the farm.

Babe the pig's foray into sheepherding sets off a string of disruptions of the way things are: Rex, infuriated at the pig's invasion of his territory, becomes vicious and must be sedated; for a while, his very identity as a sheepdog is imperiled when he is not allowed near the sheep. The sheep, involved in a new relationship with their "kinder, gentler" pig herder, become different kinds of sheep, confidently demanding that the sheepdogs treat them differently. "The way things are" changes in myriad ways as a result of the relationships the little pig forges with the animals (human and nonhuman) on Hoggett's farm.

At the center of these relationships stands the one between Farmer Hoggett–Boss, as he is known to the animals–and Babe. That relationship begins auspiciously: when Boss tries his luck at a "guess the weight of the pig contest," man and pig look into each other's eyes in a moment of recognition. The narrator observes, "The pig and the farmer regarded each other, and for a fleeting moment, something passed between them, a faint sense of some common destiny" (*Babe*).

Auspicious beginning notwithstanding, when the pig arrives on the farm, the relationship between him and the Hoggetts can be summed up in one word: "dinner." Babe is to be fattened up for Christmas. His "meat potentiality" shapes his encounters not only with Boss, but with every non-human animals on the farm as well–although out of kindness, most of the animals don't come right out and say so. As a result, Babe remains quite unaware of his fate, and as a result of *that*, he experiences considerable confusion and anxiety over just what his role on the farm *is* supposed to be. (We can understand his decision to try his hand at herding as a combination of anxiety over his apparent lack of a function on the farm; his admiration for Fly, Rex's mate and the sheepdog who mothers him; and his genuine affection for the sheep with whom he strikes up a relationship.)

Ultimately, however, Babe not only is saved from butchering, but comes to be an integral member of the working staff of the farm; his relationships with the sheep, the sheepdogs, the farmer and, yes, even the cat, transform him into a very different pig indeed than his confinement raised mother and siblings grew to be.

Of course the way Babe achieves salvation (prove to humans that you're smart, charming and useful) also makes it possible to view *Babe* as just another instance of the familiar animal-rights claim that at least some animals share with humans a capacity for intelligence, a capacity which entitles them to certain rights, including the right not to be eaten. But I think we can nest this rather parochial and human-centric argument inside a more inclusive pragmatist analysis; an analysis that understands the transformations of Babe and Boss in terms of the transformation of their relationships with each other and with the other animals on the farm. This latter approach can *account* for, but is not reducible to, the claim that Babe deserves to live because he is a rational agent and a useful working member of the farm; in a human world in which hard work on humans' behalf, and a personality pleasant to humans are valuable commodities, the relationships between pig and human have enabled the pig to develop in ways that make his meat potentiality less salient than his herding potentiality. But these and any other qualities we might choose are contingently valuable; they are not in principle the qualities that make animals inappropriate food objects. The pragmatist account allows us to see this; popularized animal rights arguments tend not to be able to.[15]

I have a suspicion that the pragmatist account also more successfully explains why the fact that we feel loyalty to *this* pig does not translate into anything like affinity for *all* pigs, and a refusal to eat any of them. Babe does not stand for all pigs; he is a pig in a very particular relationship to Hoggett and the rest of the animals on the farm. Refusing to eat him does not lead, ineluctably, to a refusal to eat bacon. While one relationship with a pig might affect a person thus, there is nothing in principle irrational about its *not* doing so–especially if it is only one of a variety of kinds of relations that one has with pigs. (Yes, eating bacon would count as a kind of relation with a pig–the kind of relation that many people would find devilishly difficult to abandon.[16]) At least this is the way that I am able to understand the fact that countless parents fearlessly let their small, suggestible children watch this movie, apparently with no worry about the way it would affect their children's meat eating. For children, on the other hand, with fewer pig-relationships to their credit, the movie might *profoundly* affect their willingness to eat meat.

"Death of a Pig"

In 1947, about a year before he began the famous children's novel *Charlotte's Web*, E.B. White wrote an essay about a pig he'd been raising to slaughter–a pig that took sick and died, thereby "avoiding" his fate.[17] Whereas *Babe* tells the fictionalized story of a pig who lives, "Death of a Pig" tells the true story of a pig who dies when he wasn't supposed to. ("This is the story of a pig who died."–*Love Story: The Squeal*). White's pig is felled by disease, not, as White planned, by a bullet between the eyes.

What interest me most about this story are two things: first, that this is a true account of a relationship between human and animal in which the human's loyalties shift very dramatically. White goes from planning the pig's demise to wishing very fervently for the pig's survival–and not simply so that he can slaughter it later. (I don't think it is naïve romanticism that leads me to assert that, had White's pig survived his illness, he would not have become bacon and ham at a later point. I think he would have become "some pig.")

Second, White's story interests me because here is a case in which the transformation in the relationship does not come about because the animal reveals itself to be agreeable, useful or similar to humans. The story does not turn on the pig's becoming more human-like, or even more attractively pig-like to us; indeed, the process of giving the pig an enema no doubt made him rather less attractive to White (though perhaps also more vulnerable, itself a quality that might endear an animal to a human). So, the story is interesting because, at the outset, it seems to rule out arguments that rely on these more familiar sorts of moves on animals' behalf.

At the beginning of his life, White's pig is destined to be food; he is a participant in "a tragedy enacted on most farms," in which "the murder. . .is quick and skillful, and the smoked bacon and ham provide a ceremonial ending whose fitness is seldom questioned" (17). Tragedy, he reports, becomes "slapstick comedy" when the pig takes sick and White finds himself playing the "farcical" role of "pig's friend and physician. . .with an enema bag for a prop" (17).

In the end, however, the story he tells is not at all a slapstick comedy; even the presence of Fred, the hilarious dachshund, cannot make it so. The relationship between White and his pig has changed swiftly, decisively–and permanently–as a result of its illness. When the pig dies, "the loss we felt was not the loss of ham but the loss of a pig. He had evidently become precious to me, not that he represented a distant nourishment in a hungry time, but that he had suffered in a suffering world" (18).

The path to the grave is three, possibly four days long; White cannot recall for certain how long it took the pig to die, so considerable is his anguish over its death. White, and his "ghoulish" dachshund spend a great deal of that time at the pig's side, trying to get it to eat, force-feeding it castor oil, giving it an enema, and, when the end seemed inevitable, trying to comfort it in whatever ways he can–with buckets of cool water, mostly. White reports that he went to bed and cried internally when the pig finally died–something he most assuredly would not have done had the pig died in the slaughter yard.

Three days of castor oil and enemas and pain and suffering transformed White's ham-in-waiting into a fellow traveler, a being whose life he cherished, and whose death he mourned.

White did not swear off pork out of loyalty to his dead pig; no doubt he raised another hog the next year. As I noted with regard to *Babe*, a single relationship with a particular does not, or need not, transform the shape and character of all the rest of one's relations with similar kinds of beings. That it might do so is a fact well explained by a pragmatist, transactionalist account.

How *do* we go from individual animals to a species? I submit that we make the move, often, because we believe there is the potential for a particular kind of interaction. We as humans tend to value or privilege certain kinds of interactions, such as communication. So, we may tend to be prejudiced *toward* a species with which we believe we can have such interactions–such as dolphins–and to be neutral or unfavorably disposed toward a species with which the interactions are unsatisfying–*e.g.* lobsters or wolves (though obviously this unfavorable impression doesn't always turn them into food!). But as I've been suggesting, we might have–and value–multiple kinds of relations with animals, including the "bacon relation." In such cases, it may not be at all as simple as saying, "look; they can communicate with us the way dolphins or dogs can. Let's remove them from the category of food." Pigs are an obvious case.

"Tails and Trotters"

"Tails and Trotters" is a song about a pig, a pig not unlike Babe or Wilbur, except that he learns his fate early in life. In this pig's case, the instrument of his disillusionment is his very own mother; in response to his question, "what'll I be when I get big?" she replies, "hush, you're gonna be a pig," and then proceeds to explain to him that "all my sons and all my daughters are hocks and hams and tails and trotters." No "motherly" protecting of him from the truth of his fate–a fate that she has been spared only because she has been turned into a piglet factory. This mother goes for brutal honesty. Why? We don't know from the song alone–it is not her story.[18] We can speculate, however. Perhaps it is because she is unsentimental. Or perhaps she wishes to steel her son to his fate. But perhaps–just perhaps–she secretly hopes that her brutality will lead him to attempt to do just what he does: escape.

In a porcine version of teenage rebellion ("I'll show you all!"), this little piggy runs to town, where his natural abilities (he can walk, talk, sing and dance), aided by some fancy clothes and jewels, enable him to assume the role of a corporate fat cat. Soon, he's "executive pork" and "nobody [*i.e.* no humans] noticed anything." (One wonders whether his mother and siblings would notice; would they recognize him in his diamond ring and tie? Or would he fool even them with his humanness?)

While it is certainly risky to build any weighty conceptual scaffolding on the small foundation of this song, I think it is interesting to draw some pragmatist transactionalist conclusions from it. Our piggy hero takes affairs into his own trotters, turning himself into a human-like creature. Stated pragmatically, his relations with humans enable him to remake himself in their (our) image so successfully that we simply assume he is one of us. All he has learned by watching us, he puts to work in interacting with us.

Part of the (decidedly black) humor of the song arises from the fact that he is mistaken for a human *by humans,* which says our capacity for recognizing the purportedly clear us-them boundary isn't very acute after all.[19] This pig isn't just showing humans that he is *like* us; he's showing that he *is* us in every relevant way. Indeed, part of the point of the tale surely is that humans are "pigs"–*i.e.* greedy and selfish. It's easy for him to act like us because we act like them. As Dewey would observe, both (actual) pig and his (presumably human) colleagues come to be more and more of a piece as a result of their mutually supportive interactions with each other–in the end, they're *all* executive pork.

Listening to this song, I can't help myself from reading into it far more than is there. In my image, he's the runt of the litter–the pig equivalent of the skinny smart kid with big glasses who's susceptible to getting beaten up on the playground for his perfect spelling scores. To survive, he assumes the character of his would-be tormenters, ending up living quite high on the hog himself. In my version, he's working in the meatpacking industry; I can almost imagine him making decisions about which of his pig siblings to butcher.

"Tails and Trotters" is a curious spin on the familiar animal rights suggestion that animals are very much like us. Whereas most versions of this suggestion involve humans recognizing animals' similarity, and according them certain treatment as a result of it, this tale involves the animal realizing how very easy it would be to become like a human–and he proceeds to develop the relationships that enable him to do so. [concluding sentence]

Conclusion

Perhaps it seems, after all, that this essay *is* an argument for vegetarianism. When I speak as my vegetarian self, I am moved to ask, why does casting our lot with these fictional pigs translate into so little change in our actions–unless we are four-year-olds, that is? Why do the relationships with animal that get turned into stories and songs remain in the category of "unusual case?" Why doesn't our affinity for these particular pigs translate into any changes in our relations with real animals? Why are we so fond of Wilber and Babe and Christopher Hogwood (the star of the latest pig tale about an animal who befriended some humans, thereby securing for himself the opportunity to die of old age)–but *also* so fond of bacon?[20] After all, we all know Wilber really wasn't "some pig"–he was just a lucky pig with a good press agent who could embroider. So why don't we realize that Wilbur and Babe and Christopher are, in fact, Everypig, and decide that we don't want to consume beings with whom such rich and rewarding relationships are possible?

These are tired old questions, but I think the pragmatists have a better answer to them than do popularized arguments for animal rights. Understood relationally, we can see that, in reading a book or watching a movie or listening to a song, very little about our context has really changed; surely not enough to make it plausible for us to change our way of eating. One interesting, cute, charming, but ultimately fictional pig hardly plays a role in one's life large enough to forsake a lifetime of BLTs on toasted white bread because of it–not if that is the only thing that changes.

But my vegetarian self isn't my only self–and my other selves have questions too. Many features of our contemporary industrial agricultural system give me reason to question the legitimacy of strict adherence to vegetarianism–and have made me quite impatient with those who believe that attending all and only to the matter of non-human animal suffering is an adequate foundation upon which to build one's eating practices. Enough humans suffer to grow, harvest and package my vegetable diet to make it clear that vegetarianism is not the path to blameworthy eating either. On the other hand, down the road from me, a guy raised twenty or thirty pigs this summer, to sell for meat. He's one of a growing number of farmers who are turning to farming methods that are much shorter on efficiency and quantity, and much longer on animal freedom and pleasure. These pigs spent the bulk of each day rooting around in a big open field, and they'd gallop alongside of me every time I went by on my bicycle or in a car–racing me, the way a dog or a little kid will. As pig lives go, I have to believe theirs were pretty deluxe. Every once in a while, I find myself thinking, "well, maybe I could eat one of *those* pigs."

For Dewey, our ultimate aim as ethical beings is to promote human growth and flourishing. I believe that, in any given situation, it is an open question what sorts of relationships with animals will best accomplish that aim. I believe, furthermore, that Dewey's transactional model provides us with an excellent framework within which to make our decisions.

Does my exploration really resolve anything. Or am I just pushing the discussion back a step? I probably am simply moving the debate–but I think the move matters. Dewey-the-experimentalist would observe here that we ought always to evaluate and reevaluate the nature of our transactions. In doing so, we may realize that characteristics such as tool use, the ability to communicate and the capacity for pain might well be criteria we want to use in making our decisions about whether to eat certain animals. We might *also* realize, however, that what count as tool use, language and pain might constantly change, as a result of new interactions–*and* that these three criteria are not the only relevant ones. Dewey's experimentalism demands that we always re-ask, "how *do* we decide? How *should* we?" These must always remain open questions.

Notes

1 There also exists a category of stories in which friends turn into food–accounts of humans eating other humans. Jonathan Swift offers us a satirical example of this genre, in his essay "A Modest Proposal," while *Alive,* by Piers Paul Reed offers a real-life account. A less well-known piece, "They're Made out of Meat," by Terry Bisson, presents humans through the eyes of a reporter from another planet; this visitor repeatedly observes to their superior that the inhabitants of our planet are, well, made out of meat. We even *think* with meat, the reporter observes. It would be illuminating to compare stories in this category to stories of creatures that make the reverse trip. Alas, I will not take up that project in this brief essay.

2 Of course many debates about vegetarianism do not arise over the question of animal intelligence, but turn on such issues as natural resource use. My discussion may not have much to say about these debates.

3 Peg O'Connor suggests that we understand those debates as struggles to make objects into subjects–or to keep objects objects. If it's food, it's an object–so if I can prove that some animals are subjects, they can no longer be considered food.

4 This is an instance of a more general claim about the impossibility of making moral decisions on the basis of isolated criteria. The belief that we will come upon such a fact strikes me as similar to the belief that there is some fact about the beginning of human life that we can discover, that will settle all debate about abortion.

5 Peg O'Connor suggests that this makes my project a kind of meta-food project. I embrace that suggestion. Maybe if we're seen as doing something akin to metaphysics, philosophers of food will start to gain a little credibility in our own country.

6 John Dewey was one of the original "big three" pragmatist philosophers, along with William James and Charles Peirce. He is perhaps best well known outside philosophy for his work on progressive education. In the past twenty years or so, interest in his philosophy–and in pragmatism more generally–has been rekindled, thanks in part to Richard Rorty, whose praise of Dewey got people reading him again.

7 Why pigs? The short answer is "because they are the animals that appeared in the examples with which I was most familiar." But no doubt there *is* something–or, more likely, a whole lot of somethings–about pigs that makes them particularly suitable for sparing. Of course pigs also "avoid" being eaten by being defined as unclean or otherwise unfit for human consumption by many persons of faith. Are these two issues connected? Perhaps–but I will not explore the matter here.

8 Does this way of thinking also have the potential to serve as a platform for a useful, consistent, and non-judgmental way to think about pets, namely, that they and we are constituted in relation, and that these relations themselves make us different kinds of animals than we would be in their absence? That my former "dogless" self is quite different from the self I'm becoming as I begin living my life with Tilde the rescued dog–just as Tilde's self is different, depending upon what sorts of human relations are in her life? It is rather a different way to view them than, say, John Berger's view, described in "Why Look at Animals?"

9 See his "Naïve Realism versus Presentative Realism."

10 In the school of twentieth century philosophy known as analytic philosophy, the so-called Naturalistic Fallacy holds that one cannot derive an "ought" statement–a moral imperative, for instance–from an "is" statement–a claim about fact. From the fact that strangulation leads to death in humans, no moral claim can be derived automatically. Notably, not only does Dewey believe that is leads to ought, but in a way, his transactionalist ontology means that, in a way, ought can lead to is. That is, our beliefs about how things should be will lead us to interact with the world in certain ways–and those ways will, in turn, have constitutive effects on both ourselves and the world.

11 For a discussion of the ethical–and educative–nature of growth, see, *e.g.*, *Democracy and Education.*

12 On the matter of producers, it is clear that what I am saying for the relation between farmer and hog *also* holds for the relation between me and the farmer. I am constituted in part through my relations with those who produce my food–both those who are visible to me and those I choose to keep out of view.

13 There is of course a trivial and obvious way in which this is true. Humans must do something to virtually every item we eat, in order to make it edible. The philosophical point is related to this, but goes beneath it. It's not a matter of

cooking–or killing, skinning, gutting, and butchering. Rather, it's a matter of categorizing.

14 Obviously they can promote our physical growth. While this isn't the sense of the term on which I mean to focus, I do think it's worth noting that these other, more intellectual and emotional forms of human growth are impossible if we are unable to thrive physically. Thus the fact that meat can be an excellent food for human physical development is a fact that cannot be ignored.

15 By "popularized animal rights arguments," I mean not the arguments of scholars, but rather the sorts of arguments that might be offered at the dinner table. Whether pragmatism is also more successful than scholarly animal rights arguments is a question I do not consider here.

16 While I have decided to forsake the bacon relation myself, I understand its power because of my own chocolate relations. Despite my firm resolve to eliminate chocolate produced by companies that do not renounce the use of child labor in their processing, and despite my having seen articles, photos, and films about particular children involved in chocolate slavery, my long and deep relations with particular chocolate products regularly proves more compelling to me in the moment than the reality of their anguish. Understanding this failure as a consequence of my relations does not excuse me from my choices–but it does point me toward a solution, namely to cultivate more three-dimensional kinds of relations with the children who produce chocolate.

17 White is of course far more well known for Wilber, the runt-turned-celebrity in *Charlotte's Web*. "Death of a Pig" was written about a year before he began writing perhaps his most famous work, but White allegedly said that the essay–and the incident that it describes–were not in his mind when he began writing it. One commentator notes, "that's awfully hard to accept.... *Charlotte's Web* gave White the chance to write a different ending" (Ott).

18 The only detail we get about her is the fact that the song she sings about the fate of her children is "mournful."

19 Gullible–or downright stupid–humans also feature prominently in *Charlotte's Web*, and in a popular recent chick flick, *Chicken Run*.

20 See Montgomery.

References

Babe. Dir Chris Noonan. Perfs. James Cromwell, Magda Szubanski, Christine Cavanaugh. Film. Universal Pictures 1995. Based on the novel *Babe: The Galleant Pig*, by Dick King-Smith.

Berger, John. 1980. "Why Look at Animals?" *About Looking*, New York: Pantheon, 1-28.

Bisson, Terry. "They're Made Out of Meat." Online at http://www.physicsforums.com/archive/topic/t-10987_They're_made_out_of_meat..html

Dewey, John. 1980. *Democracy and Education: An Introduction to Philosophy of Education.* In *The Middle Works of John Dewey, 1899-1924,* Vol.9. Ed. Jo Ann Boydston. Carbondale, Ill.: Southern Illinois University.

_____. 1981. *Experience and Nature.* In *John Dewey: The Later Works 1925-1953,* Vol 1. Ed. Jo Ann Boydston. Carbondale, Ill.: Southern Illinois UP.

_____. 1985. *Ethics.* In *John Dewey: The Later Works 1925-1953,* Vol.7. Ed. Jo Ann Boydston. Carbondale, Ill.: Southern Illinois University.

_____. 1985. "Naïve Realism versus Presentative Realism." In *John Dewey: The Middle Works.* Vol.7. Ed Jo Ann Boydston. Carbondale, Ill.: Southern Illinois University.

_____. 1986.. *Logic: the Theory of Inquiry.* In *John Dewey: The Later Works 1925-1953,* Vol. 12. Ed. Jo Ann Boydston. Carbondale, Ill.: Southern Illinois University, 32.

_____. 1988. "Time and Individuality." In *John Dewey: The Later Works 1925-1953.* Ed. Jo Ann Boydston. Carbondale, Ill.: Southern Illinois University, 1988. 110.

Goodenough, J.B. Tails and trotters 1990. [Recorded by G. Bok, A. Mayo, & E. Trickett]. On *And so will we yet* [CD]. Sharon, Connecticut: Folk-Legacy Records.

Montgomery, Sy. 2006. *The Good, Good Pig: The Extraordinary Life of Christopher Hogwood.* New York: Ballantine.

Ott, Bill. 2004. "Charlotte's Web." *Storylines New England Discussion Guide* No. 11, storylines.ala.org. Accessed 2/23/07.

Read, Piers Paul. 2005. *Alive.* New York: Harper Perennial.

Swift, Jonathan, 1729. "A Modest Proposal." Online at http://www.uoregon.edu/~rbear/modest.html.

White, E.B. 1977. "Death of a Pig." In *Essays of E.B. White.* New York: Harper and Row.

Appendix

Tails and Trotters

Judy B. Goodenough

Little piggy hollered in the middle of the night,
"Tell me now, Mama, I wanna get it right.
What'll I be when I get big?"
"Hush," said his mama, "You're gonna be a pig.

CHORUS
"That's how it is when you get older,
"You're bacon, butt, and picnic shoulder.
"All my sons and all my daughters
"Are hocks and hams and tails and trotters."

"Oh, no," said the piggy, "That's mighty hard,
"There's more to me than loin and lard.
"I can walk and talk, I'm young and strong."
"Hush," said his mama, "Not for long."

"Oh, no," said the piggy, and he started to howl,
"There's more to me than cheek and jowl.
"I'm pink and pretty, I can sing and dance."
"Hush," said his mama, "You'll never get a chance."

"Oh, no," said the piggy, "I'll show you all."
He went under the fence and over the wall.
He ran and he ran till the moon went down;
He ran and he ran till he came to a town.

With a kink in his tail and a wink in his eye,
He put on a hat and he put on a tie;
He parted his hair, bought a diamond ring,
And nobody noticed anything.

"Oh, see," said the piggy, "I'm one of you."
And everyone said, "How true, how true."
He paid his money and he lived in style;
Sometimes he sang with half a smile:

He's a big boar now, he's executive pork,
And he eats his vittles with a knife and fork;
He often thinks of the lonesome tune
His mama sang by the light of the moon:

On the Bok/Muir/Tricket album "And So Will We Yet"

Chapter 26

Elephant Images and Representations: Narratives and the Production of Cultural Signifiers in Kerala

☆ *Syam Sudhakar*

The role of elephants in the cultural sphere of Kerala, oft-discussed and perennially fascinating subject remains so, not only for Malayalees but for all those interested in animal issues. The recent focus on various 'news reports' about elephants in the Kerala media can be considered as not just reports, but may be viewed in the context of cultural tools to satisfy (or to disturb) the needs of the public. In other words, members of the public are concerned about the condition of their state animal and it is almost as if their inner psyche demands 'news' about the animal. This paper first discusses the socio-cultural status of the elephant in Kerala and how the media portrays and presents various stories of the animal to the public; then explores how the elephant is represented in the films and literary texts of Kerala; and finally it focuses on the elephant stories of *Aithihyamala*, comprising the lore and legends of Kerala. The paper also showcases the manner in which these representations reinforce an established iconoclastic and majestic image of the elephant in the minds of the people and how these images confront contemporary values and issues related to animal rights and raise questions of the problemitisation of ethics.

According to the BBC report, "India's Overworked Elephants", Kerala has the largest domesticated elephant population with nearly 700 captive elephants. Being hailed as the state animal, and featured on the official emblem of the State Government of Kerala, the elephant can be identified as an integral part of the culture and tradition of the land. There are more than 250 individual elephant owners all over the state. Ownership of an elephant has always been not just a business but a symbol of the traditional pride of the family. It is noteworthy that the Guruvayur temple in Thrissur district alone owns 66 captive elephants and more than 200 mahouts have been appointed to look after them.

Apart from a number of temple elephant camps, there are other camps that are run by the Forest Department: Konnik, Kodanad, Nilambur, Thekkady, Muthanga being some of them. The Government is taking steps to minimize the violation of animal rights, especially in the case of elephants. The implanting of a ten-digit microchip ID card behind the left ear is one of the recent measures taken by the Government. The details of the elephant will be recorded in that chip so that the owner and the mahout would be accountable for any physical problems the animal faces.

The welfare of the elephant, a revered animal, has long been a concern to the State. The eleventh chapter of *Matangaleela*, the traditional book on Indian Elephants by Thirumangalathu Neelakantan Moosath, written in about the 16th or 17th century, discusses elephant care " how the animal is to be treated and tamed gently. The well-being of the animal should be the primary concern at the time of training. According to *Matangaleela*, each animal is different and needs a different kind of attention, and training should be conducted by adhering to animal ethics. It is only by showing love and respect to the animal that the mahout can become the 'leader' of the animal. Firmness in approaching the animal should never turn into violence.

The text, which is written in 153 verses in 12 chapters, talks in detail about the good and bad signs (*Shubha - Ashubha lakshanam*), stamina, strength, measurements (*Ayur, Vayo, Maana lakshanam*) grading process (*Moolya nirnayam*), capturing and protecting (*graham, rakshadhi vicharam*) etc. Based on behavioural features, the elephant is again classified into *deva, asura, gandharva, yaksha, raksha, pishacha, sarpa and manusya*. Like the hierarchal (caste based) classification of *Manusmrithi*, *Matangaleela* also classifies elephants into *Vipra Sathvam, Kshatriya Sathvam, Vaisya Sathvam* and *Soodra Sathvam*. The final classification is a valid source to argue that elephants were considered as important as human beings in ancient times. The last chapter of the text is dedicated to the discussion of the importance of having a skilled mahout for the well-being of the animal and it deals with the techniques that could be used to make the elephant obey its 'leader'.

The existence of three translations of *Matangaleela* into Malayalam (from Sanskrit), explains the demand for such an excellent book on elephants in the cultural context of Kerala. No other animal in Kerala commands similar love and respect of the masses. It would not be an exaggeration to state that the elephant is almost a religion in the cultural sphere of Kerala. It is an inevitable part of every gathering and ceremony. No wonder the media takes up even minor issues regarding elephants and problemitises them in public spheres. Poomulli Narayanan Nambootiri, a well-known scholar in *Matangaleela* and an elephant owner/lover, in a personal interview says:

> The attitude of a Malayalee towards the elephant should be problemitised. For instance, people like different aspects of a field- some like its greenery, some look at the shape, sound fascinates some others, different aspects attract different people. Earlier, people were fond of the aesthetic quality of an elephant. We choose our elephant looking at its signs. The culture of having an elephant was in the very blood of Malayalee, just as the bull is to the Spaniard. But recently, I am afraid; this aesthetical sense gave way to mere money and pride.

Elephant and the Popular Culture of Kerala

In the West, there is a saying which has almost become a cliché now, 'the elephant in the room' meaning people are avoiding speaking out loud about something which is obvious to all but not polite or politically expedient to talk about. In my research on elephants in the Malayalam language, I have come across a number of sayings like this. "*Kattile thadi, thevarude aana*" (the wood of the forest and the elephant of the god) which satirically implies for the material benefit, we are exploiting the property of others; "*Aanavayil ambazhanga*" (*ambazhanga*, a small fruit in the mouth of the elephant) means the need is large and the amount we have is small; "*Aanakkaryatinidayilano chenakkaryam*?"(Why talk about elephant yam while talking about elephants? Which means why talking about small things in big discussions?); "*Aana melinjal thozhuthil kettilla*" (Though the elephant has become lean, we cannot tie it in the cattle shed) means though the person has lost his pride and grace now, we cannot categorize him along with the folks. These are a few adages from Kerala about elephants, which are popular and used in the day-to-day conversations. It is to be noted that in all these sayings, the elephant stands for positive qualities like grace, pride etc. The image of the elephant arises in each adage since the animal represents all these profound qualities in popular minds. It almost seems as if it has been actually happening with the elephant – that it is so large and non-understandable that people use their emotional/spiritual understanding, but do not bring this being into the intellectual realm of speech or discussion other than through these popular sayings.

The Indian Express on 1st Sept. 2010, reported the recent recommendation of considering the elephant as the national heritage animal with a financial allocation of 600 crore INR for the prevention and protection of elephants. This shows the interest of the country in elephants. The elephant population in the country was reported to be nearly 27,000, out of which more than 3,500 were in captivity. Several organisations like ELA have been formed to protect and promote the elephant in the state. Measures have been taken to reduce the stress of the animal in public/religious processions, work places like timber mills and during transportation. In this context, to reveal the popular attitude towards the animal, it is worth mentioning some interesting news about elephants reported by different televisions channels, websites and information collected from a few personal interviews from Kerala.

When one opens the window to the continuing cultural significance of the elephant, one cannot ignore Puthenkulam village in Bhoothakkulam Panchayat (Kollam Dist). The place consists of 40 captive elephants. The village has become a tourist spot because of its huge unusual number of elephants. Kottur in Thiruvananathapuram district has recently become famous because of an elephant orphanage. The orphanage is acting as a refuge for the diseased and injured elephants. There are a number of retirement centers and elephant hospitals in Kerala like the one run in North Paravur, Ernakulam district. *Elephant-Kerala* records that the news about an Elephant patrol against poaching in Wayanad, by Kerala Forest Department in association with the Forest Department of Karnataka and Tamilnadu, was also well-received by the public. Transporting forest officials through dense forest and the protection of wild animals were the major concerns.

A number of incidents were reported about the hostility between elephants and farmers near the forests. Recently, *The Hindu* on July 5, 2010, reported that a herd of elephants destroyed crops up to 12km near Palakkad. Some desperate farmers used small bombs and electric fences to frighten the animal. This sometimes leads to the death of the creature. P. Balan in one of his documentaries aptly points out that elephants started coming to human habitation only because humans first disturbed their habitat by deforestation, wild fires or by creating dams.

Death of elephants owing to the lack of proper care and attention is a problematic issue to all elephant lovers. They cannot forget the death of prominent elephants"Manisseri Rajagopal, Perumbillissery Sajprasad, Shankarankulangara Ganapathy, Avittathoor Shivaprasad" all of whom died due to food poisoning (*erandakettal*) after the ceremonial feast for elephants called *aanayoottu*. There were steps taken by the ELA (Elephant Lovers' Association) to ban the parade of elephants during the religious festivals. To some extent, such moves were successful thanks to the support of the public. The ELA called for a probe into the sudden death of a seven-year-old male elephant near Sasthamkotta and there was public rage to find out the reason behind the death of a wild elephant and calf at the Aryankavu Forest Range in Kollam district, reported *The Hindu* on March 15 and July 27, 2010. These show people's love, devotion and respect towards the animal. Aiswarya, a 25-year-old, from Thrissur, whose car was recently crushed by an elephant, is not ready to blame the animal: "It was the mistake of the people. They created a big fuss and thus the animal got frightened, and crushed my car". She continues, "... he did it unknowingly, and I don't have any problem with him. He is innocent in this issue, I am sure". In Kerala one can see a number of 'crazy' people who dedicate their lives for elephants: Vibha, the first female mahout in Kerala; Dr. Giridas, who invited around 50 elephants for fascinating models of elephants using cement, steel, paper pulp, mosaic powder etc.; Alke Riesterer, a U.S.-based researcher who came to Kerala and practiced massage to reduce the stress of the elephant; Kunjutti(Haridas) who has become a mahout at the age of 12.

Elephant Images and Representations in Context

It is here that we need to look into the representation of a being as large, historically prolific and important as the elephant in the landscape, which therefore must have always caused it to be large, omnipresent and important in the minds of the people. There are a number of movies in Malayalam where the elephant is portrayed as a major 'hero' of the story. *Guruvayoor Kesavan, Pattabishekam,*

Gajakesariyogam and *Prayikkara Pappani* are a few of them. A number of short stories, novels and poems have been written on or about elephants in Malayalam. *Ntuppuppakkoranendarnnu* by Vaikom Muhammad Basheer, *Sahyante Makan* by Vailoppilli, *Megharoopan* by Atoor Ravivarma, *Gajendramoksham* by Sugatakumari are some 'classical' works in Malayalam on elephants. *Aithihyamalai* (the Garland of Legends) the popular classic which has an indispensable role in the literary history of Kerala, written by Kottarathil Sankunni (1855-1937) dedicates a number of stories exclusively about elephants and their relationship with the contemporary society. The text is important because it reflects not only the social and cultural ethos of the geographical region called Kerala, but also acts as the symbolic reinvention of Malayalee tradition. *Aithihyamala* is accorded the status of peoples' literature and in the current scenario there is a revived interest in both print and visual media of the legends of this text. The elephant images and representations appearing in the stories of this text, thereby open a new perspective to the study of the ethical and moral attitudes of the people of Kerala towards elephants.

The elephants mentioned in *Aitihyamala* were cooperative and had a peaceful nature " Kidangur Kandangoran is one such elephant. Being an integral part of the festival, elephants enjoyed a high status in the society. Wealthy families adopted them, in accord with the elephants' nobility. As discussed earlier, there were treatises to identify, categorise and look after the health of elephants. Each elephant portrayed in the text behaves almost like a human being. Each was endowed with intellect. Every elephant had his or her personality. For instance, Kidangoor Kandangoran did not allow himself to be accompanied when his time to arrive at the temple came. He went alone and insisted those who climbed on him should use his hind leg. Only the person who held the deity's idol could use his fore leg (132). Once he met an elderly *andarjanam* (Brahmin woman) on a narrow road who fainted out of shock suddenly. He took her with his tusk and laid her on a nearby platform. She was obstructing his way (133). These were some of his 'miraculous' deeds as Sankunni observes. "Some might consider use of 'he' rather than "it" as inappropriate. But this elephant (Vaikatthu Thiruneelakandan) is not an animal by nature" (230). But echoing the public belief leads to a kind of converging of the relative status of elephants and humans, making the use of 'he' highly appropriate.

Attribution of human qualities is an important characteristic of all elephant stories in this book. People can talk to and trust them like another human being. There are instances where elephants settle disputes among themselves amicably. Koniyil Kochayyappan had two attendants and he was allowed to play with the children of the household. There is a juncture in the story where Sankunni narrates a story of an elephant proving his innocence when he was charged with a false accusation. The story of Aranmula Valiyabalakrishnan is one among them. A villainous mahout accuses Valiyabalakrishnan of destroying a sugar cane farm. But Valiyabalakrishnan, in order to prove his innocence, takes an elderly man called Kuruppu on his back and shows him the footprints on the wet floor of the farmland. His foot was bigger than the footprint and thus people realize his innocence (709).

All these stories juxtapose elephants with human beings by showing their intellectual sense and capacity to behave according to the situation. Sankunni's language while narrating elephant stories deserves special mention. He uses exaggerations like, "no one has ever heard of such a great elephant in their life time". A reader can trace the same kind of usages as he describes almost all elephants in the same vein in various stories. Critics like Ambalapuzha Ramavarma, in his introduction to *Aithihyamala*, sounds slightly critical about such statements and question their very authenticity (14). But texts like *Aithihyamala* need not be approached directly depending on its language regarding scientific accuracy, but with a more meditative and subtle approach to decipher the cultural identity of the Kerala society.

Elephant stories of *Aithihyamala* also demand a reading of caste and class issues in the cultural paradigm of Kerala society. The economic status of a family or a temple, which owns an elephant can be weighed through these stories. The relationship between the elephant owners (upper class/caste) can be problematised under this light. The Marxian concept of political economy is highly applicable in the socio-cultural paradigm of Kerala with respect to the animal called elephant. Since the maintenance of an animal like an elephant is high, it is only affordable for the upper class or a wealthy temple. Sankunni, in his story about Avanamakkal Gopalan points out that the prize of an elephant calf that belongs to a good breed is not less than Rs.5000. The story is set in 19th century Kerala, and Rs. 5000 in that age was certainly a huge amount. The food of the animal was also expensive. In a story about Kooniyil Kchayyappan, Sankunni says that the elephant used to eat nearly 2.5 *para* of rice daily and the government had increased it to 3 *para* (30 kg) as they feel that the elephant will feel hungry. Another instance appeared in the text regarding the 'wage' of Kidangur Kandankoran. He was once offered 10 stems of banana, 10 coconuts, and one kilo of jaggery to take some timber from one place to another. The palm leaves that elephants eat are also expensive. The wealth that an elephant brings to the owner should also be taken into consideration. Konniyil Kochayyappan, an elephant of the government, which was used to catch wild elephants, brought a lot of wealth to the state. He along with another elephant called Padmanabhan, says Sankunni, could easily catch the wild elephants, however big they may be. The story continues with a number of instances of captivity of wild elephants by both these 'giants'. Sankunni also observes the different nature of elephants during the time of captivity. Some elephants like Manjapra Thiruneelakandan were portrayed in a bad light because of its violent nature, and thereby brought monetary loss to the governments.

> Manjapra Thiruneelakandan was a first class taming elephant. But there was a big difference between the nature of Thiruneelakandan and Kochayyappan. Kochayyappan was able to show enough patience to the wild elephants and he always somehow managed to cage them. But Thiruneelakandan had the least patience and he killed the wild elephants when he felt that it was difficult to cage them. And because of this attitude of Thiruneelakandan there was a lot of monetary loss to the Government (365).

Though the elephant can be identified as an animal that brings wealth and fame to its owner, family, government and the society, one cannot forget its existence in a cultural realm and its role in the minds of the people. *Aithihyamala* itself describes a number of instances like the *ammaathol* (old woman from a Brahmin family) who calls their elephant, Avanamanakkal Gopalan, "Son! Son!" (486). Sankunni says: "If someone sees the interaction of *ammaathol* and Gopalan, they will think that if not in this birth, of course in the previous birth, Gopalan must have been her son" (486). Sankunni continues, "She has affection towards Gopalan like for a son and Gopalan does love and respect her like his own mother". The conversation of *Valiamma*, a woman in the house of Ranniyil Karthavu, calls Kochayyappan, "My son". Moolayil Govindassar, the mahout of Vaikkathu Thiruneelakandan used to call him, "*makane*" (son) or by the pet name 'Sanku' (231).

All the elephant stories in *Aithihyamala* in one way or the other point out that there was love and respect towards animals in the minds of the masses. People were highly ethical in their attitudes and the government took necessary steps to preserve the rights of the animal. Texts like *Aithhyamala* are capable of opening a new paradigm of the cultural, political, social and economical dimensions of the animal through its representations. It is only primarily as a result of the importance given to the animal by a society that one can trace the cultural paradigm through the society's literary and other cultural representations over time.

References

"Articles on Kerala Elephants." *Kerala Travel and Tour Portal Providing Tourism Information about Kerala India*. 9 July 2010. Web. 23 Aug. 2010. <http://enchantingkerala.org/kerala-elephants.php>.

Babu, I.V. 2001. *Keraliya Navodhanavum Namboothirimarum(The Rennaissance of Kerala and the Namboothiris)*. Kottayam: Sahithya Pravarthaka Co-operative Society Ltd.

Balan, P. "The Silenced Witness." *YouTube*. 4 Dec. 2008. Web. 20 Aug. 2010. <http://www.youtube.com/watch?v=H4ZTYVH1GD8>.

Bhaskaranunni, P. 2005. *Keralam Irupatham Noottandinde Aarambhathil* (*Kerala in the Beginnings of Twentieth Century*). Thrissur: Kerala Sahithya Academy.

Carla Jodey, Castricano. 2008. *Animal Subjects : An Ethical Reader in a Posthuman World Environmental Humanities*. Canada: Wilfrid Laurier UP.

"ELA Calls for Probe into Elephant's Death." *The Hindu : Home Page News & Features*. 15 Mar. 2010. Web. 12 Sept. 2010. <http://www.thehindu.com/news/states/kerala/article245885.ece>.

"Elephant to Be Declared National Heritage Animal." *Latest News, India News, Breaking News, India Latest News Online - The Indian Express*. 1 Sept. 2010. Web. 12 Sept. 2010. <http://www.indianexpress.com/news/Elephant-to-be-declared–national-heritage-animal-/675653>.

"Elephant-Kerala.com." *Elephants of Kerala*. Web. 12 Sept. 2010. <http://www.elephant-kerala.com/index.htm>.

Engels, Friedrich. 1978 . "The Origins of Family, Private Property and the State". *The Marx- Engels Reader*, 2nd ed, Ed. by Robert. C. Tucker. New York: W.W Norton.

Ganesh, K. N. 2008. *Kerala Samooham: Innu, Nale*. Thrissur: KSSP.

"India's Overworked Elephants." *BBC News*. 4 Mar. 2010. Web. 12 Sept. 2010. <http:/ /news.bbc.co.uk/2/hi/8533776.stm>.

"Interviewing Aiswarya S." Personal interview. 18 Aug. 2010.

"Interviewing Poomulli Narayanan Nambootiripad." Telephone interview. 12 Aug. 2010.

Kemmerer, Lisa. 2006. *In Search of Consistency: Ethics and Animals*. Vol. 3. Bosten: Brill.

Logan, William. 1887. *Malabar*. New Delhi: Asian Educational Services.

Mathew, Akhel. "Church and Temple Vow Not to Parade Elephants." *Gulfnews*. 21 Oct. 2007. Web. 12 Aug. 2010. <http://gulfnews.com/news/world/india/church-and-temple-vow-not-to-parade-elephants-1.207386>.

Narayanan, T.C. 2009. *Lore and Legends of Kerala*. New Delhi: Oxford University Press.

Narayanan Namboothiri, Cheruvalli. 2007. *Matangaleela*. Kodungallur: Devi Books.

Radhakrishnan, MG. "Living: Woman's Own Country." India Today, 22 Nov. 1999. Web. 27 Aug. 2010. <http://www.india-today.com/itoday/19991122/living.html>.

Raghunathan Nair, A. B. 1989. *Kottarathil Sankunni: Jeevithavum Krithikalum*. Trivandrum: The State Institute of Languages.

Sankunni, Kottarathil. 2008. *Aithihyamala*. Kottayam: DC Books.

Sathya Narayanan, Kundu, and Nair Naveen. "Kerala Govt Sets up Elephant Training Camps for Mahouts." *CNN-IBN, Live India News,Top Breaking News,World,India,Business,Sports, Entertainment & Health News*. 19 Oct. 2008. Web. 12 Sept. 2010. <http://ibnlive.in.com/news/kerala-govt-sets-up-elephant-training-camps-for-mahouts/76221-3.html>.

"Task Force on Elephants to Meet April 30." *DNA*. 23 Apr. 2010. Web. 12 Aug. 2010. <http://www.dnaindia.com/india/report_task-force-on-elephants-to-meet-april-30_1374651>.

Varner, Gary E. 1998. *In Nature's Interests? : Interests, Animal Rights, and Environmental Ethics*. New York: Oxford University Press.

"Wild Elephant and Calf Found Dead." *The Hindu : Home Page News & Features*. 27 July 2010. Web. 20 Aug. 2010. <http://www.thehindu.com/news/states/kerala/article536764.ece>.

"Wild Elephants Raid Farms Again." *The Hindu : Home Page News & Features*. 5 July 2010. Web. 16 Aug. 2010. <http://www.thehindu.com/news/states/kerala/article501534.ece>.

Chapter 27

Morality, Ethics, and Animal Rights in Romantic Poetry and Victorian Thought

☆ *Johansen Quijano-Cruz*

Notions of the superiority of man over animals, or brutes as they were commonly referred to by philosophers John Locke and Immanuel Kant, were heavily internalized during the Enlightenment. This led to practices like animal exploitation and torture, as well as vivisection – surgery performed on animals for experimental purposes. Despite this era being characterized for its emphasis on the importance of logic and reason, notions that, as Kant argues, humans did not owe any moral duties to animals come from the Scala Naturae, or the Great Scale of Being – a medieval hierarchy that indicates humans as superior beings than animals. Certainly, thinkers of the Age of Enlightenment tried to use reason to explain their superiority. In *Lectures on Anthropology*, Kant wrote that human beings can rise above other beings on earth. He further stated that humans are "a being altogether different in rank and dignity from things, such as irrational animals, with which one may deal and dispose at one's discretion" (Kant, LA, 7, 127). This rationalization, however, presupposes a construct based on faith and accepts it as truth without any sort of higher thinking. In this issue reason was not used to question already existing knowledge that had no logical basis, but to reinforce it. This, however, was not the case with every thinker of the Enlightenment. Sammuel Johnson, considered by many as the most intelligent man of the eighteenth century, stated in *The Idler* that

> Among the inferior Professors of medical knowledge, is a race of wretches, whose lives are only varied by varieties of cruelty; whose favorite amusement is to nail dogs to tables and open them alive;

> and try how long life may be continued in various degrees of mutilation, or with the excision or laceration of the vital parts; to examine whether burning irons are felt more acutely by the bone or tendon; and whether the more lasting agonies are produced by poison forced into the mouth or injected into the veins (94-5).

This condescending tone that Johnson takes with the medical doctors of the time shows a certain level of disdain towards the widespread animal experimentation and cruelty which was much widespread during the eighteenth century. However, even Samuel Johnson himself is, on some level, accused of engaging in animal cruelty. In *The Life of Samuel Johnson*, James Boswell writes that "it is told, that, when a child of three years old, he [Johnson] chanced to tread upon a duckling, the eleventh of a brood, and killed it" (Boswell, 25). Furthermore, Bergen Evans, lexicographer and professor of English at Northwestern University, states in his introduction to *The Life* that Samuel Johnson was seen on ocassion "smiling with a soft, humorous satisfaction as he pushed a dead cat over a waterfal" (Boswell, xi). Certainly, during the Enlightenment there were a number of individuals who advocated for the notion of animal rights; however, for the most part society was largely disconnected from such issues and was rather content with engaging on the standard, torture-like practices of the time thanks to their internalized medieval notions of superiority. During the Romantic period, however, the tone of this conversation would shift public opinion regarding animal rights drastically.

Romanticism was largely a rebellion against eighteenth century thought. As Marshall Brown, professor of English at the University of Washington, states, it can be agreed that Romanticism was an all-encompassing revolutionary movement "in the sense in which a revolution gathers up and recollects, as it sweeps all with it toward the future" (Brown, 47); and despite several scholars disagreeing on the principal ideals behind Romanticism, and even in the exact dates that the Romantic period encompasses, most Romantic scholars agree that whatever else can be said about Romanticism, two things are certain: (1) "no other intellectual or artistic movement has had comparable variety, reach, and staying power since the end of the Middle Ages", and (2) "poetry mattered during the Romantic period in a way that has not mattered since" (Curran, 218). Because of the shift in ideals caused by the Romantic movement, an unprecedented amount of writing advocated kindness to animals in England during the second half of the eighteenth century. David Perkins, literary critic and author, effectively describes the changes in attitudes surrounding animals that led up to the establishment of The Society for the Prevention of Cruelty to Animals in 1824. As Perkins, and several other authors, show, these changes in attitudes are best observed in Romantic poetry. Indeed, poetry communicated the ideals of the Romantic period just as philosophy and essays communicated ideas in the previous century. With poetry being the primary medium for communicating ideas during this period, then it is to its poetry that one must turn to in order to understand the Romantic mindset towards issues of importance.

Some critics, such as M.H. Abrams, argue that William Wordsworth is the Romantic poet *par excellence* and claim that Wordsworth is "the great exemplary poet of his age" (Abrams, 14). Other critics, such as Northrop Frye, argue that William

Blake is the one who best embodies the ideas and notions of the Romantic Movement, and as such it is to Blake who one must turn to in order to fully understand the scope and magnitude of Romantic Sensibility. He was rightly called "the futuristic prototype of visionary romanticism" (Eaves, 239). Certainly, Blake and Wordsworth are the seminal Romantic poets; yet any discussion of the morality and ethics of animal rights in Romantic poetry that would not include John Keats, Percy Bysshee Shelley, or Samuel Taylor Coleridge would be an incomplete one. This chapter will take a critical look at the poetry of the five aforementioned writers in order to ascertain that with the shift in attitudes towards animals during the Romantic period people stopped seeing animals as things towards which no moral or ethical duty was owed and began considering them as divine gifts having a spirit of their own.

William Wordsworth, along with Coleridge, helped launch the Romantic Age in English literature. He describes animals as agents of pleasurable memory. In his poem *Lines Written in Early Spring* (1798), Wordsworth recalls how when sitting under the shade of a certain tree:

> The birds around me hopped and played,
> Their thoughts I cannot measure: -
> But the least motion which they made
> It seemed a thrill of pleasure.

In these few verses Wordsworth is assaulting eighteenth century ideas that the only worth animals had was for food and experimentation. By stating that birds have a certain pleasurable quality Wordsworth is expanding his readers' mindset. He also suggests the same in his poem *To The Cuckoo* (1804), where he writes:

> I hear thee and rejoice:
> O Cuckoo! Shall I call the bird,
> Or but a wandering voice?
> Thrice welcome, Darling of the Spring!

In these passages Wordsworth is, again, emphasizing the pleasurable feelings of joy felt when in the presence of a bird's song. This kind of ode to nature can easily be spotted in several of Wordsworth's poems. His argument for the acceptance of animals, however, does not end at simply stating that birds are pleasurable. In *The Redbreast Chasing the Butterfly* (1807), Wordsworth writes:

> ART thou the bird whom Man loves best,
> The pious bird with the scarlet breast,
> Our little English Robin;
> [...]
> The bird, that by some name or other
> All men who know thee call their brother,
> The darling of children and men?

In these two passages Wordsworth is furthering his argument by suggesting not only that birds are pleasurable animals, but also by implying a type of kinship with it. By urging the reader to see the Robin as more than just a bird, but as their bird, he is strengthening the bond between human and animal. Furthermore, by suggesting that men call this bird their brother he is giving the animal a certain humane property that works on the reader by redefining his preconceived notions of what animals are. In these powerful verses Wordsworth is suggesting that animals are not below humans, as the Scala Naturae would suggest, but creatures similar to humans with which a certain degree of kinship is shared.

Some might argue that the reason for Wordsworth's and other poets' fascination with the bird and its song lies in their privileged position in relation to other animas in the Scala Naturae. In this medieval construct of hierarchical moral worth, birds were considered as inferior to humans due to their lack of soul, but higher than earthbound animals, as they have wings and, thus, are closer to God. Wordsworth's thoughts on the morality of animal rights, however, did not end with the admiration of birds. In *Hart-Leap Well* (1888), Wordsworth makes a strong criticism of the practice of hunting. The first part of this two-part poem presents Sir Walter, a gallant knight, engaging in the popular English pastime for the aristocrats of the Romantic era: hunting. During this first part the reader becomes acquainted with the knight and becomes sympathetic towards his emotions. During the second part of the poem, however, Wordsworth opens with "The moving accident is not my trade. To curl the blood I have no ready arts", a statement that has been taken by many critics, including Perkins, as a critique against the inhumane treatment of animals in the sport of hunting. His descriptions of the once-beautiful lands encompassing Hart-Leap Well, which are now cursed because of the hunt, would lend further weight to Wordsworth's anti-hunting argument. In his essay on Wordsworth and the Polemic Against Hunting, David Perkins reminds his readers that Sir Walter embodies a young and immature version of Wordsworth, one that is obsessed with a sense of power and dominance over nature. The older, more mature self, Wordsworth as the Poet, embodies a voice that speaks out against animal cruelty. Furthermore, when one considers the act of killing an animal as one that might lead to intense solitude, the representation of the knight of the past as a hunter and the poet of the present and future as a type of animal rights activist can be taken as a representation of the eighteenth century values that would use animals as sport and material for experimentation and their shift towards the more animal-friendly morale of the Romantic era.

Several other poems of Wordsworth advocate for the recognition of animal rights. In *Last of the Flock*, (1888) a grown man mourns the loss of his flock of sheep, which, at times, he claims to have loved more than his own children. A number of poems on the restorative qualities of butterflies on memory show that in Wordsworthean lyric even insects had their rightful place worth respecting. Some might still be skeptical and offer commentary that a poetry advocating for animal rights is only to be expected of the writer often referred to as The Nature Poet, but this style of poetry that, directly or indirectly, advocated a moral duty towards animals was a trend that other romantics followed.

In *A Lay Sermon* on the *Unitarian Creed* (1852), Samuel Taylor Coleridge wrote that "for if man be no other or nobler creature essentially, than he is represented in their system, the meanest reptile, that maps out its path on the earth by lines of slime, must be of equal worth and respectability, not only in the sight of the Holy One, but by a strange contradiction even before man's own reason" (Coleridge, 1852, 209). In this quote Coleridge is stating that humans, who have the power of reason and aspire to spiritual and moral greatness, has as much worth in the eyes of God as the animal considered of least worth by the eighteenth century mindset: the lizard. By equating man with the reptile, a symbol of sin and temptation in English literature, Coleridge is arguing that God loves all creatures the same. This would imply that it may be possible for God, or Spirit as it is more commonly referred to in Romantic poetry, to punish man, a thinking, rational being with a moral responsibility towards other animals, when man decides to abuse or kill animals without a purpose. This implication is, perhaps, not as strong in the *Sermon* as it could have been, but it is explicit in Coleridge's seminal poem *Rime of the Ancient Mariner*.

In *Rime of the Ancient Mariner* (1798), the mariner, the impersonation of the poet within the poem, tells a story of how when he sailed through the ocean and found himself and his crew in trouble after sailing through a number of storms, an albatross came to the rescue. Coleridge writes:

> At length did cross an Albatross,
> Thorough the fog it came;
> As if it had been a Christian soul,
> We hailed it in God's name....

By equating the albatross to a Christian soul Coleridge is not only heightening the moral level of the animal to that of humans, but even making it a higher being. The pious albatross is more than just an animal to the sailors – it is an ideal representing hope and salvation. However, the mariner killed his albatross with his crossbow. This event of animal cruelty for sport is one that, according to Coleridge, is terrible enough to torture the soul of the abuser. He further argues that those who hunt animals for sport as the Mariner hunted the albatross will be ostracized from society. Coleridge writes:

> And I had done a hellish thing,
> And it would work 'em woe
> For all averred, I had killed the bird
> That made the breeze to blow.
> Ah wretch! Said they, the bird to slay
> That made the breeze blow!

As with Wordsworth's Hart-Leap Well, it can be considered that this mariner who shot the albatross is the mindset of the eighteenth century ideology regarding moral duty towards animals. Just as many of the eighteenth century thinkers subscribed to the Lockean notions of animal rights, so did the mariner. The shifting attitudes towards animals can be seen in the poem when, after suffering through a number of supernatural ordeals, the albatross was hung on the neck of the mariner. In Coleridge's work the albatross takes a mythical form. Being considered as an elemental spirit of good omen sent by God, the connotation is not only that animals have a divine purpose, but that also help humans become better. If the poem is to be interpreted as a representation of the changes in attitudes regarding moral duty towards animals, then the mariner's road to redemption is a reference to the realization that animals are living, breathing beings that share the planet with humans. Furthermore, the mariner's desire to captivate wedding guests with what is described in the poem as "his glittering eye" to tell them the story of how killing an animal cost him dearly can be seen as a warning to future generations to not stray off the path of moral righteousness.

While Coleridge uses animals as religious symbols, he is not the only one to do so. In "The Lamb" and "The Tyger," William Blake gives religious meaning to two animals representing contrary states of being. Blake opens his poem "The Lamb" by asking "Little Lamb who made thee? Dost thou know who made thee?" After praising the tender voice and wooly, pure clothing of the lamb Blake proceeds to answer his own question by giving the lamb a mythical aspect. Blake writes:

> Little Lamb, I'll tell thee,
>
> Little Lamb, I'll tell thee.
>
> He is called by thy name,
>
> For He calls Himself a Lamb.

Here Blake is ascribing a holy nature to the animal. Once again, as in Coleridge, we see that the animal transcending the animalistic and human state of being in order to become something more abstract. By seeing animals as holy entities Blake is giving them a heightened sense of being and placing the burden of being morally dedicated to the wellbeing of animals on humans. Blake, however, argues that just as the lamb is God, so humans are God. Blake continues by writing how "He became a little child – I a child and thou a lamb, we are called by His name". By establishing a direct relationship between God and the individual and another between God and the animal, Blake is implying that both animal and human are part of the same larger entity. In doing this he asserts that animals are not things for experimentation or food supply, but beings towards whom we have a moral duty. In "The Tyger" Blake asks "what immortal hand or eye could frame thy fearful symmetry?" and concludes, after a series of rhetorical questions that admire the tiger, by asking "did he who made the Lamb make thee?" In doing so he acknowledges some higher being, the same that created the lamb and the child, also created the tiger. Here Blake is acknowledging that even in exotic animals there is beauty to behold.

It is obvious that in "The Lamb" and "The Tyger", while Blake might not have necessarily arguing for the notion of a moral responsibility towards the acceptance of animal rights as a universal value, he was certainly imbuing these animals with a certain supernatural quality, similar to Coleridge's albatross, in order to provoke some sort of change in the mindset of the reader. Still, Blake's thoughts on the rights of animals can be seen in his poem *Auguries of Innocence*. Published in 1863, it is the last Romantic poem published dealing with the issue of animal rights. This poem is full of couplets that make powerful statements on animal rights, such as "a horse misused upon the road – calls to heaven for human blood" and "each outcry of the hunted hare – a fiber from the brain does tear" (Blake, 1863). To look at the entire poem in a full would take an entire volume to compile, and so it will not be dealt with in this chapter. It will be mentioned, however, that this poem is the strongest statement for arguing in favor of animal rights and moral responsibility towards animals.

Even though Blake can be credited with having written the strongest poem advocating for animal rights and a moral duty towards animals, it can be argued that Percy Bysshe Shelley was more passionate about the issue. He wrote several essays on the issue of animal rights and on the moral duty of humans towards animals. He takes the issue a step further by arguing that animals should not even be used for food. Despite having written several poems, such as "The Revolt of Islam", "Alabaster" and "Queen Mab," Shelley's strongest statements regarding animal morality are found in his essays. In *A Vindication of Natural Diet*, Shelley clearly states that "man resembles no carnivorous animal" (Shelley, 1813). Shelley reminds his readers that "comparative anatomy teaches us that man resembles frugivorous animals in everything, and carnivorous in nothing; he had neither claws wherewith to seize his prey, nor distinct and pointed teeth to tear the living fiber" (Shelley, 1813). By pointing out that man is like animals, he is stressing that bringing harm to animals is akin to bringing harm to oneself. He further explains that humans are not carnivorous because "it is only by softening and disguising dead flesh by culinary preparation, that it is rendered susceptible of mastication of digestion; and that the sight of its bloody juices and raw horror does not excite intolerable loathing and disgust" (Shelley, 1813). If animals are not meant to be used as a source of food, then there is no need to harm them. By comparing the physical properties and dietary habits of humans to those of orangutans to make an argument that not only that man should treat animals kindly, but that man should also refrain from eating animals, Shelley manages to propose what could be considered as a predecessor to Darwin's work.

It was towards the end of the Romantic period when people and poets alike began realizing the flaws of the Romantic ideal. The poem *Ode to a Nightingale*, by John Keats, demonstrates such a shift in attitudes better than any other Romantic poem. This poem presents a triangular structure in which the reader is taken from a perspective of reality to a perspective of a world of ideal beauty while trying to escape the harsh realities of the real world. However, the poem's shift to the loss of pleasure and inevitable death describe what according to Claude Finney, describes "the inadequacy of the romantic escape from the world of reality to the world of ideal beauty" (Finney, 632). It is through Darwin's work on evolution, particularly *On the Origin of Species*, which was written for non-specialist readers, that the Victorians realized these "inadequacies". By succesfully disceminating the conept and nature of evolution and adaptation, and by making scientific studies on the anatomy of humans and animals, Darwin managed to spark scientific, philosophical, and religious debates regarding both human and animal nature which uninevitably led to the promotion of scientific naturalism and a larger debate on issues regarding animal rights.

References

Abrams, M.H. (Ed.). 1975. *English Romantic Poets*. New York: Oxford University Press.

_____. 1973. *Natural Supernaturalism*. New York: Norton.

_____. 1971. *The Mirror and the Lamp*. New York: Oxford University Press.

Bentley, G.E. 2001. *A Stranger from Paradise: A Biography of William Blake*. New Haven: Yale University Press.

Bloom, Harold. (Ed.) 1970. *Romanticism and Consciousness*. New York: Norton.

Boswell, James. 1965. *The Life of Samuel Johnson*. Abridged Modern Library College Edition. New York: Mc Graw Hill.

Brown, Marshall. 2004. "Romanticism and Enlightenment". In Curran, Stuart (Ed.). *The Cambridge Companion to British Romanticism*. (p. 25 – 47). United Kingdom: Cambridge.

Curran, Stuart. (Ed.) 2004. *The Cambridge Companion to British Romanticism*. United Kingdom: Cambridge.

De Man, Paul. 1984. *The Rhetoric of Romanticism*. New York: Columbia University Press.

Denenholz Morse, Deborah and Danahay Martin A. 2007. *Victorian Animal Dreams: Representations of Animals in Victorian Literature and Culture*.Portland, OR: Ashgate Publishing.

Eaves, Morris. (Ed.) 2006. *The Cambridge Companion to William Blake*. United Kingdom: Cambridge.

Erdman, David. 1977. *Prophet Against Empire*. New Jersey: Princeton University Press.

_____. (Ed.) 2008. *The Complete Poetry and Prose of William Blake*. Los Angeles: University of California Press,.

Finney, Claude. 1936. *The Evolution of Keats's Poetry*. Cambridge, MA: Harvard University Press.

Frye, Northrop. 1974. *Fearful Symmetry*. New Jersey: Princeton University Press.

Johnson, Robert. *Kant's Moral Philosophy*. Stanford Encyclopedia of Philosophy. <http://plato.stanford.edu/entries/kant-moral/>

Johnson, Samuel. 1761. *Medical Professors Experiments*. The Idler. London.

Kant, Imanuel. 1997. *Lectures on Anthropology*. Berlin: Akademie-Textausgabe.

Kenyon-Jones, Christine. 2001. *Kindred Brutes: Animals in Romantic-Period Writing*. Aldershot, England: Ashgate Publishing.

Lovejoy, Arthur. 1960. *Essays in the History of Ideas*. New York: G. P. Putnam's Sons.

_____. 1964. *The Great Chain of Being: A Study of the History of an Idea*. Cambridge, Massachusetts: Harvard University Press.

Perkins, David. 2003. *Romanticism and Animal Rights.* Cambridge: Cambridge University Press.

_____. "Wordsworth and the Polemic Against Hunting: Hart-Leap Well." *Nineteenth Century Literature.* 52.4 (1998): 421 – 445.

Shelley, Percy Bysshe. 1813. *A Vindication of Natural Diet*. PDF Version. London. <http://www.animal-rights-library.com/texts-c/shelley01.htm>

Simpson, David. 2006. "Blake and Romanticism". In Morris, Eaves (Ed.). *The Cambridge Companion to William Blake.* (p. 169 – 190). United Kingdom: Cambridge.

_____. 2004. "Romanticism, Criticism, and Theory". In Curran, Stuart (Ed.). *The Cambridge Companion to British Romanticism.* (p. 1 – 24). United Kingdom: Cambridge.

Taylor Coleridge, Samuel. *The Rime of the Ancient Mariner.* Poems and Poets. <http://www.everypoet.com/archive/poetry/Samuel_Taylor_Coleridge/samuel_taylor_coleridge_rime_contents.htm>

_____. 1852. *"On Unitarian Creed"*, in *A Lay Sermon. Lay Sermons.* London.

Wordsworth, William. *Poems by William Wordsworth.* Poems and Poets.

<http://www.everypoet.com/archive/poetry/William_Wordsworth/william_wordsworth_contents.htm >

Chapter 28

Issues of Survival in *Life of Pi* and *A Tiger for Malgudi*

☆ *S.S.V.N. Sakuntala*

> "Richard Parker has stayed with me. I've never forgotten him...............I miss him.....................I still cannot understand how he could abandon me so unceremoniously, without any sort of goodbye, without looking back even once. That pain is like an axe that chops at my heart' (Martel, 6).

Is it believable that Richard Parker is a royal Bengal tiger and the monologue is of a 16 year old boy who lived with Richard Parker on a lifeboat in the Pacific Ocean for 227 days? Amazing survival strategies exhibited by Pi Patel and Richard Parker make Yann Martel's *Life of Pi* an astonishing world of magic realism. On the other hand R.K.Narayan's *A Tiger for Malgudi* stands as an epitome of the anthropomorphic presentation of the Tiger possessing the soul of an enlightened human being. In both the novels a tiger is the central character along with his counterparts or the social animals. Both are narrated in first person singular – *Life of Pi* by a human and *A Tiger for Malgudi* by the Tiger himself. In both the novels, one written by a western writer and the other by a writer of the orient, the Tiger is portrayed as a complement to the human characters and is referred to as 'he' not as 'it'. This itself is an evidence of acceptance of animals as fellow creatures, in fictional world if not in reality. Both novels dwell upon and probe into aspects such as animal world, animal rights, atrocities against animals, offensive and defensive attitude of the animals and attitude of the animals towards humans and vice versa. The journey of Narayan's Tiger is from Jungle to Zoo; that of Martel's tiger is from zoo to jungle. However some of the aspects related to animals and humans, their habits, habitats, habitation and struggle for survival reveal the underlying and unseen nexus that exists between the human and animal world. They both survive in a world that is independent and interdependent, each species complementing the other. It is a universally accepted fact that the co-existence of humans and animals is inevitable and indispensable. Unfortunately human beings, being animals imbued with knowledge and reason, think and feel themselves superior and exhibit their hegemony over the fauna. From time immemorial all religions and scriptures have been preaching kind and considerate attitude towards fellow creatures. But the unethical treatment meted out to animals is suggestive of their strained relationship with humans. The concept of animal rights has become current in the legal scenario as it aims at making the humans see animals as "fellow creatures". This is an idea that is extended and based on the concept of "aesthetic and environmental well-being, like economic well – being, are important ingredients of the quality of life in our society….." (*Harvard Law Review*, p.1215)

Animals, especially wild animals, have always been targets of exploitation for the entertainment and enjoyment of human beings. In the process many atrocities and varied forms of violence are inflicted upon them. Their freedom is restricted and they are confined to closed enclosures either in a Zoo or in a circus or in films. It is a mere coincidence that the stories of Narayan's Raja and Martel's Richard Parker begin at Zoo. Raja, awaiting death like an ascetic, accepts the human cries and stones thrown at him by the visitors whereas Richard Parker is introduced as a young and ferocious three year old Bengal tiger that was brought to the Pondicherry Zoo as a cub. Raja reaches the Zoo with willingness and resignation whereas Richard Parker is brought to the Zoo after the death of his mother and is unaware of the wild life and law of the jungle. For him the enclosure in the Zoo becomes a dwelling and a world of its own. As Martel rightly points out through Pi Patel:

> Animals are territorial……………animals are, conservative, the smallest changes can upset them. In the wild, animals stick to the same paths for the same pressing reasons, season after season. In a Zoo, if an animal is not in its normal place, in its regular posture at the usual hour, it means something (Martel, 16-17).

One is alarmed and surprised to notice Richard Parker's behaviour on the lifeboat against these observations. Behind the conservative attitude regarding territory and boundary the basic instinct of struggle for existence seems to be apparent. Martel's description of Richard Parker's attempts to save his life when the ship sinks create a breath – taking effect. In such circumstances survival becomes the only priority either of humans or of animals. How strategic they become in this struggle for existence is realistically depicted through Pi Patel and Richard Parker . They both are caught in the oceanic wilderness from which there is no probable and possible escape.Though momentary Pi shows concern for his fellow creature.

> "Richard Parker! Don't give up, please. Come to the lifeboat…swim, swim!... He looked panic – stricken. He looked small and helpless…In a few seconds You'll be aboard and we'll be together…Together? …Have I gone mad?" (*Ibid*: 97)

But destiny destines them to stay together in the same lifeboat for 227 days. Such co-existence whether factual, fictional or mythical was earlier referred to by Kipling in his *Jungle Book* and long ago by Kalidasa in his *Abhignana Sakuntalam* which occur as passing references, when we think of Pi Patel and Richard Parker. Where as Narayan's Raja willingly surrenders to the ascetic power as the last resort to save his life and spends the 'Sanyasa" or the final stage of human life in the holy company of the man who rescues him. Raja as a fictional character possesses the ability to understand human speech and is a typical representative of the oriental belief and life. What about animals who lack the faculty of speech and are the victims of human indifference? As Narayan puts it

> "Man assumes he is all important, that all else in the creation exists only for his sport, amusement, comfort or nourishment" (Narayan, 18).

Such self-centred attitude of human beings is well analysed and portrayed by Narayan who seems to suggest that animals too possess traits such as selfishness, altruism and nepotism.

> "We the denizens of the jungle can communicate, without words, exactly as human beings do-we are capable of expressing to each other. Sympathy, warning, abuse, irony, insult, love and hatred exactly, in the manner of human beings..." (*Ibid*: 14).

Raja's love and affection and concern for his cubs and his attempts to protect them exemplify the typical fatherly concern. Whereas human beings treat animals as creatures imbued with wild strength sans emotions. This hegemonic attitude leads humans to attack wild animals for no reason and encroach upon their habitats. Animals attack humans only when their existence is endangered. Man in his pride and self esteem forgets that animals are creatures that possess ego, values and feelings. The law of the jungle seems to operate on the principle of the survival of the fittest. Even there the territories are earmarked. Survival of the weak and the strong animals is conveniently and skillfully organized following the basic principle "Live and Let Live."Human world has only one motto-live and don't let live. Raja, the uncrowned king of the jungle assumes that he is omnipotent. Later he realizes,

> "I had thought that there could' never be any creature stronger than a tiger. I was mistaken. A human being may look small, without prominent teeth or claws, but he is endowed with some strange power, which can manoeuvre a tiger or an elephant as if they were toys?" (*Ibid*: 28)

That is what exactly Pi does. He analyses and studies the nature and behaviour of Richard Parker as Pi happens to be the son of a Zoo curator and has been warned to be careful with wild animals since his childhood. Though he is sure to be attacked by Richard Parker after it had finished the hyena as its prey, he could recognize the sounds made by Richard Parker as signs of acceptance and approval.

The relationship between Pi Patel and Richard Parker is more of compassionate and friendly companionship. No traces of cruelty on either part are apparent as they both live with a mutual - love - fear complex in a strange situation. Each seem to be afraid of the other and at the same time each becomes a companion of the other.

Rather it can be called kind of survival strategy "better adapted for immediate local environment", as Darwin explained it. A situation where individual struggle has been reduced to its narrowest limits and the practice of mutual aid had attained the greatest development. Before reconciling to survive with Richard Parker, Pi Patel does think of ways and plans to kill it – typical of human mentality though the animal is on friendly terms :

> "During those long cold, dark hours...I held on to one thought: Richard Parker. I hatched several plans to get rid of him so that the lifeboat might be mine" (Martel, 157).

What selfishness, monopoly and hegemony? Isn't it another being, another animal struggling for life just like him? When revelation downs upon him that Richard Parker is a 450 pound, living fierce animal his attitude changes and he realizes that it is better to come to terms with Richard Parker as it is difficult to outlive him. Pi Patel adapts this strategy when the truth that he is the weaker than the tiger dawns upon him. The fear of being outlasted by an animal makes him resolve to continue his struggle for survival on a friendly note.

At that moment Richard Parker makes a sound known as "Prusten: the quietest of tiger calls, a puff through the nose to express friendliness and harmless intentions" (*Ibid*: 163-4). Pi Patel, son of a Zoo curator distinguishes the sound from other sounds made by tigers and is full of fearful wonder. He decides to tame Richard Parker and to continue the comradeship.

> "It was not a question of him or me, but of him *and* me........... We would *live* or we would *die-together*" (*Ibid*: 164).

Selfishness leads to hegemony, hegemony to realization, realization to fear, fear to wonder and wonder to comradeship. This attitude of togetherness results in co-operation rather than competition. It also makes Pi accept Richard Parker's presence as a source of outliving despair and loneliness. But for Richard Parker, Pi's existence would have become a kind of solitary confinement which is unbearable for human beings who are regarded as social animals.

This feeling of togetherness makes Pi feel grateful to Richard Parker:

> "....if he died I would be left alone with despair, a foe even more formidable than a tiger... Without Richard Parker, I wouldn't be alive today to tell you my story (*Ibid*: 164).

A magnanimous gesture on the part of Richard Parker and a truthful confession of Pi Patel. Richard Parker, unusual of his species, becomes a peace marker and a lovable companion in a crisis. Rather the critical situation in which both are placed demanded a strategic co-existence. What if Pi Patel had a powerful weapon with him? Had he allowed Richard Parker to survive just as Richard Parker expressed his friendliness though imbued with a more powerful weapon (his paws) remains a million dollar question. Given a chance, how cruel humans can be towards animals is realistically portrayed by R.K.Narayan in narrating the plight of Raja. Raja is forced to enter the human habitation when his cubs are captured. The brutish and devilish behavior of the villagers turns Raja into a vengeful creature. As a result, he gets trapped and becomes a victim of the heinous acts of the circus captain. The physical and psychological agony of Raja remains unnoticed by the captain as human beings are ignorant of the emotional disturbances of animals.Raja confesses,

> "I can hardly describe that kind of suffering, an emptiness, a helplessness, and a hopelessness behind the bars" "...now this combination of man and metal subdued me..."(Narayan: 47).

Narayan shows remarkable skill in comparing the hard heartedness of man to a metal. How unkind and unethical to keep an animal for days without food and water leaving it to suffer in loneliness. Can ever a human being imagine himself in the place of Raja? No wonder humans commit such shameful and violent acts against fellow humans for no specific reasons. Whereas animals attack their prey only for two reasons – to satisfy their hunger and for self defense: "Tigers attack only when they feel hungry unlike human beings who slaughter one another without purpose or hunger..."(*Ibid*:117). Raja is made to undergo rigorous training and perform feets beyond reason and comprehension. In the process he gets whipped, injured and understands that he gets enough of stale pieces of meat if only he entertains the humans. In course of time Raja starts admiring the capabilities of his captain:

> "I began to admire him – a sort of worshipful attitude was developing in me. I had thought in the jungle that I was supreme. Now that was gone. I was a defeated king, and captain was the unquestioned Suzerain" (*Ibid*: 53).

Such an appreciation of an animal by a human being is far beyond imagination either in real or in the fictional world. Pi Patel is gracious enough to express gratitude for what Richard Parker had done to him. Simultaneously his hegemonic attitude emerges:

> I pushed myself off not forgetting to blow the whistle hard several times, to remind Richard Parker of who had so graciously provided him with fresh food (Martel, 186).

This smug complacency soon recedes as Richard Parker raises its paws when Pi Patel attempts to occupy his territory. Pi prepares a shield with a turtle shell and saves himself from the minor attacks.

Richard Parker appears to be the living proof for its compassionate attitude while Raja loses its temper and resorts to cruelty by killing the captain of the circus. One has to be sympathetic with Raja as he narrates his agony and shame when he is made to act in a film, which ultimately enrages him to kill his captain. "A creature needs the support of four legs for stability.. ...while they tempted me with a bait to stand up, the camera followed my action the act repeated till I was sick of it". (Narayan p.112).The captain of the circus is an epitome of nepotism. For the sake of his survival and entertainment he inflicts physical and psychological pain on Raja. Disobeying him results in starvation and isolation to Raja for days. Such inhuman treatment leads Raja to think, "Why should I fear this creature no bigger than my tail."

Animals can be compassionate like Richard Parker; they can be cruel like Raja. There are instances where Richard Parker becomes cruel – when he kills another

Sailor on the boat' and Raja becomes kind and compassionate after listening to the words of the ascetic. Therefore it is a mutual or give and take relationship or bond between humans and animals that can bring peace and harmony. It is the same bond that makes Pi Patel await Richard Parker's arrival when he escapes from the island. It is the same bond or the feeling or responsibility for others that makes Pi Patel feel orphaned when Richard Parker finally leaves him. The association and companionship between Pi and Parker appears to be broken as they reach the shore. The human nature of Pi anticipates a ceremonial farewell gesture from Richard Parker. Richard Parker's departure into its natural and peaceful world makes Pi Patel disheartened. The throngs of separation create feelings of helplessness, despair and hopelessness in him which is unusual.

> "I was truly alone, orphaned not only of my family, but now of Richard Parker, "I wept like a child....I couldn't have done it without you" (Martel, 285-6).

Will an animal ever be able to understand this gratitude of humans as animals are aware only of the cruel, unethical and inhuman attitude of humans? That is why Pi Patel warns Richard Parker "watch out for Man. He is not your friend" (Martel 286) Pi Patel also refers to the fact".............. the most dangerous animal in a Zoo is Man" (*Ibid*:29). There is no limit to the heights of sadism of men who visit Zoo and inflict pain on animals. It is an unbelievable fact that many animals of zoos were killed being fed foreign bodies by visitors- apples with small nails, ballpoint pens, paper clips, combs, coffee spoons, broken bottles etc.

Many laws have been framed regarding animal rights. Do the animals know their rights? Can they understand the laws made for them by humans? Who are humans to frame laws for beings created on par with them by God? All these questions can be answered and change in the state and status of animals can be possible with a change in the mind set of humans. To put it in Raja's words "People talk ill of us all the time. Its our misfortune that neither the denizens of the jungle nor those of the towns would trust us...," "...and that left alone a tiger could be as innocent and harmless as a cow or a chimp? (Narayan, 96-7). As long as humans fail to understand and treat animals with a compassionate and humanitarian attitude and as co-survivors animals have only one option – to cry in unison

> Let us hope and pray
>   To have a word to say
> And wait for a day
>   Full of freedom and gay
>
>   or

to keep themselves away from the human jungle of vicious thoughts to derive vicarious pleasure. The sound and fury of the human world no doubt disturbs the serene and quiet world to which the animals belong. However, Richard Parker and Raja owe their gratitude to their creators for christening them and making them immortal, not as animals but as characters with psychological insight worth studying. Both Martel and Narayan excelled in making them the real heroes of their fictional world.

References

Martel, Yann. 2002. *Life of Pi.* New Delhi: Penguin Books India.

Narayan, R.K. 2006. *A Tiger for Malgudi.* Chennai: Indian Thought Publications.

Harvard Law Review. Vol. 122, No. 4. February 2009.

http://en.wikipedia.org/wiki/Survival_of_the_fittest, 8th September, 2010

Chapter 29

The Bengal Tiger, the Irrawaddy Dolphin, an Humble Fisherman and Lone Researcher: A Tale of Compassion, Connection and Foreboding

☆ *Denise Russell*

The Hungry Tide by Amitav Ghosh takes us into the interwoven lives of animals and humans in the most extraordinary environment: the Sundarbans of West Bengal. Rivers, creeks, islands, mudflats and large tidal differences dominate, creating challenges for humans and animals and the setting for Ghosh to raise some critical questions for the twenty first century. What if anything is the human obligation to consider the needs and interests of animals? How do we weigh these needs and interests against those of humans?

Given the likely severe impact of climate change on the Sundarbans, and their inhabitants, who should have responsibility for addressing this problem? The novel also engages with stories of human relationships and politics that will not be the focus here.

The central human figure is Piya, an Indian born American dolphin researcher visiting the Sundarbans on a research grant to study the Irrawaddy and Gangetic dolphins. It is her story and her relations to animals that are used by Ghosh to raise the above questions. The key animal players are dolphins and tigers but crocodiles, fish, and crabs also have significant roles.

Interactions Between Humans and Animals

Benign as well as hostile interactions are explored in *The Hungry Tide*. It is a commonly reported phenomenon that the Irrawaddy dolphin (*Orcacella brevirostris*) will work co-operatively with humans to drive fish into nets (even though humans are not always so co-operative in sharing 'their' dolphins with other humans) (Watson 257). Piya witnessed this in waters near Mandalay on another research trip. Several dolphins worked as a team when called to a fishing boat by the beating of a stick on the gunwale. Later the fishermen rattled weights on the net, the signal for two of the dolphins to break off from the pod. They swam in circles herding the fish close to the boat. Then one of the fishermen called out indicating to the dolphins to move off and he cast the net in the centre of the area where the dolphins had been patrolling. This was followed by all the dolphins in the pod working together to keep the fish in the net. A great number of fish were caught but many ended up in the mud and the dolphins harvested them. This phenomenon is not unique to Irrawaddy dolphins. It has been documented in waters off Laguna in Brazil with bottlenose dolphins. Even one such dolphin, using speed and acrobatics can conduct the herding of fish towards the nets and in these activities, the dolphin signals to the fishermen when to pull in the net. The dolphin is paid in fish (Connor and Peterson 64-5).

Many species of dolphin will make use of humans by for instance following a fishing boat for a feed either from fish escaping from nets or the disposal of by-catch (Connor and Peterson 55-64). Sometimes however they offer humans a gift. A National Geographic photographer was snorkelling in Hawaiian waters when a dolphin swam up to him with a large mahi-mahi in its mouth. The dolphin left the fish in the photographer's face and backed away (Connor and Peterson 133). Two people were sitting on a beach on Hinchenbrook Island in Australia when 'a dolphin silently swam up and threw a fish to them, then just as quietly disappeared' (Hast 13).

The Irrawaddy dolphin has some fishing techniques which don't involve humans but may require cooperation of the pod. One remarkable activity is mentioned by Ghosh. Here the dolphins herd fish into shallow water. The fish bury themselves in the mud trying to avoid their pursuers. Then 'much like rabbits uprooting a harvest of carrots' (167), the dolphins pick up the fish from the riverbed. These dolphins may use other fishing techniques as well that don't involve co-operation. Piya had 'a couple of beautiful sightings in which they shot water from their mouths, creating fountains in the air' (136). *Orcacella brevirostris* also sometimes throw their prey in the air, in the manner of their relative, the orca whale. The former may be to confuse their prey, the latter to stun, but the animals could simply be playing. At least one study counts 'tossing fish' as play behaviour in the bottlenose dolphin (Bel'kovich *et al.*, 74).

In *The Hungry Tide* dolphins provide company until a boat arrives, for a man marooned on an island with no human inhabitants. He is afraid of the tigers there; tigers who are quite capable of attacking on land or in the water or even up a tree if it would support their weight. Historically stories abound of humans being saved by dolphins, either by their actions as guardians fending off attacks by predators such as sharks or by actually pushing drowning people to shore (Connor and Peterson 129). Dolphins may also guide swimmers to a lost partner, lost object, or an anchor line, or lead a yacht past an obstacle (Doak 51-2).

Ghosh reinforces the link between humans and dolphins by mirroring actions and mental attributes. Piya is rescued from a dangerous situation on the water by Fokir. They make their way slowly in his small fishing boat to 'Lusibari', a small village in the Indian Sundarbans. Ghosh portrays the gentleness of the relationship between Fokir and his young son, Tutul, as Fokir teaches his son the ways of the boat and waters. This mirrors the close educative bond between a mother dolphin and her calf encountered on a few occasions. One evening Piya looks upon Fokir and Tutul curled up peacefully together sleeping in their fishing boat and 'the rhythm of their breathing reminded her of the pair of dolphins she had been watching earlier' (138). Sadly one of each pair dies in the novel.

Ghosh suggests that dolphins also mirror humans in that they make a choice about habitat. He claims that Irrawaddy dolphins may live in salt or fresh water, but this isn't explained by any anatomical difference. It is the expression of a preference for one habitat over another by particular groups. Such a clear cut division is confounded however by the Sundarban dolphins who seem to exist in fresh and salt water. In the novel and, in general, when dolphins relate to humans it is generally in a friendly manner though humans are responsible for threatening dolphins in various ways indicated below. By contrast human interactions with tigers are generally hostile.

The tigers in the Sundarbarns of West Bengal are treated in *The Hungry Tide* as objects of fear. Yet in the background we are told about attempts to set aside tiger reserves that indicates a high level of concern for their welfare given the demands on space for human habitation and exploitation. There is one incident in the novel where a connection is forged between Piya and a tiger when they are both trying to survive a storm. During the eye they notice each other across several hundred metres. The tiger looks at her and makes no movement other than a twitch of its tail. Piya reflects that 'if she had been able to put her hand on its coat she would have been able to feel the pounding of its heart' (389). Then the storm returns and the connection is lost.

Piya realizes that she has an indirect connection with crabs in that they are the reason why Fokir, who is fishing for crabs, is on the waters which interest Piya. While he was catching crabs, Fokir became very familiar with the ways of dolphins and is able to be of invaluable assistance to Piya. The crabs help humans in hidden ways, 'they kept the mangroves alive by removing their leaves and litter; without them the trees would choke on their own debris' (142). In fact, Piya reflects that they are the keystone species of the entire ecosystem. Perhaps even her destiny is ruled by the crab.

The harmful interactions between humans and animals are many. The Indian Sundarbans were settled by humans in the early 20th century and there have been regular deaths from crocodiles, estuarine sharks, snakes and tigers. Tigers have been squeezed out of their territory and looking for food threaten human settlements. The attempts to set aside reserves for tigers have not always worked. The demand for space has on occasion led to violent disputes between those who want the reserves and those who want the land. One such conflict occurred on Morichjhanpl island which had been set aside as a protected forest reserve. In 1978 many refugees initially from neighbouring Bangladesh attempted to establish a settlement there. This led to increasingly determined efforts by government forces to evict the settlers that eventually succeeded in 1979. According to some reports the forcible evacuation led to thousands of human deaths. As the story is relayed in *The Hungry Tide* it is the human tragedy which moves us. The long-term tragedy of tigers becoming extinct is not weighed against this terrible loss. In fact we are made to sympathize wholeheartedly for one of the inhabitants who comments on the announcements of the officials trying to evict the inhabitants. She says that they 'say our lives, our existence, was worth less than dirt or dust .... "This island has to be saved for its trees, it has to be saved for its animals, it is part of a reserve forest, it belongs to a project to save tigers, which is paid for by people from all around the world"' (262). She then asks

> Who are these people ...who love animals so much that they are willing to kill us for them? ... it seemed to me that this whole world has become a place of animals, and our fault, our crime, was that we were just human beings, trying to live as human beings always have, from the water and the soil (262).

The people evicted from Morichjhanpl Island and in the surrounding areas were desperately poor and had limited choice as to where to live. The government offered no assistance in relocation. Yet should we just ignore the plight of the tiger? Left unasked here at this point in the novel are the questions: what will happen to the tiger as a species if they have no place to live? Should we as humans stand by as they become extinct?

The novel pushes back against this rush of human sympathy generated by the Morichjhanpl story in the heartrending account that follows of a tiger cornered in a livestock pen who is pounded with sticks and then burnt alive. Piya is horrified and tries in vain to stop the attack. Fokir participates to some extent and drags Piya away. The villagers believed that this tiger had been responsible for two human deaths and had also killed some livestock. Piya's response is to say 'This is an animal ... You can't take revenge on an animal' (294). Trying to calm Piya, Fokir says 'when a tiger comes into a human settlement, it's because it wants to die' (295). However it is far more likely that it simply wants food.

The dilemma here is elaborated in a conversation between Piya and Kanai, another visitor to 'Lusibari'. Kanai remarks that nearly every week people are killed by tigers in the Sundarbans but that it goes unremarked. This is a horror – 'that we can feel the suffering of an animal, but not of human beings' (301). Furthermore he accuses Piya of being complicit because it is people like her 'who made a push to protect the wildlife here, without regard for the human costs' (301). He criticizes himself for complicity also, because it is people like him who choose to hide these costs, 'basically in order to curry favour with ... Western patrons. It's not hard to ignore the people who're dying – after all they are the poorest of the poor' (301). Piya's response is to argue that the world would be worse off if humans decided that they were the only species who mattered. Besides if it is acceptable to kill other species that will lead to a callousness in regard to other humans. This Kantian twist fails to impress Kanai who returns to the fundamental shame of the loss of human life. Piya's retort is that she would not always put the human above the animal. In fact she declares that she would give up her life if it 'might make the rivers safe again for the Irrawaddy dolphin' (301-2). This then is one of the crucial dilemmas explored in the novel and there is no resolution. Yet there has to be some way of moving beyond the 'save the tiger' vs 'save the human' thinking. I will return to this issue.

Irrawaddy and Gangetic dolphin numbers are down in most places. Piya conjectures that it is dramatic deterioration in the habitat which has led to a drop in the number of dolphins in these waters where they were once plentiful. She regards it as a task of utmost importance to document the decline in numbers. Fishing practices are likely to threaten the food supply of these dolphins. Ghosh indicates that while fish are abundant now in the Sundarbans there are predictions that they will be gone within 15 years with new nets that have a fine mesh which catches fish eggs. The nets are used to catch tiger prawn spawn which is a lucrative business. Collisions with power boats also cause dolphin deaths and the calf that Piya felt connected to dies in this way. Fokir indicates that such a fate is not uncommon.

Tales of crocodiles attacking sailors and threatening boats are included in the novel. In one incident Piya is saved by the quick action of Fokir when a crocodile approaches their small boat. We are told that even crabs may be seen as a threat despite their beneficial qualities mentioned above, as in 'untold multitudes' they burrow into the sand levies and weaken them. Yet it is also noted that the multitudes are diminishing along with the birds. We are told that mosquitoes, snakes, scorpions and even fish may turn up in beds of the unprepared. At least for me the latter would be such a novelty that I would be charmed. The koimachh or tree perch is able to use spiny fins to drag itself overground for short distances.

The threat of humans to animals is both local and global. As we see from the novel, humans may move animals out of their territory, compete for resources such as fish, or simply kill the animals. It is surprising that poaching is not mentioned as, according to Ruth Padel, in Tigers in Red Weather, it is a significant problem in the Sundarbans of West Bengal (82). When we as humans indulge in activities which generate climate change we also harm the animals by bringing about the conditions for habitat change detrimental to their existence.[1]

Connections Between the Natural World and Animals

Apart from the explicit human/animal interactions in *The Hungry Tide* there are other ways in which the links between nature, humans and animals are explored. For example, during a storm the tide is 'rising, rising, gnawing at the mud and sand [which the people] had laid down to hold the river off. Imagine what went through their heads as they watched this devouring tide eating its way through the earthworks, stalking them wherever they were. There was not one among them ....who would not rather have stood before a tiger than have looked into the maws of that tide' (203). Also writing about the wreck of various ships in the nineteenth century off the Sundarban coast, Ghosh says 'today on those sites nothing is to be seen; nothing escapes the maw of the tides; everything is ground into fine silt' (225).

The dolphins' normally serene movements and soft breath match the generally calm waters. Yet when the atmospheric pressure drops before a storm their behaviour alters. Piya notices that 'they were surfacing with unusual frequency, with barely a minute or two separating their exhalations. And more than once, along with the breathing, she heard a sound not unlike a squeal.' (366). Their swimming patterns seem to be attuned to the tides. Piya conjectures that these patterns are related to salinity levels, the Irrawaddy dolphins preferring low saline levels. She is fired with enthusiasm to pursue this idea, especially as she thinks that it could have implications for conservation of the species.

Connections Between the Natural World and Humans

The Ganges and the Brahmaputra rivers converge in the tide country and two humans, Piya and Kanai come together on a train heading for the Sundarbans or tide country of West Bengal. They are from very different social backgrounds. Piya grew up in Seattle though she was born in Kolkata (Calcutta) from Indian parentage. Kanai is from New Delhi. Piya has dedicated her life to researching dolphins in different parts of the world and 'getting her hands dirty' in doing so. Kanai lives the life of the mind, an interpreter and translator of six languages. His head is full of conjugations and abstractions. Piya has lived a life of sexual austerity while Kanai has focussed on relationships with a sequence of women. In describing himself, Kanai says that he 'has loved in six languages' (353).

They converge in mutual attraction and Kanai invites Piya to stay at his aunt's place, 'Lusibari' in the Sundarbans where he is heading. When Piya's research does not begin well she takes him up on this offer. Kanai is uninterested in dolphins calling the Irrawaddys 'ridiculous porcine little things' (304). However his desire to be with Piya leads him to assist her in practical ways with her research. Curiously, they are similar in another way too. When confronted with natural dangers they panic – Piya flounders in the water when she falls off a boat and needs to be rescued. She also simply freezes when a crocodile attacks the small boat she is in. Kanai flails about in panic when be believes a tiger is nearby and he has no ready escape.

One thought that these stories prompt is that both Piya and Kanai lack a strong connection to place, Piya because of her travels and Kanai because of the abstract sort of life he leads. It is Fokir, the humble fisherman, deeply connected to place, who really excels in these dangerous situations. He jumps into the water and rescues Piya. He calmly hits the crocodile on the head with a paddle thus warding off the attack. He handles the presence of tigers by falling back on his spiritual beliefs connected to the tiger's habitat and in particular the thought that he will be protected by deities because he is a good man. The main deity who Fokir puts his faith in is Bon Bibi who rules over the tigers, crocodiles and other animals and may be called upon to forestall attacks.

The lives of Piya and Kanai come together in the Sundarbans where the two great rivers join. Piya reflects on the way that she lives as 'an alibi for a life' perhaps because of the lack of connection to anywhere or anyone. Towards the end of the novel she returns to the Sundarbans to take up a long-term study of the dolphins and talks of this as 'coming home'. It is her commitment to the dolphins which has given her the possibility to drop the alibi. There is a possibility of a deeper connection with Kanai too as we are told that he will also return.

Parallels Between Human/Animal Connections in Myth and Reality

When Kanai visited 'Lusibari' in 1970 as a young boy, he befriended the young girl Kusum whose father was kiiled by a tiger while seeking firewood in a nearby island forest. This story is relayed just after the telling of a myth about a spirit in the form of a tiger stalking a boy lured into the tiger's domain by the promise of wealth from honey. After loading up the boats he was left onshore. The boy called upon the aid of Bon Bibi who offered him protection. Kusum appeals to this deity when her father is in trouble but unsuccessfully. The tiger's mind according to Ghosh is full of judgment and intention. Friends and relatives are screaming warnings on the opposite shore but the wind means that the sound does not carry.

> [The tiger] was skilled in dealing with the wind and it knew that the people on the other bank were powerless against these gusts. So great was its confidence that in the last stretch it actually broke cover and went racing along the shore, in full view of the far shore; intent on its prey, it no longer cared about concealment. This was in itself an astonishing sight, almost without precedent, for the great cats of the tide country were like ghosts, never revealing their presence except through marks, sounds and smells. They were so rarely seen that to behold one, it was said, was to be as good as dead – and indeed the sight caused several of the women on the embankment to lose consciousness (108).

The same man who is killed by the tiger is supposed to have been saved from a tiger attack in an earlier situation by the intervention of Bon Bibi's messengers: dolphins who stayed with him until with the rising tide he could be rescued. Fokir also relates how his mother told him that the dolphins would bring back news of the rivers to Bon Bibi. They were her eyes and ears.

In a letter recounting the past beliefs, we are told that the river dolphin was originally from the sea but its mother Tethys, left the dolphin as a legacy to the two rivers, the Ganga and the Sindhu who 'nurtured it and made it their own' (182).

Future Prospects

It is the interconnected web of all life and spirit that is the central concern of *The Hungry Tide* but the novel also encourages us to ask, what will happen in the future? Will the tigers and dolphins simply become extinct? Will the whole of the Sundarbans succumb to sea level rise? There are serious attempts to set aside areas where the tigers can live in five national parks in West Bengal. The Sundarbans National Park covers about 2585 square kilometres with a 1330 square kilometre core. The rest is a forest buffer zone (Tilson and Seal 429). There are varying estimates of the number of tigers now in the reserves but as a species they are threatened. Measures to allow humans to safely gather timber, extract honey or to fish have been employed. No exploitation is allowed within the core of the Park but limited exploitation is allowed in the buffer zone. The tiger attacks on humans are correlated with the salinity of the creek water. So fresh water ponds have been dug for the tigers with good results (*Ibid,* 430-1).Tiger attacks on humans tend to be on the right nape, so workers in the forest buffer zone 'are instructed to carry strong Ceriops sticks on their right shoulders' (*Ibid,* 430). Protective fiberglass head gear with iron spikes on the back has been found to deter attacks. Attempts to make tigers fear humans have been tried with human dummies that will administer a mild electric shock with attacked. These measures have led to a significant drop in man-eating (*Ibid,* 433). With the continuation of these practices and the control of poaching it may yet be possible for tigers to continue to exist in the wild, peacefully with the human inhabitants.

In many of their traditional habitats around Asia, the Irrawaddy dolphin numbers are rapidly declining. However in a remarkable discovery by the Wildlife Conservation Society in 2009, researchers found 6,000 Irrawaddy dolphins swimming in the Sundarbans of Bangladesh and adjacent waters of the Bay of Bengal. (Wildlife Conservation Society 1). They are at risk of being caught in fishing nets and the Society is working with the Ministry of Environment and Forests to establish a protected area 'to secure the dolphin stronghold' (Wildlife Conservation Society 2).

The future of the Sundarban habitat is at risk because of climate change. Already wild storms and tidal surges lash this area as we see in the novel. Ghosh also mentions the storm of 1970 which devastated villages, smashed boats, killed 300,000 people and unrecorded numbers of animals. We are likely to see an increased number of storms with increased intensity and bigger tidal surges with the predicted changes in climate (Intergovernmental Panel on Climate Change). Nirmal, a school teacher in 'Lusibari' reflects on how rapidly the land in the Sundarbans is being reclaimed from the sea in a period of a few decades. He asks, 'What would it take, to submerge the tide country? Not much – a minuscule change in the level of the sea would be enough.' (Ghosh 215). Who should be shouldering the responsibility of mitigating the effects? This is not just a local question. It is hardly a local question at all – except if re-location of humans and animals is a possibility. While not much space is given in the novel to the exploration of this issue, if climate change continues unabated then it may well be too late to wonder about human relations with tigers or dolphins.

Notes

1 For a description of the likely effects of global warming on West Bengal see G.M. Polya. *Jane Austen and the Black Hole of British History: Colonial rapacity, holocaust denial and the crisis in biological sustainability*, 2nd ed. Melbourne: La Trobe University, 2008.

References

Bel'kovich, V.M., A.V. Agafonov, O.V. Yefremenkova, L. B. Kozarovitsky, and S.P. Kharitonov. "Dolphin Herd Structure". *Dolphin Societies: Discoveries and Puzzles.* Ed. Karen Pryor and Kenneth S. Norris. 1991. Oxford: University of California Press. 17-79.

Connor, Richard C. and Dawn Micklethwaite Peterson. 1994. *The Lives of Whales and Dolphins*. Sydney: Reed Books.

Doak, Wade. 1988. *Encounters with Whales and Dolphins*. Auckland: Hodder & Stoughton.

Ghosh, Amitav. 2004. *The Hungry Tide*. London: Harper Collins Publishers.

Hast, Fai Chivell. 1989. *Thar* (Is it Correct?) *She Blows! Whales and Dolphins in Australia*. Melbourne: Magistra Pub. Co.

Intergovernmental Panel on Climate Change, accessed 26 Aug. 2010, <http://www.ipcc.ch/about/index.htm>

Padel, Ruth. 2005. *Tigers in Red Weather*. London: Little, Brown.

Sanyal, P. 'Managing the Man-Eaters in the Sundarban Tiger Reserve of India – A Case Study'. In Tilson, Ronald L. and Ulysses S. Seal. *Tigers of the World: The Biology, Biopolitcs, Management, and Conservation of an Endangered Species*. New Jersey: Noyes Pubs., 1987. Pp. 427-434.

Seidensticker, John. 'Managing Tigers in the Sundarbans: Experience and Opportunity'. In Tilson, Ronald L. and Ulysses S. Seal. present petition.16-426.

Watson, Lyall. 1988. *Whales of the World: A Complete Guide to the World's Living Whales, Dolphins and Porpoises*. London: Hutchinson.

Wildlife Conservation Society, accessed 26 Aug. 2010, <www.wcs.org/saving-wildlife/ocean-giants/Irrawaddy-dolphin.aspx>

Chapter 30

The Poetry of Inclusion: Tukaram, Animals, and the Cosmic One

☆ *Thomas Crowley*

Introduction

In exploring the ideological roots of the current environmental crisis, scholars have highlighted the ecological shortsightedness of various mainstream Western traditions, from Christianity (see White) to Cartesian dualism (see Plumwood; Devall and Sessions; Curry) to Platonism (see Smith). While some have tried to find alternative, eco-friendly traditions within the Western canon (for instance, in Heidegger's phenomenology; see Zimmerman), others have looked East for green inspiration. Hinduism has appeared particularly attractive to many eco-critics, as evidenced by the publication of weighty anthologies about Hinduism and ecology in 1998 and 2000 (Chapple and Tucker; Nelson). An annotated bibliography on Hinduism and ecology, published in 2004, lists hundreds of articles and books on the subject (Chapple and Forum on Religion and Ecology).

Much of the green Hinduism literature focuses on the religion's major traditions and texts, from the great Sanskrit epics (see Lee; Lutgendorf) to the Vedas (see Patton) to the Bhagavad Gita (*see* Dwivedi). However, as other eco-critics have pointed out (sometimes in the same anthologies!), many Hindu traditions seem, at their heart, anti-ecological, lauding transcendence of the material, earthly world (see Nelson; Sherma).[1] Further, much of mainstream Hinduism is steeped in hierarchical thinking, which often places humans above animals and plants, in opposition to the egalitarian emphasis of contemporary environmental ethics.

Nelson makes clear the hierarchical nature of classical Vedanta, one of the most influential Hindu philosophies. "The university is not a community, but a hierarchy," Nelson writes of Vedanta, "in which gods enjoy great happiness, human beings experience moderate happiness and pain, and animals suffer 'extreme misery'" (67). Nelson also shows how this hierarchy enters the human sphere in the form of caste and gender. In the Vedantic universe, "the best possible birth is that of a human being. Better still, a human male. Better still, a male Brahmin, for such can become renouncers and escape the entire process [of birth and death]" (68).

Despite the growing emphasis on social justice in the environmental movement, the debate on Hinduism and ecology has largely skirted issues of hierarchy in general and caste in particular. In the anthology *Hinduism and Ecology: The Intersection of Earth, Sky, and Water*, only three of the twenty-one articles make any mention of caste, and none of the articles seriously investigate how caste could complicate a green interpretation of Hinduism (Chapple and Tucker). For instance, Larry D. Shinn recognizes that the ancient Hindu notion of an interdependent cosmos is largely predicated on the notion of caste, but he does not address the problematic nature of basing a modern environmental ethic on a cosmos so conceived.[2] An even more striking example: two articles listed in the Hinduism and ecology bibliography speak highly of the ecological guidelines laid down in the *Manusmriti*, an ancient law book, not mentioning the foundational role of the book in systematizing the caste system.[3]

Perhaps, then, it is time to look beyond the dominant Hindu texts and practices. There are many traditions within Hinduism that question the hierarchical nature and the otherworldly focus of conventional Hindu thought. The *varkari* sect of Mahararashtra exemplifies such non-mainstream traditions. (*Varkari* literally means pilgrim, marking the importance of this sect's annual pilgrimage (*vari*) to the town of Pandharpur.) The *varkaris* are part of the widespread but disparate *bhakti*, or devotional, movement in India, which stresses direct access to the divine, unmediated by priests or scholars. The *varkari* tradition has been propagated by a line of poet-saints, of whom Tukaram is perhaps the most popular. Tukaram is particularly deserving of critical attention because of his respect for the animal world and his questioning of caste distinctions, both of which point to his more general rejection of hierarchy.[4]

Further, since Tukaram is part of a living tradition, an ecological interpretation of his work is more likely to have a practical impact. Thousands of devotees still gather for the *vari*, which – as Chitre notes – is rooted in an ecological appreciation of the monsoon rains ("Earth and Bhakti"). Tukaram plays a central role in the present-day *vari*. One of the most important *palkhis* (covered litters) in the pilgrimage comes from Dehu (Tukaram's birthplace) and carries Tukaram's *paduka* (sandals). The Dehu devotees sings Tukaram's poems as they walk hundreds of kilometers to their final destination, ensuring that Tukaram's legacy lives on.

Outside of Maharashtra, Tukaram is known to the English-speaking world largely through the translations of Dilip Chitre. His book *Says Tuka – 1*, a reprint of an earlier collection of translations, was meant to be the first installment of a three-volume work; however, Chitre was unable to complete the final two volumes before his death in 2009. Although *Says Tuka – 1* includes only about 250 of Tukaram's poems, it is still the best resource for English-speakers wishing to understand the spirit of Tukaram's poetry. The most complete translation of Tukaram's poetry into English (3721 poems in all) was completed in 1915 and, unlike Chitre's artful translation, it is made up of prose paraphrases that do not capture the poetic nature of the original (Chitre, "Introduction" xxiii).

Animals make many appearances in *Says Tuka – 1*. These appearances provide a useful lens for viewing Tukaram's rejection of hierarchy and his ethics of inclusivity. A close reading of *Says Tuka – 1* reveals not only Tukaram's familiarity with the animal kingdom, but also his deeply holistic worldview, which stresses the continuity and the divinity of all life.

Tukaram's Biography as Context for his Poetry

Tukaram's unorthodox worldview emerged, in part, from his unusual life experiences. A short summary of Tukaram's life will help place his poetry, and his treatment of animals, in context. Most accounts of his life are more hagiography than biography, and much of the biographical information we have is from his own poems; as a result, we know more about the legends than the actual details of Tukaram's life. However, some basics are well established (see Chitre, "Introduction" and More).

Tukaram was born in 1608. His caste background reflects the complexity of social classification in India. His parents were affluent traders from the Maratha *jati*.[5] Tukaram often identifies himself as a merchant, a typical occupation of the Vaishya *varna*, while today Marathas are seen as part the Kshatriya *varna*. In Tukaram's time, however, Brahmins considered everyone outside their *varna* as Shudras. This last categorization was most important in Tukaram's life as a poet-saint.

Since Tukaram's elder brother had chosen a path of renunciation at an early age, Tukaram was groomed by his father to take over the family business, which included not only trading, but also farming and money-lending. When he was still in his teens, his mother and father died in quick succession, and his elder brother left for a pilgrimage and never returned. So at eighteen, Tukaram found himself in charge of a large, profitable business, which he handled ably.

Just three years later, a major famine hit Maharasthra. Tukaram's first wife and only son died in the famine, and Tukaram lost all desire to run the family business. Like the Buddha, Tukaram saw suffering, and his agony pushed him onto a spiritual path. He began to spend long periods of time in the wilderness, and he used what money he had to renovate his family's shrine to Vitthal, the god at the center of the monotheistic *varkari* tradition. (Vitthal is also called Pandurang by Tukaram and other devotees; although the god's origins are obscure, by the time of Tukaram, he had been incorporated into Vaishnavite tradition and was worshipped as Vishnu.)

One night, as the legend goes, Tukaram had a dream in which Vitthal and Namdeo, a fourteenth-century devotee of Vitthal, told him to "stop fooling around" and start writing poems (Tukaram 3). This was the start of Tukaram's career as a poet-saint. As the years went by, his body of work grew larger and larger, as did his legions of followers. This rise to fame occurred despite widespread opposition from

Brahmins, who believed that a Shudra like Tukaram had no place composing religious verse.

At the height of his popularity, Tukaram disappeared. Some claim that Tukaram simply vanished into thin air while leading a session of devotional singing; other followers say that he was taken up to heaven by Vishnu's eagle mount, Garuda. Some darkly claim that he was murdered by jealous Brahmins, although this claim has been thoroughly discredited. Whatever the case, Tukaram left behind a vast body of work – between 5,000 and 8,000 poems (Chitre, "Introduction" xiv).

Animals in Tukaram's Grounded Poetics

Tukaram's poems are striking for their down-to-earth, sometimes profane, nature. Writing in the vernacular (Marathi) as opposed to the usual language of high religion (Sanskrit), he talks of his desire to be a pot to receive the Lord's shit (Tukaram 87); he jokes about a man trying to bugger a donkey (149); and he excoriates the man who farts too much, with his "arse bared to public view" (172). He also talks extensively about the soil, the water and the animal kingdom, from glow-worms to tigers. Even his mystical imagery is corporeal; he speaks of the Vast One's "hairs/ Of infinity" (27). He is far removed from the ethereal overtones often associated with the religious; he tells Vitthal, "I am/ The soil/ Touched by/ Your feet" (88).

In its accessibility, Tukaram's poetry marks a deepening of a trend that began with Jnandev, the *varkaris'* first major poet-saint.[6] Jnandev (also known as Jnaneshwar) was the first influential religious poet to write in Marathi. In contrast to Vedantic philosophers, Jnandev held that "the phenomenal world was not illusory or unreal, but a unique and unrepeated expression of creativity underlying everything that comes into being" (Chitre, "Notes and Glossary" 267). So, from the very beginning of the *varkari* poetic tradition, there was an emphasis on the sensuous particulars of day-to-day life.

However, Tukaram's poems have a folksy, colloquial feel that Jnandev's poetry lacks. Jnandev's works are abstract and philosophical, while Tukaram's emphasize lived experience. In the words of Dilip Chitre, "to [Tukaram], all poetry was empirical and so was all religion" ("Introduction" xx). Tukaram himself puts it bluntly: "Life is too short for me to read books" (159).[7] Refusing to stray too far into abstraction, Tukaram keeps his poems deceptively simple; he bases his holistic, egalitarian ethics on his first-hand experiences and frames it in the language of animals and agriculture.

Tukaram's frequent invocation of animals (as well as his more general groundedness) should be seen in the context of his historical setting and his caste background. As a Shudra in a pre-industrial society, Tukaram interacted frequently with animals and depended on them in immediate ways.[8] By contrast, Brahmins, even in pre-industrial times, promoted an ideology that separated them from the day-to-day concerns of tending animals, growing food and, more generally, producing the material conditions necessary for a smoothly-running society. As Kancha Ilaiah (2007) argues, "The Brahmin mind…is prohibited from interacting with nature or with production tools, indeed with any of the forces of production. Human beings are not supposed to relate to nature and to other human beings, they must only relate to the 'other world.'" (44). Not limited by this restrictive worldview, Tukaram strongly identifies with the animal world.

The role of animals in Tukaram's grounded poetry deserves careful examination. Because Tukaram is a poet, not a systematic philosopher, his reflections on animals (among other topics) cannot be neatly summarized; they are sometimes contradictory, they shift in tone and emphasis, and they defy easy categorization. Yet, through the tangle of meanings and subtexts, some patterns emerge in the poems of *Says Tuka – 1*. Of the many ways that Tukaram uses animals, two are of particular note. First, Tukaram frequently uses animals as parts of metaphors or allegories. While such metaphorical invocation surely shows Tukaram's familiarity with the animal kingdom, it does not prove his deeper ethical respect for animals. This proof can only be found in Tukaram's second, less frequent, use of animals: not as part of literary devices, but as members of his all-inclusive ethical framework.

Animal Metaphors – and Beyond

Tukaram's most frequent references to animals are metaphorical. Like Jesus, he is fond of parables; he speaks in fables and allegories as he tries to make deeper spiritual points. Just as "for a bedbug, the bed is a mountain/ So hard to ascend" (Tukaram 150) and "insects in a fig/ Cannot imagine/ Worlds other than the fig" (27) so too – says Tukaram – the human mind has trouble grasping the greatness of God. In another poem, a clever tiger and a trickster jackal represent self-serving egoists (151); in yet another, those receptive to Tukaram's message are described as thoroughbreds, while those who aren't are common horses (156).

There are many more examples. Of those who only pretend to worship God, Tukaram says, "The spark of a glow-worm/ Cannot kindle a flame" (175). Speaking of "the art of seeking," Tukaram notes, "The empty space between a bird and a fruit/ Is never an obstacle for the bird" (173). Instructing his listeners to follow their own thoughts and beliefs, he asks "Who tells an unborn snake to hide itself/ As soon as it emerges from an egg" (163)? Such rhetorical tools show Tukaram's awareness of the animal world, and they would have resonated with his contemporary listeners, most of whom worked the land.

In addition, Tukaram uses similes and metaphors to suggest a certain commonality between humans and the rest of the animal kingdom. He often compares himself to animals. Sometimes, this is to express his spiritual anguish: he is "like a stubborn/ Bullock" (37), "like a monkey/ In a tree" (45), "a moth unable to reach the flame" (61), or a tortured, twisting fish "parted from water" (79). In a more cheerful mood, he speaks of himself as "God's own dog" (99-106) and a rooster crowing (162). Even more than his animal allegories, Tukaram's human-animal metaphors suggest an identification with and sympathy for the animal world; at one point, Tukaram directly says, "We too are creatures like ants" (161). By comparing humans to lowly ants, Tukaram is both expressing humility and identifying the commonality of all God's creatures.

However, it is important to note that his metaphorical use of animals does not necessarily imply any deep ethical or spiritual recognition. One would not call George

Orwell an animal rights activist because he penned *Animal Farm*. The portrayal of animals in literature often reveals ambivalence rather than straightforward ethical respect (see Soper 81-90). It may seem obvious that simply talking about animals does not imply respect for them; however, the two notions are commonly conflated in the literature on religion and ecology. For instance, Puri, arguing for a green interpretation of Sikhism, uses a common Sikh animal metaphor that hardly seems sympathetic to the embodied world: "Man's mind is like a wild elephant wandering bewildered in the forest of Maya delusion; worldly material objects drive him about" (14). While, in other parts of his book, Puri does give convincing arguments for the ecological sensitivity of Sikhism, his larger thesis is not helped by his references to metaphors that compare animals to unenlightened states.

Writing about a different Indian tradition, Rukmani interprets yoga as an eco-friendly practice because many of the common *asanas* (postures) are named after animals. Despite this superficial attention to animals, yogic philosophy, at its core, appears deeply un-ecological. Classical yoga is based on dualistic Samkhya philosophy, which separates the *purusha* (the seer, the true self) from *prakriti* (primal matter or nature), and urges the yogi to separate himself from impure nature (see Sherma 95-102).[9] Alerted by these examples, we should be cautious while drawing connections between Tukaram's poetic use of animals and his ethical respect for them. Still, it is important to recognize that awareness of animals is a necessary (although not a sufficient) condition for a deeper respect. Without being aware of animals, their behavior and their interactions, it is impossible to build any practical ethical system around them. Tukaram's environmental ethic, then, is rooted in his everyday connections to real-life animals.

Animals as Part of an Inclusive Ethics

Tukaram's ethical respect for animals derives from his more general embrace of an interconnected cosmos. Ironically, Tukaram had to escape from human society before he could realize the connection between humans, animals and the divine. Like many great religious figures, Tukaram went into wilderness to find what transcends the human.[10] Tukaram's linking of wild space and divine revelation could suggest a potential divide between mundane human culture on the one hand and ecstatic wilderness on the other; however, the very enlightenment Tukaram reaches in the wilderness allows him to transcend the divide between city and forest and see divinity in the unified whole of life. In his enlightened state, Tukaram exclaims: "The forest and the city are identical:/ There are no divisions in my kind of space" (240). This earthly unity is explicitly linked to God's all-pervasive presence; as Tukaram says, his poetry "will find otherness nowhere/ In crowds and forests I find Pandurang" (28). Emphasizing our participation in the divine, he says, "God is a vast jig-saw puzzle of which we are/ Mere tiny bits," (157) and "there is Pandurang/ In every thing" (235).

With God suffusing all worldly things (or, to put it more poetically, with God "the very/ Being of light/ Contained/ In every ray" (24)), Tukaram experiences continuity and community everywhere. Speaking of his divine revelations, he equates "absolute being" with "continuity" (208), saying, "those who experience/ Cosmic

Being/...See all things as equal.../ The whole argument/ About difference/ Ends here" (184). Since all beings share in this continuity, Tukaram emphasizes the importance of community. He enjoins his listeners to "blend with all/ You'll join/ A global family" (177). And: "being one with all is being beyond all bonds" (55). Plants and animals have a prominent place in the cosmic community; Tukaram says, "Trees, creepers and the creatures of the forest/ Are my kith and kin./ And birds that sweetly sing" (182). Before he has reached enlightenment, Tukaram prays that, when he is blessed by Vitthal, "these trees/ These shrubs/ These stones/ Will relate to my blood" (192). Thus, the mystical communion he strives for is ecologically sensitive, evoking a deep sense of connectedness not just with living beings but the geological world that supports them.[11]

Despite this sensitivity, some phrases in Tukaram's poetry seem, at first glance, to be arguing against the value of the body and the material world. For instance, Tukaram urges his listeners to "give up [their] bodily concerns" (95); he rants, "To hell with the body/ To hell with its bonds" (85). He even says that "life in this world is unreal" (83). However, seen in the context of Tukaram's overall philosophy, these statements should not be read as simple echoes of Vedanta, with its nature-denying stance that all we see is an illusion. (Tukaram himself said, "I just do not allow the language of Advaita [Vedanta] to come to my ears" (qtd. in More 6).) Rather, these statements express Tukaram's concern with going beyond the narrow, ego-bounded body and seeing the divinity in all things.

In modern religious terms, Tukaram is a panentheist: he holds that God is both transcendent and immanent. He states that God "informs every living being/ Yet He is Being itself, apart from all its forms" (157). By reflecting on the nature of the divine in this world, Tukaram suggests, we can catch glimpses of the divine spirit that transcends all.[12] With his panentheistic view of the divine, Tukaram exhorts his followers to, "Scatter your own ego to the winds,/ Then you will pass the crucial test" (169). When we abandon our ego, we realize that "We are blind./ You [Vitthal] are our legs./ You are our thoughts./ You are our walk" (73).

Statements like this reinforce the idea that Tukaram is not criticizing the earthly world as such, but rather the earthly world divorced from divinity. Claiming that all pursuits except the worship of God are "unreal" (167), he argues "God is not uniform:/ What makes Him/ Homogenous/ Is purity of heart/...You must come/ Out of your involvements/ And vested interests/ to grasp Him/ Whole" (75). That is, we see differences only when we are wrapped up in our "vested interests;" when we have left our egos behind, we can see the continuity and pervasiveness of God.

Tukaram practices what he preaches, recognizing, "It is/ The Cosmic one/ Making me speak" (12). He describes his transformation: "I was confined to the sole territory of my ego/ Having renounced it, I enjoy the abundance of it all/... The whole world is His imprint/ And I wear it like a tattoo on my body" (223). Far from denigrating the body, Tukaram is celebrating it as a reflection of a divine imprint.

Tukaram is explicit about the ethical implications of the ego-less, expanded sense of self he lauds. One remarkable poem on this subject deserves to be quoted in full:

Don't kill a snake
Before the eyes of saint
For the saint's being
Includes all living things
And he's easily
Hurt

A single hair
Plucked from one's body
Causes instant pain
And the soul that perceives
Life as a community
Always suffers

To those who equate
Their being
With everything
Even the other
Is a sensation
Of the self

Says Tuka
I give you
The very root
Of ethics
Be happy to worship
The Lord of All Being
In your own being (Tukaram 191).

In this poem, Tukaram moves from concern about a specific animal to a concern for all life, emphasizing that a saint quite literally feels the pain of other creatures. From his metaphysical assertions about God's omnipresence and an expanded sense of self, he derives "the root of all ethics."[13] Since Tukaram feels that his being "includes all living beings," he stresses the importance of mutual love and non-exploitation. "Don't let the bond of love break" (155), he says in one poem. In another: "Don't crush a flower/ To possess its fragrance/... O people/I'm telling you a principle" (164). Thus, Tukaram urges his listeners to develop compassion and ethical respect for all in the cosmic community.

While this article has focused on Tukaram's portrayal of animals and plants, his inclusive ethics also extends to other groups outside the mainstream. He has a keen sense of social justice and speaks in defense of the downtrodden, especially lower castes. He says, "He who identifies/ With the battered and the beaten/ Mark him as a saint/ For God is with him" (185). He states in no uncertain terms: "The *Brahmin* who flies into rage at the touch of a *mahar* [from a so-called "untouchable" caste]/ – That's no *brahmin*/ The only absolution for such a *brahmin*/ Is to die for his own sin" (140).

Tukaram's inclusive ethics also extends to issues of gender. In keeping with the *varkari* tradition, he emphasizes the feminine aspects of the divine. Unlike Yoga and Vedanta philosophies, which associate masculinity with the true self and femininity with illusory nature, the *varkaris* view both the immanent and transcendent aspects of God as a union of male and female. As Chitre notes, *varkaris* refer to the poet-saint Jnandev as *Maooli*, or Mother, "a significant use of the feminine gender that conveys their belief that their Cosmic Parent is an indivisible couple" ("Notes and Glossary" 279). Tukaram refers to Vitthal as both Vithoba (Father Vitthal) and Vithabai (Lady Vitthal). In addition to referring to God as Narayana (a name of Vishnu that literally means "son of man" or "son of waters"), he often invokes the divine simply as Mother. In one poem, he even calls God "my Father, my Mother" (188). What is more, he refers to himself in the feminine: "'She's crazy for God!'/ Says the world of me" (213). He feels a deep, corporeal connection to the feminine divine, saying, "You are my mother" (84) and pleading, "Suckle/ My soul/ And/ That's/ No metaphor" (20). In opposition to the male-heavy Brahminism dominant in his time, Tukaram and the *varkaris* unabashedly embrace the feminine as an equal partner of the masculine.

Tukaram, in his inclusive ethics, questions the value-laden divide between masculine and feminine, low caste and high caste, humans and nature. His radical egalitarianism is the direct result of his ethical holism, in which all beings, human and non-human, are part of the "Cosmic One." This sense of divine community extends to Tukaram's own self-conception; he sees himself as a vessel through which God flows. Summarizing his life, he says, "My only assets in this world are/ The poems Pandurang made me speak" (256).

Tukaram Today

A defender of the feminine divine, a hero of the downtrodden, a companion of trees and birds, Tukaram is a truly progressive figure. Defying the religious authorities of his day, he wrote poems that were at once brutally honest and beautifully mystical. Throughout his work, he stresses the importance of community and inclusion. Because of his compassionate, inclusive spirit, as well as his down-to-earth poems, Tukaram developed a huge following in his life; on many occasions, large crowds gathered to hear Tukaram sing his poems and to join in with the choruses. To this day, *varkaris* come together to sing Tukaram poems in gatherings open to all, a reflection of the power of song to build and sustain religious communities.

Recently, *varkaris* have been invoking Tukaram's name in a much more political context. In early 2008, activists in Maharasthra began protesting the construction of a Dow Chemicals R&D facility in Shinde Vasuli, a village not far from Tukaram's birthplace. In India, Dow is largely known as the company that bought Union Carbide, which was responsible for the 1984 Bhopal disaster. Dow has consistently refused to give adequate compensation to the victims of the Bhopal tragedy, despite evidence that (as of 2009) the site was continuing to leak toxic materials (Jayaraman). Further sullying Dow's reputation, a 2003 study found that Chlorpyrifos, a Dow insecticide, was found in 29 of 34 samples of bottled water from India; the amount of the pesticide in the water ranged from 3 to 370 times the legal limit (Mathur et al. 22).

In this context, the villagers of Shinde Vasuli were understandably worried when the government repeatedly failed to provide information about the upcoming R&D facility, despite many requests. The protests against the facility – originally organized by left-wing activists – received little attention until an influential *varkari* leader, Banda Tatya Karadkar, decided to join in. He brought many *varkaris* with him, and the protests grew. As villagers became increasingly frustrated by the lack of government response, a small group of protesters gathered at night and burned down part of the construction site. Karadkar took responsibility for the arson; he was arrested, but because of his popularity in rural Maharashtra, police soon decided to release him. For the next several months, the future of the facility remained up in the air. This was largely for political reasons; elections were approaching and no party wanted to alienate the politically strong *varkaris*. Finally, after the elections, due to continuous pressure from the anti-Dow movement, construction of the R&D facility stopped.

Perhaps the most remarkable part of the protest (leaving aside the problematic use of arson) was its invocation of Tukaram.[14] *Varkaris* sang Tukaram poems at anti-Dow gatherings. Karadkar, in an interview with the press, said, "Tukaram wrote his world famous spiritual treatise which also urged the masses to revere and nurture Mother Nature with all its grandeur near the village where the 'polluting plant' was coming up" (Gonsalves). More generally, the protest was framed as an effort to save the land of Tukaram from pollution ("Anti Dow Chemicals Movement").

The anti-Dow movement is a compelling example of the modern-day relevance of Tukaram. The poet-saint's ecological sensitivity still resonates with the farmers who work the land of Maharashtra. However, given that *varkaris* have also, in recent years, voiced nativist and anti-Muslim sentiment, Tukaram is also a reminder that even the most progressive religious figures can be co-opted by forces of intolerance. Still, with his connection to plants and animals, his strong critiques of social injustice and his balance between masculine and feminine notions of the divine, Tukaram remains an inspiration.

India is facing massive environmental problems, from extensive water shortages to widespread soil erosion to dangerous levels of air pollution. Some have attempted to develop a homegrown Hindu environmentalism to combat these problems; however, these attempts have too often failed to grapple with the hierarchical nature of mainstream Hindu thought. A shift away from the mainstream – to figures like Tukaram – can bring attention to more egalitarian traditions within Hinduism. Such traditions are powerful because they resonate with a majority that has long been marginalized by the socio-religious elite. The people who make up this majority are both the most connected to the land and the most severely affected by environmental degradation, and are thus in a unique position to lead a socially-conscious environmental movement. If an egalitarian Hindu sect can energize a grassroots movement that triumphs over the world's second-largest chemicals company, there's much cause to celebrate non-mainstream traditions – in India and beyond.

Notes

1 Even the articles that have focused on lived traditions largely pick mainstream practices, such as Durga *puja* (goddess worship – see Khanna) and *kolam* drawing (ritual drawings made of grain, left to be eaten by animals – see Nagarajan). These very practices are becoming un-ecological, with clay Durga idols being replaced with Plaster-of Paris, chemically-dyed ones (as Khanna recognizes), and the grain for *kolam* being replaced with inedible but cheap synthetic powder (as I personally witnessed when traveling through Tamil Nadu).

2 Shinn's article focuses on Gandhi's reinterpretation of ancient Hindu thought. While Shinn acknowledges Gandhi's strong opposition to untouchability, he does not adequately explore Gandhi's overall support for the caste system (which drew bitter criticism from anti-caste leaders like Ambedkar).

3 The Dalit leader Ambedkar famously burned this book as part of an anti-caste protest (see Omvedt 151). (Dalit, literally meaning "the oppressed" in Marathi, refers to castes formerly known as "untouchable.")

4 As Chitre suggests, anti-Brahminical movements in India, from Tukaram's *varkari* tradition to Budhhism and Jainism, have often been the most vocal defenders of non-violence and animal rights ("Introduction" xxvi).

5 The English word "caste" generally conflates two Sanskrit terms: *varna* and *jati*. The *varna* system (expounded in major Hindu texts like the Manusmriti and the Bhagavad Gita) divides society into major social classes, most commonly into four categories: Brahmins (priests and scholars), Kshatriyas (warriors and administrators), Vaishyas (merchants and artisans) and Shudras (laborers). *Jati* is a much more fine-grained term; it describes the thousands of endogamous groups in India. *Jatis* are more fluid and region-specific, and serve as a better description of on-the-ground reality in India; however, each *jati* is often assigned a place in the broader *varna* system.

6 Jnandev's background is further evidence of the complexity of Indian socio-religious life. According to folk history, Jnandev was born a Brahmin, but was excommunicated from his caste because of the transgressions of his father (who became a *sannyasin* (renunciate) then went back on his vows and resumed family life). After his parents committed suicide, Jnandev was initiated in the heterodox Shiva-worshiping Nath sect. He then went on to become a main saint of the Vishnu-worshiping *varkaris*.

7 This statement must be taken with a grain of salt; Tukaram, in his other poems, shows an impressive mastery of many religious texts. However, this mock-modest

statement makes clear his preference for *bhakti* (devotion) over intellectual exchange.

8 This is not to say that humans in say, modern cities, are totally independent from animals; however, the connection is felt much less viscerally if one buys one's meat pre-cooked in a restaurant.

9 This is not to say that a green defense of yoga cannot be constructed; however, it should recognize the problematic dualism of traditional yogic thought. On a practical level, many modern yoga studios are promoting a "green yoga"; of course, the exercises performed in these studios have very little resemblance to the yoga of classical Indian philosophy.

10 In addition to the Christian example of Jesus in the desert, the practice of finding transcendence outside of the bounds of society is common in Hindu lore (see Lutgendorf). In Tukaram's case, the retreats into the wild seem like more than legend; the available historical evidence suggests that Tukaram frequently visited two hills on the outskirts of his town (Chitre, "Introduction" xx); these hills are still visited by his devotees.

11 Tukaram also sees divinity in water. According to folk history, as well as Tukaram's own poems, some Brahmins forced Tukaram to submerge his manuscripts in the Indriyani River. After thirteen days, the manuscripts emerged from the river undamaged. Of this experience, Tukaram tells Vitthal, "You were with me and deep inside the river, too" (198).

12 Some of Tukaram's more powerful poems address the difficulty of worshiping a being that transcends normal human reason; he says, for instance, "We sing songs addressed to You,/ though You are way beyond words" (113). Some of his poems (including one with that begins "Water has turned/ Into the sky" and ends "Sunrise is sunset" (212)) are reminiscent of Zen koans, which are intended to shock the listener into an intuitive understanding of transcendental truths.

13 In deriving ethical principles from a metaphysical notion of enlightened selfhood, Tukaram follows a line of thinking that eco-philosopher Arne Naess takes up centuries later. (See Naess 103; Naess was quite influenced by Gandhi's interpretation of Hinduism; to complete the circle, Gandhi wrote, "Tukaram is very dear to me" (qtd. in "Tukaram and Gujarat").)

14 Karadkar invokes Tukaram even to justify the arson, quoting a Tukaram poem: "To a good man, we will even give our loin clothes. But for a bad man, we will have our sticks ready" (qtd. in "Warkari Leader Blames Govt"). Karadkar is used to being at the center of controversy; before his involvement in the anti-Dow movement, he was known for his anti-Islamic sentiments. The left-wing organizers of the anti-Dow movement took some flak for joining with Karadkar, but decided that, in the case of the Dow plant, their goals and Karadkar's were the same. Progressive groups see the anti-Dow movement as the first step in moving the varkaris away from their recent association with the hard-right nativist Shiv Sena, and towards their roots as defenders of tolerance and inclusivity.

References

"Anti Dow Chemicals Movement." *Yuva Bharat*. N.p., n.d. Web. 7 Aug. 2010.

Chapple, Christopher Key and Forum on Religion and Ecology. 2004 "Hinduism and Ecology Bibliography." *The Yale Forum on Religion and Ecology*. Forum on Religion and Ecology. Web. 23 Jul. 2010.

Chapple, Christopher Key and Mary Evelyn Tucker, eds. 2000. *Hinduism and Ecology: The Intersection of Earth, Sky, and Water*. Cambridge: Harvard University Press.

Chitre, Dilip. "Earth and Bhakti." *Tukaram.com*. Tukaram.com, n.d. Web. 2 Aug. 2010.

_____. 2003. Introduction. *Says Tuka – 1. The First Volume of the Selected Poems of Tukaram*. By Tukaram. Trans. Dilip Chitre. Pune: Sontheimer Cultural Association. xiii-xli.

_____. Notes and Glossary. 2003. *Says Tuka – 1. The First Volume of the Selected Poems of Tukaram*. By Tukaram. Trans. Dilip Chitre. Pune: Sontheimer Cultural Association. 259-298.

Curry, Patrick. 2006. *Ecological Ethics: An introduction*. Cambridge: Polity Press.

Devall, Bill, and George Sessions. 1985. *Deep Ecology: Living as if nature mattered*. Salt Lake City: Gibbs M. Smith.

Dwivedi, O. P. "Dharmic Ecology." Chapple and Tucker 101-126.

Gonsalves, Michael. "Judges Join Protest against Dow Centre in Maharashtra." *MyDigitalFC.com*. Financial Chronicle, 30 Jul. 2008. Web. 25 Jul. 2010.

Ilaiah, Kancha. 2007. *Why I Am Not a Hindu*. 2nd ed. 2005. Kolkata: Samya.

Jayaraman, Nityanand. "Bhopal: Generations of Poison." *CorpWatch*. CorpWatch, 2 Dec. 2009. Web. 4 Aug. 2010.

Khanna, Madhu. "The Ritual Capsule of Durga Puja: An Ecological Perspective." Chapple and Tucker. 469-498.

Lee, David. "The Natural History of the *Ramayana*." Chapple and Tucker 245-268.

Lutgendorf, Philip. "City, Forest, and Cosmos: Ecological Perspectives from the Sanskrit Epics." Chapple and Tucker 269-290.

Mathur, H.B. et al. "Analysis of Pesticide Residues in Bottled Water." *Centre for Science and Environment*. Centre for Science and Environment, Jan. 2003. Web. 3 Aug. 2010.

More, Shreedhar. "Tukaram: A Brief Biography." *Tukaram.com*. Trans. Vijay Lele. Tukaram.com, n.d. Web. 1 Aug. 2010.

Nagarajan, Vijaya. "Rituals and Embedded Ecologies: Drawing *Kolams*, Marrying Trees and Generating Auspiciousness." Chapple and Tucker 453-468. Print.

Naess, Arne. Interview by David Rothenberg. 1993. "Everything Really Important is Dangerous: A Discussion with Arne Naess." *Wisdom in the Open Air: the Norwegian Roots of Deep Ecology*. Eds. Peter Reed and David Rothenberg. Minneapolis: University of Minneapolis Press. 99-111.

Nelson, Lance E. "The Dualism of Nondualism: Advaita Vedanta and the Irrelevance of Nature." Nelson 61-88.

_____. ed. 1998. *Purifying the Earthly Body of God: Religion and Ecology in Hindu India.* Albany: State University of New York.

Patton, Laurie L. "Nature Romanticism and Sacrifice in Rgvedic Interpretation." Chapple and Tucker 39-58.

Plumwood, Val. *Environmental Culture: The ecological crisis of reason.* London: Routledge, 2002.

Puri, Gopal Singh. 1995. *Man: Environment Crisis and Sikh Faith.* Delhi: Falcon Books.

Rukmani, T. S. "Literary Foundations for an Ecological Aesthetic: *Dharma,* Ayurveda, the Arts and Abhijnanasakuntalam." Chapple and Tucker 269-290.

Shinn, Larry D. "The Inner Logic of Gandhian Ecology." Chapple and Tucker 213-244.

Sherma, Rita DasGupta. "Sacred Immanence: Reflections of Ecofeminism in Hindu Tantra." Nelson 89-132.

Smith, Huston. 1973. "Tao Now: An Ecological Testament." Earth Might Be Fair. Ed. Ian G. Barbour. Englewood Cliffs, NJ: Prentice-Hall. 62-81

Soper, Kate. 2000. *What is Nature? Culture, Politics and the non-Human.* 1995. Oxford: Blackwell.

Tukaram. 2003. *Says Tuka – 1. The First Volume of the Selected Poems of Tukaram.* Trans. Dilip Chitre. Pune: Sontheimer Cultural Association.

"Tukaram and Gujarat." *Tukaram.com.* Tukaram.com, n.d. Web. 3 Aug. 2010.

"Warkari Leader Blames Govt Apathy for Violence at Dow Chemicals." *ExpressIndia.com.* The Indian Express Limited, 6 Aug. 2008. Web. 30 Jul. 2010.

White, Lynn. "The Historical Roots of Our Ecological Crisis." *Science* 155 (1967): 1203-1207.

Zimmerman, Michael. "Martin Heidegger: Antinaturalistic Critic of Technological Modernity." *Minding Ecology: The Philosophers of Ecology.* Ed. David Macauley. New York: The Guilford Press. 59-81.

Chapter 31

Representational Ethics: Dog Narrators and Focalisers in Patrice Nganang's *Dog Days: An Animal Chronicle* and Paul Auster's *Timbuktu*

☆ *Wendy Woodword*

Traditionally, nonhuman animals have been figured as narrators of folktales, fables and allegories. These tellers of stories are entirely accepted within such genres; the nonhuman animals speak in human language and often embody a wisdom inaccessible to their human counterparts. The animal speakers are agentive, directing the readers to new points of view, which may be part of the didactic and ethical function of the story. But when contemporary writers have dogs as narrators, critics and readers tend to be discomfited, either dismissing the story as sentimental or, possibly, denying the very dogness of the narrator, thus occluding the ethical substance of such writing strategies. A number of contemporary novelists have chosen to have dogs as partial narrators or focalisers or as one point of view in the narrative, for example Orhan Pamuk's *My Name is Red*, John Banville's *The Infinities*, David Wroblewski's *The Story of Edgar Sawtelle*, Kerstin Ekman's *The Dog*. But Patrice Nganang in *Dog Days: An Animal Chronicle* and Paul Auster in *Timbuktu* go further than the above-named novels: Nganang has Mboudjak the dog as a narrator throughout while Auster has the canine Mr Bones as a focaliser rather than a narrator *per se* with the novel, almost entirely, filtered through his point of view.[1]

Centrally incorporating the points of view of dogs suggests that the writers, in their very different ways, are recognising dogs as "knowing subjects", an acknowledgement which challenges the basic tenets of humanism with its exclusive focus on humans as epistemologically superior beings. Having dogs as narrators and focalisers is inherently paradoxical however: linguistically and cognitively they are, of course, an impossibility, but imaginatively and ethically, I would argue, they can be justified. Such tensions exist in both novels but if, conventionally, readers willingly suspend disbelief about the capabilities of a human narrator and the transparencies of human minds, then a reader willing to play with notions of human-animal subjectivities and epistemologies is more likely to be convinced by the fiction of a dog as narrator or focaliser. Such a reading accords animal roles and subjectivities the ethical attention they deserve, and considers, as well, the (im)possibility of representing a dog as a central consciousnesses in fiction.

A posthumanism which extends epistemological subjectivity to animals in ways that are not merely pluralist is a useful concept to harness for the essay which follows. Cary Wolfe, in his analysis of posthumanism, implicitly critiques the humanist constitution of the "knowing subject" (569) and he enlists Derridean philosophy to dismantle hubristic notions of the "'auto'" in relation to "autonomy,.. agency .. authority over one's own autobiography" (570). Thus the "embodied finitude that we share with nonhuman animals" (570) is extant on a physical and mortal level and in a discursive realm which renders us "radically other ... in the materiality and technicity of a language that is always on the scene before we are, as a radically a human precondition for our subjectivity, for what makes us human" (571). To be entirely comfortable, then, with the authority of human narrators is to ignore their own "finitude" within language but such a fiction goes largely unrecognised in our reading of conventional novels.

Ironically, even as these novels aspire to the posthumanist, their very rendering of dog consciousness may have its limitations; Donna Haraway, for example, is adamant that the (animal) subaltern, in general, should not be made to speak (20). Adam Begley is dismissive of *Timbuktu* as mere "sentimental whimsy" for, in his opinion, "Auster can't quite dodge the silliness inherent in transcribing canine cogitation." (http:www.guardian.co.uk/books/1999/may/29/fiction/paulauster). Ruediger Heinze, sounding like a latter-day Descartes, is adamant that "the reader has no frame of reference against which to judge the experientiality of a dog or a machine" (n.p.). Wendy Doniger, on the other hand, asserts that "[n]o one can prove that someone else does *not* know how animals feel" (103).

That dogs should understand and be capable of wielding human language does, of course, challenge the bounds of conventional realism, but the project of having a dog as a narrator or focaliser does not have to be rejected on realistic grounds. My reading of these novels, while it may touch on mimetic issues, stresses the imperative for the reader to accede to some epistemological unsettling and the ethics inherent in a writer's imaginative representation of nonhuman animals. In her analysis of poetry, Elizabeth Costello speaks of the "sympathetic imagination" (quoted in Coetzee 35) which may be without boundaries as she stresses the "faculty [o]f *sympathy*, that allows us at times to share the being of another"(34). Helen Tiffin celebrates literature which has animals speaking in the dominant discourse, regarding this writing strategy as "focal in the restitution of an animal subalternity" because it engages with what she terms "the vexed issue of the species boundary"(17). Martha Nussbaum interprets any projection of the imagination into the life of another being as an ethical act, suggesting that "imagining the lives of animals makes them real to us in a primary way, as potential subjects of justice" (355). Read within such a theoretical framework, the novels have a feasibility of their own. The novelists under discussion not only reject the conventional dualism of the species boundary but undermine other dualistic assumptions: the dog narrators may be both subaltern and hero, artist and animal, part of culture, within which they are firmly ensconced, as well as nature.

Having dogs as narrators by necessity engages with the fraught issue of animal consciousness, one which J.M. Coetzee has Elizabeth Costello belittle in *The Lives of Animals* as "Descartes's key state, which has an empty feel to it: the feel of a pea rattling round in a shell"(33). Western philosophical discourse, in which Descartes is a central figure, "reduces every thing to [human] subjective experience and consciousness" as Atterton and Calarco maintain (xix). Thus Georges Bataille can claim that if we try to imagine the "universe" through the gaze of an animal "we can only call up a vision in which we see nothing" (34). He proposes instead poetic discourse which "describes nothing that does not slip toward the unknowable" in order to try to convey "immanent animality."(34-35). The fiction under discussion contradicts such definitions. While both Auster and Nganang depict animality as embodied, the former has animality as transcendent and the latter has dog experience as knowable and intellectualised. Auster's Mr Bones is an artist, constructing comforting fictions about his beloved human companion's death and dreaming intricately about him as he aspires to a milieu of the spirit. Nganang's Mboudjak is a social scientist, an observer and researcher of humanity, who remains, mostly, in the quotidian, and we see much through their points of view.

If nonhuman animals' inability to speak human language has meant that they have been categorised as lacking in rationality or consciousness, then having the dogs engage in human language as both writers do, conversely misses the potential to imagine other modes of canine expression or communication. In "Eating Well" Derrida suggests:

> If one defines language in such a way that it is reserved for what we call man, what is there to say? But if one reinscribes language in a network of possibilities that do not merely encompass it but mark it irreducibly from the inside, everything changes ... These possibilities or necessities, without which there would be no language, are themselves not only human (116).

For a writer, even one ethically motivated, to represent animal awareness through this "network of possibilities" is immensely challenging. How best to inhabit animal consciousness with some degree of 'reality' in relation to animal embodiedness and knowledges is addressed by Vicki Hearne, trainer of dogs and horses, poet and philosopher. In her reading of Auster's novel, she implies that his sympathetic imagination has not gone far enough in *Timbuktu*. In her opinion, here "we find ourselves without a language by which dogs can live in literature" although she allows for the novelist's "fate" of "writing in the impoverishment we now take for speech" (http://www.slate.com/id/2000087/entry/1002990/). She draws attention to the "issue of truth" in relation to language, claiming that "[I]t is art's job to be true *to* the world"(http://www.slate.com/id/2000087/entry/1002990/). According to Hearne, writers should be more attuned to dog senses and sensuality in their rendering of canine experience in human language.

In her poetry, Hearne strives to represent ontological aspects of dogness while reflecting on the stilted anthropocentrism of human language and its inability to convey dog-experience. In "Young Dog, Grass, and More," for example, she celebrates the merging of the young joyous dog with her environment while critiquing rational patterns of codifying the "truth" of a dog:

> ... Her clear gallop
>
> Divulges noon pouring through,
> Her eyes to gather and sow
> Knowledge, as expert as joy
>
> Residing elsewhere for no
> Reason but reason's dancing
> Refusals to mark out what
> Cannot be marked (52).

Hearne's fluidly lambent poetry attempts to convey the drama of canine heightened sense experiences which tumble over each other, as opposed to the rigidity of the boundaries of reason. "Dog" by South African poet Ruth Miller which reflects on Rilke's musing on God creating dog while "'watch[ing] it under the influence of its first embarrassments and inspirations'"(54) also has dog consciousness as joyous, creative, and even poetic:

> Under the influence of its first inspirations
> What dog could not discover and create –
> A stick in his mouth, a stone in a pond, a feather
> Becoming bone to be buried, splash to be harried and all the weather
> Blowing on a blue day. Nothing but was recreated
> In the imagery of a poet; he knew gardens were there to be looted
> And that no deliciousness occurred without a reason
> Whether smells from a distant cooking pot or a bitch in season (55).

The poem glows with the momentousness of the dog's joy and with the momentum of his discoveries. If Bataille's "immanent animality" is in evidence, both poems extend beyond this frontier as Miller and Hearne embody the forms of feeling and sensuousness of a dog's point of view through conveying the immediacies of dog-delight in motion and smell, in the case of the former, and the ecstasy of merging with the environment in the case of the latter.

Joseph Smith, in his lyrical novel *The Wolf,* has the eponymous protagonist interpret the world visually and impressionistically and, like Hearne's and Miller's dogs, through a sophisticated sensory immanence that also goes beyond what Bataille imagines for animal points of view. Animals communicate effortlessly with each other through pictures, as the wolf engages with potential prey:

> [The beast's] eyes tell me that she is old, but still strong. I see in them a flash of images, her memories, the things she wants me to see. I see where she has come from and where she has been, that she has raised many calves, that she has fought with many wolves and never been brought down to be torn open and eaten. She is telling me that she is not ready for death ... (5).

Smith implicitly extends the awareness of animals beyond the boundaries of their embodiment into an unbounded telepathic realm of visual connections, which Temple Grandin and Catherine Johnson would surely endorse, given their thesis in *Animals in Translation* that both nonhuman animals and autistic humans think in pictures.

Smell is, indubitably, the paramount canine sense, however, and both Nganang and Auster attempt to figure dog-being partly through smell. Nganang has Mboudjak speak of the bliss he encounters in a "mountain of garbage" behind the Party Headquarters:

> I rediscovered my joie de vivre ... I let myself get carried away and wandered in the unknown, the Infinite of Art ... I crouched, I even lay right down in the stench, amid the flies' never-ending song. In the depths of the garbage I discovered a shoe with tantalizing leather ... and finally a piece of cloth that smelled like the feast of the decade (103).

Auster also playfully represents the sense of smell in relation to art, in having Willy, the primary human companion of Mr Bones, contemplate and then scheme an "olfactory art" for dogs. He laboriously constructs a Symphony of Smells for Mr Bones, an invention, he sometimes feels, which "would turn them into millionaires many times over" (5). But Mr Bones has his reservations:

> For once, Mr Bones was glad that he had not been endowed with the power of human speech. If he had, he would have been forced to tell Willy the truth, and that would have caused him too much pain. For a dog, he would have said, for a dog, dear master, the fact is that the whole world is a symphony of smells. Every hour, every minute, every second of his waking life is at once a physical

> and spiritual experience. There is no difference between the inner and the outer, nothing to separate the high from the low (53-54).

The Symphony of Smells is an artwork made by Willy at his craziest, during a winter-long obsessive phase. The family friend he visits in an attempt to market his invention cruelly furnishes the voice of reason as he reduces the grandiosity of the symphony to what it is–a cardboard box of dirty rags.

If Auster mocks such efforts at alternative representation of an animals' embodied point of view, Jessica Chalmers and Una Chaudhuri, in an article entitled "Sniff Art," describe the installation-theatre "On the Scent" which recalls the inspiration behind Willy's art for Mr Bones. They suggest, via Freud, that paying attention to smell challenges the human epistemological favouring of sight as the most significant sense, even while "olfaction has proved itself maddeningly resistant to art" (77); inevitably, "[e]ven the words we use for smells ... do nothing at all to actually produce the sensory experience they refer to" (79). Implicitly, the writers under discussion here have both Mr Bones and Mboudjak recognise this.

Mr Bones believes absolutely in the notion of *Timbuktu*–the novel title comes not from the venerated African city but from Willy G. Christmas's assigning the name to " 'an oasis of spirits'" (60). As the novel opens on the streets of Baltimore, Mr Bones, the loyal companion and protector of Willy G. Christmas, is painfully aware that his 'master' is dying. Willy, born Gurevitch in New York of parents who were holocaust survivors, dropped out of his own life at a young age and had a drug-induced psychotic episode. In recovery, he heard Santa Claus speaking to him from the television set and took on the name Christmas as he styled himself a poet-saint, spending summers on the streets of New York and other cities, and the winter months with his mother before she died (some years before the narrative begins). Now, abandoned by Willy's death, Mr Bones struggles to live on the streets alone until he encounters the young Henry Chow who keeps him in a box in an alley until his restaurant owner father discovers the deception. Terrified of being relegated to an item on a Chinese menu, Mr Bones runs for three days until he ends up in North Virginia and is adopted by the middle class Jones family. When they go on holiday he escapes from the boarding kennels, a seriously ill dog. Unable to resume his wandering or to manage the walk back to the Jones home, he chooses to dice with death in the traffic on an eight lane highway, confident that he will meet Willy in Timbuktu before the day ends.

Dog Days has a broader sweep and is a social and political satire, with the dog Mboudjak the deeply moral chronicler and observer of human foibles in vibrant and populous Madgascar, a working class quarter of Yaounde in the Cameroons. He analyses how suffering is due, largely, to government incompetence and corruption. When government workers and civil servants do not get paid, for example, they spiral down into debt and poverty. Mboudjak belongs to Massa Yo. Initially a civil servant and a "Big Man" at the start of the narrative, he is forced to open a bar, The Customer is King, when he loses his job. His wife Mama Mado and son Soumi disappear from the narrative when the marriage breaks down due to Massa Yo's sexual profligacies. Unlike *Timbuktu*, the novel has a large cast of characters, including that of the corrupt and gun-toting police commissioner, but few are developed. The Crow, as people suspiciously call him, is, a putative "member of the opposition" who studies the bar patrons in order to write his book *Dog Days*. Mboudjak decides in disgust to try living on the streets in the interest of freedom from his hyper-masculine "master" but the narrative is without ongoing drama and the action is picaresque rather than plot-driven. Finally Mboudjak returns to the bar and the novel ends with some hope as protests against the regime begin to take hold in Madagascar.

Both writers are highly conscious of the ethical resonance of their foregrounding of dog points of view as a close, comparative reading of the two novels will show. Mr Bones and Mboudjak have reason and the logos, which very dramatically contradicts human exceptionalism, even as it undermines the embodied nature of dog ontologies. *Dog Days* opens with Mboudjak proclaiming "I am a dog" and then, astutely aware of the "finitude" of language, he refuses its connotations as he explains that it is now "Nothing more than a word, a noun" (7). In his youth he was "wounded" by human words:

> Because for me the word [dog] signified the arrogance with which men name the world, assigning a place to each thing, and ordering them to be silent. Each and every time it was used to refer to me, the word let me know I was an object in the human universe, that I had stopped being what I really was, and that I had no right to speak (7).

Now, having shifted from his aggressive response to humans at the prospect of being given a rabies injection, he answers to the name Mboudjak "which means 'the outstretched hand'"(8). Nganang stresses the linguistic abilities of the dog who is far more erudite than the humans he observes. Auster, in the opening of the novel, does not have Mr Bones so self-consciously aware of language. Instead he emphasises the dog's epistemological capabilities:

> Mr Bones *knew* that Willy wasn't long for this world ..... Every time one [of Willy's] attacks started, Mr Bones *half expected* Willy's body to explode from the rockets of blood pressure bursting against his rib cage. He *figured* that blood would be the next step, and when that fatal moment finally occurred on Saturday afternoon, it was as if all the angels in heaven had opened their mouths and started to sing. Mr Bones *saw* it happen with his own eyes...(3, emphasis added).

Not only does Mr Bones have a rich repository of knowledge about Willy and the universe they inhabit, he is a visionary who sees angels. With Mr Bones as focaliser rather than as narrator, his words and perceptions blur with those of the narrator. Consider the following:

> ... it was more than just love or devotion that caused Mr. Bones to dread what was coming. It was pure ontological terror. Subtract Willy from the world, and the odds were that the world itself would cease to exist (4).

The first and second sentence quoted here suggest extradiegetic comments: sensitive and attuned to life and death as Mr Bones is, he would not use terms like "ontological" unlike the intellectual Mboudjak. The final sentence above is more likely to be the free indirect speech of Mr Bones. Adept at analysing the particular, Mr Bones is not a wise political commentator, having accepted Willy's not-always-sane world-view. Mboudjak's abilities as a political analyst, on the other hand, emanate from his own natural visceral awareness of human-animal hierarchies, rather than from a human teacher.

Auster has his dog focaliser humorously aware of his canine deficits when it comes to human language. Mr Bones is "at home in the native lingo" because Willy " a genuine, dyed-in-the-wool logomaniac" speaks to him "from the instant he opened his eyes in the morning until he passed out drunk at night"(7). But "biology was against him":

> what with the configuration of muzzle, teeth and tongue that fate had saddled him with, the best he could do was emit a series of yaps and yawns and yowls, a mooning muddled sort of discourse. He was painfully aware of how far from fluency these noises fell, but Willy always let him have his say, and in the end that was all that mattered (7).

Mr Bones knows that he is unable to enunciate human language; when Henry Chow asks him what his name is, Mr Bones thought "Ha …[a] little wise guy. And how does he think I'm supposed to answer that one?" (124). Still, this does not stop him from barking out "a perfect anapest, with each syllable of his name accorded the proper stress, balance, and duration"(124-125).

Auster has Mr Bones able to understand what he calls Ingloosh in which "he had thoroughly mastered the ins and outs of syntax and grammar" (6) and able to recognise that certain words "like *shelter* meant trouble" (5) for here "the so-called humane societies" (5) would quickly dispatch him. Mr Bones' knowledge is useful in his daily life, but Auster critiques the ridiculousness of experimenting on dogs and language in the story he has Willy tell (an experience of his college roommate called Anster or Amster, he can't remember which) about Thomas Mann's daughter teaching an English setter to type on a specially modified typewriter the sentence " 'Ollie is a good dog'" (103). The reason Willy, as he is dying, berates himself for not having taught Mr Bones to read, is that a literate dog could have negotiated his way through the streets to find refuge with his school teacher from high school. The Dolittle fantasy of speaking with nonhuman animals recurs in both novels, even as it is undermined. In *Timbuktu*, Mr Bones dreams of conversing with the late Willy who gives him advice when he is on the run, but it is only in dreams that such conversations are mooted. In *Dog Days*, Mboudjak suddenly finds that he is speaking with the bottle collector with whom he has teamed up in his travels. But the "garbage genie" is not to be trusted and, in a drunken state teases Mboudjak that " 'where [he] come[s] from we eat dogs'" (129).

Humans in both novels have the power of life or death over Mboudjak and Mr Bones. If neither Nganang nor Auster imagine utopian solutions to such human-dog hierarchies, they do have the dogs define themselves in what Carol J. Adams, in Neither Man nor Beast, calls "relational epistemologies" (155). Epistemology, as Marsden reminds us, requires "a priori distinction between subject and object" (39) but a relational epistemology contradicts such binary thinking. Tragically, apart from Willy who addresses Mr Bones as an equal, humans do not de-centre their subjectivities in relation to dogs in the novels. When Mboudjak barks " ' you are my brother'" (15) to Soumi, Massa Yo's fickle ten-year old son the boy does not understand him. For Soumi, Mboudjak's "four-legged condition defined the limits of [his] being; it signified that [he] was condemned by fate to crawl flat on [his] belly like a reptile" (16). The boy, like his father, maps the world dualistically and is devoid of any sympathetic imagination:

> This difference in our souls created an invisible barrier between us that even my side-stepping of his anger couldn't destroy: camaraderie's unbridgeable gap ...even at the end of a day spent playing together, Soumi used the cut-and-dried tone of a master to speak to me (16).

Donna Haraway's ideal of dogs and humans reciprocally connecting into "multispecies knots" (35) does not feature in *Dog Days* although Mboudjak longs, alternately, for more connection with humans (or complete freedom from them). Although his very name means "the outstretched hand," in his experience the hand could also be a violent one, as in the case of Massa Yo, whose predominant discourse with Mboudjak is tyrannous; Soumi, goes even further, attempting to murder the dog.

In "Reflections" Barbara Smuts theorises an alternative to the hierarchies of human-animal pet configurations in her lived relationships with Safi, her dog, whom she regards as a person and "personhood connotes a way of being *in relation to others*" (118). Certainly, both writers play with such notions of non-hierarchical connections as ideal. If Mr Bones' "personhood' is taken for granted by Willy, however, in *Dog Days* both father and son refuse to regard Mboudjak in this light. Still, humans in his environment cannot deny some inter-twining of their identities. Massa Yo barely feeds Mboudjak and taunts him: " 'A dog you feed isn't really a dog ...It's a child'" (56) but the latter rejects the demeaning role of a pet animal who is always already dependent and inferior. Mboudjak critiques domestication and what it has done to his subjectivity: "I no longer have the wolf's reflexes or the hyena's wicked wit. As a result of observing men and doing nothing else but, maybe I don't even have my canine attributes anymore" (30). Mark Rowlands, in *The Philosopher and the Wolf*, his non-fiction narrative of his relationship with his putatively domesticated and resident wolf, interprets this dependence more positively, arguing that dogs' reliance on humans has developed canine capabilities and that a dog may "use us to solve its various problems, cognitive and otherwise...We humans are part of the dog's extended mind ... And to use us, [dogs] have to be able to read us ... But a wolf will not do this" (30-31).

Auster and Nganang engage quite extensively with the ramifications of pet-keeping and its conventions in North America and West Africa respectively, implicitly undermining Deleuze and Guattari's mockery of pets "as individuated animals, family pets, sentimental Oedipal animals each with its own petty history, 'my' cat, 'my' dog" (240). When Mama Mado transforms Mboudjak "into a poodle" (67) in order to boast about the money she is making as a street vendor, Nganang does not represent the dog as a substitute object but as a being capable of analysing the homologies of speciesism and racism. While Mboudjak suffers the humiliation of being washed and having his coat put into curlers at a salon, he feels that a woman who grumbles that dogs should not be permitted is a "racist". Later in the narrative Mboudjak is shocked into developing a "much clearer sense of what the communist dog often referred to as 'race consciousness'"(108) when the brothel keeper Mini Minor throws hot water over the wall onto him.

Animality and race, Nganang seems to be saying, are elided, with the dogs as oppressed subalterns. Humorously, Massa Yo has an essentialised idea of what "canitude" involves, wishing that his dog would present him with slaughtered chickens but "canitude" is also socially constructed as the narrative shows. As Mama Mado and the newly coiffed dog return home along the streets, some passers-by comment " 'That poor dog, alienated from his canitude!'" and Mboudjak thinks "Afrocentric intellectuals to be sure" (70). Nganang has it both ways here, mocking the actual notion of "canitude" (and negritude) while critiquing any tampering with dogness through the figure of the grotesquely primped dog. If Mboudjak is initially seduced by his transformation, his pride in his appearance soon changes when he realises that he and his mistress are sexually objectified by passers-by and that his new appearance is easily satirised with Massa Yo's mocking comment that he looks like a ground squirrel. Canine identity then, is indeterminate, multi-faceted and even contradictory; dogness and negritude have comic parallels, with foreign and inappropriate standards of beauty undermining colonised identities. Mama Mado's triumphal notions of canine utopian identity are, however, monolithic and fixed:

> Mama Mado let it be known throughout the quarter that I was the dog of the future. No more starving dogs. No more dogs chased by flies. No more dogs with their teats sweeping the streets. No more cadaverous dogs. No more vampirish dogs. No more three-legged communist dogs. No more dogs scarred with mange … Me, Mboudjak, my time had come …(70).

Such radical ideals are rendered unattainable by the political status quo which keeps humans and animals in ongoing states of poverty.

Auster does not represent Mr Bones as so astutely class consciousness. In *Timbuktu*, dogness, rather than being a political issue, is one of personal or species identity although Mr Bones does take on the class status of his owner whether it is the vagrant Willy or the bourgeois Jones family. As Rob Gordon reminds us in "Fido: Dog Tales of Colonialism in Namibia," dogs are classed and raced along with their owners and the well-fed Mboudjak reciprocally reflects Massa Yo's initial status as a civil servant. The regular event of a dog being walked by his 'master' is the source of much astonishment and mirth for the street dogs who mock Mboudjak for being a cosseted sex slave. When Massa Yo plummets in status and income after the loss of his job, Mboudjak has to perform as a watch dog so that Massa Yo can "become a man" (24) or, at least, retain some masculine status.

Mr Bones never has such a clear role in his relational epistemology, but is unable to construct or imagine existing without human relationships, and is filled with terror at the prospect of not being cared for. Like Mboudjak, Mr Bones is highly sensitive to the performativity of being a dog, with the need to please humans and act as humans expect. He is a bodyguard and companion to Willy, a fellow sportsman to some young boys he encounters, a confidante to Henry Chow, an intelligent trick-dog for the Jones family, a fellow protagonist in Alice Jones's fantasy world. Mboudjak, at times, prefers the prospect of freedom to Massa Yo's abuse. His various sorties into the city to get away from Massa Yo, are fruitless, however. After he follows the garbage picker, he is overwhelmed by horror: he had "entered the very heart of fear: into the belly of whispers, the arteries of fantasy, the intestines of the insanity that is Yaounde" (126). Ultimately, he rejects the life of a stray, which involves suffering from hunger, infestations of parasites and being vulnerable to human and canine violence. When he returns to The Customer is King, the hens who come to hear his stories are incredulous closet revolutionaries:

> Seeing me tied up, they lifted their beaks in surprise: who can prefer chains to wandering? They didn't believe me when I said I'd come back because I missed my master's bar. How so?! They believed me even less when I said that life on the road is no life at all, and that in order to survive, you have to give up your intelligence. They repeatedly raised their voices: 'Freedom has no price' (166).

Mboudjak is a pragmatist then, aware of the compromises he has to endure in order to live. Nganang satirises the class choices open to Mboudjak: the street dogs whom he has been forced to mix with taunt him for being a " 'capitalist dog'" because he never embraces their free but tenuous ways of living and he ignores the three-legged "communist dog"(106) finally preferring to be a bar dog than to be homeless and starving.

The dogs in these narratives also serve as metaphors for the human condition–as in Mboudjak's fluid insertion into class structures. When Mr Bones is on the run from Baltimore and Mr Chow's restaurant, Willy appears to him in a dream, reprimanding him for his self-pity and telling the dog of his mother, Mrs Gurevitch, being " 'hunted .. down like a dog' " during the holocaust: "'People get treated like dogs, too, my friend'" he says (150). Groups of humans who have been subjected to oppression or even genocide throughout history have been cast as animals so that the perpetrators can condone their violence.[2] Thus Njabulo S. Ndebele writes, in his essay "The Year of the Dog: A Journey of the Imagination," of the discursive construction of humans as "dogs" by those who wish to justify violence against them. The apartheid regime evicted Africans from their land "like dogs" in 1913 and fired on protesting schoolchildren "like dogs" in 1976; communities murdered innocent women, or criminals murdered police "like dogs" (252-253).

Nonhuman animals themselves have always been regarded as lacking in ethical capabilities. Emmanuel Levinas, for example, philosophises that "[t]he being of animals is a struggle for life. A struggle of life without ethics" (50) but recent ethological research challenges such reductive suppositions. Marc Bekoff and Jessica Pierce contend, in *Wild Justice: The Moral Lives of Animals,* that nonhuman animals have "moral intelligence" as animals are quite capable of expressing themselves ethically. For Ndebele, dogs embody the characteristics of "intelligence, empathy, loyalty, dependability, courage, protectiveness, sensitivity and caring," (256) ethical qualities which contrast shockingly with the absence of ethics in his imagining of a vengeful crowd which kills a dog. Similarly, in *Dog Days,* Nganang has Mboudjak as the observer and critic of unethical human behaviour. When the police commissioner unjustly and brutally arrests the cigarette vendor, only Mboudjak and the Crow protest: the former "barked out [his] indignation, but couldn't rouse the local men from their torpor"(95) and the latter questioned whether the commissioner has a warrant for the arrest. Then voices protest and Mboudjak hears "the word 'justice', then the words 'injustice,' 'dictatorship,' and 'new deal,' the name 'Biya,' and finally the sentence, 'Cameroon is Cameroon.'"(96) When the Crow is about to be hauled off by the police, Mboudjak protests "wildly ... outraged that all the men in the quarter let themselves be dominated by one man, even if he did have a gun in his hand" and bites the commissioner's calf so severely that the latter fires randomly and threatens to shoot the " 'mutt'" (97). When the Crow finally returns from prison, he greets Mboudjak as " 'A real hero ... This quarter's only hero!'" (111) but he vilifies the inhabitants of the bar, calling them " 'cowards ... getting drunk on indifference'" (113).

If Mr Bones's ethical behaviour in which he attempts to help people as Willy did could be construed as self-interested, he has the qualities of Ndebele's dogs and acts, often, like a "messenger" to use Luce Irigaray's term (196) as though he embodies compassionate wisdom for the humans he encounters. Willy muses, in free indirect speech about his dog companion, "If God had sent his son down to earth in the form of a man, why shouldn't an angel come down to earth in the form of a dog?" (44). Elsewhere, Willy cogitates on the linguistic and denotative connections between dog and god. Certainly, Mr Bones acts as a kind of dog-saint in his caring for people, particularly in his humouring of children, even though he worries that they do not have the power to nurture him as he would wish. To Henry Chow "an only child whose parents worked long hours and had steadfastly refused to allow a pet in the apartment, Mr Bones was the answer to his prayers" (126). For young Alice Jones he embodies "her partner, her fellow protagonist, her male lead" in her "world of imaginary beings" (191-192).

Humorously, Nganang and Auster have both dogs anxious not to be thought of as unethical beings who enact the "dog eat dog" proverb and as quite aware of the limitations of cultural constructions of "dog." On one of his travels, Mboudjak comes across the cadaver of the three-legged "philosopher" dog, eaten by maggots with a crowd of flies doing a "mourning dance" (135) around him. Nauseated by the fate of the "philosopher dog" who had been "run over and left like an offering in the sanctuary of the genie of the slums" Mboudjak attempts to "pay [his] final respects" but a child shouts that he is eating his "'brother'" and another speaker agrees, disgusted, that "'dogs are savages!'" (135) When Mr Bones stays with Henry Chow, he is seduced by the restaurant left-overs, even though he fears, following Willy's prejudices, that he might be eating "a fellow dog": " in an effort to assuage his guilty conscience, he would tell himself that if this was to be his fate as well, he only hoped that he would taste as good as the thing he had just eaten" (131).

In relation to fellow canines then, Mr Bones is pragmatically compassionate. In his commitment to a human being, however, he moves onto another level telling the story of his beloved Willy's death and dreaming about him in ways which exemplify not only his capabilities as a visionary artist but his longing for guidance from a known, expansive presence. Jacques Derrida notes in "The Animal that Therefore I Am" that, indisputably, some animals dream which raises "a relation among consciousness, sub-conscious and unconscious, as well as representation and desire" (62) that is illustrated in *Timbuktu*. When Willy is picked up by an ambulance from his prone position outside Edgar Allan Poe's house in Baltimore and taken to hospital, Mr Bones imagines that he is a fly who clings to the side of the vehicle and accompanies Willy until he dies. This 'fly on the wall' perspective has a more literary echo perhaps in Emily Dickinson's "I Heard a Fly Buzz–When I Died–"(39) but it also a profound strategy for Mr Bones to imagine what happens to Willy. Dream and reality merge as Mr Bones divides himself into two; in the quotidian "the dog part of him"(87-88) remains on the corner, but in the dream the fly accompanies Willy. The dream/tale Mr Bones tells himself is worthy of a sentimental Hollywood production: Willy's old high school teacher who had nurtured his writing talent appears miraculously in Our Lady of Sorrows Hospital, Willy dies a painless, effortless death and his last words concern Mr Bones and his future. But as the dog wakes, Willy is still with him on the pavement speaking torrentially, as he did in Mr Bones' dream. The dream is futuristic, then, a story with which Mr Bones comforts himself.

The dog has further dreams of Willy subsequently; one a paranoid vision of his late companion as a "vengeful and sarcastic Willy, a devil Willy"(209) but another is profoundly healing on a metaphysical level. Willy speaks of the time coming when Mr Bones will go to Timbuktu:

> "You mean dogs are allowed?"
>
> "Not all dogs. Just some. Each case is handled separately."
>
> "And I'm in?"
>
> "You're in."
>
> "Don't kid me, master. If you're joking now, I don't think I could stand it."
>
> "Believe me, pooch, you're in. The decision's been made" (221-2).

In spite of the light tone, Auster represents a dog's interest in his life after death and implies that dreams are potentially a means to work out large epistemological issues. Mr Bones' fly dream, for example, is prefaced by the extradiegetic narrator commenting:

> because [Mr Bones] was a biased witness to the proceedings, we should not begrudge him the consolation of last-minute hopes...But what do flies know? And what do dogs know? And what, for that matter, do men know? It was in God's hands now, and the truth was that there was no turning back (89).

Through these cosmic questions, the narrator gestures towards the limitations of dualistically categorising human and other–whether animal or insect. And the question of whether animals dream is, according to Derrida,

> in its form, premises, and stakes, at least analogous to the questions 'Does the animal think?' 'Does the animal produce representations?' a self, imagination, a relation to the future as such? Does the animal have not only signs but a language and what language? Does the animal die? (63)

Timbuktu has many of the same philosophical preoccupations (more obviously than *Dog Days*) and that Auster has Mr Bones as an artist producing representations, even if they are within his dreams, confirms the dog focaliser's well-defined self and imagination.

In modernity, generally, because an animal does not have a self s/he does not die in the same way as a human being. Auster implicitly contradicts such anthropocentrism: Mr Bones has a soul, he has a knowledge of dying and looks forward to joining Willy in Timbuktu through making a choice about his death. The dog rationalises that he is not committing suicide but making a sublime choice by playing the sport of "dodge-the-car" in which "[t]he moment you lost, you won" (226-227). Sick and in pain, he galvanises himself for the last time– for a hero's dash onto the highway which had appeared to him "as a spectacle of pure radiance, a field of overpowering light" (226). Vicki Hearne objects to this ending for Mr Bones, arguing that Auster cannot convince her that the old, woeful dog becomes a "dancer" capable of "playing for truth in a way that only a handful of human poets and painters and sculptors and writers even dare to consider" Still, she concedes, if somewhat condescendingly, that Auster's "sporting attempt to lift a street dog to the sublime slips, but it breaks new ground" ((http://www.slate.com/id/2000087/entry/1002990/).

My reading of the ending is less critical. Auster, I would argue, has prepared the reader for the "truth" (to use Hearne's term) of Mr Bones through the tragedy of his abortive quest to find a real home, his desire to be with Willy, his pondering of metaphysical notions of being and the entirely relational construction of his epistemology. My objection to Auster's narrative lies in the occasional jarring rhetorical questions posed by the extradiegetic narrator to the reader: "What was a poor dog to do?" (4) "What difference did it make if he didn't fully understand? He was a dog, wasn't he?" (51) or the description of Mr Bones as "mere dog" (109,195) and being "only a dog" (218). Such questions and statements coyly exclude Mr Bones who is, contrary to these negative sentiments, represented as a being with a strong sense of self and one who is compassionate, agentive and creative. He tells stories so that he can understand the world and his place in it, but the narrator's questions and statements paradoxically detract from the "truth" of his focalising, almost as though Auster is uncomfortable about sustaining such a positive representation of a nonhuman animal. Adam Begley appositely critiques the collusion set up between "a highly sophisticated writer and a reader aware of the sophistication" but he finds even the attempt at a "doggy voice" misjudged and neglects the ethical dimensions of the novel as he denigrates "the amusing but not particularly urgent doggy activity depicted on the page" http:www.guardian.co.uk/books/1999/may/29/fiction/ paulauster accessed 3/8/09.

Dog Days also includes Mboudjak's refrain that he is "only a dog"(202) with the narrative belying the negativity of the statement. Nganang points out repeatedly the disparity between human underestimation of dogs and the nuances of the dog's sensibilities and knowledges, a strategy made possible by having Mboudjak as narrator.

Earlier, when he attacked the police commissioner, Mboudjak ignored the man's threats to kill him because of the passion of his righteous rage. But his master assumed that it was because he was too stupid to realise his life was being threatened. A bar customer agrees, saying:

> dogs don't recognize guns ... dogs don't even know what death is until they're already dead, adding that we're like chickens who've never caught on about the hand that rings their neck every Christmas. He concluded by saying that my master shouldn't hold it against me because I was just an animal (98).

In Nganang's novel, any collusion between writer and reader is obviated by having Mboudjak as narrator. The irony is directed more often, as it is here, against a human character who does not have the sensibilities to recognise the nuanced intelligence of the canine narrator. Through his narrator, Nganang raises serious issues in relation to human perceptions of dogs, but without being didactic or high serious.

The ending of *Dog Days* has human and canine united in action and voice. With Madgascar finally galvanised into action after the shooting of a child as well as protesters, students and black marketeers, the chant "BIYA MUST GO! ..could only be barked out"(205). The narrator is ecstatic as he runs to the forefront of the "cyclone in the streets":

> United we were, Man and me, in the spasmodic rush of our language: our barks. We marched, not only to bring somebody else's child back to life, but above all and foremost to chase out the crazed lion. We marched, hunters in the urban jungle (205-6).

Political action, a denial of the cowardice that paralysed the quarter, seems linked to a primeval, posthuman connection between human and nonhuman animals here. And the last sensory perception is accorded to the dog who "sniffed, re-sniffed, re-re-sniffed and ... re-re-re-sniffed ... the scent of [the child's] blood [which] was still hot."(206). The dog is the ultimate witness, then, able to establish the evidence of murderous violence.

Nganang's novel demonstrates here and throughout the narrative, as does Auster's, that a focus on the lives of dogs derives not from a sentimental, apolitical pursuit favouring nonhuman animals over human ones, but from an ethical analysis of the interconnections of forms of otherness and the potential kinship between humans and other animals. Both novels broaden the reader's suppositions of human-canine connections and the place of dogs in modernity. If Mboudjak is a political animal, an observer and social scientist, Mr Bones is a visionary, an artist motivated by an ethics of care and spiritual compensation. Paradoxically, for the reader to be alerted to such putative canine capabilities, the dogs have to be humanized to some extent–through their cognitive faculties and their facility with human language, even as the "finitude" of language is foregrounded. Both novels could broadly be defined as posthumanist and ethically resonant: the dogs tend to be represented as "knowing subjects" who, along with the humans in their worlds, are embodied and mortal. Both novels are brave attempts within the constraints of such discourses to encourage the readers to reconfigure the word in posthumanist ways. They gesture towards the discursive limits of human language, in their ethical representations of dogs as narrators or focalisers. Thus, in radical ways, the novels under discussion extend the conventions of the contemporary novel which has, predominantly and, it seems, unproblematically until recently, inhabited a human point of view.

Notes

1 *King: A Street Story* by John Berger is not analogous to these two novels. Although King, narrates the novel almost in its entirety, his identity is uncertain. Whether he is canine or human remains undefined

2 See Yann Martel's *Beatrice and Virgil* for a chilling narrativising of this homology.

References

Adams, Carol. J. 1995. *Neither Man nor Beast: Feminism and the Defense of Animals*. New York: Continuum.

Atterton, Peter and Matthew Calarco. 2004. "Editors' Introduction: The Animal Question in Continental Philosophy. "In *Animal Philosophy: Ethics and Identity*, Peter Atterton and Matthew Calarco (eds.). London: Continuum.

Auster, Paul. 1999. *Timbuktu*. London: Faber and Faber.

Banville, John. 2009. *The Infinities*. London: Picador.

Bataille, Georges. 2004. "Animality." In *Animal Philosophy: Ethics and Identity*.

Begley, Adam. "Mutt from New York."

http://www.guardian.co.uk/books/1999/may/29/fiction.paulauster, accessed 3 August 2009.

Bekoff, Marc and Jessica Pierce. 2009. *Wild Justice: The Moral Lives of Animals*. Chicago: Chicago University Press.

Berger, John. 1999. *King: A Street Story*. London: Bloomsbury.

Chalmers, Jessica and Uma Chaudhuri. "Sniff Art." *The Drama Review* 48 (2004): 76-80.

Coetzee, J.M. 1999. *The Lives of Animals*, edited and introduced by Amy Gutmann. The Princeton Center for Human Values Series. Princeton: Princeton University Press.

Deleuze, Gilles and Felix Guattari. 1987. *A Thousand Plateaus: Capitalism and Schizophrenia*. Trans Brian Massumi. London, Minneapolis: University of Minneapolis Press.

Derrida, Jacques. 2008. *The Animal That Therefore I Am,* edited by Marie-Louise Mallett, translated by David Wills. Perspectives in Continental Philosophy. New York: Fordham University Press.

_____. 1991. "Eating Well." In *Who Comes After the Subject?* edited by Eduardo Cadava *et al*. New York: Routledge.

Dickinson, Emily. 1973. "I Heard a Fly Buzz when I Died." In *Norton Anthology of Modern Poetry*, edited by Richard Ellmann and Robert O'Clair. New York: Norton.

Doniger, Wendy. 1999. "Reflections." In *The Lives of Animals* by J.M. Coetzee, edited and introduced by Amy Gutmann. The University Center for Human Values Series. Princeton, New Jersey: Princeton University Press.

Ekman, Kerstin. 2009. *The Dog*. Translated from the Swedish by Linda Schenck and Rochelle Wright. London: Sphere.

Gordon, Rob. 2008. "Fido: Dog Tales of Colonialism in Namibia." In *Canis Africanis: A Dog History of Southern Africa,* edited by Lance van Sittert and Sandra Swart. Human-Animal Studies Series. Leiden: Brill.

Grandin, Temple and Catherine Johnson. 2005. *Animals in Translation: Using the Mysteries of Autism to Decode Animal Behaviour*. Lonodn: Bloomsbury.

Haraway, Donna J. 2008. *When Species Meet*. Posthumanities Vol 3. Minneapolis: University of Minnesota Press.

Hearne, Vicki. 2007. "Young Dog, Grass and More." In *Tricks of the Light: New and Selected Poems,* edited and introduced by John Hollander. Chicago: University of Chicago Press.

_____. *The White Bone* and *Timbuktu*: A Dog's Life.

http://www.slate.com/id/2000087/entry/1002990/ accessed 3 August 2009.

Heinze, Ruediger. 2009. "Violations of Mimetic Epistemology in First-Person Narrative Fiction." *Narrative* (2008) 16 (3): Web. 9 Oct..

Irigaray, Luce. 2004 . "Animal Compassion." In *Animal Philosophy: Ethics and Identity*.

Levinas, Emmanuel. 2004. "The name of a dog, or natural rights." In *Animal Philosophy: Ethics and Identity*.

Martel, Yann. 2010. *Beatrice and Virgil*. New York: Spiegel and Grau.

Miller, Ruth. 1990. "Dog" In *Poems, Prose, Plays,* edited and introduced by Lionel Abrahams. Cape Town: Carrefour.

Ndebele, Njabulo S. 2007. "The Year of the Dog: A Journey of the Imagination" In *Fine Lines from the Box: Further Thoughts about our country*. 251-256. Roggebaai: Umuzi.

Nganang, Patrice. 2006. *Dog Days: An Animal Chronicle,* translated by Amy Baram Reid. Charlottesville: University of Virginia Press.

Nussbaum, Martha. 2006. *Frontiers of Justice: Disability, Nationality, Species Membership.* Cambridge, Massachusetts: Belknap Press of Harvard University Press.

Pamuk, Orhan. 2001. *My Name is Red.* Translated by Erdag M. Goknar. London: Faber and Faber.

Rowlands, Mark. 2008. *The Philosopher and the Wolf: Lessons from the Wild on Love, Death and Happiness.* London: Granta.

Smith, Joseph. *The Wolf.* London: Jonathan Cape.

Smuts, Barbara. 1999. "Reflections." In *The Lives of Animals* by J.M. Coetzee, edited and introduced by Amy Gutmann. The University Center for Human Values Series. Princeton, New Jersey: Princeton University Press.

Tiffin, Helen. 2007. "The Speech of Dumb Beasts." Abstract *In Animals and Society II: Considering Animals Conference HandBook.* Conference organised by the Animals and Society Study Group (Australia) and the University of Tasmania. Hobart: Conference Design, Tasmania, 17.

Wolfe, Cary. 2009. Human, All Too Human: "Animal Studies" and the Humanities. *PMLA.* 124 (2): 564-575.

Wroblewski, David. 2009. *The Story of Edgar Sawtelle.* New York: Ecco.

Chapter 32

Modern Media and Animal Rights: How Training your Dragons and Pokémon Influence Contemporary Animal Ethics

☆ *Johansen Quijano-Cruz*

This text looks at how different fields of study throughout various periods have portrayed the issue of animal rights. This debate, which became especially controversial during the Victorian period, still rages on today. With animal rights activist groups like People for the Ethical Treatment of Animals (PETA) advocating for animal rights and various farmer and scientist groups strongly debating that animals should have no rights, the debate will go on forever. Scientists, politicians, and media personalities have all too often made their opinions publicly known in order to influence the mindset of individuals one way or another. However, it is worth noting that, in today's visually charged world, film and visual media have an incredible persuasive potential that is only starting to become integrated into the larger discourse regarding animal rights. As any moviegoer will be all too happy to testify, it is worth noting that film has an unquestionable impact on how individuals think, as they have the potential to reinforce already existing tendencies or ideas and criticize new trends, as well as the ability to question previously existing assumptions by making them part of a larger national or international conversation. When this is taken into consideration, it is worth taking a critical look at some of the major animal-related media hits in recent years in order to figure out what they are saying about morality and animal rights. This chapter will look at how the hit animation and videogame series *Pokémon* and the popular film by Dreamworks called *How to Train*

Your Dragon deal with the issues of animal rights, and how these visual representations might influence today's younger generations.

Pokemon, a popular media franchise owned by Nintendo, was released in 1996 and has since become one of the most lucrative media franchises in the world. In the *Pokemon* world, humans coexist with over five hundred species of animal-like creatures after which the show is named –*Pokemon*. Although Pokemon are fictional creatures – electrically-charged rodents and fire-breathing salamanders are hardly real, after all – it is obvious that Satoshi Taijiri drew inspiration for his *Pokemon* designs from Butterfree's cocoon-to-butterfly evolution to Pidgeot's bird form, straight from real animals. His inspiration to create *Pokemon*, in fact, comes from his childhood fascination with insect collecting. His original idea was to create a world that could depict the collection and companionship of his childhood hobby, and in doing so he created a narrative revolving around universally accepted principles of teamwork and strong bonds of friendship; however, he also created a narrative that could be interpreted by some as one that advocates animal cruelty.

The *Pokemon* franchise boasts thirteen films, an animated series, and over 20 videogame titles, some of them with various versions available. Even though each narrative follows different characters through distinct adventures, the overall concept in each of the narratives is fairly constant: collect as many Pokemon as possible and become the best Pokemon trainer. It is in this concept that, as many *Pokemon* critics and fans have admitted, that "If there is any aspect of Pokémon that has been controversial, it is the perceived animal cruelty inherent in the concept" (Hlin, 2010). As previously stated, in the world of *Pokemon*, the only way to become a Master Pokemon Trainer is to capture as many Pokemon as possible. In order to do this, the Pokemon trainer must summon a Pokemon, an animal-like creature that could be as massive as a car, from the Pokeball, a red and white ball the size of a baseball, in which it is stored. Once the Pokemon is temporarily released from its storage the trainer instructs it to battle. Even though the Pokemon has been released from its prison, it has not been released from captivity. By following the trainer's instructions, the captive Pokemon fights the wild Pokemon into submission. Once the wild Pokemon has been subdued, it is magically stored in a Pokeball to become part of the trainer's Pokemon collection. This kind of collection system, universal in all the *Pokemon* narratives, has caused animal rights advocates to speak against *Pokemon*, as it may instill values of animal abuse in younger, more impressionable children, some even going as far as asking the question "If *you* were kidnapped from your family and forced to beat up your own kind, would *you* like it?" in order to make a point.

Not everyone, however, sees the act of subduing a Pokemon as an act of coercion. Many *Pokemon* fans have argued that, according to the *Pokemon* canon, "Pokémon are generally happy with the current arrangement of training", which involves being summoned at a moment's notice to battle, and that "there is plenty of evidence that contradicts the idea that they are coerced at any point in the process" (Hlin, 2010). However, the fact is that in order for a Pokemon to join a trainer, it must first be weakened by another Pokemon in battle – this is the textbook definition of coercion. In the very first episode of the popular *Pokemon* TV series we see an excellent example of this Pokemon coercion happen. Pikachu, Ash's first Pokemon, is an electric mouse who dislikes being inside the Pokeball. Once Ash receives this Pokemon from his mentor, Professor Oak, Ash decides that it is "the best Pokemon"; however, Professor Oak deems it as a "defective Pokemon". Pikachu's distaste for being trapped inside a Pokeball quickly labels him as a "weird" Pokemon. Lacking any other Pokemon to fight Pikachu and subdue it, in order to take Pikachu in his quest to become a Pokemon Master, Ash has to tie it with a rope and drag it around while wearing rubber gloves to prevent electrical charges. This situation, where a ten year old boy ties a Pokemon down and drags it around in order to make it follow orders, is an obvious depiction of animal abuse. Furthermore, when he realizes that Pikachu is a stubborn Pokemon, Ash unties him and sets off to capture another Pokemon. After running into, Pidgey, a baby bird Pokemon, Ash excitedly exclaims "enjoy your last moments of freedom, Pidgey, 'cuz you're mine!". He then proceeds to throw his Pokeball at it. It is here that his Pokedex, a handheld Pokemon encyclopedia, tells Ash that "in order to capture a Pokemon, you usually have your other Pokemon fight it" (I Choose You, 1998). Furthermore, despite how some have commented that Pokemon are not really forced to fight other Pokemon as they are more powerful than their trainers and could, therefore, attack the trainer or simply refuse to fight and leave, it is trainers who have the final say on whether a Pokemon is able to go free or not, as, with very few exceptions, they can simply recall the Pokemon back into the pokeball whenever they see fit. In the end, the act of capturing / collecting Pokemon is nothing more than an act of subversion where the Pokemon retains its freedom if it's strong enough or is captured if it is not. Yet collecting Pokemon is only half of what makes a Pokemon Master. In order to become a Pokemon Master, the main characters in the *Pokemon* narratives must engage in Pokemon battles with other Pokemon trainers. In these battles each trainer tells his or her Pokemon what attacks to use in order to defeat the other trainer's Pokemon. This practice resembles organized animal fights, such as cockfights and illegal dog fights.

Certainly, there is more to *Pokemon* philosophy than animal collecting and fighting. It is worth noting that in all of the *Pokemon* narratives, elements such as teamwork, human-Pokemon cooperation, and perseverance are outlined, and the interactive titles further encourage reading and strategy planning. Furthermore, there is no denying the fact that many of the actions which the characters engage in, such as Ash saving Pikachu from a flock of wild bird Pokemon, are actions that would inspire trust and friendship. Throughout all the *Pokemon* narratives there is a common factor that trainers who love their Pokemon and treat them with respect are "the good guys" and those who abuse their Pokemon are "the bad guys". Even the fact that the world of *Pokemon* is filled with Pokemon centers – hospitals that treat injured Pokemon for free – and that there are extremely few – if any – hospitals for humans should be an indicator that the *Pokemon* creator did not mean to create a narrative that advocated for animal cruelty. When this is taken into consideration, it becomes obvious that the overall narrative is well-meaning and does not wish to purposefully send a message that animal fighting is a good practice. Indeed, it would be ridiculous to suggest that a child would, after watching *Pokemon*, run out, attempt to capture a small animal, and teach it to "bite" and "tackle"; however, this is exactly what Pokemon trainers do.

Many *Pokemon* fans have argued that the central morale of the *Pokemon* narratives is the teamwork element between humans and Pokemon. Some have even claimed that those who speak out against *Pokemon* "argue that all the talk about love and trust is just sugarcoating a dark, bleak world that is actually abusive and cruel" (Hlin, 2010). However, this is not so. The argument that concerned individuals raise is not that talk about morality, friendship, and teamwork in *Pokemon* exist to gloss over some dark agenda of animal abuse, but that despite the talk of morality, friendship, and teamwork, the *Pokemon* narratives portray acts of coercion and animal abuse, and the one message regarding animal morality and animal rights that *Pokemon* does not portray is that it is imperative to remember that while humans may be the ruling species on the planet, one must always remember that harming any living creature for sport is unfair and cruel. Indeed, if it is content with having children learn values of trust, friendship, and teamwork at the cost of developing a certain degree of sensitivity towards animals, then *Pokemon* should be as acceptable as narrative as any other, but of moral values towards animals are a major concern, then there are other narratives that one could turn to. One of them is Disney's *How to Train your Dragon.*

The setting in *How to Train you Dragon* is Berk, a "miserable village" in a fictional island in the icy north Atlantic regions. The Vikings who inhabit this area have been in a nearly eternal war against dragons who inhabit a nearby volcanic region and regularly raid the Viking village for food. This film, however, is not the traditional tale of dragon-hunting Vikings in quest for glory. Hiccup, the protagonist of the narrative, is a Viking teenager whose physical attributes are less than impressive. Although at the beginning of the film he is eager to slay his first dragon in order to impress his father, Stoic the Vast, who is the leader of the Vikings, his lack of physical strength and apparent clumsiness prevent him from doing so. In order to compensate for his lack of physical attributes, Hiccup apprentices under the town blacksmith and creates a mechanic device that allows him to trap a Night Fury – a mythical type of dragon which no Viking has ever seen. However, when faced with the opportunity to kill the Dragon, Hiccup has a change of heart and decides to set it free. Despite having been told throughout all his life that all dragons are ruthless creatures that will "kill you on site", Hiccup notices that the Night Fury is just as afraid of him as he is of the dragon. In a moment of empathy, Hiccup frees the dragon. Hiccup's inner struggle between two choices – that of being the first Viking to kill a Night Fury and that of setting another living, breathing, sentient being free – allow the viewer to recognize a certain degree of respect for the dragons represented on-screen; something that *Pokemon* fails to do when its main character ties up a small rodent and drags it around on the ground.

The aspect of the film which more strongly advocates for animal rights is when Hiccup bonds with the Night Fury. Despite having been freed, the Night Fury is unable to return to its lair, as it hurt its tail when it was trapped by Hiccup's mechanical contraption. It is here when Hiccup tries to befriend the Night Fury. Although in their review of the film the United States Conference of Catholic Bishops writes that some audiences may find troublesome "the assumption that dragons ought to be domesticated", and further remark that one "does not have to be an animal-rights activist to feel uneasy about the degree of servitude implied" (USCCB, 2010), a closer look at the interaction between Hiccup and the Night Fury, whom Hiccup calls "Toothless", will reveal that the dragon-human relationship portrayed through Hiccup and Toothless is not one of domesticated servitude, but one of real friendship and teamwork. Unlike *Pokemon*, where "training" consists of coercion, the "training" of a dragon in Disney's film is closer to a process of getting to know someone. Hiccup's initial approach and Toothless' initial wariness about Hiccup's intentions, as well as the distance they keep from each other, show a degree of mutual respect. Once the initial fear of the other disappears, curiosity creeps into the scene. In order to satisfy this curiosity both Hiccup and Toothless engage in practices distinct to the other. In order for Hiccup and Toothless to form a bond, Hiccup finds himself having to eat a raw fish that Toothless had just regurgitated. Toothless, in exchange, proceeds to draw on the sand, just as he sees Hiccup do. Despite the obvious implication of the domestication of a naturally wild species, specially when the title is taken into account, the actual narrative shows two living, breathing, sentient beings of different species getting to know each other and developing some form of companionship rather than one superior species coercing the other into subservience.

The notion of a relationship based in teamwork is further strengthened when Hiccup decides to help Toothless by fixing his tail. Using his natural mechanical skills, Hiccup manages to create a tail extension to fit Toothless. Although he eventually develops enough trust in the dragon to attempt to saddle and ride it, Toothless' first impulse once it is equipped with the proxy tail is to fly off on its own. It is only after the trust between the dragon and the human fully develops that Toothless allows Hiccup to ride on its back. Furthermore, all of Hiccup's actions throughout the film are not geared towards the eventual capture of all dragons, but towards the eventual coexistence of dragons and humans. It is through the events encompassing the adventures of Hiccup and Toothless that one is able to understand the underlying message of the narrative: animals are not mindless beasts to be hunted for sport and abused for comfort; they are living, breathing, sentient beings whom humans must learn to coexist with.

Although the main message of *How to Train your Dragon* is certainly noble in nature, it does not mean that the film does not portray instances of animal abuse. In Hiccup's Viking society dragons are kept imprisoned and are often used as training objects. During training, dragons are released unto a caged ring in order to be killed by the young Vikings in training. Stoic's generation of Vikings sees this practice as perfectly acceptable. However, Hiccup does not. When forced to partake of the training exercises, which he does reluctantly, he uses tricks which he learns with Toothless in order to force dragons to sleep instead of using techniques to kill them. It is indeed worth noting that while in dragon training older Vikings teach younger Vikings skills of cooperation to fight dragons, when Hiccup befriends Toothless he learns of the possibilities of human-dragon cooperation and ends up saving both the human village and the dragons.

The climactic battle in the narrative takes place in the dragon's nest island, where, Hiccup has discovered, a giant beast forces dragons into servitude. The giant dragon overlord forces dragons to find food for it, for if they don't they will be eaten alive. Once Stoic the Vast and the other Vikings realize this, they join Hiccup, Toothless, and the other dragons in a battle for the freedom of all dragons. It is during this climactic battle that Hiccup is wounded. Just as Toothless lost half a tail in the opening battle sequence, Hiccup looses one of his legs in the final battle. This conveys the message that violence, regardless of how just a cause, will have consequences. The film's conclusion portrays a Berk which is inhabited by humans and dragons. Both species now live together, sharing the same space, in an almost symbiotic manner. With dragons aiding humans with heavy everyday tasks and humans providing dragons with the necessities for survival, the message that *How to Train your Dragon* offers is not one of animal domestication and subservience, nor is it one of unwarranted human dominance, but one of peaceful coexistence.

By looking at these two examples, the conclusion that can be drawn is that although media does, in fact, make allusions to desired behaviors by representing moral values, the moral values in question vary from title to title. *Pokemon*, which at first glance might seem like an animal-friendly narrative of teamwork and cooperation, does indeed teach values of teamwork and cooperation, but at the expense of animal rights, as the system by which Pokemon are trained is coercive in nature and the rules which govern the success of a trainer are based on a system of animal abuse. On the other hand, *How to Train your Dragon*, which at first glance might seem to advocate cruelty towards animals, actually offers an ideology of mutual acceptance and peaceful coexistence with animals. Certainly, each film producer has his or her own vision of what is correct and not, and they will undoubtedly project their views unto their work and, by extension, the viewer. It is because of this that one must be aware of the messages present in these media. Indeed, film has the power to alter old beliefs and offer new ones. Because of its unique ability of representing ideas through visual and aural means, its persuasive power may indeed be superior to that of any other medium. With film becoming more important to philosophical and ideological debate worldwide, there is no doubt that in the future there will be groundbreaking contributions that will attempt to shift perspectives on already existing conversations through interesting narrative conventions that are inherent to film.

References

Cua-Eoan, Howard. *Beware the Pokemania*. Time Magazine Online. November 14, 1999. <http://www.time.com/time/magazine/article/0,9171,34342-1,00.html>

Gee, James P. 2006. *Social Linguistics and Literacies*. New York: Routledge.

Handa, Carolyn. 2004. *Visual Rhetoric in a Digital World*. New York: Bedford/St. Martin's.

"I Choose You." -Pokemon. Satoshi Taijiri. W.B. 8 September, 1998.

Lim, Andrew. *The Pokemon Craze*. Epinions. December, 2010. <http://www0.epinions.com/game-review-5207-5039337-39D3BD67-prod1>

Smith, Lee. *How to Train your Dragon*. WLTX. April, 2010. <http://www.wltx.com/news/story.aspx?storyid=85740>

Stammitti, Rob. *How to Train your Dragon. Week in Review.* March 27, 2010. <http://www.weekinrewind.com/2010/03/how-to-train-your-dragon-movie-review.html>

Steiff, Josef and Tamplin, Tristan. 2010. *Anime and Philosophy.* Chicago: Open Court.

USCCB. *How to Train your Dragon.* Catholic News Services Media Review Office. 2010. <http://www.usccb.org/movies/h/howtotrainyourdragon.shtml>

Vilhjálmsdóttir Önnudóttir, Hlín. *Animal Cruelty. The Cave of Dragonflies.* June 9, 2010. <http://www.dragonflycave.com/animalcruelty.aspx>

Wartenberg, Thomas. 2008. *Philosophy of Film.* Stanford Encyclopedia of Philosophy. <http://plato.stanford.edu/entries/film/>

Chapter 33

Dogs of Istanbul

☆ *Aiswarya S. Babu*

Animal rights and ethics are often problematised thanks to ambivalent conceptions and want of neatly laid down rules. From the Biblical idea of man's supremacy over all other creatures to the Buddhist concept of equality among all beings, religion too has varying parameters for animal rights. These uncertainties continue today as well. For example, rights activists of PETA (People for Ethical Treatment of Animals) condemn the breeding and buying of animals, equating it to killing while many animal lovers are proud owners of pedigreed dogs. Jan Fennel, whose humane training methods for dogs have popular acclaim was a breeder herself and owns a pack of pedigreed dogs. It is in this respect that animal lives and rights depend on and are affected by their representation.

From talking dogs of comic books and animated movies, to those in serious literature; these dogs could affect our attitude towards this canine member of our or the neighbour's family. This paper looks at the evolution of the lives and rights of dogs and its popular representations. Jan Fennel's *Dog Listener* and the conventions put forward by PETA and some representative pieces on rights of dogs are used for this study. In the light of these, dogs in Orhan Pamuk's texts and their rights are read against this context. The texts used are, *My Name is Red, Istanbul:Memories of a City*, and *Other Colours:Essays and a Story*.

Dogs are the oldest domesticated animals in history. Their co-operation can be traced back to the evolution of homosapiens. Dogs, *canis familiaris* evolved from wolves, *canis lupus* around the same time, at about 150, 000 BC. The social hierarchies in the wolf pack similar to the human society helped this domestication. The dog which is used to the Alpha's leadership in a wolf pack could choose to be the subordinate or the protective leader of the hunting group and act as guardians of the livestock and the household as it guarded its pack. It could respond to orders and find its place and function well within the social hierarchy as demanded by its human companions.

With the evolution of human societies, dogs were assigned specific roles and selective breeding that produced dogs with specific qualities started. One of the earliest instances according to Jan Fennel was in Mesopotamia in 7000BC with the Arabian Desert Wolf into the earliest pure breed, Persian Grey Hound (12). With selective breeding, the status of a dog in the social structure began to change. They were not supposed to be just dogs anymore. If not for a few functional ones like the police dogs and companion dogs for differently-abled, they were reduced to decorative pieces, show animals, and toys. Their rights too suffered a similar fall.

As the Dog in Pamuk's *My Name is Red* states, migration of people from villages to cities brought in the concept of civilised pets and the idea of dirtiness got attached to dogs (13). Before the advent of pedigrees, dogs were unfussy creatures that were part of households. A stray in the street elicited a friendly glance and some grub, not fear and suspicion unless it is rabid. They were not left in the streets to wander and contribute to the 'dog menace' instead got old and died at home.

The dog as a playmate or toy was expected to stay indoors and adhere to the human etiquette for dogs. Cleanliness meant not just dogs free of fleas and rabies. Dog showing emphasised shiny mane, well formed features, accessorising as well as familiarity with human commands. The earlier shepherd dogs were overlooked in favour of pedigrees and there was a growing insecurity about the untrained dogs with no owners. Like the marginal human population, local dogs too acquired a stigma and were eyed with fear and suspicion as they were not part of the ordered world of pets. When it was instilled in the pet dogs not to bite, street dogs still fought threats from humans and other dogs with their teeth. From these developed the concept of "the dog menace".

The local dogs displaced by the breeds as pets, were left to fend for themselves and roamed the streets. Yet the street dogs one finds on the road are not always stray mongrels. Abandoning of pet dogs in the instance of its old age, disease or the owners' migration adds to the street dog population. The pent up aggression once suppressed through obedience training often resulted in bouts of anger, the dog displaying fighting instincts. A change of place, introduction of a new family member or a new pet too can make it feel threatened and awake the community instinct to fight. The dog is said to "have turned", from the obedient pet to a snappy threat. It often results in euthanasia or in the dog being abandoned for there is no place for instinctual/pack behaviour inside the four walls of contemporary homes. The inbreeding that focuses on certain preferred qualities also gives rise to medical conditions: Dalmatians and Pomeranians are prone to deafness, Alsatians to hip dysplasia, bull dogs to heart disease, and Dachshunds to spine dislocation. A blatant violation of the animal's right to a healthy life, it also pushes some of them to the streets.

Animal rights conventions adopted by various countries and organizations like PETA advocate certain rights and ethical treatment towards dogs. Ensuring access to clean food and healthy atmosphere is a basic premise. Many breeders and kennels as well as government run shelters economise on them while dog owners lack the awareness. This is one more reason why animal lovers advocate adopting a stray than buying a breed. Dogs have a right to an abuse-free existence, be it from the owner or from other social elements. Animal lovers endow the government with duties to provide proper medical care to dogs, at least in the recognised and funded shelters. Euthanasia too should be humane and professionally done: while an intravenous injection of sodium pentobarbital administered by a trained professional is the preferred method, electrocution, beheading and shooting amount to cruel violation of the rights. Breeding is condemned as animals without desired features are killed or abandoned by the breeders. Dog shows that emphasise on certain preferred features are equally guilty in accentuating this crime. The pure breeds become completely dependent on the owners and do not possess the survival qualities of mixed breeds. If abandoned, they are prone to be hit by cars or to die of starvation. The obedience training (pet training for obeying human commands like fetching) these pure breeds undergo call for the use of force–except in a few cases like Jan Fennel's Amichien Bonding–and pressure that may amount to cruelty towards dogs.

The most problematic issue in animal rights is the lack of education and representation of these rights. Literature and media hardly represent the cause while propagation of tilted representations contributes to animal abuse. PETA states in "The AKC and Dog Breeders: Partners in Crime" how a breed in a Hollywood movie can increase its demand in the market and will result in abandoned pups that do not exactly resemble it: "Every time Hollywood makes a dog movie, the breed goes to hell," says one caretaker of Bouvier des Flandres dogs. The telecom advertisement in India with the pug following the boy everywhere sure increased the popularity of the service provider. It also increased the demand for pugs in the market and breeding mills went on churning out the pug with the desired colour and other features. If *101 Dalmatians* meant hell for that bred of dogs, the math can be quite shocking in this case too.

Literature too has a similar effect on animal rights. Be it Nana from *Peter Pan* or the Hound from *The Hound of Baskervilles*, literature tend to fantasise and glamorise the idea of dogs as faithful companions or as images of terror. The shadowy world of animal cruelty remains in the shadows. Coetzee's *Disgrace* is a commendable deviation from these texts. The abandoned dogs constantly waiting for adoption while moving towards the euthanising table, is an uncommon sight in a novel. The underfunded and closed down shelter that operates on Bev's kindness and meagre donations is the picture of many animal welfare facilities. The human abusers abusing animals as well; the animal owners' hesitance to opt for euthanasia for a suffering animal; inadequately supplied animal facilities; the humane treatment at the euthanising table and at the crematorium apart from death from suffering and salvation are aptly represented by Coetzee in the novel, sans the drama.

If *101 Dalmatians* gave "hell" to the breed, if *Disgrace* threw light on the last days of abandoned mongrels, then what does Orhan Pamuk's texts and the street dogs attempt to do to this centuries old characteristic of Istanbul streets? In the light of the rights, practises and representations, the lives of Istanbul's dogs (those of the text and of the city) are discussed. It problematises the literary presence of the street dogs, their representation, the attitude and ethics evident from the texts and how they reflect on the rights and lives of the street dogs that are synonymous with Istanbul streets.

According to Linda Taal who supports and runs dog shelters in Turkey, there are approximately 150,000 stray dogs in Istanbul and almost 30,000 living in shelters. The Turkish Animal Welfare Bill of 2004, implemented since June 2006 asks the municipalities "to solve their stray dog problems by means of catch, neuter, vaccinate and return", she told *Abolitionists-Online* in an interview. But most do not comply with this law or do not follow it properly. The 33 shelters in Istanbul run by the municipality are overcrowded, has scarce food, and no medical aid. Taal cites an instance of a pesticide company contracted for spaying or neutering. The dogs were transported in non-ventilated cars, were not neutered professionally or humanely, were starved and then dumped in unfamiliar territories. A Turkish animal welfare association, EHDKD took the Istanbul Metropolitan Municipality to court and the municipality took up the job themselves. The visible change according to Taal was, "only one difference: dogs are no longer dumped in unfamiliar areas in the cities, but in desolated areas (forest, highways) without anything to eat or drink". The dogs were supposed to be returned to the same territories as they are attuned to the provisions of survival there. Animal protection websites like *straydogsinturkey.com* and videos posted in YouTube like, "Animal Cruelty in Turkey" and "Police Brutality towards Dogs" tell similar stories of abuse and abandonment. They also mention the fear of the people, the prejudice of the religion, cruelty of some social elements to dogs, lack of awareness and absence of adequate funds. It is in this context that Orhan Pamuk's representation of street dogs acquires improved significance.

"What must it be like to be a dog?" asks Orhan Pamuk in "What I Know about Dogs" (*Other Colours* 48). Be it the aggressive narrator of *My Name is Red*, the Dog, the brown stray in *Other Colours:Essays and a Story*, the resilient street dogs of *Istanbul*, the snoopy dogs shadowing Ka in *Snow* or those in his Nobel lecture and newspaper columns completing his picture of Turkey, it is normative for dogs to walk into Pamuk's writings.

Orhan Pamuk identifies Istanbul's street dogs as integral part of the city, "Only one of the city's idiosyncrasies has refused to melt away under the Western gaze: the packs of dogs that still roam the streets" (*Istanbul* 219). "They look alike, their coats all the same colour for which no one has a name", he opines (*Istanbul* 39). Apart from these postcolonial and nationalist notions, these dogs point to a flawed system in place for the street dogs in Turkey.

The campaigns that failed to "drive away" stray dogs from the streets, the military generals who "mentions the dog menace", the state and the school system that "have launched campaign after campaign"; all emphasise the resilience of the dogs for Pamuk (*Istanbul* 39). They, in fact establish the ineffectiveness of the attempts of spaying or neutering of dogs. The dogs that have an instinct for procreation were multiplied through centuries and their rehabilitation programmes too were ineffective. "After the Constitutional Monarchy, there was another 'reform' drive, this one aided by the gypsies, but the dogs removed one by one to Sivriada managed to find their way triumphantly back home". (Pamuk, *Istanbul* 219) Dogs removed from familiar places tend to find their way back thanks to the unavailability of food and problems of survival at these unfamiliar places. They "cling to their old turf" for rear of being stripped off its food sources and shelter. If one is to take Taal's comments into account, the improved neutering and rehabilitation drive is just the old wine in a new bottle. And there were attempts to poison dogs to control their population, a blatant violation of these animals' right to kind death.

While talking about the attitude of the public, the fear and suspicion towards stray dogs is a problem. "Fearsome as they are...", Pamuk qualifies them in *Istanbul* (39). This fear is shared by the people as well agrees GAS-Der Straydogs Workgroup. The public's fear can be traced back to the abundance of strays and the lack of effective awareness campaigns. Pamuk carries this fear into his texts: "And the dogs that have followed me all my life are portrayed as creatures that cause alarm and suspicion in the heroes to whom I feel close" (*Other Colours* 50). These representations could only help in accentuating the fear for dogs rampant in the milieu. The menacing Dog with long and pointed teeth, for whom, "nothing is as satisfying as sinking his teeth into his miserable enemy in a fit of instinctual wrath" can send shivers down the spine (Pamuk, *My Name is Red* 10). Though attributed to a personal prejudice, fearsomeness of Pamuk's stray dogs put their counterparts in a poor light. Pamuk further complicates it as he links the attack on him by eight to nine dogs in Maka Park to his shadowed representations of them in his novels. He terms it an act of vengeance by these mongrels in "A Note on Poetic Justice" (*Other Colours* 50). He could only tarnish the perception about dogs further with, "but there are always dogs waiting to catch the vengeful poet alone at a corner and sink their teeth into him" (*Other Colours* 51). If a doting Dalmatian from a movie can produce love for the animal, a strained representation could very well produce hostility.

Yet aggressiveness and biting are instinctual to dogs. Their life in the wolf pack and early role as guardians and hunting companions of humans was in tally with this quality. But pet dog decorum has no place for these qualities, except in the case of selected breeds of guard dogs. So the stray dogs with their own will and habitual instincts intact induce fear in the public. Yet Pamuk seems to have a certain regard for their resilience.

The most vociferous of the street dogs in Orhan Pamuk's works is the Dog, one of the nineteen narrators of *My Name is Red*. A part of the Sultan's great book of paintings, it finds its way to a scrap paper at a coffee shop, supplying to the plot while telling its own story like the other narrators. Fearsome, aggressive and gigantic, it is pretty chatty about the life of stray dogs and their pedigreed companions, the prejudices of Islam and the people, and the changes in the status of dogs in the social structure. The writer's strain of admiration for the analytic, rational and proud dog despite its menacing qualities is contagious.

There is certain disapproval for the repression and dismissal of packs and communities of dogs in "I am a Dog" chapter of *My Name is Red*. The Dog looks down on the European dogs who have owners, accessorised in human clothing and are on leashes, "fettered like most miserable of slaves and dragged around in isolation"(2002:14) The Dog instead is proud of,

> Dogs who roam the streets of Istanbul freely in packs and communities, the way we do, dogs who threaten people if necessary, who can curl up in a warm corner or stretch out in the

> shade peacefully, and who can shit wherever they want and bite whoever they want. (2002:14).

Pamuk here is putting together a picture of Istanbul's street dogs. It connotes a sense of approval for dogs living their lives as dogs, with their own free will. They are not owned properties but free members of dog "packs and communities". The pack life calls for a degree of aggression where erring members are taught their place with a snip or bite. If there is a human threat identified to their territory or self, the dogs react in the same mode. It is one of the reasons behind the attacking streaks in otherwise well-behaved pets. The bottom line is, the problem lies not with the dogs who follow their instincts but with the humans who intend to suppress them. The problem lies with the authorities who fail to carry out proper spaying and neutering which not only controls the population but is proven to reduce aggression in dogs. "A spayed or neutered dog is healthier, lives longer, is less aggressive, is less likely to roam and get lost or hit by a car, and won't suffer from cancers of the reproductive system", observes PETA in "The AKC and Dog Breeders: Partners in Crime".

Another prejudice Pamuk's Dog represents and addresses is the centuries old religious prejudice which ensured suffering and cruelty for dogs. Taal and GAS-Der mentions how some followers of Islam disapproves of the animal. Pamuk's Dog represents and confronts this issue well. It is critical of religious reservations towards its counterparts. The religious hostility is attributes to a misunderstanding of the Prophet's act of kindness to a cat. The mention of a dog in the chapter, "The Cave" of *Koran* is brought in to condemn their barring from mosques and beatings by caretakers. "...The slander that a pot licked by a dog must be thrown away or retinned" is dismissed as a creation of tinsmiths (2002:13). It also refers to the "months of dog" prevalent before Islam. The dog lover in Pamuk is more rational and less sarcastic while putting forward the case. Yet he makes their case worth listening by mingling an unnerving canine and coffeehouse humour.

Taal too has a balanced approach to offer and quotes Ayat (verse in the *Koran*) 6:38, "All the beasts that roam the earth and all the birds that wing their flight are but communities like your own. We have left out nothing in the Book. They shall all be gathered before their Lord", she quotes in her interview to *Abolitionist-Online* to affirm that the holy text also offers a kinder interpretation of the rights of dogs.

Pamuk's dogs are shaped by the author's attitude towards and encounter with dogs. They take after their city and are essentially Turkish in the sense that they reflect the moods and atmosphere of the city. Be it the resilience towards changing regimes, the refusal to be wiped away, the haunting of the shadowy streets or the very colour of their mane, they are typical to the country. No wonder they found a place in the Ottoman Sultan's precious miniature painting in *My Name is Red* and a Turkish Nobel laureate's texts. Pamuk's Dog displays a great sense of analogy, a way with words, a strong sense of pride and freedom as well as a clear and rational mind. But he is a window to the lives of the dogs of Istanbul. Pamuk's dogs may be strays but they represent the cause of their fellow beings aided by an author whose fear of them do not deter him from giving them a voice. Pamuk's representation confronts the rights violations and animal sufferings that go back to centuries. The street dogs are integral to his texts for a reason and not just to induce suspicion and fear in the heroes. The dogs of Istanbul do talk to Pamuk and he does listen. The world certainly is talking more about Turkey since Pamuk. If the sufferings of the stray dogs of Istanbul can make people take notice, as the human rights issues did, then their problems need not wait long for redressal.

References

"The AKC and Dog Breeders: Partners in Crime". *Peta Media Center*. People for the Ethical Treatment of Animals. 2010. Web. 14 Aug. 2010.

Coetzee, J.M. *Disgrace*.1999. London: Vintage-Random, 2000.

Fennel. Jan. *Dog Listener*. 2000. New York: Harper Resource-Harper Collins, 2004.

Pamuk, Orhan. 2006. *Istanbul: Memories of a City*. Trans. Maureen Freely. London: Faber and Faber.

_____. 2002. *My Name is Red*. Trans. Erdag M.Göknar. 2001. Newyork: Vintage-Random.

_____. 2008. *Other Colours: Essays and a Story*. Trans. Maureen Freely. 2007. London: Faber and Faber.Print.

"Stray Dogs and Cats in Turkey". Akyaka.org. GAS-Der Straydogs Workgroup.Web. 11 Aug. 2010.

Taal, Linda. Interview. "Turkish Stray Dogs". By Claudette Vaughan. *Abolitionist-Online*. 7(2008): n.pag. Web. 15 Aug. 2010.

Chapter 34

The 'Silent' Listener in Anton Chekhov's "*Heartache*"

☆ *D. Venkataramanan*

The 'animal in humans' and the 'human in animals' has come to be viewed as a complementary phenomenon. Literature, History, Philosophy – in short – the Arts, has gone about vociferously portraying the junctures when human rights are violated. The Arts abound in works that reveal this. Interestingly, this has been extended to animals also. Our concern here is only with the manner in which Literature views animal rights. World Literature showcases several literary works that deal with the animal kingdom. The Western literary sphere flaunts Classics like *Black Beauty*, *Animal Farm* and *Moby Dick*. In most of the works from the Western side, the portrayal of the animals has a technical purpose. For example, George Orwell uses animals in his *Animal Farm* to satirise the Russian political feud involving Stalin and Trotsky. In contrast, the world of animals in Russian literature is a part of the human world. Both spheres co-exist. Much of Russian literature has been a direct product of suffering. The Russians mostly do not see the animal as the 'other'. A peculiar kind of closeness is seen between the characters and the animals. Several literary works from the erstwhile Soviet Union reflect this complex but meaningful relationship between animals and human beings.

This paper, contextualizing this situation of human-animal nexus, explores how Anton Chekhov (1860-1904), a master of the Russian short story genre, has presented the 'animal sensibility' in his story *'Heartache'* (1886), also translated as *'Misery'* or *'Lament'*. It also shows how Chekhov, instead of moralizing, tries to humanize the animal kingdom and highlights how animals 'communicate' where human beings fail.

'Heartache' deals with the mental trauma of Jonah Potapov, a cab driver who is desperate to share his grief over the recent death of his son Kuzma and unable to do

so with human beings, ends in pouring it out to his horse, a 'silent listener'. This two-page short story presented a few hours in the life of an ordinary cab driver. This kind of approach characterizes most Russian literary works, notably Solzhenitsyn's novel *One Day in the Life of Ivan Denisovich,* which describes a day in the life of a camp inmate. Most of the Russian stories deal with the dilemma of existence and World Literature deals with this theme on a large scale. "Existentialism", a potential movement celebrates this philosophy. Mostly, there is a note of pessimism. It is interesting to note that a majority of Russian writers find ways of engaging with this dilemma of existence. For example, Solzhenitsyn's Ivan Denisovich is described as working strenuously amidst inclement weather. Ivan does not give in to the hardship but endures and even finds a sense of fulfillment in his labour. The pessimistic feature of existentialism is transformed in most of Russian literature. This definitely takes place in Chekhov's *'Heartache'.*

The setting, the language, the description, the pace, the positions taken up by Jonah, the various passengers and the mare – all come together in Chekhov's hands to become a masterful narrative that touches readers and finally makes them also to think deeply. The story introduces Jonah Potapov as sitting "huddled", "motionless" and "white as a ghost" (186). The same words "white" and "motionless" are also used to describe Jonah Potapov's "little old horse". Though this parallelism seems innocuous at a surface reading, Chekhov's placing of a 'human being' and an 'animal' on the same plane reveals his belief in 'animal ethics'. Chekhov's description of the horse is mundane but he does not stop with that. He writes, "In all probability it [the horse] is plunged in thought" (186). Chekhov attributes the ability of human feelings to the horse. Here, readers are not prone to rationalize whether animals can think or not. Chekhov subtly argues, "For one who has been torn from the plough, from the familiar greyness of the landscape, and thrust into this whirlpool of monstrous lights, incessant tumult and surging humans, it is impossible not to think" (186-187). The parallelism continues with Chekhov pointing out that "For a long time Jonah and his horse have not stirred" (187). It is not merely the physical world that seems to be common to both but almost the mental world too. Chekhov sprinkles his story with such subtleties and he reverses the roles towards the end. It is here that to Jonah Potapov, the other humans become the 'other' and the traditional 'other (the animal) replaces the position of the human beings.

It is almost a week since the death of Jonah Potapov's son. Jonah is lost in grief, "stares round like a lunatic" and "doesn't know where he is and why he is there" (187). His attempts to share his grief with all his passengers on his carriage fail. His anguish filled statement, "This week, sir my son … um … passed away" (187) is met with an indifferent reply, "Indeed? What did he die of?" (188). Another passenger says, "That's what we shall all do" (189). Jonah tells another passenger:

> These days my only wife is the damp earth … ho-ho-ho! … the grave, that is … My son's gone and died, and I'm still alive and kicking … A queer business: death came in at the wrong door … Instead of calling on me, he called on my son …. (189).

The passengers are indifferent and happy that they have reached their destination. No one is "disposed to listen" (188).

Chekhov brilliantly captures that moment of deep sorrow that forces one to question the purpose of existence:

> Were Jonah's breast to burst and his sorrow come gushing forth, it seems it would flood the whole world; but nevertheless it is unseen. It has managed to conceal itself in so insignificant a shell that you couldn't discover it by broad daylight with a lantern .... (190).

This tends to move towards a pessimistic note with regard to the plight of Jonah Potapov. As readers, we tent to sympathise with Jonah's predicament and begin to think that this will be the status quo of Jonah. Jonah realizes that there is "No longer any point in speaking to people" (190). Here comes the reversal. Jonah feeling "a stab of pain" and unable to withstand it he drives home. Jonah's horse "as though reading his thoughts starts off at a jog-trot" (190).

At home, Jonah makes another desperate attempt to speak to a young cab driver but the latter dozes off. Jonah is desperate to speak about so many details, "how his son fell ill, how he suffered, and what he said before he died, how he died ... Describe the funeral, the journey to the hospital to collect his son's belongings ... (191). Jonah even contemplates about sharing his thoughts with women. He feels, "To talk to women would be best of all. They may be fools, but say two words and they're already in tears" (191). At this point Jonah remembers his horse. He begins by talking to his horse about oats and his having become "too old to drive" and goes on to speak of his son who ought to have been helping him. Jonah addresses his horse:

> That's how it is, my dear ... Kuzma my son is no more ... Passed away ... Went and died, and for what? ... Now let's take you; you've got say, this foal goes and dies ... You'd be dory, wouldn't you? (191)

Anton Chekhov describes the response of the mare thus: "The mare chews, listens, [emphasis added], breathes on her master's hands" (191). He ends his story with the words, "And Jonah, carried away, [emphasis added] tells her the whole story ... "191).

We see that human communication fails miserably. People are shown as indifferent to the feelings of fellow beings. This is illustrated through the depiction of Jonah Potapov's predicament. Then Jonah realizes the companionship of his horse in the throes of an existentialist trauma. It is here that we understand how Russian existentialists differ from their European counterparts by not completing negating life but by seeking alternatives. This is what Jonah Potapov does. He is unable to get anyone to listen to him. He listens to himself and feels sorry that he had not earned enough. He feels, "A fellow who know his business ... who's well-filled himself and his horse too, is always at peace" 190). He remembers his horse and goes to it. The horse fills the void in Jonah Potapov's emotional schema and performs the role that human beings are supposed to take on - the role of a listener in a dire situation.

Jonah's mare, the 'silent listener', becomes a catalyst for a letting-out of pent up grief – Jonah's heartache.

Chekhov's *'Heartache'* does not idealise the horse. It rather describes it in conventional term but Chekhov ascribes human feelings to it. The core of the phenomenon of 'Animal Ethics' concerns itself with' feelings' towards animals emphasizing its rights to live as human beings do. There seems to be no question go 'rights'. When human beings are desperate to live with self-dignity, then it is more a question of 'existence', irrespective of qualities like living, non-living, animate, non-animate or communicative, non-communicative. So, the 'right to live' is a common, unstated rule of existence. It is laudable that Russian literature have seen deep into this aspect and represented animals as a part of one's life, a companion and not an intruder. Chekhov's *'Heartache'* highlights this. The horse does its duty, is considerate towards its master, is aware of his grief and responds to it. It "breathes on her master's hands" (191) brings in a breath of new life into the master's life. There is no human language involved here. No words of solace are spoken. This mere gesture of this silent listener makes Jonah Potapov unburden his grief to his mare.

Chekhov's short story is no overt treatise on 'animal ethics' but the absence of 'ethics' that should exist in a human being. When an animal responds instead of human beings, it assumes an ethical significance. In such a context, the unsaid becomes more potent. We, as readers, are made to think of our attitude towards animals, 'how we treat them' and 'how we should treat them'. Literary works such as Chekhov's *'Heartache'* exemplify such a progressive move. We are led to introspect this phenomenon of ethics vis-à-vis the animal kingdom.

Many literatures uphold and portray the function and significance of animal life. The 'What' and 'How' of this portrayal makes all the difference. Chekhov's *'Heartache'* is a work that is bound to not only move readers and touch them emotionally but also start them on an inner journey of self-revelations. 'Silence' is a much more powerful language that animals are proficient in. We need to follow 'Animal Ethics' and practise it 'silently' respecting and acknowledging that 'silent power of language' that animals possess.

References

Chekhov, Anton, "Heartache". Trans. Walter Morrison. *A First Series of Representative Russian Stories – Pushkin to Gorky*. Ed. Janko Lavrin. Westhouse, London: 1946. 186-191.

_____. *One Day in the Life of Ivan Denisovich*. 1962. Trans. Ralph Parker. Intro. Marvin L. Kalb. Foreword Alexander Tvardovsky. New York: Signet Classic, 1963.

Chapter 35

Beauty, Spirit and Power: Divine Animals in the Writings of Louise Erdrich

☆ *K. Subapriya*

The world, a perfect combination of both living and non-living things is classified into broader groups as Homo sapiens (man), plants, animals and reptiles. We, the modern men/women try to exercise absolute control over all the other categories denying the fact that they are in no way inferior to us. The opinion of ancestors varied from the present situation, where the least attention is paid towards the survival and needs of every species. The man/woman of previous age strongly believed in the quality of all the species. Therefore, they praised and worshipped religions and scriptures that promoted the concept of sharing love. The air of superiority has enforced the present generation to deny the importance of animals. But animals are wonderful creatures which have synchronized with the life of humankind. The majority of the present generation fails to recognize this strong bond.

People during the ancient times were aware of the importance of animals which provided them food, shelter and clothing. Hence, they worshipped them as Gods. Through out the globe, every religion manifest the notion of love for animals and Hindu religion further emphasizes this ideology. According to Hindu myth, animals are not a separate entity but one among us. It is interesting to note their efforts to get off their Karma to attain salvation. They have the goal to attain liberation and hence they seek the human company to achieve this, stated the well renowned sage Sankaracharya. Diana Tierney in her article "Hindu Sacred Animals" discusses the reverence attributed to cows and monkeys in Hinduism. These two animals have been given primary importance and apart from these many other animals like snakes, crows, cocks and dogs are worshipped in Hindu religion. These kinds of worship exhibit the importance of animals.

However, the importance and uniqueness of animals has disappeared in due course of time with the scientific advancements and disbeliefs about nature and religion. Attempts are being made by philanthropists and writers like Lorri Houston, Ric O' Barry, Jarid Manos, Jonathan Safron Foer and Charles Seibert to retain the valuable ideas of treating all the species with equality. In ancient times, the situation prevailed, where animals gained the superior strata over human beings by serving as a source of inspiration, a motivating and guiding spirit and a symbol of love and beauty, as seen in the writings of Louise Erdrich – a Native American Renaissance writer. The fact that Native American culture preserves its ideology about animals is evident through a study of Louise Erdrich's *The Birch Bark House, Four Souls, Tracks* and *Love Medicine.* These texts serve as a testimony to the ability and superiority of animals over humankind. Her works symbolize the energy and spirit derived from animals. One can feel the power and the charm of animals in her works. This paper attempts to explore the representation and power of animals over humankind, by revealing a close bond between Native Americans (Ojibway) and animals through select texts of Louise Erdrich.

Native Americans are well known for expressing their extraordinary love and attachment towards animals. A number of myths encapsulate the close relationship between animals and the Native Americans. In their culture, animals are not mere creatures with five senses but are elevated to the level of Gods. Animals are living Gods among human beings who dwell amidst them to provide a remedy for their troubles. Sometimes, these animals are even considered as spirit forms that protect the chosen spirit.

According to Native tradition, every individual after a certain age has to dream in search of a spirit. Spirits related to the family/ individuals come forward to guide and protect them until their end. Louise Erdrich's *The Birch Bark House, Love Medicine, Four Souls* and *Tracks* aptly illustrates and validates this celestial relationship between humankind and animals, The beauty, power and spirit - both in abstract and concrete forms - are merged in the writings of Erdrich.

The intertwined relationship between animals and human beings perfectly blend in Native American culture. Therefore, it is rather difficult to realize the function and relation of animals thriving as an individual form. So, these unnoticed aspects of the animals can be traced in Louise Erdrich's novels where animals excel over humankind at various situations and guide them.

To begin with, her children's work *The Birch Bark House* is a tale about Omakayas and her family, a well renowned medicine woman in the Ojibway community. Erdrich is a descendant of the Ojibway community and therefore all her works unveil their tradition and culture. In this children's work, three major characters (animals) perform a major task. The bear, crow, and the frog dominate the story. However, there are reference to horse, deer, fish, dogs and other birds.

The very name of the protagonist Omakaya means little frog. She was named so "because her first step was a hop" (*The Birch Bark House* 5). Therefore, it is evident that animals are paid due respect where people are named after the animals. There is yet another character whose name is 'fish tail'. Naming the humankind with the name of an animal is a form of reverence shown to the animals. As people in India are named after Deities, so the natives are named after animals, which show the superiority of animals over human beings.

Nevertheless, it cannot be denied that the Natives are basically hunters whose main source of livelihood is through hunting, but they never kill animals for sport. They request the respective animals for food and state that they 'need them' for survival. They offer a pinch of tobacco (sign of respect) after hunting the animals, to appease the spirit of the killed. Though this act of hunting raises the question of violence against the animals in the context of Native culture, killing the animals for one's needs after paying due respect reveals the ethics of Natives. They do follow ethics in killing. The act of killing therefore is not mere destruction but is transformed into a different act combined with a purpose. The Natives hunt animals only for food. In general, the reverence/violence does not go together and therefore Natives may be blamed for hunting animals. However, in every culture, human beings are dependant on animals for food. Even in India, we worship animals and at the same time kill them for food. But Native Americans are far better as they pay homage to animals and then kill them. Except this reason for the act of hunting, one can observe the cordial relationship between animals and the Natives. This does not distort but only reiterates and highlights the Natives' friendly attitude towards animals.

Hunting for sport is rather different from hunting for need. Killing a fellow human being in a war is different from murder. It is considered as patriarchy and proper homage is paid to the soldiers for killing. The aspect of killing animals is not justified but in the native context, this cannot be considered as violence. It is done for a purpose. And the Natives relationship with animals cannot be merely restricted to the level of hunting. It is much profound. By providing food, animals reach the level of God and no wonder Gods are thanked for fulfilling needs. An animal by providing food for the Natives attains the position of a deity. Animal as a deity both serve and guide them. Therefore, the Natives love animals and pay due respect to them.

In *The Birch bark House* during the winter when food cannot be obtained from crops, the spirit guides humankind towards their food. When the family of Omakayas runs short of food, her grandmother dreams and tells the exact place of animal dwelling, which can end their starvation. Deydey accurately follows her guidance and finally finds a one horn (a variety of deer) and shoots him dead. He offers tobacco to the spirit of the deer and thanks him for favouring his family.

Animals are treated with respect throughout the text and Erdrich never fails to address them as "him/her" instead of "it". The bear takes up a position in this text when compared to all the other animals. It is the spirit of bear (bear people) who becomes the guiding sprit of Omakayas. Omakayas' mother puts charcoal on her forehead before going to bed. Initially, Omakayas could not find her guiding spirit through dreams in the first two attempts. Later, she dreams of a motherly woman who declares, "I'm going to help you"... "I feel sorry for you, and I love you. Just remember to give me tobacco. I'm the bear spirit woman. I knew your great-grandma and great-grandpa. They lived with me now" (*The Birch Bark House* 170).

Omakayas narrates the dream to her mom and she confirms that it is the bear people who extend their favour. However, the strong bond between Omakayas and the bear is seen right in the beginning of the story. She feels a sort of attraction towards the bear cubs, plays with them and when the mother bear tries to attack Omakayas to defend her cubs, she explains her innocent attitudes to the mother bear. Her unharming intention is sensed by the mother bear, which leaves her calmly.

During the first meeting Omakayas gives a piece of advice and warning about the human beings. Her oneness with the bear is established through the smell, "Perhaps the bear smelled the touch of Grandma and mama's bone-and-spruce wood comb, her baby brother's cuddling body, the skins and mats she had slept in, and Little Pinch who had whined and sobbed the night before…fish from the night before last night. The berries she was eating. The bear smelled all" (*The Birch Bark House* 30-31). Her knowledge about the bears gained previously from her grandma with regard to the similarities between them and the human beings enables her to consider those two bear cubs as her brothers.

The image of the bear and their voice haunts her often. She is able to sense the presence of the bear even after coming home. Her newly established relationship with the bear in the woods is explicit in the following lines, "She knew the bear had followed her home. She knew that when she needed the bear she would be able to call on the bear. The bear had understood something she had said and she had understood something the bear had thought, and although she couldn't tell exactly what, Omakayas turned back to her task with her head clear and her hands cheerful" (*The Birch Bark House* 36).

Her second meeting with the bear cubs and the mother bear happens when she goes in search of her another pet Andeg (crow) along with her mother. Omakayas' mother is able to sense the close relationship between the bear and her daughter. Even her grandmother anticipates her to be a medicine woman in the future. Her third meeting with the bear finally helps her to evolve as a healer. She meets them jovially, offers her usual advice to be aware of human beings and then places tobacco to ask for something but she does not know what to ask for. She hears voices but is unable to decipher its meaning. Finally, when she reveals her problem to grandma, she realizes the destiny of Omakayas to be a healer. Immediately, an incident follows, where Big Pinch's legs are hurt and Omakayas treats him like an experienced healer. Nevertheless, the relation between Omakayas and the bear is eternal.

Along with the bear, yet another character who flies in and out of the story leaving a strong impression in the minds of readers is Andeg (crow). Andeg becomes the friend of Omakayas in a strange way. In the process of hunting, a little crow is left behind without killing which later turns out to be the best friend of Omakayas. Apart from becoming one in the family, it gains name for its speaking skill. It learns to tell two words -"Gago Pinch" (stop pinch) and ahneen (a greeting, or sometimes it is a form of question). Andeg also serves as a comic relief in the story.

This little bird gets resurrected in the hands of Omakaya with her decision to set the last bird free. However, when she finds the inability of the bird to fly with its hurt leg, she puts it in her sack and carries it to home. Later, she forgets about the bird. The bird reveals itself in a way, which evokes a burst of laughter among the family members. After feasting on the hunted birds, Deydey narrates a ghost story. At the end of the ghost story, Deydey offer gifts to everyone, which he had brought during the trip. He appreciates Omakayas for her better preparation of makazins (footwear usually made of tanned moose hide or deer skin, often trimmed with beads and/or fur). After receiving Deydey's old gun barrel as the gift, suddenly she notices the movement of the makazins that had been laid in a corner. Louise Erdrich narrates the incident with a tinge of humour as follows:

> Suddenly, her heart thumped. Her throat shut. Owah! Deydey's makazin. Oh, she was sure she saw it move. It had moved. She was sure of it. Again!
>
> "Neshkey!" Her voice trembled. She pointed. The others looked.
>
> ...They were all frozen in shock. The makazin sat quite, still. Nokomis leaned over, poked at it with a stick. The makazin twitched and then squawked. Omakayas jumped up, for she remembered instantly that she'd left her carrying sack right next to the door with the makazins. Sure enough, her bird popped its head out, bright, and in the sound of everyone's laughter it blinked with interest, cheerful, greedy, hunger, unafraid. (*The Birch Bark House* 69-70).

Omakayas realization of the bird with its head popping out from the sack left behind the makazins invokes great laughter among the family members. More than being in a shelter, it takes refuge on Omakayas' shoulder and head. Andeg dwells with them, serves the family and finds food for them during winter. When Andeg is punished by Omakayas' mother, believing the words of Pinch for eating the berries, it flies far away from them.

When they realize their folly and go back to call, Andeg feels afraid of Omakayas' mother. However, the relationship once again continues until one day Andeg flies off with its flocks. Though its behaviour changes wild after some time, Omakayas does not blame Andeg for it. She is quite aware of separate kingdom of animals. She is filled with hope about the return of Andeg, after listening to Lapautre's description of a bird. Lapautre who never knew Andeg before narrates the story of a strange bird who could tell "Ahneen". He expects cheer and amazement from this family after hearing the strange story of a bird with talking skill, but to his amazement the family remains passive. No wonder the family easily identifies Andeg from Lapautre's description.

Fish, horse, dog and rabbit form a part and parcel of their life. Fishes and Rabbits fill their belly. Rabbit stew serves both as a healthy recipe filled with medicinal qualities and as delicious food. Horse for its part highly charms Deydey. He is so carried away by the charm of horse that even in the game of chess, his most favourite is the horse. He loves the curves of the horse.

Dog is another character which seems to behave in an indifferent way towards Omakayas. The encounter between the dog and Omakaya is highly captivating. For the first time readers come across a situation, where right from the beginning to the end, the dog snares at the girl with a cruel intention. The hunting dog, for no reason, snares at the girl. Old Tallow specially warns the hunting dog, that least it attempts to touch Omakayas, it has to face a terrible death. But despite of the warning, at the end when the dog attempts to catch hold of her, it meets the mortal death at the hands of old Tallow. Old Tallow hates to kill the dog but for the love of Omakayos she kills him. This act also evokes distaste in the mind of Omakayos. Omakayos avoids old Tallow after this incident. She could not bear to see the yellow dog dying in the hands of its own master because of her. Omakayas love for animals becomes explicit through this incident. She is able to show a deep affection even towards those animals that act indifferently.

Above all these, the origin myth of the world as narrated by grandmother is a crown to all the other animal myths. The Ojibway myth brings out the power of an animal in creating a new world. Nanabozho seeks the help of a beaver, an otter and a muskarat to bring at least a speck of mud from the bottom of the sea. Beaver and otter fail in their attempts but Nanabozho gathering the mud from muskarat throws it over the water thus creating a new world. It is the tradition to worship the creator and the one who helped in creation. Hence, the animals are prone to worship who has played a major role in procuring them a piece of land to live with all resources.

In *Four souls*, though Louise Erdrich does not refer much to animals as such, their presence is felt throughout the text. The characters of humankind are qualified with the characters of the animals. This adds to the credit of the animals. The animal characters are attributed to human kind to show the resemblance between the two in many ways and at times even animal excels human beings. Therefore, when Fleur Pillager bounces on John to kill him, he would know "the wolf in her couldn't kill him on instinct and the woman in her could not destroy him out of sheer intrigue" (*Four Souls* 73). The partial presence of animal is felt in all human beings.

When John regrets his injustice, he wants to marry Fleur and surrender his spirit and body at her service like a dog. He tells, "My spirit is meant to be the slave of your spirit. I will make you my wife and give you everything I own. And more than that, I will love you no matter what you do to me, as a dog does. My spirit is meant to be g'dai, your animal, to do with as you wish, let live or kill" (*Four Souls* 46). John Mausser's attempts to surrender spirit like a dog establish the faithful quality of dog.

If God is the provider of food, clothing and shelter, then animals provide all these things to the Natives. It provides them food; their skin is used for clothing and shelter. For example, the quills of porcupine and dyed hairs of moose keep them warm. So, animals deserve due respect, space to live and worship.

Apart from the noble qualities of animals, other factors draw one's attention towards animals in Native culture. Love medicine is a common factor, which appears both in the *Four Souls* and *in Love Medicine*. In *Four Souls* Nanapush prepares a love medicine to attract his wife Margaret but the poor Shesheeb's dog swallows it quickly without the attention of Nanapush. Nanapush preparation of love medicine seems to turn futile. However, later it helps him to get out of the trap laid by Shesheeb. The dog comes to the rescue of Nanapush when Shesheeb tries to seduce him mistaking him for woman.

In *Love Medicine*, Lamartine's preparation and content of love medicine is different. He plans to secure a pair of geese heart and provide them for his grandpa and grandma. The consuming of a female heart by grandma and male heart by grandpa is expected to have a good result, because "If it's true that the higher feelings of devotion get lodged in the heart like people say, then we'd be home free" (*Love Medicine* 242).

Metaphorically speaking, Love Medicine is considered as a central metaphor of Erdrich works where "Medicine" represents "interdisciplinary network of religion, philosophy, spirituality, health care, psychology, ecology, natural history and more" (*Native American Literatures* 100). This medicine combined with the element of love helps to provide a healthy environment is the basic conception of Erdrich. But the unique feature is that the usage of Love Medicine in the text is made through animals. Either animals are used in preparation of love medicine or animal gets interfered in the preparation of Love Medicine.

As stated early, the human characters are attributed with animal qualities. Men and women are described with the animal characters. For example, Lulu Nanapush in *Love Medicine* calls herself as wild. Even the features of human beings are explained in the context of animals. In *Tracks* when Eli examines the hand of Pauline and discords with interest, Erdrich explains it as, "he looked at my hand with curiosity, no intent, and then like a fish too small to keep, he threw it back" (*Tracks* 77). The sense of taste and the dressing style is also related with animals. The lines like "a smell coming off her like duck feathers, oiled and sly" (78) and "Her skirt was spread like a trap for fish" (79) serves as a best example. When Fleur struggles to deliver the child and Eli is left at a loss, it is the drunken bear, which comes to the rescue of Fleur to deliver the child safely. By frightening her to push the baby out, she delivers the baby. Similarly, phrases like "wicked serpent" and "skinny old dog" are used to blame. To praise someone people say, "shrewd young bird" "her fragrance like a horse" and for pathetic condition they use "trapped like a fox". So, in the entire situation the emotions and feelings of human beings are best expressed through animal imagery.

Thus, this study reiterates how Louise Erdrich's works lend themselves as a viable source for exploring the extraordinary power of animals. In the spheres of beauty, spirit and power they are not merely an abstract divine who can be felt and not seen but the concrete superpower in midst of humankind. They always favour humankind and the merging of animals with the souls of men. This effectively gets reflected through her works which signifies the strong bond between the Natives and the animals. Along with stating the relationship between the Natives and the animals, a new mode of ethics in treating animals in Native culture has been brought into focus. Louise Erdrich works thus contributes a new literary aspect in human animal relationships that following certain ethics in treating animals will help to restore the status of animals in society.

References

Erdrich, Louis. 1999. *The Birch Bark House*. New York: Hyperion publishers.

_____. 2004. *Four Souls*. New York: Harper Collins publisher.

_____. 1993. *Love Medicine*. New York: Harper Perennial.

_____ 1988.. *Tracks*. New York: Henry Holt and Company.

Lundquist, Suzanne Evertsen. 2004. *Native American Literatures: An Introduction*. New York: Continuum publishers.

Tierney, Diana. "Hindu Sacred Animals". Suite 101.com. Web. 22 Mar 2010.

Glossary

Ahneen: A greeting and sometimes it can also be a form of question.

Deydey: Daddy

Gago: Stop

Makazins: Footwear made out of deer skin or tanned moose-hide.

Sankaracharya: The well renowned philosopher and sage who promoted the concept of Brahmam and the man as of one essence. He motivated the people to achieve oneness with Brahmam.

Ojibway: Native American tribe and the people are called as Ojibwa

Chapter 36

Myth Unmasked: Eating Your Relatives with a Clear Conscience

☆ *Lisa Kemmerer*

"Native"

The Navajos, who came from the far north, arrived in the Southwestern United States shortly before the first Europeans, somewhere between 1200 and 1400 CE (Brown 19). While most other "native" groups were settled in their perspective areas many centuries earlier, they too came to the Americas from elsewhere. Evidence indicates that all human residents of North America originated elsewhere, and migrated to the Americas when land formations were very different from contemporary continents.

As these immigrants moved about, and grew in number and size, they squabbled over hunting rights and land. Those who were overtly defeated in battle, crowded out, or who simply moved in search of more hospitable environments, settled in areas their ancestors had not known.

Since early European immigrants applied the misnomer "Indian" to peoples living in the Americas, we have sampled a variety of better ways of referring to this diverse group of people: American Indian, Amer-Indians, native Americans, and First Nations. While it has become most common to refer to these various peoples as "native," the term is not appropriate, since they are not native to the lands on which they live. "Earlier immigrants in North America" (EINA) seems a more factually correct referent, and it is therefore the term I use throughout this paper.

Distant Time Stories

The Koyukon, an Athabaskan people in the interior of Alaska and Canada, provide a representative concrete example of traditional worldviews transmitted through myth. Koyukon myths are encompassed in a body of oral knowledge passed down as Distant Time Stories that "provide the Koyukon with a foundation for understanding the natural world."[1] Distant Time Stories are the sacred word of the people, best understood in relation to the historic position of the Bible in Western societies.[2] Myths tie traditional Koyukon people to their past while continuing to govern daily life.[3] Myths are historical records, a directive of traditions and customs that offer insights,[4] social cohesion, and entertainment.[5]

The telling of myths was particularly important during the long, dark, cold winter months.[6] The myth discussed in this article ends, "...the winter had just begun and now I've chewed off part of it" (Attla 109), suggesting both the entertainment value of myth, and the much more weighty importance of myth–sustenance itself. While many contemporary Koyukon are indifferent to the myths of their past, for others the power of language and of ancient mythology continues to be actualized in daily life.

Distant Time Stories explain how animals came to have their present characteristics, and how people ought to relate to nature to maintain balance and harmony.[7] Myths instruct people in a vast array of rituals that guide interactions with the natural environment and safeguard human sustenance. Ritual actions permeate traditional Koyukon lifeways so that almost every act is an act of worship, a religious act. Through rituals traditional Koyukon discover the Distant Time Stories perpetually asserted in the world around them; life is a religious act ever renewed in the daily rituals of living.[8]

Distant Time Stories reveal that Raven created a harmonious universe where all that live both control and are controlled in equal measure. Mythology records how Raven created all beings, that the natural universe rules, and that the role of human beings is to serve nature, the dominant force.[9] Raven's world is one where spiritual powers are pervasive, and constant reverence is required to avoid misfortune brought on by insulted spirits.[10]

Consequently, the Koyukon associate with other animals *as spiritual powers.*[11] Just as there is no ruler amongst the Koyukon, so there is no almighty ruler of the universe.[12] For the Koyukon, "Nature is not governed by God, nature *is* God."[13]

Eagle Man Who Carried People Far Away

Eagle Man Who Carried People Far Away is a Koyukon Distant Time Story that begins with a woman picking berries. She sees an eagle "gliding around high in the sky" and wishes she could be high up in the sky with the eagle. (Alternatively, the eagle wills her to have this wish.) Eagle Man then abducts her, and her husband spends the autumn crying and searching for his lost wife. A powerful spirit-woman reveals her lodge and sends out a go-between to offer help to the man-in-search-of-his-wife, but first he must provide for her to prove his sincerity. She then offers him three tools and explains how he might locate his wife and defeat Eagle Man, "a cane, the old woman's caribou pants, and a wolverine skin." The spirit-woman instructs him as to how and when to use these powerful tools:

> Whenever you are ready to spend the night, plant the cane in the snow. Go in the direction toward which it falls...When you start wrestling, wrap [the caribou pants] around him, and then you will be able to throw him down...When everybody is leaving the hall, he will block your path with a stump. When this happens, wrap the soaked wolverine skin around yourself (Attla 95).

The man-in-search-of-his-wife follows her instructions, travels far to find and Eagle Man, who is holding his spouse, defeats Eagle man in a fierce and bloody battle, then returns home with his wife. At the close of the myth he returns spirit-woman's possessions, and spirit-woman's lodge disappears.[14]

Analyzing Eagle Man Who Carried People Far Away

Animals are represented in five distinct but interconnected ways in *Eagle Man Who Carried People Far Away*:

- ☆ Through the power of parts of their bodies,
- ☆ As indistinct from human beings,
- ☆ As characters in their own right,
- ☆ As a concrete means of subsistence, and
- ☆ As part of the natural surroundings.

What can these animal representations teach us about animal ethics among traditional Koyukon?

Subsistence and Cooperation

Nonhuman animals are central to *Eagle Man Who Carried People Far Away.* Nonhumans beings are constantly noted as part of the natural surroundings, and they are perpetually noted as essential to the subsistence of the Koyukon. The myth begins with reference to putting up fish, and the man-in-search-of-his-wife pays for the old woman's services with furs and dried flesh. He subsists on Spruce Grouse, and after defeating Eagle Man, man-in-search-of-his-wife brings in caribou for the village for which Eagle Man had provided. The complexity of the relationship between humans and nonhumans, between kin who help one another survive, is critical to understanding the worldview and ways of the Koyukon.

Old spirit-woman brings animals into the story as positive powers. By way of helping man-in-search-of-his-wife, she offers him a cane, which brings Spruce Grouse to provide him with food: In the evenings, when he "planted" this cane in the snow, a "spruce grouse would land nearby... and he would shoot it with an arrow."[15]

In Western culture we might tend to view Spruce Grouse as a spirit-bird, but the Koyukon worldview maintains an unrelenting physical reality that is reflected in the world of myth. For the Koyukon, spirits live. They live for the old spirit-woman in the Distant Time Stories, they live and provide food for the man-in-search-of-his-wife, and they continue to be manifest in every element of the natural environment for many traditional Koyukon of today. Each time the man eats, he eats a very real, individual Spruce Grouse.

Understanding the relationship between the spirit-woman, man-in-search-of-his-wife, and the various Spruce Grouse is central to understanding the traditional Koyukon view of human-nonhuman relations. Through a well-lived life, a Koyukon gains the power necessary to influence the subsistence environment. Koyukon people who follow taboos and maintain rituals dutifully believe that they will be able to influence animal spirits. This belief has led to myriad rituals and rules, rooted in respect for hunted and fished animals, that ultimately offered a sense of control for human beings in dealing with the capricious animals on which they depend for survival. "Koyukon subsistence is more than just an economic pursuit – it is manifestly bound to traditional ideology and ritual practice."[16] Koyukon people ask other animals for help, and use parts of other animals as amulet to increase their hunting and trapping catch. Traditional Koyukon believe that, if respect is shown – if an animal is approached and treated in the proper manner[17] - that animal will permit the human to use them for subsistence. Thus, a hunter's (trapper's/fisher's) skill does not determine catch. For success, the spirits of animals must be *willing* to die. When the Koyukon suffer food shortages they are not likely to say, "I cannot kill deer anymore," but rather, "Deer don't want to die for me."[18]

In this light it is easier to understand how the old spirit-woman and man-in-search-of-his-wife gain the cooperation of many spruce grouse in securing food. Each spruce grouse willingly comes to the man-in-search-of-his-wife, by the power of the old spirit-woman – and due to the man's proven worthiness – to provide food. A close reading of *Eagle Man Who Carried People Far Away* reveals this consent from Spruce Grouse. When the man is returning home with his wife, the story reports that "a spruce grouse landed in the tree for them" – *for them* specifically. The bird is purposefully fulfilling the man's needs at the request of the old woman, given the worthiness of both the spirit-woman and the man-in-search-of-his-wife. The bird's spirit will live on, the body is donated by way of looking after human needs. When humans are hungry, nonhuman animals share – as kin ought to do.

The Presence and Power of Nonhuman Animals

The Koyukon, like most earlier immigrants to North America, believed that animals, plants, and the natural world were "endowed with spirits and with spiritually based power" (Nelson 228). The Koyukon knew that they depended on other animals for survival. Their rituals and beliefs stemmed from this understanding, and were a constant reminder of their need, and of the generosity of other animals in providing for needy human beings – who were understood to be kin. They believed that the understanding and knowledge of animals was beyond that of human beings, and they taught their young "never to bother anything unless we really needed it" (Nelson 225).

Traditional, ancient cultures tend to treat other animals respectfully *because* they believed that non-humans possessed "consciousness, will, and other capacities... superior to those of humans" (Harrod 159). Traditional Koyukon believe that each animal has a spirit, and that many animals have very powerful spirits. The Koyukon considered other animals sacred and powerful–quite capable of preventing a hunter from having success (Nelson 231). Perhaps out of respect for the terrifying power invested in non-human animals, traditional hunters upheld "a code of moral and social etiquette" that encompassed all creatures (Nelson 228). Many myths of earlier immigrants reveal the power of non-human animals, and the respect with which they ought to be treated; Koyukon myths are no exception.

In the Koyukon worldview death is not sudden. Awareness continues in the deceased as the remains "vibrate with spirituality."[19] An animal is not considered completely dead just because biological life is gone. Traditional Koyukon maintain certain rituals and practices because they believe that animals remain powerful after death. For example, wolverine pelts are not completely dead until one year later; while geese should not be prepared for eating until one or two days after they have been shot.[20]

Lingering power is also manifest in flesh, even after a body has been prepared for consumption. Traditional Koyukon butcher and prepare animals in prescribed ways in order to show deference to the power of lingering spirits. In some cases, "their names should not be mentioned, nothing should be burned lest the smell offend the spirits, metallic noises should be avoided, and even if it is unfrozen and skinned the carcass should be kept indoors overnight."[21] Such taboos and rituals indicate which animals traditional Koyukon hold in the highest regard.[22] The use of pelts of spiritually potent animals, such as Bear, Wolf, and Wolverine, require elaborate rituals and entail many taboos. A memorial potlatch, often held for deceased human beings, is also held at the death of a bear as "a kind of funeral ritual, a final act of deference and respect."[23] A freshly killed wolverine is ritually wined and dined, and praised as a guest of honor, when killed and brought back to the village.

Traditional Koyukon believe that those who come into contact with an animal, even long after it has been dead, may gain that animal's traits, including physical attributes, temperament, appearance, or motor function. Consequently, a tremendous number of taboos (especially for children and women of childbearing age) surround the use of certain animals. For example, to gain grace and beauty, young girls are encouraged to eat swallows. Clumsy water birds are taboo in the diets of all children. SuckerFish has a bad reputation for stealing in the Distant Time (the mythological history of the people), and few Koyukon will eat the meat of this fish "lest they acquire its thieving personality."[24]

These beliefs are evident in *Eagle Man Who Carried People Far Away*. Both the second and the third items that the spirit-woman gives the man-in-search-of-his-wife are parts of deceased animals: Caribou pants and a wolverine skin. These animal's skins retain their power. The caribou pants, when wrapped around Eagle Man, allow the man-in-search-of-his-wife to throw Eagle Man to the floor in a wrestling match–though Eagle Man is huge and has killed many more weighty opponents. Spirit-woman offers the searching man Wolverine's skin because wolverines are one of the most spiritually powerful animals in the traditional Koyukon worldview. Wolverine's skin transforms the man into a viscous wolverine, providing him with the necessary fierceness and prowess to defeat Eagle Man. Meanwhile, Eagle Man transforms into an eagle, and the two engage in a fierce battle of claws and teeth, of digging and flying, of blood and guts and feathers (103-104). Having put on the

Wolverine skin, the man-in-search-of-his-wife, "being a wolverine," chews up the house, engages in a fierce battle with Eagle, tears out the great raptor's insides, then eats him. Only a few feathers remain on the floor of the house when Eagle's wives return (104-5). No mere human could do so well!

Spirit-woman offers Wolverine's pelt only after the man-in-search-of-his-wife shows sincerity in his quest for his wife by giving fur and flesh to the spirit-woman. Humans must be worthy of the powers of other animals, and only through tenacity and displays of respect does man-in-search-of-his-wife prove himself to be worthy of the powers held in deceased nonhuman animal parts. In the lives of traditional Koyukon, only through perpetual diligence and the proper performance of tradition is the delicate balance of survival maintained. Traditional Koyukon believe that these rituals and taboos originated in the Distant Time, and only through the maintenance of these ancient ways is it possible for the Koyukon to survive.

Interpenetrability

At the outset of *Eagle Man Who Carried People Far Away*, the wife is picking berries, looking up, and watching an eagle soar in the sky. The eagle is portrayed as a natural and normal part of the landscape–as it would be. Yet before the page is turned the eagle appears before the woman as a "very handsome man." Eagle Man is neither all eagle, nor all man, he speaks with humans, and he has human wives. Similarly, the man-in-search-of-his-wife is transformed into a wolverine.

Distant Time Stories reveal a Koyukon worldview where animals are not distinctly separate from human beings. These connections between animals and people–this lack of distinction between what is exclusively human and what is animalian–is embedded in the Koyukon worldview.

> The distinction between humans and animals [in the Koyukon worldview] is less sharply defined than in Western thought. Based on Distant Time Stories and their own interpretations of animal behavior, the Koyukon attribute many characteristics to animals that Western thought reserves for humans. For example, animals were human in the Distant Time, they still understand human speech, they have spirits, and they must be treated according to a code of moral and social etiquette. Thus the closeness between humans and animals is based more on animals' possessing human qualities than on humans' possessing animal qualities.[25]

Distant Time Stories perpetually reveal the lack of clear distinction between people and other animals. Species are interpenetrable. In the mythological Distant Time of the Koyukon, all living beings shared the same characteristics.[26] They once looked like humans, lived like humans, and talked like humans.[27] Many Koyukon myths involve animals who talk and interact with people. Myths of Lynx and Wolverine tell how these animals emerged from human form and how Wolf once hunted with the Koyukon people in the Distant Time.[28]

Kinship

Traditionally, EINA tended not to see any concrete division between other-animal and human-animal (Erdoes 389). An animal trickster, Raven, created the Koyukon world. All animals (people included) were envisaged as relatives. Myths reveal a belief that nonhumans share "human" qualities, and that a social and spiritual relationship exists between people and non-human animals (Gill: 121). Consequently, how traditional Koyukon interact with nature is critical–they believe that humans and other natural entities share a "constant spiritual interchange that profoundly affects human behavior" (Nelson 229).

Traditional Koyukon view animals not as separate and distinct, but as part of a living bridge that unites all living beings. For centuries, traditional Koyukon lived closely with other animals, and considered these other living beings to be their kin. The traditional Koyukon relationship with other living beings, as revealed in Koyukon myths, goes beyond subsistence into the realms of interpersonal relations and spirituality that permeates every fiber of Koyukon daily life.[29]

Traditional people knew and shared streams, woodlands, tundra, and lakes, with myriad wild creatures. Koyukon subsistence depended on specialized knowledge of animal behavior–where they sleep, feed, and when certain species come and go with the seasons.[30] Today, through generations of Distant Time Stories, Koyukon people continue to "remember" and identify human characteristics in non-human animals. The close relationship between the Koyukon and the animals that live near them is expressed in unexpected ways, even today. One Koyukon notes: "Sometimes people will hunt the loon, but me, I don't like to kill it. I like to listen to it all I can and pick up the words it knows."[31] Koyukon people see the traits of other animals in themselves, and they see their human traits in other animals:

> You know, beavers work hard all fall and winter, so every spring they take a vacation. They drift down the river someplace or they wander around the rough lakes. Then, later on in summer they start heading back home, and it really makes them feel happy. So that's when they like to sing the most, while they swim along toward home.[32]

A contemporary Koyukon journal is filled with similar appreciation for the lives of nonhuman beings: The inner nature of Mink's personality, the magic of watching Otter in the water, and the joy of having Fox in camp sharing the remnants of a hunt. Common Koyukon phrases play with the closeness and similarities envisaged between people and animals: A greedy person may be enlikened to a gull, a hard working wife to a female beaver, and a child who stares at others is "Just like a mink at its hole."[33] Koyukon people continue to rediscover themselves in the activities and perceived attitudes of other animals.[34]

Through hunting, Koyukon observed the secret worlds of other animals, witnessed their pain and fear when confronted with death, and long after that terrible moment they continued to feel the soft fur of that animal around the hoods of their parkas. Traditional Koyukon held a "friendly, kin-like relationships with animals, which, by tradition and necessity they were forced to kill."

The crucial importance of game and knowledge of game animals is graphically demonstrated in the content of Athapaskan mythology and religion. Recognition of a close, kin-like relationship with game animals was in some respects, however, at variance with the importance of hunting and fishing for survival.[35]

Many Koyukon continue to refer to other animals by kinship terms–Bear as grandfather, Wolf as brother.[36] In Koyukon villages, using a dog's skin is likened to using the clothes of a deceased relative; the owner of a dog is referred to as that dog's grandfather or grandmother. Traditional hunters, trappers, and fishers suffered anxiety, and no doubt a good deal of sadness, over the "forced hostility" inherent in subsistence hunting.[37] By way of expressing these concerns, Koyukon sometimes note the "striking and somewhat frightening" resemblance of a skinned corpse of a bear to that of a human body. Some do not like to kill certain animals, such as wolves, because they are "too much like people."[38]

Myths from around the world suggest an understanding of human-nonhuman animal relations as kinship (McLuhan 56, 99), and they also express the conflicted feelings of those who must kill their kin for survival are evident in myths from peoples around the world. World mythology suggest that indigenous communities struggled emotionally and ethically with killing other animals for their sustenance, animals they tended to be viewed as kin (Serpell 29). This discomfort surfaces in ancient mythology, often in the form of a Great Peace that once existed between all living beings, and a story line that explains how and why this peace was broken. For example, The Cherokee remember a time of universal peace, when "humans and animals freely communicated" as one community (Bierlein 115). Their myths also recall a time when humans greedily began to kill animals for fur and food: "It was easy to do at this time, as the animals were completely unprepared to be hunted and they walked up to human beings, trusting them. Then the animals became angry" (*Ibid* 115). The bears met in council, and decided on open warfare (Spence 249). Then the deer also met in council, then fish and reptiles, and finally birds and insects. They were all angry with humans, and inflicted humans with various punishments, such as nightmares and disease (Bierlein 115).

Eagle Man Who Carried People Far Away is part of this body of mythology that eases the conscience of those who kill their kin. This myth (like many others from a host of peoples) teaches people that hunting, fishing, and trapping are acceptable *to nonhuman animals*–hunted, trapped, and fished animals *choose* to die for the Koyukon people–and even through this sacrificial death nonhuman animals do not lose their power.

Taboos and rituals that surround the use of nonhuman animals are critical to the process of mitigating the pain that are likely to accompany the killing of one's kin. When these rituals and taboos are broken, it is no longer acceptable to kill kin. A contemporary Koyukon hunter commented: "My brother doesn't follow hutlaanee (taboos). When he cut a beaver's throat in front of me, it just *hurt* me to see that happen."[39] By maintaining ways attributed to the Distant Time, Koyukon people ease the pain of killing. When taboos are broken, the pain returns. By repeatedly reenacting ritualized reverence and respect for the nonhuman animals whom they depend on for sustenance, the traditional Koyukon make peace with hunting kin.

Ancient myths such as *Eagle Man Who Carried People Far Away* continue to reaffirm the Koyukon belief that nonhuman animals are complicit in this system of giving and taking life–humans cannot kill without the help of nonhuman animals. Through a belief in the living spirits of the deceased, traditional Koyukon focus their attention on the expectation of rebirth, rather than on the death of a living being.[40] They view those hunted as ever-returning to those who are respectful both toward the living and the "dead." Koyukon mythology brings meaning and peace to the daily process of killing to live–killing animals that the Koyukon cannot help but view as old kin and community.

Interestingly, humans and other animals are also bound by the same ethic. Kingfisher is held in some disdain for being wasteful with fish, while Osprey is looked down upon for being miserly. In *Eagle Man Who Carries People Far Away*, Eagle is too powerful and takes the wives of others–an imbalance of power that needs to be rectified. To bring the powerful Eagle Man into line, Spirit-Woman and many nonhuman animals are necessary.

Revisiting Koyukon/EINA Animal Ethics in Contemporary North America

The Koyukon once struggled to survive in an environment that was much bigger and more powerful than they. Koyukon people no longer struggle for survival. They buy sugar and coffee from other parts of the United States; they travel by snowmobile, cars, and small aircraft; and they use high-powered weaponry and sophisticated fishing equipment. Koyukon people, like EINA across North America, enjoy products from across the country and around the world. There are perhaps a few remaining EINA who depend solely on subsistence hunting, fishing, and trapping, but earlier immigrants no longer *need* to hunt, trap, or fish for survival. Rather than import and purchase ammunition, weapons, traps, and fishing gear, EINA can buy brown rice and potatoes, whole grains and fresh veggies. *There is currently no need for the Koyukon, or any other EINA, to hunt, fish, or trap for survival–there is no need to eat animal products at all, not even in the most remote areas of North America.*

Today, whether hunting, trapping, and fishing are practiced by Koyukon who believes an animal has given his or her life, or by a European who cares nothing for nonhuman animals, *that creature is needlessly killed*. Though I might lay down my life for my kin if my death were necessary for their survival, I would be sorely disappointed if this generosity was later used to justify killing me when the need had passed. Traditional EINA, and many other earlier immigrants in North America, believe that killing without necessity is dangerous and taboo – it is certainly contrary to ancient traditions and lifeways. For example, a contemporary Waswanipi Cree hunter (Canadian Subarctic), while noting that technology and modern techniques allow humans to take life beyond what is necessary, cautions against doing so: killing anymals "for fun or self-aggrandizement" demonstrates "the ultimate disdain" for other creatures. Like the Koyukon and many other EINA, traditional Cree believe that animals decide whether or not to be hunted, and such disdain would surely lead to starvation; no anymal would return to such a hunter. At the end of the day, whether one wines and dines the deceased, or simply shoots a hole through the animal and drags it off by the leg, if an individual is killed unnecessarily, then that individual is killed disrespectfully. While Westerners can continue such crass practices with impunity, traditional EINA cannot treat their powerful nonhuman kin with such disrespect, and still be traditional.

Given that hunting, trapping, and fishing are no longer necessary, earlier immigrants who continue to hunt or fish breach traditional values, traditions, and cultural practices. *Given that subsistence no longer requires us to kill other animals at all, hunting, trapping, and fishing are incongruous with respect for life.* Consuming flesh is wasteful, and for traditional Koyukon and other traditional EINA, disrespectful, because it is unnecessary. While hunting and fishing might remain the *preferred* lifestyle for some EINA, earlier immigrants cannot maintain that they kill with respect – in the traditional way – when *there is no need to kill at all.*

Conclusion

In *Eagle Man Who Carried People Far Away*, Eagle soars in the sky with grace and intrigue, then reappears in human form to speak to a woman who is picking berries. The man-in-search-of-his-wife, through the power of Wolverine, becomes a wolverine and destroys Eagle. Spruce Grouse comes whenever needed, offering her flesh to the man-in-search-of-his-wife.

By knowing Distant Time myths, when a traditional Koyukon sees an eagle, she sees power. She sees both the beauty and the majesty of a bird high in the sky *and* the handsome Eagle Man who speaks to her people in human form. She believes that the many things that live around her–wolverines, spruce grouse, and eagles–are kin who once shared a language, and continue to share many "human" attributes. Species are interpenetrable, and form no barrier in the storyline of the Distant Time.

The Koyukon, like many indigenous peoples, viewed other animals as their kin. Traditional earlier immigrants consider wild animals to be their relatives, and they believe that their relatives have given their lives to respectful, appreciative, and needy human beings – their dependant brothers and sisters – so that humans might survive. Consequently, those who kill other animals do so with a clear conscience–animals *wanted* to die – and *needed* to die – so that humans might live.

Instead of identifying the raw act of killing as an EINA cultural heritage, we would do well to rediscover the spirituality *behind* these practices: All living beings are kin, and killing is only acceptable in the confines of respect, which requires that such killing be necessary. In *Eagle Man Who Carried People Far Away*, Spruce Grouse gave her life *because* there was no other way for the man-in-search-of-his-wife to survive, and because his life exemplified respect for animal kin. One cannot respect other animals and kill them when there is no need to do so. Today, the ancient teachings of the Koyukon, and other earlier immimgrants in North America, indicate a vegan diet and lifestyle. Concurrently, these body-parts hold tremendous power to smite those who do not show due respect. Parts from other animals are not thought to be dangerous, but are kept by the Koyukon simply for good luck, such as the skin of Flicker, Lemming, or Loon. Some children carry dried salmon-tails in their pockets for protection because the power of these animal spirits is believed to continue after death.

Notes

1 Richard Nelson, *Make Prayers to the Raven.* (Chicago: University of Chicago Press, 1983), pp. 18 and 227.

2 Rev. Jules Jette, "On Ten'a Folk-Lore," in "Journal of the Royal Anthropological Institute of Great Britain" (London: Anthropos, 1908), 38: 298-367.

3 Chad Thompson and Helen Thompson, A Teacher's Guide to Bakkaatugh Ts'uhuniy, Stories We Live By (Fairbanks: Yukon Koyukuk School District and Alaska Native Language Center, University of Alaska, 1989).

4 Jette, "On Ten'a Folk-Lore," 38: 298-367.

5 Clark McFadyen, "Koyukon," in Handbook of Native American Indians, Volume Six, Subarctic (Washington: Smithsonian Institution, 1981), p. 595.

6 Jette, "On Ten'a Folk-Lore," p. 298.

7 VanStone, Athapaskan Adaptations; Hunters and Fishermen of the Subarctic Forest. (Chicago: Aldine, 1974), p. 61.

8 Rev. Jules Jette, "On Superstitions of the Ten'a Indians (Middle Part of the Yukon Valley, Alaska) in "Journal of the Royal Anthropological institute of Great Britain" (London: Anthropos, 1895).

9 *Ibid.*, p. 88.

10 Nelson, Make Prayers to the Raven, p. 240.

11 VanStone, Athapaskan Adaptations, p. 65.

12 Jette, "On Superstitions of the Ten'a Indians," p. 97.

13 Richard Nelson, Make Prayers to the Raven, KUAC-TV, (Fairbanks: University of Alaska Fairbanks, N.D.) Video #1 "Passage of Gifts."

14 Catherine Attla, Bakkaatugh Ts'uhuniy, Stories We Live By (Fairbanks: Yukon Koyukuk School District and Alaska Native Language Center, University of Alaska, 1989).

15 Attla, Bakkaatugh Ts'uhuniy, Stories We Live By, p. 97.

16 Nelson, Make Prayers to the Raven, p. 234.

17 VanStone, Athapaskan Adaptations.

18 Robert Heizer and Albert Elsasser, The Natural World of the California Indians (Berkeley: University of California Press, 1980).

19 Richard Nelson, Make Prayers to the Raven, KUAC-TV, (Fairbanks: University of Alaska Fairbanks, N.D.) Video #1 and #2.

20 *Ibid.*, pp. 92 and 152.

21 Nelson, Make Prayers to the Raven, p. 24.

22 VanStone, Athapaskan Adaptations, p. 65.

23 Nelson, Make Prayers to the Raven, pp. 181 and 191.

24 Nelson, Make Prayers to Raven, p. 76.

25 *Ibid.*, p. 228.

26 VanStone, Athapaskan Adaptations, p. 61.

27 *Ibid.*, p. 16.

28 *Ibid.*, pp. 103-104, 136, 143-159.

29 Nelson, Make Prayers to Raven, 79.

30 VanStone, Athapaskan Adaptations, p. 101.

31 *Ibid.*, 87.

32 *Ibid.*, p. 136.

33 Nelson, Make Prayers to Raven, p. 144.

34 Nelson, Make Prayers to Raven, p. 227.

35 VanStone, Athapaskan Adaptations, p. 122.

36 McFadyen, Subarctic, p. 591.

37 *Ibid.*, p. 88.

38 Nelson, Make Prayers to Raven, pp. 159, 179, 191.

39 Nelson, Make Prayers to the Raven, p. 137.

40 Jette, "On Ten'a Folk-Lore," p. 101.

References

Attla, Catherine. 1989. *Bakkaatugh Ts'uhuniy, Stories We Live By*. Fairbanks: Yukon Koyukuk School District and Alaska Native Language Center, University of Alaska.

Bierlein, J.F. 1994. *Parallel Myths*. NY: Ballantine.

Brown, Joseph Epes. 1991. *The Spiritual Legacy of the American Indians*. NY: Crossroads.

Buege, Douglas J. "The Ecologically Noble Savage Revisited." *Environmental Ethics*. 18 (1996): 71-88.

Clark, Wissler. 1966. *Indians of the United States*. NY: Anchor.

Erdoes, Richard and Alfonso Oritz, eds. 1984. *American Indian Myths and Legends*. NY: Pantheon.

Gill, Sam D. 1983. *Native American Traditions: Sources and Interpretations*. Belmont: Wadsworth.

Harrod, Howard L. 1987. *Renewing the World: Plains Indian Religion and Morality*. Tucson: U. of AZ.

Heizer, Robert and Albert Elsasser. 1980. *The Natural World of the California Indians*. Berkeley: U of CA.

Jette, Rev. Jules. "On Superstitions of the Ten'a Indians (Middle Part of the Yukon Valley, Alaska)." *Journal of the Royal Anthropological Institute of Great Britain in Anthropos*. Vol. 6. Wien, Osterreich: "Anthropos," 1911.

_____. "On Ten'a Folk-Lore." *Journal of the Royal Anthropological Institute of Great Britain. Anthropos*. Wien, Osterreich: "Anthropos," 1911.

Kwiatkowska-Szatzscheider, Teresa. "From the Mexican Chiapas Crisis: A Different Perspective for Environmental Ethics." *Environmental Ethics*. 19 (1997): 267-278.

McFadyen, Clark A. 1981 . "Koyukon." *Handbook of Native American Indians*. Volume 6, *Subarctic*. Washington: Smithsonian Institute.

McLuhan, T. C. 1971. *Touch the Earth*. N. Y: Promontory.

Nelson, Richard. 1983. *Make Prayers to the Raven*. Chicago: U of Chicago.

_____. *Make Prayers to the Raven*. KUAC-TV. Fairbanks: University of Alaska, Fairbanks, N.D. Videos #1, 2, 3, and 5.

Serpell, James. 1992 . "Animal Protection and Environmentalism: The Background." Ed. Richard D. Ryder. *Animal Welfare and the Environment*. Melksham: Duckworth. 27-39.

Spence, Lewis. 1985. *North American Indians: Myths and Legends*. London: Studio.

Thompson, Chad and Helen. 1989. *A Teacher's Guide to Bakkaatugh Ts'uhuniy, Stories We live By*. Fairbanks: Yukon Koyukuk School District and Alaska Native Language Center, University of Alaska.

VanStone, James W. 1974. *Athapaskan Adaptations; Hunters and Fishermen of the Subarctic Forest*. Chicago: Aldine.

White, Lynn Jr. "The Historical Roots of Our Ecologic Crisis." *Science* 155 (1967): 1203-1207.

Index

www.ingramcontent.com/pod-product-compliance
Ingram Content Group UK Ltd.
Pitfield, Milton Keynes, MK11 3LW, UK
UKHW041133071026
15587UKWH00009B/452

9 789351 240877